1,001 Pre-Calculus Practice Problems For Dummies

by Mary Jane Sterling

1,001 Pre-Calculus Practice Problems For Dummies®

Published by: **John Wiley & Sons, Inc.,** 111 River Street, Hoboken, NJ 07030-5774, www.wiley.com

Published simultaneously in Canada

For general information on our other products and services, please contact our Customer Care Department within the U.S. at 877-762-2974, outside the U.S. at 317-572-3993, or fax 317-572-4002. For technical support, please visit www.wiley.com/techsupport.

Wiley publishes in a variety of print and electronic formats and by print-on-demand. Some material included with standard print versions of this book may not be included in e-books or in print-on-demand. If this book refers to media such as a CD or DVD that is not included in the version you purchased, you may download this material at http://booksupport.wiley.com. For more information about Wiley products, visit www.wiley.com.

Library of Congress Control Number: 2014936392

ISBN 978-1-118-85332-0 (pbk); ISBN 978-1-118-85281-1 (ebk); ISBN 978-1-118-85334-4 (ebk)

Manufactured in the United States of America

V10014433_100419

Contents at a Glance

Table of Contents

Introduction

Pre-calculus is a rather difficult topic to define or describe. There's a little bit of this, a lot of that, and a smattering of something else. But you need the mathematics considered to be pre-calculus to proceed to what changed me into a math major: calculus! Yes, believe it or not, I started out as a biology major — inspired by my high school biology teacher. Then I got to the semester where I was taking invertebrate zoology, chemistry, and calculus (yes, all at the same time). All of a sudden, there was a bright light! An awakening! "So this is what mathematics can be!" Haven't turned back since. Calculus did it for me, and my great preparation for calculus made the adventure wonderful.

Pre-calculus contains a lot of algebra, some trigonometry, some geometry, and some analytic geometry. These topics all get tied together, mixed up, and realigned until out pops the mathematics you'll use when working with calculus. I keep telling my calculus students that "calculus is 60 percent algebra." Maybe my figures are off a bit, but believe me, you can't succeed in calculus without a good background in algebra (and trigonometry). The geometry is very helpful, too.

Why would you do 1,001 pre-calculus problems? Because practice makes perfect. Unlike other subjects where you can just read or listen and absorb the information sufficiently, mathematics takes practice. The only way to figure out how the different algebraic and trigonometric rules work and interact with one another, or how measurements in degrees and radians fit into the big picture, is to get into the problems — get your hands dirty, so to speak. Many problems given here may appear to be the same on the surface, but different aspects and challenges have been inserted to make them unique. The concepts become more set in your mind when you work with the problems and have your solutions confirm the properties.

What You'll Find

This book contains 1,001 pre-calculus problems, their answers, and complete solutions to each. There are 16 problem chapters, and each chapter has many different sets of questions. The sets of questions are sometimes in a logical, sequential order, going from one part of a topic to the next and then to the next. Or sometimes the sets of questions represent the different ways a topic can be presented. In any case, you'll get instructions on doing the problems. And sometimes you'll get a particular formula or format to use. Feel free to refer to other mathematics books, such as Yang Kuang and Elleyne Kase's *Pre-Calculus For Dummies*, my *Algebra II For Dummies*, or my *Trigonometry For Dummies* (all published by Wiley) for even more ideas on how to solve some of the problems.

Instead of just having answers to the problems, you'll find a worked-out solution for each and every one. Flip to the last chapter of this book for the step-by-step processes needed to solve the problems. The solutions include verbal explanations inserted in the work where necessary. Sometimes, the explanation may offer an alternate procedure. Not everyone does algebra and trigonometry problems exactly the same way, but this book tries to provide the most understandable and success-promoting process to use when solving the problems presented.

How This Workbook Is Organized

This workbook is divided into two main parts: questions and answers. But you probably figured that out already.

Part I: The Questions

The chapters containing the questions cover many different topics:

- **Review of basic algebraic processes:** Chapters 1 and 2 contain problems on basic algebraic rules and formulas, solving many types of equations and inequalities, and interpreting and using very specific mathematical notation correctly. They thoroughly cover functions and function properties, with a segue into trigonometric functions.
- **Graphing functions and transformations of functions:** Functions and properties of functions are a big part of pre-calculus and calculus. You work with operations on functions, including compositions. These operations translate into transformations. And all this comes together when you look at the graphs of the functions. Transformations of functions help you see the similarities and differences in basic mathematical models — and the practice problems help you see how all this can save you a lot of time in the end.
- **Polynomial functions:** Some of the more familiar algebraic functions are the polynomials. The graphs of polynomials are smooth, rolling curves. Their characteristics include where they cross the axes and where they make their turns from moving upward to moving downward or vice versa. You get to practice your equation-solving techniques when determining the x-intercepts and y-intercept of polynomial functions.
- **Exponential and logarithmic functions:** You're not in Kansas anymore, so it's time to leave the world of algebraic functions and open your eyes to other types: exponential and logarithmic, to name two. You practice the operations specific to these types of functions and see how one is the inverse of the other. The applications of these functions are closer to real-world than most others in earlier chapters.
- **Trigonometric functions:** Trigonometric functions take being different one step further. You'll see how the input values for these functions have to be angle measures, not just any old numbers. The trig functions have their own rules, too, and lots of ways to interact, called *identities*. Solving trig identities helps you prepare for that most exciting process in calculus, where you get to find the area under a trigonometric curve. So keep your eye on that goal! And the trig applications are some of my favorite — all so easy to picture (and draw) and to solve.
- **Complex numbers and polar coordinates:** Complex numbers were created; no, they aren't real or natural. Mathematicians needed to solve problems whose solutions were the square roots of negative numbers, so they adopted the imaginary number i to accomplish this task. Performing operations on complex numbers and finding complex solutions are a part of this general arena. *Polar coordinates* are a way of graphing curves by using angle measures and radii. You open up a whole new world of curves when you practice with these problems dealing with polar graphs.
- **Conic sections:** A big family of curves belongs in the classification of *conics*. You find the similarities and differences between circles, ellipses, hyperbolas, and parabolas. Exercises have you write the standard forms of the equations so you can better determine individual characteristics and create reasonable sketches of the graphs of the curves.

- **Systems of equations and inequalities:** When you have two or more statements or equations and want to know whether any solutions are common to both or all of them at the same time, you're talking about solving systems. The equations can be linear, quadratic, exponential, and so on. You'll use algebraic techniques and also use matrices to solve some of the linear systems.
- **Sequences and series:** Some problems cover the basic arithmetic and geometric series. And, as a huge bonus, you'll use the binomial theorem and Pascal's triangle to expand binomials to fairly high powers.
- **Limits and continuity:** The basics of limits and continuity are covered — analytically and graphically. This point is actually the launching spot for calculus — where pre-calculus finishes, calculus begins.

Part II: The Answers

This part provides not only the answers to all the questions but also explanations of the answers. So you get the solution, and you see how to arrive at that solution.

Beyond the Book

This book is chock-full of pre-calculus goodness, but maybe you want to track your progress as you tackle the problems. Or maybe you're stuck on a few particularly challenging types of pre-calculus problems and wish they were all presented in one place where you could methodically make your way through them. No problem! Your book purchase comes with a free one-year subscription to all 1,001 practice problems online. Track your progress and view personalized reports that show where you need to study the most. And then do it. Study what, where, when, and how you want.

What you'll find online

The online practice that comes free with this book offers you the same 1,001 questions and answers that are available here, presented in a multiple-choice format. The beauty of the online problems is that you can customize your online practice to focus on the topic areas that give you the most trouble. So if you aren't yet a whiz at exponential and logarithmic functions, you can select these problem types and BAM! — just those types of problems appear for your solving pleasure. Or, if you're short on time but want to get a mixed bag of a limited number of problems, you can plug in the quantity of problems you want to practice and that many — or few — of a variety of pre-calculus problems appears. Whether you practice a couple hundred problems in one sitting or a couple dozen, or whether you focus on a few types of problems or practice every type, the online program keeps track of the questions you get right and wrong so that you can monitor your progress and spend time studying exactly what you need.

You can access this online tool by using a PIN code, as described in the next section. Keep in mind that you can create only one login with your PIN. After the PIN is used, it's no longer valid and is nontransferable. So you can't share your PIN with other users after you've established your login credentials.

This book also comes with an online Cheat Sheet full of frequently used formulas and more goodies. Check it out for free at `www.dummies.com/cheatsheet/1001precalculus`. (No PIN is required. You can access this info before you even register.)

How to register

To gain access to additional tests and practice online, all you have to do is register. Just follow these simple steps:

1. **Find your PIN access code:**
   - **Print-book users:** If you purchased a print copy of this book, turn to the inside front cover of the book to find your access code.
   - **E-book users:** If you purchased this book as an e-book, you can get your access code by registering your e-book at `www.dummies.com/go/getaccess`. Go to this website, find your book and click it, and answer the security questions to verify your purchase. You'll receive an email with your access code.
2. **Go to `Dummies.com` and click** Activate Now.
3. **Find your product (*1,001 Pre-Calculus Practice Problems For Dummies* (+ *Free Online Practice*)) and then follow the on-screen prompts to activate your PIN.**

Now you're ready to go! You can come back to the program as often as you want — simply log on with the username and password you created during your initial login. No need to enter the access code a second time.

For Technical Support, please visit `http://wiley.custhelp.com` or call Wiley at 1-800-762-2974 (U.S.), +1-317-572-3994 (international).

Your registration is good for one year from the day you activate your PIN. After that time frame has passed, you can renew your registration for a fee. The website gives you all the important details about how to do so.

Where to Go for Additional Help

The written directions given with the individual problems are designed to tell you what you need to do to get the correct answer. Sometimes the directions may seem vague if you aren't familiar with the words or the context of the words. Go ahead and look at the solution to see whether it helps you with the meaning. But if the vocabulary is still unrecognizable, you may want to refer to *Pre-Calculus For Dummies, Algebra II For Dummies,* or *Trigonometry For Dummies,* all published by the fine folks at Wiley.

You may not be able to follow a particular solution from one step to the next. Is something missing? This book is designed to provide you with enough practice to become very efficient in pre-calculus topics, but it isn't intended to give the step-by-step explanation of how and why each step is necessary. You may need to refer to the books listed in the preceding paragraph or their corresponding workbooks to get more background on a problem or to understand why a particular step is taken in the solution of the problem.

Some pre-calculus topics are sometimes seen as being a bunch of rules without a particular purpose. Why do you have to solve for the exponent of that equation? Where will you use the fact that $\tan^2 x + 1 = \sec^2 x$? All these questions are more apparent when you see them tied together and when more background information is available. Don't be shy about seeking out that kind of information. And all this practice will pay off when you begin your first calculus experience. It may even be with Mark Ryan's *Calculus For Dummies!*

Part I
The Questions

Visit www.dummies.com for great (and free!) Dummies content online.

Part I

The Questions

In this part . . .

You find 1,001 pre-calculus problems — many different types in three different difficulty levels. The types of problems you'll find are

- Basic algebraic rules and graphs as well as solving algebraic equations and inequalities (Chapters 1 through 5)
- Properties of exponential and logarithmic functions and their equations (Chapter 6)
- Trigonometry basics and solving trig identities (Chapters 7 through 11)
- Complex numbers, polar coordinates, and conic sections (Chapters 12 through 13)
- Systems of equations, sequences, and series (Chapters 14 and 15)
- Limits and continuity (Chapter 16)

Chapter 1

Getting Started with Algebra Basics

The basics of pre-calculus consist of reviewing number systems, properties of the number systems, order of operations, notation, and some essential formulas used in coordinate graphs. Vocabulary is important in mathematics because you have to relate a number or process to its exact description. The problems in this chapter reacquaint you with many old friends from previous mathematics courses.

The Problems You'll Work On

In this chapter, you'll work with simplifying expressions and writing answers in the following ways:

- Identifying which are whole numbers, integers, and rational and irrational numbers
- Applying the commutative, associative, distributive, inverse, and identity properties
- Computing correctly using the *order of operations* (parentheses, exponents/powers and roots, multiplication and division, and then addition and subtraction)
- Graphing inequalities for the full solution
- Using formulas for slope, distance, and midpoint
- Applying coordinate system formulas to characterize geometric figures

What to Watch Out For

Don't let common mistakes trip you up; keep in mind that when working with simplifying expressions and communicating answers, your challenges will be

- Distributing the factor over every term in the parentheses
- Changing the signs of all the terms when distributing a negative factor
- Working from left to right when applying operations at the same level
- Assigning points to the number line in the correct order
- Placing the change in y over the change in x when using the slope formula
- Satisfying the correct geometric properties when characterizing figures

Identifying Which System or Systems a Number Belongs To

***1–10** Identify which number doesn't belong to the number system.*

1. Which is not a rational number?

$4, -\frac{3}{8}, -2, \frac{\sqrt{3}}{7}, 0$

2. Which is not a rational number?

$591, -\frac{163}{164}, -\frac{4}{1}, 3i+1, 11^2$

3. Which is not a natural number?

$83, 27, 0, 14, 2^3$

4. Which is not a natural number?

$1, 1.9, 49, 101, 15$

5. Which is not an integer? $-3, \sqrt{20}, 0, -153, 9^4$

6. Which is not an integer?

$-23, \frac{49}{7}, -\frac{100}{6}, 27^3, 111$

7. Which is not an irrational number?

$\sqrt{25}, \sqrt[3]{81}, \sqrt[4]{624}, \sqrt{28}, \sqrt{7}$

8. Which is not an irrational number?

$\sqrt{101}, \sqrt{0}, \sqrt[5]{15}, \sqrt[3]{300}, \sqrt{2}$

9. Which is not an imaginary number?

$3i, 2i+1, i^2, 2-3i, 4i+i^2$

10. Which is not an imaginary number?

$6i, 2+i^4, 3-2i, 4i+4i, i^3$

Recognizing Properties of Number Systems

***11–20** Identify which property of numbers the equation illustrates.*

11. $14+11=11+14$

12. $-19+(9-4)=(-19+9)-4$

13. $\left(\frac{1}{4}\cdot 3\right)\cdot\frac{1}{3}=\frac{1}{4}\cdot\left(3\cdot\frac{1}{3}\right)$

14. $a+b+0=a+b$

15. $-3(x+7)=-3x-21$

16. $-11\left(-\frac{1}{11}\right)=1$

17. $(x+3)(y+2)=(x+3)\cdot y+(x+3)\cdot 2$

18. $9(20+3)=(20+3)9$

19. If $x=3$ and $y=x$, then $y=3$.

20. $\frac{8}{13}+\left(-\frac{8}{13}\right)=0$

Simplifying Expressions with the Order of Operations

21–30 *Simplify the expression by using the order of operations.*

21. $4(3)^2$

22. $\frac{6+18}{6}$

23. $\frac{\frac{1}{3}+\frac{1}{6}}{\frac{5}{7}+\frac{19}{21}}$

24. $\frac{\frac{13}{6}-\frac{1}{2}}{\frac{1}{6}+\frac{1}{3}}$

25. $\frac{-4+\sqrt{4-4(2)(-12)}}{4}$

26. $\frac{-20-\sqrt{400-4(-5)(-15)}}{-10}$

27. $\dfrac{\sqrt{18^2+4(19)}}{|-7+3|}$

28. $\dfrac{\sqrt{25-16}+\sqrt{9(6-2)}}{\sqrt{61^2-60^2}}$

29. $\dfrac{|-4(3)+2(7)|}{6\left(17-3^2\right)}$

30. $\dfrac{\left|3^2-4^2\right|-\left|4^2-5^2\right|}{\left(5^2-6^2\right)^2}$

Graphing Inequalities

31–40 Graph the inequality.

31. $x+y\geq 7$

32. $2x-y\leq 4$

33. $4x+5y>20$

34. $3x-2y<12$

35. $x>6$

36. $x\leq -2$

37. $y\geq -4$

38. $y\leq 0$

39. $y\leq 3x-7$

40. $x\geq 2y+1$

Using Graphing Formulas

41–50 *Solve by using the necessary formula.*

41. Find the slope of the line through the points (–2, 3) and (4, 9).

42. Find the slope of the line through the points (–4, –3) and (–6, 2).

43. Find the slope of the line through the points (4, –3) and (4, –7).

44. Find the slope of the line through the points (–2, –9) and (2, –9).

45. Find the distance between the points (–8, –1) and (–2, 7).

46. Find the distance between the points (0, 16) and (7, –8).

47. Find the distance between the points (6, –5) and (–4, 3).

48. Find the midpoint of the segment between the points (–5, 2) and (7, –8).

49. Find the midpoint of the segment between the points (6, 3) and (–4, –4).

50. Find the midpoint of the segment between the points $\left(\frac{1}{3}, -\frac{1}{2}\right)$ and $\left(\frac{5}{6}, -\frac{1}{4}\right)$.

Applying Graphing Formulas

51–60 *Use an appropriate formula to compute the indicated value.*

51. Find the perimeter of triangle ABC, whose vertices are A (1, 1), B (1, 4), and C (5, 1).

52. Find the perimeter of the parallelogram DEFG, whose vertices are D (0, 10), E (9, 13), F (11, 7), and G (2, 4).

53. Find the center of the rhombus HJKL, whose vertices are H (0, 3), J (4, 6), K (8, 3), and L (4, 0).

54. Determine which type(s) of triangle ABC is if the vertices are A (1, 1), B (4, 5), and C (9, –5).

55. Determine which type of triangle ABC is if the vertices are A (0, 0), B (0, 12), and C$\left(6\sqrt{3}, 6\right)$.

56. Find the length of the altitude of triangle ABC, drawn to side AC, with vertices A (0, 0), B (5, 12), and C (21, 0).

57. Find the length of the altitude of triangle DEF, drawn to side DF, with vertices D (2, 3), E (2, 12), and F (42, 3).

58. Find the area of the parallelogram PQRS with vertices P (4, 7), Q (7, 12), R (15, 12), and S (12, 7).

59. Find the area of circle A if the endpoints of its diameter are at (6, 13) and (–8, 21).

60. Find the area of triangle ABC with vertices A (0, 0), B (5, 12), and C (14, 0).

Chapter 2

Solving Some Equations and Inequalities

The object of solving equations and inequalities is to discover which number or numbers will create a true statement in the given expression. The main techniques you use to find such solutions include factoring, applying the multiplication property of zero, creating sign lines, finding common denominators, and squaring both sides of an equation. Your challenge is to determine which techniques work in a particular problem and whether you have a correct solution after applying those techniques.

The Problems You'll Work On

In this chapter, you'll work with equations and inequalities in the following ways:

- Writing inequality solutions using both inequality notation and interval notation
- Solving linear and quadratic inequalities using a sign line
- Determining the solutions of absolute value equations and inequalities
- Taking on radical equations and checking for extraneous roots
- Rationalizing denominators as a method for finding solutions

What to Watch Out For

Keep in mind that when solving equations and inequalities, your challenges will include

- Factoring trinomials correctly when solving quadratic equations and inequalities
- Choosing the smallest exponent when the choices include fractions and negative numbers
- Assigning the correct signs in intervals on the sign line
- Recognizing the solution $x = 0$ when a factor of the expression is x
- Determining which solutions are viable and which are extraneous
- Squaring binomials correctly when working with radical equations
- Finding the correct format for a binomial's conjugate

Using Interval and Inequality Notation

61–70 Write the expression using the indicated notation.

61. Write the expression $x < 6$ in interval notation.

62. Write the expression $-5 \le x < 8$ in interval notation.

63. Write the interval notation $(4, 16]$ as an inequality expression.

64. Write the interval notation $[-8, \infty)$ as an inequality expression.

65. Describe the graph using inequality notation.

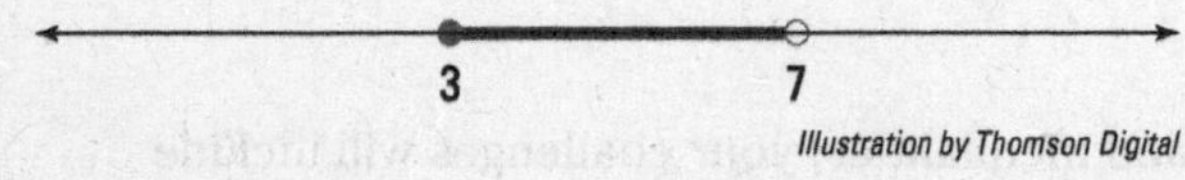

Illustration by Thomson Digital

66. Describe the graph using inequality notation.

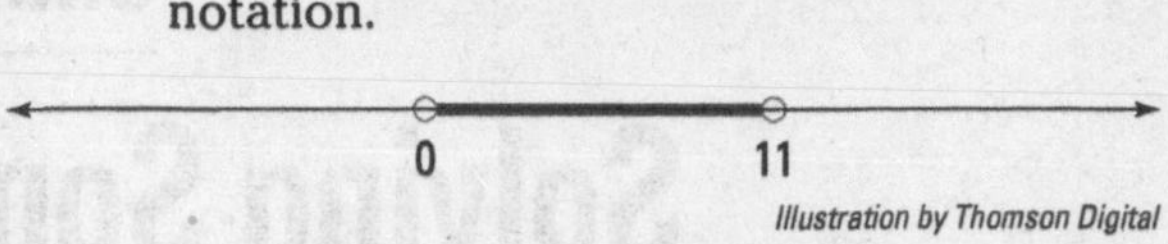

Illustration by Thomson Digital

67. Describe the graph using inequality notation.

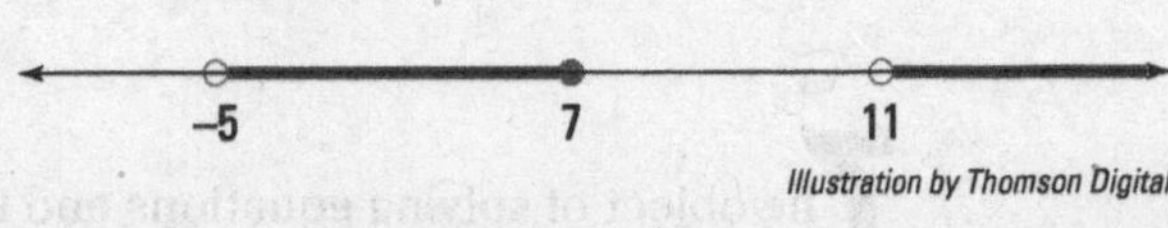

Illustration by Thomson Digital

68. Describe the graph using interval notation.

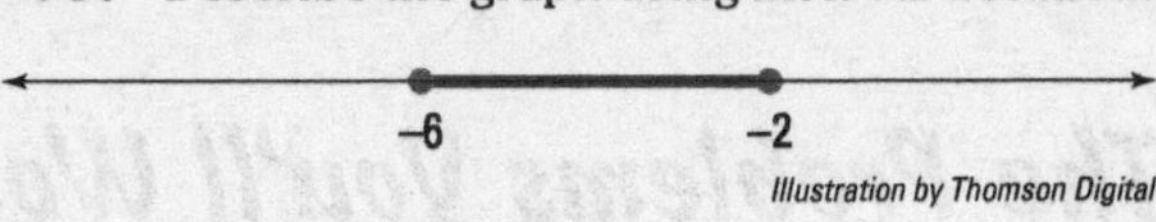

Illustration by Thomson Digital

69. Describe the graph using interval notation.

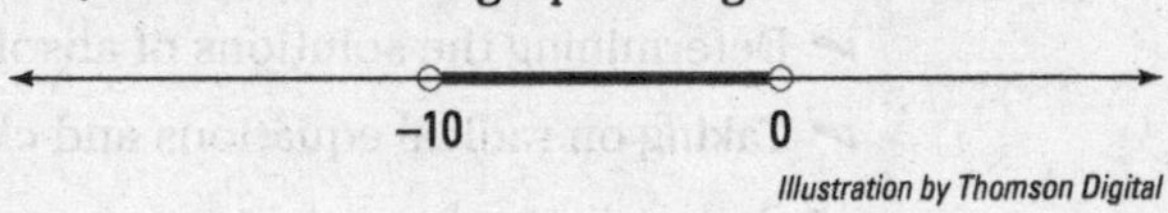

Illustration by Thomson Digital

70. Describe the graph using interval notation.

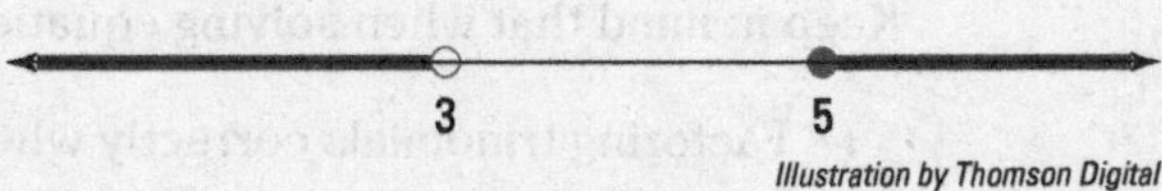

Illustration by Thomson Digital

Solving Linear Inequalities

71–74 *Solve the inequality for x.*

71. $4x-3>9$

72. $5x-6\leq 7x+8$

73. $16-3x<4x+2$

74. $-\frac{5}{3}x-3\geq 1-\frac{1}{3}x$

Solving Quadratic Inequalities

75–80 *Solve the nonlinear inequality for x.*

75. $x^2-7x-8\geq 0$

76. $24-10x+x^2>0$

77. $x^3-9x\leq 0$

78. $x^3-6x^2-x+6<0$

79. $\frac{x-3}{x+4}\geq 0$

80. $\frac{x+7}{x^2-25}<0$

Solving Absolute Value Inequalities

81–85 *Solve the absolute value inequality for x.*

81. $|x-4|\leq 3$

82. $|2x-3|>5$

83. $4|2x+1|<20$

84. $3-|x+5|\geq 1$

85. $5+|3x-1|\leq 2$

Working with Radicals and Fractional Notation

86–89 Change the expression using rules for fractional exponents.

86. Rewrite the radical expression $\sqrt[3]{x^4}$ with a fractional exponent.

87. Rewrite the radical expression $\frac{1}{\sqrt{x}}$ with a fractional exponent.

88. Rewrite the fractional exponent $x^{5/3}$ as a radical expression.

89. Rewrite the fractional exponent $x^{-3/4}$ as a radical expression.

Performing Operations Using Fractional Exponents

90–94 Simplify the expression.

90. $x^{2/3}\cdot x^{4/3}$

91. $x^{1/5}\cdot x^{-6/5}$

92. $\frac{x^{2/3}}{x^{1/3}}$

93. $\frac{x^{1/2}}{x^{-1/2}}$

94. $\frac{x^{-3/5}}{x^{1/10}}$

Factoring Using Fractional Notation

95–100 *Factor the greatest common factor (GCF) from each term and write the factored form.*

95. $2x^{1/2}+4x^{3/2}$

96. $-9y^{2/3}+4y^{1/3}$

97. $6x^{2/5}-4x^{-3/5}$

98. $15y^{-3/5}+10y^{-8/5}$

99. $4x^{1/3}+8x^{2/3}-5x$

100. $-16y^{-3/4}-32y^{-7/4}+48y^{-11/4}$

Solving Radical Equations

101–110 *Solve for x. Check for any extraneous solutions.*

101. $\sqrt{4x+5}=5$

102. $\sqrt{10-2x}=4$

103. $4+\sqrt{3x-2}=9$

104. $\sqrt{4-7x}-3=2$

105. $\sqrt{6x-5}=2x-5$

106. $\sqrt{7-2x}=x+4$

107. $\sqrt{5x+14}-6=2x$

108. $\sqrt{3x+10}-x=4$

109. $\sqrt{2x+3}-\sqrt{x+1}=1$

110. $\sqrt{11-5x}+\sqrt{x+6}=7$

Rationalizing Denominators

***111–120** Simplify by rationalizing the denominator.*

111. $\dfrac{4}{\sqrt{6}}$

112. $\dfrac{10}{\sqrt{5}}$

113. $\dfrac{6+\sqrt{2}}{\sqrt{3}}$

114. $\dfrac{10-\sqrt{14}}{\sqrt{2}}$

115. $\dfrac{5}{1+\sqrt{2}}$

116. $\dfrac{12}{3-\sqrt{3}}$

117. $\dfrac{8}{\sqrt{5}+3}$

118. $\dfrac{\sqrt{6}}{4+\sqrt{10}}$

119. $\dfrac{\sqrt{6}-\sqrt{10}}{\sqrt{2}+\sqrt{5}}$

120. $\dfrac{\sqrt{14}-\sqrt{3}}{\sqrt{7}-\sqrt{3}}$

Chapter 3

Function Basics

A *function* is a special type of rule or relationship. The difference between a function and a relation is that a function has exactly one output value (from the range) for every input value (from the domain). Functions are very useful when you're describing trends in business, heights of objects shot from a cannon, times required to complete a task, and so on. Functions have some special properties and operations that allow for investigation into what happens when you change the rule.

The Problems You'll Work On

In this chapter, you'll work with functions and function operations in the following ways:

- Writing and using function notation
- Determining the domain and range of different types of functions
- Recognizing even and odd functions
- Checking on whether a function is one-to-one
- Finding inverses of one-to-one functions
- Performing the basic operations on functions and function rules
- Working with the composition of functions and the difference quotient

What to Watch Out For

Don't let common mistakes trip you up; keep in mind that when working with functions, your challenges will include

- Following the order of operations when evaluating functions
- Determining which values need to be excluded from a function's domain
- Working with negative signs correctly when checking for even and odd functions
- Being sure a function is one-to-one before trying to determine an inverse
- Correctly applying function rules when performing function composition
- Raising binomials to higher powers and including all the terms

Using Function Notation to Evaluate Function Values

121–125 *Evaluate the function for the given value.*

121. Given $f(n)=4n^3+n^2-6n+2$, find $f(-3)$.

122. Given $g(x)=|x-3|+2$, find $g(-6)$.

123. Given $h(y)=-2^{2y+2}$, find $h(3)$.

124. Given $g(x)=\begin{cases} x^2+2 & \text{if } x\le -4 \\ |x| & \text{if } -4<x\le 0, \\ 2x^2-1 & \text{if } x>0 \end{cases}$

find $g(-5)$.

125. Given $f(x)=2x^2-3x$, find $f(x+h)$.

Determining the Domain and Range of a Function

126–135 *Find the domain and range for the function.*

126. $f=\{(5,3),(2,4),(-2,6),(1,3),(0,2)\}$

127.

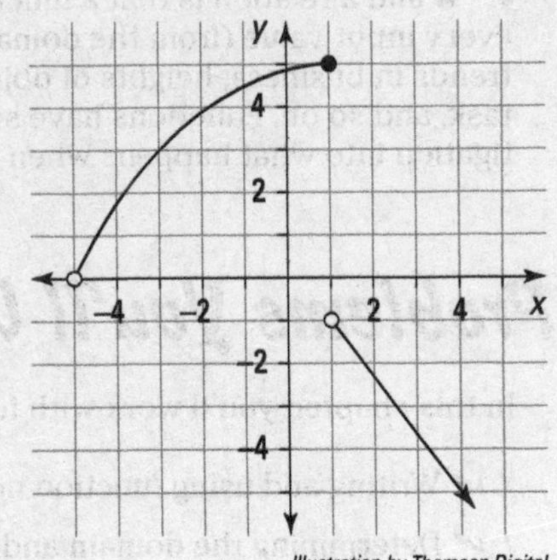

Illustration by Thomson Digital

128. $f(x)=-3x+4$

129. $f(x)=2x^2+1$

130. $h(t)=128t-16t^2, 0\le t\le 8$

131. $r(x) = x^3 - 6$

132. $g(x) = \sqrt{5-2x}$

133. $h(x) = \sqrt[3]{2x+9}$

134. $g(x) = \dfrac{x-5}{2x+6}$

135. $f(x) = |x-4| + 3$

Recognizing Even Functions

136–137 *Determine which function is even.*

136. $f(x) = -3x^3 + 4$, $f(x) = -3x^2 + 4x$, $f(x) = -3x^2 + 4$, $f(x) = -3x^3 + 4x$, $f(x) = 3x^2 - 4x$

137. $g(x) = 1 - x^2 - 2x^4 + x^6$, $g(x) = 1 + x^3 - 2x^4 + x^6$, $g(x) = -1 + x^3 - 2x^4 + x^6$, $g(x) = x - x^2 + 2x^4 + x^5$, $g(x) = -x - x^2 + 2x^4 + x^6$

Identifying Odd Functions

138–139 *Determine which function is odd.*

138. $f(x) = -4x^3 - x^2$, $f(x) = -4x^3 + x^5 + 3$, $f(x) = -4x^3 + x^5$, $f(x) = 4x^3 - x^2 + x$, $f(x) = -4x^3 - x^2 + x$

139. $g(x) = x^2 + 2x^3 - x^5$, $g(x) = x + 2x^4 - x^5$, $g(x) = x + 2x^3 - 3$, $g(x) = x + 2x^3 - x^5$, $g(x) = x^2 + 2x^3 - x^4$

Ruling Out Even and Odd Functions

140 *Determine which function is neither even nor odd.*

140. $h(x) = 2x^5 - 3x$, $h(x) = 2x^6 - 3$, $h(x) = x^2 + 2$, $h(x) = x^2 - 2x$, $h(x) = 2x^4 - 4x^2 - 3$

Recognizing One-to-One Functions from Given Relations

141–143 *Determine which of the given relations is a one-to-one function over its domain.*

141. $\{(1,a),(2,d),(3,b)\}$, $\{(1,b),(2,c),(3,b)\}$, $\{(1,a),(2,c),(3,b),(1,d)\}$, $\{(1,b),(1,c),(1,d),(2,a)\}$, $\{(1,d),(2,d),(3,d)\}$

142. $\{(2,3),(4,5),(1,5),(3,4)\}$,
$\{(2,3),(4,2),(2,5),(3,4)\}$,
$\{(2,2),(4,2),(1,2),(3,2)\}$,
$\{(2,3),(4,2),(1,5),(3,3)\}$,
$\{(2,3),(4,2),(1,5),(3,4)\}$

143.

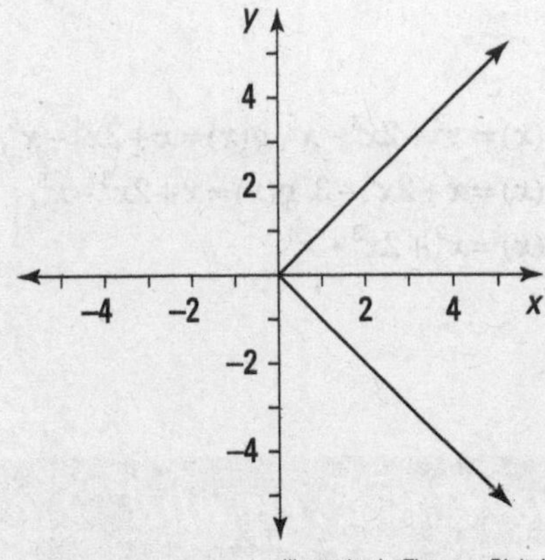

Illustration by Thomson Digital

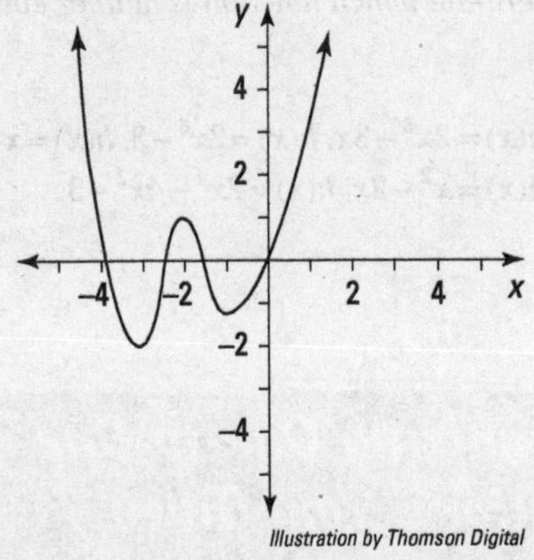

Illustration by Thomson Digital

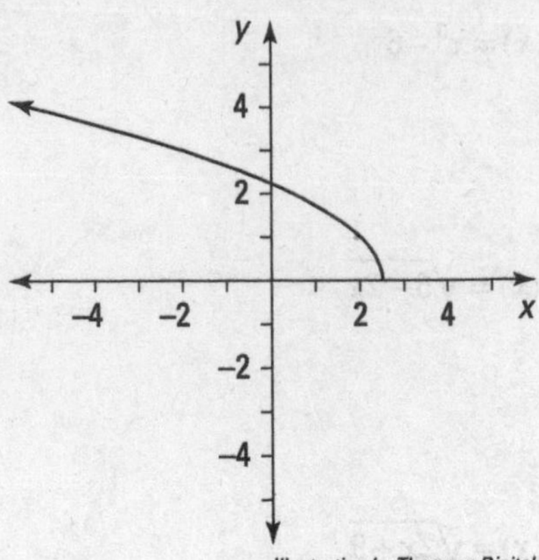

Illustration by Thomson Digital

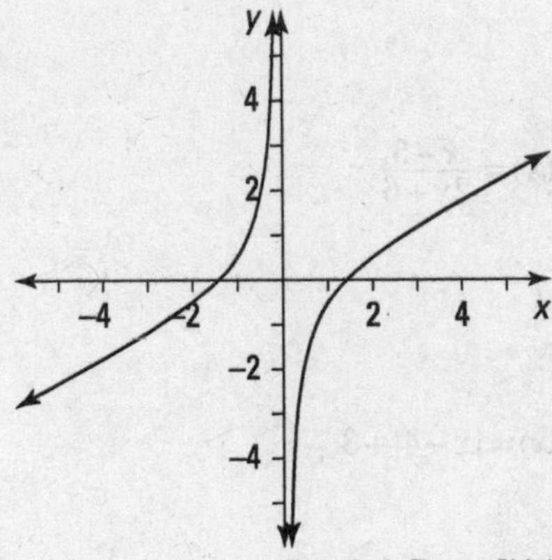

Illustration by Thomson Digital

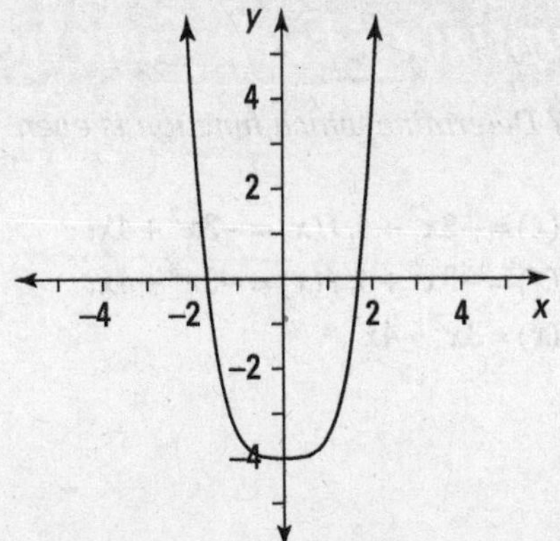

Illustration by Thomson Digital

Identifying One-to-One Functions from Equations

144–145 *Determine which of the functions is one-to-one over its domain.*

144. $f(x)=x^2+2$, $f(x)=-x^3$, $f(x)=x^2+x$, $f(x)=8$, $f(x)=x^4$

145. $g(x)=x^2-2x+1$, $g(x)=|x+2|$, $g(x)=\frac{x-3}{x^3}$, $g(x)=\frac{2x+1}{x}$, $g(x)=\frac{3x^2+2}{x-1}$

Recognizing a Function's Inverse

146–150 *Determine which pair of functions are inverses of each other.*

146. $f(x)=3x-1, g(x)=\frac{x}{3}+1$,
$f(x)=3x-1, g(x)=x+\frac{1}{3}$,
$f(x)=3x-1, g(x)=\frac{3x+1}{3}$,
$f(x)=3x-1, g(x)=-3x+1$,
$f(x)=3x-1, g(x)=\frac{x+1}{3}$

147. $f(x)=4+\frac{3}{2}x, g(x)=\frac{2}{3}x-4$,
$f(x)=4+\frac{3}{2}x, g(x)=\frac{2}{3}x+4$,
$f(x)=4+\frac{3}{2}x, g(x)=\frac{2}{3}(x-4)$,
$f(x)=4+\frac{3}{2}x, g(x)=\frac{3}{2}(x-4)$,
$f(x)=4+\frac{3}{2}x, g(x)=4-\frac{2}{3}x$

148. $f(x)=-\frac{2}{x}-1, g(x)=\frac{2}{x+1}$,
$f(x)=-\frac{2}{x}-1, g(x)=-\frac{2}{x+1}$,
$f(x)=-\frac{2}{x}-1, g(x)=\frac{x}{2}+1$,
$f(x)=-\frac{2}{x}-1, g(x)=-\frac{2}{x}+1$,
$f(x)=-\frac{2}{x}-1, g(x)=\frac{2}{x}+1$

149. $f(x)=\frac{x-2}{x+2}, g(x)=\frac{x+2}{x-2}$,
$f(x)=\frac{x-2}{x+2}, g(x)=-\frac{x+2}{x-2}$,
$f(x)=\frac{x-2}{x+2}, g(x)=\frac{2x+2}{x-1}$,
$f(x)=\frac{x-2}{x+2}, g(x)=-\frac{x-2}{x+2}$,
$f(x)=\frac{x-2}{x+2}, g(x)=\frac{-2x-2}{x-1}$

150. $f(x)=-(x-2)^3, g(x)=2+\sqrt[3]{x}$,
$f(x)=-(x-2)^3, g(x)=-2-\sqrt[3]{x}$,
$f(x)=-(x-2)^3, g(x)=-2+\sqrt[3]{x}$,
$f(x)=-(x-2)^3, g(x)=2-\sqrt[3]{x}$,
$f(x)=-(x-2)^3, g(x)=\sqrt[3]{x}-2$

Determining a Function's Inverse

151–160 *Find the inverse of the function.*

151. $f = \{(-4,1),(-3,2),(0,0),(1,10),(2,3)\}$

152. $f(x) = -2x + 3$

153. $f(x) = -\frac{3}{5}x + 2$

154. $f(x) = 1 - \frac{4}{x}$

155. $f(x) = \sqrt[3]{x-5}$

156. $f(x) = \frac{1}{x+3}$

157. $f(x) = \frac{4}{-x-2} + 2$

158. $f(x) = \frac{3}{x^3 - 2}$

159. $f(x) = \frac{x-7}{4x+5}$

160. $f(x) = \sqrt{1-x^2},\ 0 \le x \le 1$

Executing Operations on Functions

161–165 *Perform the indicated operation using the given functions.*

161. $f(x) = x^2 + 2x + 1$
$g(x) = -x - 5$
Find $(f+g)(x)$.

162. $h(n) = -3n - 3$
$k(n) = n^2 + 5$
Find $(k-h)(n)$.

163. $g(a) = 3a - 2$
$h(a) = a + 7$
Find $(h \cdot g)(a)$.

164. $f(x)=x^2-3x-4$
$g(x)=x+1$

Find $\left(\frac{f}{g}\right)(x)$.

165. $f(x)=8-3x$
$g(x)=x^2-2x$

Find $(g\cdot f)(-2)$.

Performing Function Composition

166–169 *Find $f \circ g$.*

166. $f(x)=3x+1$
$g(x)=x^2$

167. $f(x)=4x+3$
$g(x)=x-2$

168. $f(x)=x^2+2x-3$
$g(x)=x+1$

169. $f(x)=x^2-3$
$g(x)=\sqrt{x+3}$

Doing More Function Composition

170–176 *Perform the indicated operation.*

170. $f(x)=3x$
$g(x)=2^x$

Find $g \circ f$.

171. $f(x)=-x^2-2$
$g(x)=3x+5$

Find $g \circ f$.

172. $f(x)=5x-3$
$g(x)=\frac{2x-1}{3x+2}$

Find $g \circ f$.

173. $f(x)=x^2-2x$

Find $(f \circ f)(x)$.

174. $f(x) = -6x + 3$
$g(x) = 3x - 9$

Find $(f \circ g)(12)$.

175. $f(x) = \sqrt{x+7}$
$g(x) = x^2$
Find $(f \circ g)(-3)$.

176. $f(x) = x + 1$
$g(x) = x^2 - 2$
$h(x) = 4x$

Find $(h \circ g \circ f)(x)$.

Using the Difference Quotient

177–180 *Evaluate the difference quotient, $\frac{f(x+h) - f(x)}{h}$, for the given function $f(x)$. Assume $h \neq 0$.*

177. $f(x) = 4x + 3$

178. $f(x) = x^2 - 4$

179. $f(x) = 2x^2 + 5x - 3$

180. $f(x) = \frac{2}{x+1}$

Chapter 4

Graphing and Transforming Functions

You can graph functions fairly handily using a graphing calculator, but you'll be frustrated using this technology if you don't have a good idea of what you'll find and where you'll find it. You need to have a fairly good idea of how high or how low and how far left and right the graph extends. You get information on these aspects of a graph from the intercepts (where the curve crosses the axes), from any asymptotes (in rational functions), and, of course, from the domain and range of the function. A good working knowledge of the characteristics of different types of functions goes a long way toward making your graphing experience a success.

Another way of graphing functions is to recognize any transformations performed on basic function definitions. Just sliding a graph to the left or right or flipping the graph over a line is a lot easier than starting from scratch.

The Problems You'll Work On

In this chapter, you'll work with function graphs in the following ways:

- Graphing both a function and its inverse
- Determining the vertices of quadratic functions (parabolas)
- Recognizing the limits of some radical functions when graphing
- Pointing out the top or bottom point of an absolute-value function graph to establish its range
- Solving polynomial equations for intercepts
- Writing equations of the asymptotes of rational functions
- Using function transformations to quickly graph variations on functions

What to Watch Out For

When graphing functions, your challenges include the following:

- Taking advantage of alternate formats of function equations (slope-intercept form, factored polynomial or rational functions, and so on)
- Determining whether a parabola opens upward or downward and how steeply

- Graphing radical functions with odd-numbered roots and recognizing the unlimited domain
- Recognizing when polynomial functions don't cross the x-axis at an intercept
- Using asymptotes correctly as a guide in graphing
- Reflecting functions vertically or horizontally, depending on the function transformation

Functions and Their Inverses

***181–190** Find the inverse of the function. Then graph both the function and its inverse.*

181. $f(x)=3x-5$

182. $f(x)=-\frac{2}{3}x+5$

183. $f(x)=x^3$

184. $f(x)=x^5$

185. $f(x)=x^3-3$

186. $f(x)=\sqrt[3]{x-4}+7$

187. $f(x)=\frac{1}{x-2}$

188. $f(x)=\frac{x+3}{x-4}$

189. $f(x)=\frac{2}{\sqrt[3]{x-3}}$

190. $f(x)=\frac{3}{\sqrt[3]{4-x}}$

Sketching Quadratic Functions from Their Equations

***191–195** Sketch the graph of the quadratic function.*

191. $f(x)=\frac{1}{2}x^2-1$

192. $f(x)=x^2+6x+8$

193. $f(x) = -(x+2)^2 + 9$

194. $f(x) = 4\left(x + \frac{1}{2}\right)^2 + 1$

195. $f(x) = -7x^2 + 14x$

Writing Equations from Graphs of Parabolas

196–200 *Given the graph of a quadratic function, write its function equation in vertex form,* $y = a(x-h)^2 + k$.

196.

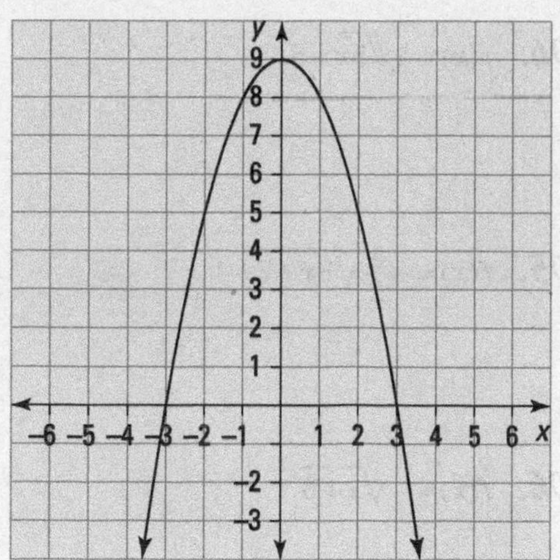

Illustration by Thomson Digital

197.

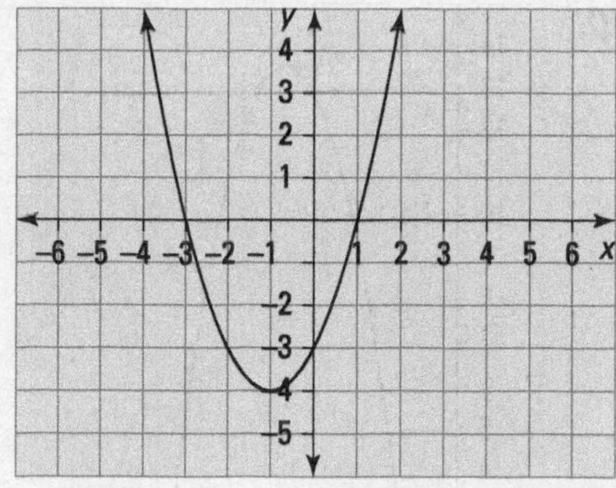

Illustration by Thomson Digital

198.

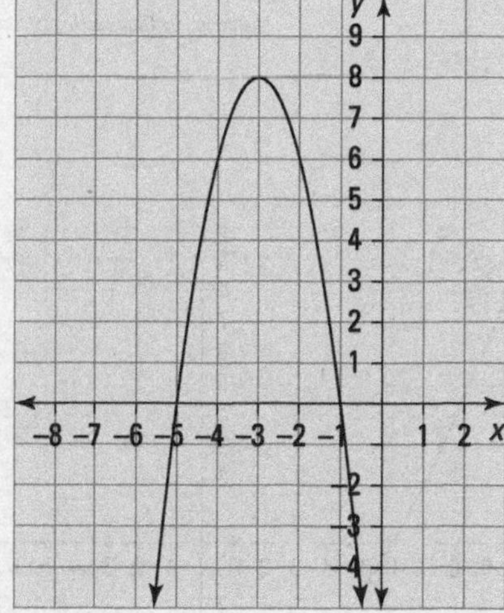

Illustration by Thomson Digital

199.

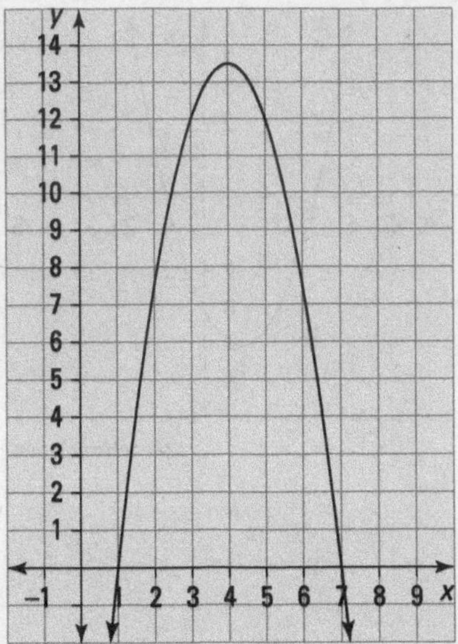

Illustration by Thomson Digital

200.

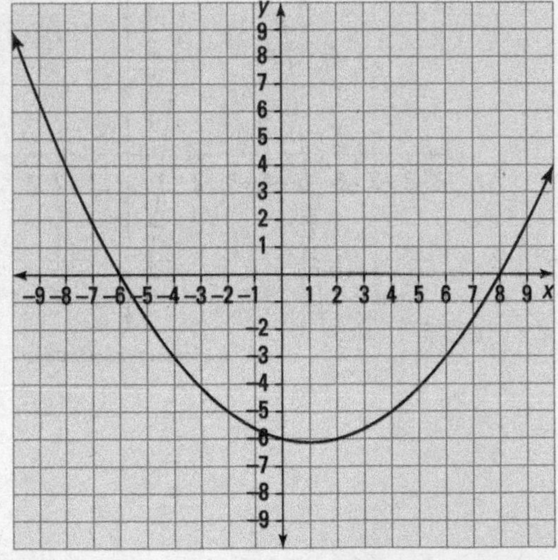

Illustration by Thomson Digital

Investigating and Graphing Radical Functions

***201–210** Identify the domain of the radical function and any intercepts. Then sketch a graph of the function.*

201. $f(x)=\sqrt{x+4}$

202. $f(x)=-\sqrt{2x-3}+1$

203. $f(x)=3\sqrt{x}-6$

204. $f(x)=4\sqrt{5x}-8$

205. $f(x)=-\frac{1}{2}\sqrt{3x-7}+4$

206. $f(x)=-\sqrt[3]{x+3}$

207. $f(x)=\sqrt[3]{3x-4}-3$

208. $f(x)=\frac{1}{2}\sqrt[3]{6x+2}+5$

209. $f(x)=3\sqrt[3]{x}-7$

210. $f(x)=-2\sqrt[3]{x-2}+1$

Investigating Absolute Value Functions

***211–215** Find the range of the absolute value function. Then sketch the function's graph.*

211. $f(x)=|2x-6|+3$

212. $f(x)=3|5x|-7$

213. $f(x)=|x^2-4|$

214. $f(x)=|x^3+1|$

215. $f(x)=|4\sqrt[3]{x-3}|$

Investigating the Graphs of Polynomial Functions

***216–223** Determine the intercepts of the graph of the polynomial. Then sketch the graph.*

216. $f(x)=x(x-4)(x+5)$

217. $f(x)=(x-3)^2(x+2)^2$

218. $f(x)=(x+4)^2(x-3)$

219. $f(x)=x^3(x-2)^3$

220. $f(x)=(x^2-4)\,(x^2-25)$

221. $f(x)=x\,(x^2-9)\,(x^2-36)$

222. $f(x)=x^2\,(x^2-16)$

223. $f(x)=(x^2-25)\,(x+1)^2$

Investigating Rational Functions

224–230 *Determine any asymptotes of the rational function. Then graph the function.*

224. $f(x)=\frac{x-2}{x+3}$

225. $f(x)=\frac{x^2-9}{x^2+3x-4}$

226. $f(x)=\frac{x+3}{x^2-25}$

227. $f(x)=\frac{x^2-5x}{x^2-25}$

228. $f(x)=\frac{x^2-2x-15}{x^2-7x+10}$

229. $f(x)=\frac{x^2-4}{x-1}$

230. $f(x)=\frac{x^3-2x^2+x-2}{x^2-1}$

Transformation of Functions

231–235 *Describe the transformations from the f function to the g function.*

231. $f(x)=x^2,\ g(x)=-(x-5)^2-7$

232. $f(x)=x^3,\ g(x)=\frac{1}{2}(x+4)^3$

233. $f(x)=|x|,\ g(x)=-|x+3|+2$

234. $f(x)=x^4,\ g(x)=3x^4-8$

235. $f(x)=\sqrt{x},\ g(x)=4\sqrt{-x}+3$

Transforming Selected Points Using Functions

236–240 *Transform the points (3, 4), (–2, 6), and (5, –1) using the function description of the transformations.*

236. $f(x+2)-4$

237. $5f(x)+6$

238. $-f(x-3)+5$

239. $\frac{1}{2}f(x+5)-8$

240. $f(-x)-3$

Sketching Graphs Using Basic Functions and Transformations

241–245 *Find the vertex of the given function, which is a transformation of* $f(x)=x^2$. *Then sketch the graph of the new function.*

241. $f(x)=2x^2-3$

242. $f(x)=\frac{1}{3}(x-2)^2$

243. $f(x)=(x-3)^2+4$

244. $f(x)=-3(x+1)^2+6$

245. $f(x)=-\frac{1}{4}(x-4)^2+2$

Sketching More Graphs Using Basic Functions and Transformations

246–250 *Find the vertical asymptote of the given function, which is a transformation of* $f(x)=\frac{1}{x}$. *Then sketch the graph of the new function.*

246. $f(x)=\frac{1}{x-3}$

247. $f(x)=\frac{2}{x+4}$

248. $f(x)=-\frac{3}{x+1}$

249. $f(x)=\frac{1}{5-x}+1$

250. $f(x)=\frac{6}{1-3x}-2$

Chapter 5

Polynomials

Polynomial functions have graphs that are smooth curves. They go from negative infinity to positive infinity in a nice, flowing fashion with no abrupt changes of direction. Pieces of polynomial functions are helpful when modeling physical situations, such as the height of a rocket shot in the air or the time a person takes to swim a lap depending on his or her age.

Most of the focus on polynomial functions is in determining when the function changes from negative values to positive values or vice versa. Also of interest is when the curve hits a relatively high point or relatively low point. Some good algebra techniques go a long way toward studying these characteristics of polynomial functions.

The Problems You'll Work On

In this chapter, you'll work with polynomial functions in the following ways:

- Solving quadratic equations by factoring or using the quadratic formula
- Rewriting quadratic equations by completing the square
- Factoring polynomials by using grouping
- Looking for rational roots of polynomials by using the rational root theorem
- Counting real roots with Descartes's rule of signs
- Using synthetic division to quickly compute factors
- Writing equations of polynomials given roots and other information
- Graphing polynomials by using end-behavior and the factored form

What to Watch Out For

Don't let common mistakes trip you up; keep in mind that when working with polynomial functions, your challenges will include

- Watching the order of operations when using the quadratic formula
- Adding to both sides when completing the square
- Remembering to insert zeros for missing terms when using synthetic division
- Recognizing the effect of imaginary roots on the graph of a polynomial

Using Factoring to Solve Quadratic Equations

***251–260** Find the solution set for the equation by factoring.*

251. $x^2+2x=0$

252. $x^2-x-3=9$

253. $4x^2-25=0$

254. $49x^2-70x+25=0$

255. $3x^2-9x-60=24$

256. $7x^2-15x+2=0$

257. $5x^2-13x-6=0$

258. $35x^2-69x+28=0$

259. $x^6+19x^3-216=0$

260. $x^4-10x^2+21=0$

Solving Quadratic Equations by Using the Quadratic Formula

***261–265** Use the quadratic formula to solve the equation.*

261. $4x^2-7x-15=0$

262. $x^2=x+4$

263. $4x^2+8x=1$

264. $2x^2-5x-5=0$

265. $(x-4)^2=11$

Using Completing the Square to Solve Quadratic Equations

266–270 *Solve the equation by completing the square.*

266. $x^2+2x-8=0$

267. $x^2+11x=-18$

268. $x^2+4x=3$

269. $5x^2+3x=2$

270. $3x^2-16x-2=-7$

Solving Polynomial Equations for Intercepts

271–276 *Find the real roots (x-intercepts) of the polynomial.*

271. $-3x^6+15x^4=0$

272. $8x^3-24x^2+18x=0$

273. $15x^4+26x^3+7x^2=0$

274. $x^4-16=0$

275. $4x^4+20x^3+6x^2=0$

276. $27x^3+8=0$

Using Factoring by Grouping to Solve Polynomial Equations

277–280 Find the real roots (x-intercepts) of the polynomial by using factoring by grouping.

277. $x^3-x^2-5x+5=0$

278. $3x^3+2x^2-3x-2=0$

279. $x^3-9x^2-4x+36=0$

280. $25x^3+5x^2-30x-6=0$

Applying Descartes's Rule of Signs

281–285 Use Descartes's rule of signs to determine the possible number of positive and negative real roots of the function.

281. $f(x)=6x^3-5x^2-3x+1$

282. $f(x)=18x^6-24x^4-25x^2+60$

283. $f(x)=7x^6-5x^4+2x^3-1$

284. $f(x)=2x^4-3x^3+8x^2-7x+5$

285. $f(x)=x^5+x^4+x^3+x^2+x+1$

Listing Possible Roots of a Polynomial Equation

286–290 Use the rational root theorem to list the possible rational roots of the function.

286. $f(x)=7x^3-18x^2+17x-7$

287. $f(x)=3x^3+2x-5$

288. $f(x)=4x^3-7x^2+4x-1$

289. $f(x)=4x^3-5x^2+3x+8$

290. $f(x)=5x^3+2x^2-10x-8$

Dividing Polynomials

291–295 *Perform the division by using long division. Write the remainder as a fraction.*

291. $(x^2+10x+15)\div(x+5)$

292. $(3x^3-7x^2-10x+5)\div(3x-1)$

293. $(35x^2-24)\div(7x+7)$

294. $(4x^4+3x^3+2x+1)\div(x^2+x+2)$

295. $(2x^3+2x^2-x-5)\div(2x^2-1)$

Using Synthetic Division to Divide Polynomials

296–298 *Use synthetic division to find the quotient and remainder.*

296. $(x^2+6x-15)\div(x+5)$

297. $(2x^3-13x^2+27x+20)\div(x-4)$

298. $(x^3-20)\div(x+3)$

Checking for Roots of a Polynomial by Using Synthetic Division

299–300 *Use synthetic division to determine which of the given values is a zero of the function.*

299. $f(x)=3x^2-11x+6$

Possible zeros: –3, 3, –6, 6, 1

300. $f(x) = x^4 + 2x^3 - 9x^2 - 2x + 8$

Possible zeros: –8, –2, –1, 4, 8

Writing Polynomial Expressions from Given Roots

301–305 *Find the lowest order polynomial with 1 as the leading coefficient with the listed values as its zeros.*

301. –3, 2, 3

302. –4, –3, 0, 1, 2

303. $1 + i, 1 - i, -1, 2$

304. $-\sqrt{2}, \sqrt{2}, -1$, and –4

305. $2 - \sqrt{3}, 2 + \sqrt{3}, 1, 2$

Writing Polynomial Expressions When Given Roots and a Point

306–310 *Find the lowest-order polynomial that has the listed values as its zeros and whose graph passes through the given point.*

306. Zeros: –3, 1, 2

Point: (–1, 24)

307. Zeros: –4, –2, 3

Point: (2, 48)

308. Zeros: –3, –2, 1, 2

Point: (0, 36)

309. Zeros: $-\sqrt{3}, \sqrt{3}, -2, 1$

Point: (2, 16)

310. Zeros: $2 + i, 2 - i, -3, 2$

Point: (–1, 120)

Graphing Polynomials

311–316 *Graph the polynomial by determining the end behavior, finding the x- and y-intercepts, and using test points between the x-intercepts.*

311. $f(x) = -2x^3 + 8x$

312. $f(x) = 3x^3 + 10x^2 - 27x - 10$

313. $f(x) = x^4 - x^3 - 6x^2$

314. $f(x) = x^4 + x^3 - 9x^2 + 11x - 4$

315. $f(x) = -x^4 + 9x^3 - 31x^2 + 49x - 30$

316. $f(x) = -x^5 - x^4 + 5x^3 + 5x^2 - 4x - 4$

Writing Equations from Graphs of Polynomials

317–320 *Write an equation for the given polynomial graph.*

317.

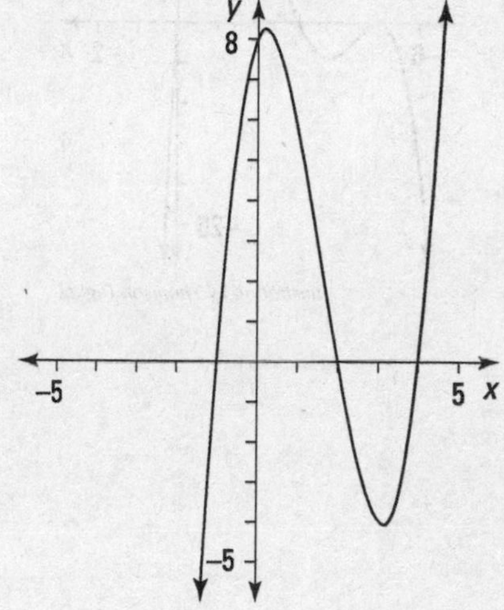

Illustration by Thomson Digital

318.

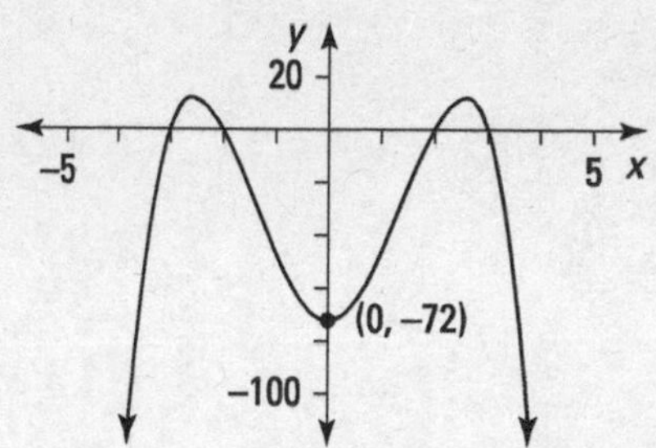

Illustration by Thomson Digital

319.

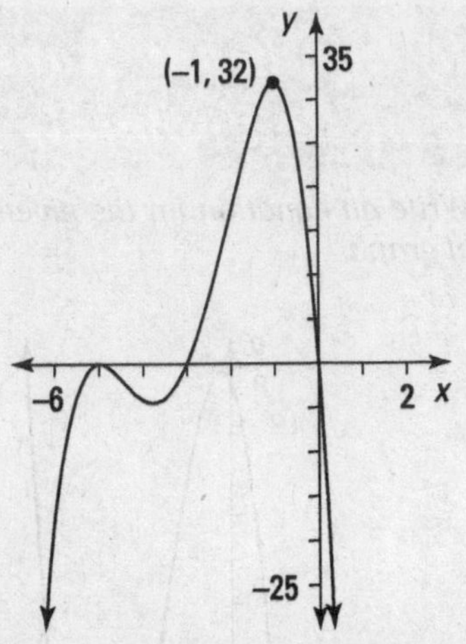

Illustration by Thomson Digital

320.

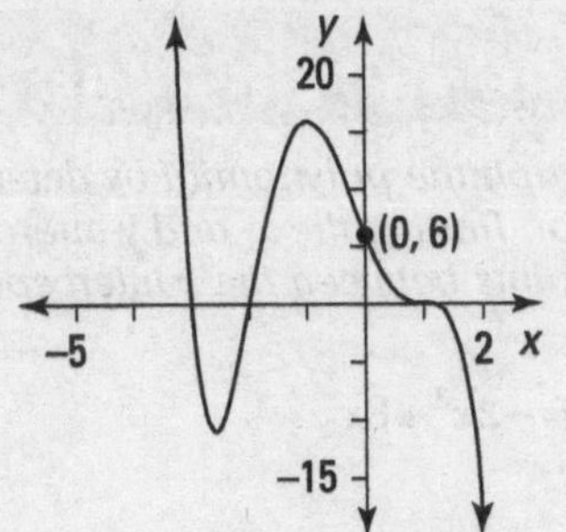

Illustration by Thomson Digital

Chapter 6

Exponential and Logarithmic Functions

Exponential and logarithmic functions go together. You wouldn't think so at first glance, because exponential functions can look like $f(x)=2e^{3x}$, and logarithmic (log) functions can look like $f(x)=\ln\left(x^2-3\right)$. What joins them together is that exponential functions and log functions are inverses of each other.

Exponential and logarithmic functions can have bases that are any positive number except the number 1. The special cases are those with base 10 (common logarithms) and base e (natural logarithms), which go along with their exponential counterparts.

The whole point of these functions is to tell you how large something is when you use a particular exponent or how big of an exponent you need in order to create a particular number. These functions are heavily used in the sciences and finance, so studying them here can pay off big time in later studies.

The Problems You'll Work On

In this chapter, you'll work with exponential and logarithmic functions in the following ways:

- Evaluating exponential and log functions using the function rule
- Simplifying expressions involving exponential and log functions
- Solving exponential equations using rules involving exponents
- Solving logarithmic equations using laws of logarithms
- Graphing exponential and logarithmic functions for a better view of their powers
- Applying exponential and logarithmic functions to real-life situations

What to Watch Out For

Don't let common mistakes trip you up. Here are some of the challenges you'll face when working with exponential and logarithmic functions:

- Using the rules for exponents in various operations correctly
- Applying the laws of logarithms to denominators of fractions

- Remembering the order of operations when simplifying exponential and log expressions
- Checking for extraneous roots when solving logarithmic equations

Understanding Function Notation

321–325 *Evaluate the function at the indicated points.*

321. Evaluate the function $f(x)=\left(\frac{1}{3}\right)^x$ at $f(-2)$ and $f(1)$.

322. Evaluate the function $f(x)=-2^x$ at $f(-3)$ and $f(2)$.

323. Evaluate the function $f(x)=5^x-2$ at $f(0)$ and $f(3)$.

324. Evaluate the function $f(x)=-3\left(\frac{1}{2}\right)^x+2$ at $f(-3)$ and $f(4)$.

325. Evaluate the function $f(x)=\frac{1}{2}(-2)^x+1$ at $f(1)$ and $f(5)$.

Graphing Exponential Functions

326–330 *Graph the exponential function.*

326. $f(x)=-3^x$

327. $f(x)=-5^x+10$

328. $f(x)=6\left(\frac{1}{2}\right)^x$

329. $f(x)=-4\left(\frac{2}{3}\right)^x-3$

330. $f(x)=\frac{3}{4}(3)^x+5$

Solving Exponential Equations

331–345 *Solve the exponential equation for x.*

331. $e^{4x} = e^{3x+6}$

332. $2^x = \frac{1}{32}$

333. $\left(\frac{1}{e}\right)^{2x} = e^3$

334. $3^{3x+4} = 9$

335. $\left(\frac{1}{5}\right)^{4x-6} = \frac{1}{625}$

336. $2e^{3x+8} = 2e^{-9x-1}$

337. $12^{-3x+2} - 44 = 100$

338. $3^{-x/2} + 80 = 809$

339. $\left(\frac{3}{4}\right)^{7x} - \frac{1}{2} = \frac{1}{16}$

340. $6\left(6^{-4x-9}\right) = 1{,}296$

341. $2\left(8^{6x-3}\right) + 24 = 1{,}048$

342. $\left(\frac{1}{49}\right)^{3x-8} + 25 = 368$

343. $(64)^{-2x-5} + \frac{1}{64} = \frac{1}{32}$

344. $4^{x^2-40} = \left(\frac{1}{4}\right)^{3x}$

345. $2^{x^2+8x-15} = 4^{4x+5}$

Using the Equivalence $b^x = y \Leftrightarrow \log_b y = x$ to Rewrite Expressions

346–348 *Rewrite the exponential expression as a logarithmic expression.*

346. $10^x = 100$

347. $y^3 = \frac{1}{27}$

348. $4^3 = z$

Using the Equivalence $\log_b y = x \Leftrightarrow b^x = y$ to Rewrite Expressions

349–350 *Rewrite the logarithmic expression as an exponential expression.*

349. $\log_8 128 = x$

350. $\log_{64} y = \frac{1}{2}$

Rewriting Logarithmic Expressions

351–358 *Write the expression in a new form using the laws of logarithms.*

351. $\log(4x)$

352. $\log_6\left(\frac{24}{y}\right)$

353. $\log_8 x^2 + \log_8 8$

354. $\log_5 5^2 + \log_5 1$

355. $3\log_6 x + \log_6 8$

356. $\ln 10 + 2\ln y$

357. $\log 100 - \log 1{,}000$

358. $5\log_4 x - 2\log_4 y$

Rewriting Logs of Products and Quotients as Sums and Differences

359–365 *Write the expression as an expanded logarithm.*

359. $\log_8\left(\frac{wx^2}{64}\right)$

360. $\ln\left(\frac{10e^{-4}}{y^2}\right)$

361. $\log 1000x^2y^3$

362. $\log_6\left(\frac{1}{36xy^5}\right)$

363. $\ln\left(\frac{6e^{-4}y^7}{7e^5z^2}\right)$

364. $\log_3\left(\frac{9x^4y^{-5}}{3z^2}\right)$

365. $\log\left(\frac{1{,}000y^{-2}z}{8x^{-4}}\right)$

Solving Logarithmic Equations

366–375 *Solve the logarithmic equation for x.*

366. $\log_6(4x-8)=2$

367. $\ln(8x-1)=4$

368. $\log(3x-5)=-1$

369. $\log_3(5x+2)+\log_3(x-4)=3$

370. $\log_5(4x+3)-\log_5(2x+5)=-2$

371. $\ln(6x-8)-\ln 10=3$

372. $\log_2(x+5)+\log_2(x-1)=4$

373. $\log(x^2-8x)=\log(4x+45)$

374. $\ln(x^2 + 16x) = \ln(4x - 32)$

375. $\log_3(3x) - \log_3(x+5) = \log_3 2$

Applying Function Transformations to Log Functions

376–380 *Find the best choice for a function rule of the transformation of $f(x) = \log(x)$ shown in the graph.*

376.

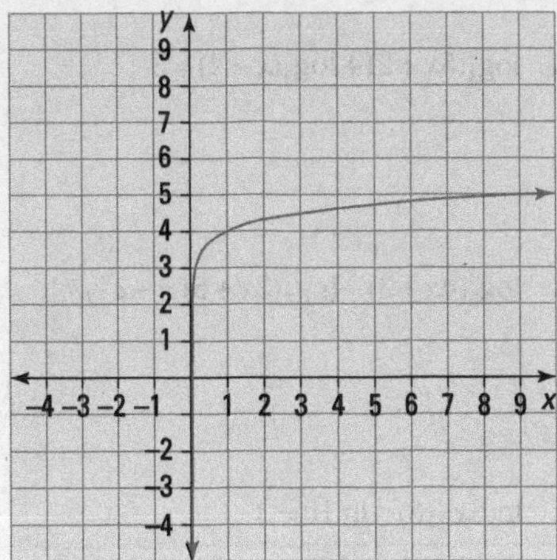

Illustration by Thomson Digital

(A) $f(x) = \log(x+4)$

(B) $f(x) = \log(x-4)$

(C) $f(x) = \log(x) + 4$

(D) $f(x) = \log(x) - 4$

(E) $f(x) = \log(4x)$

377.

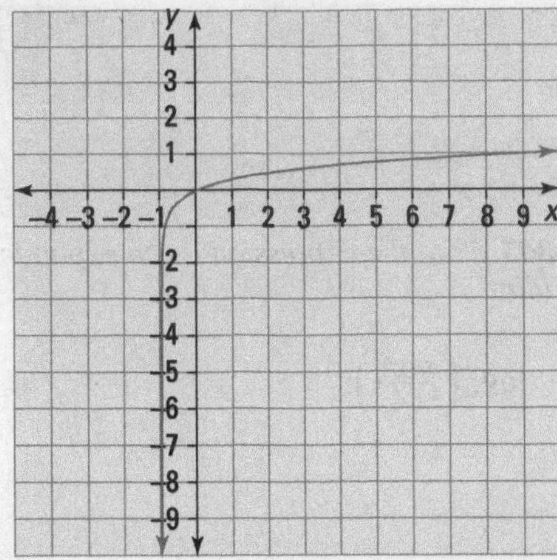

Illustration by Thomson Digital

(A) $f(x) = \log(x+1)$

(B) $f(x) = \log(x-1)$

(C) $f(x) = \log(x) + 1$

(D) $f(x) = \log(x) - 1$

(E) $f(x) = \log(-x)$

378.

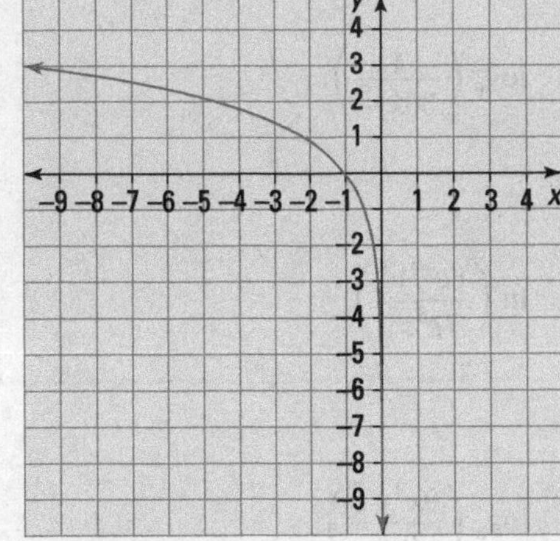

Illustration by Thomson Digital

(A) $f(x) = \log(x+3)$

(B) $f(x) = \log(x-3)$

(C) $f(x) = -3\log(-x)$

(D) $f(x) = -3\log(x)$

(E) $f(x) = 3\log(-x)$

379.

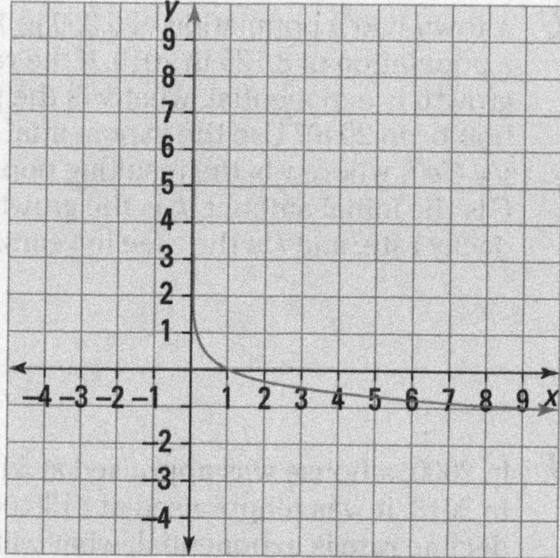

Illustration by Thomson Digital

(A) $f(x)=\log(-x)$

(B) $f(x)=-\log(-x)$

(C) $f(x)=1-\log(x)$

(D) $f(x)=\log\left(\frac{1}{x}\right)$

(E) $f(x)=\log\left(\frac{10}{x}\right)$

380.

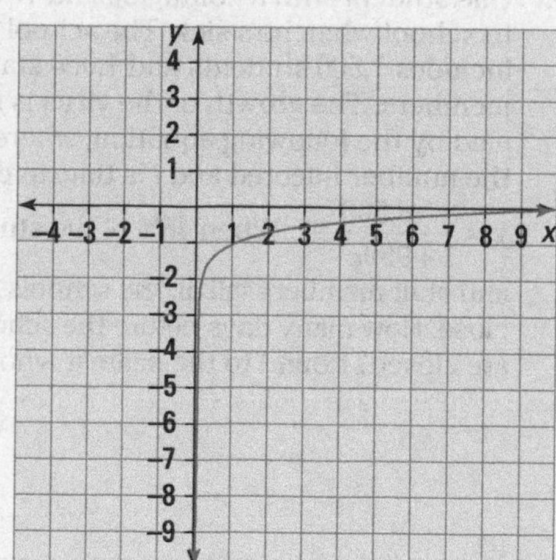

Illustration by Thomson Digital

(A) $f(x)=\log(-x)$

(B) $f(x)=-\log(x)$

(C) $f(x)=1-\log(x)$

(D) $f(x)=\log\left(\frac{1}{x}\right)$

(E) $f(x)=\log\left(\frac{x}{10}\right)$

Applying Logarithms to Everyday Life

381–390 *Use the given formula and properties of logarithms to answer the question.*

381. Orange juice has a hydrogen ion concentration of 0.001 moles/liter. What is the pH of orange juice? Use the pH formula $pH=-\log(H^+)$, where H^+ is the hydrogen ion concentration in moles/liter.

382. Ammonia water has a pH of 11.6. What is the hydrogen ion concentration (in moles/liter)? Use the pH formula $pH=-\log(H^+)$, where H^+ is the hydrogen ion concentration in moles/liter.

383. You invest $5,000 in an account at an annual rate of 1.475% with interest compounding quarterly. Find out how much is in the account (to the nearest cent) after 3 years by using the compound interest formula $A=P\left(1+\frac{r}{n}\right)^{nt}$, where A is the total amount in the account in dollars, P is the principal (amount invested), r is the annual interest rate, n is the number of times compounded per year, and t is the time in years.

384. If you invest \$1,000 at a rate of 3.25%, compounded continuously, how many years would it take for your money to double? Round to the nearest hundredth of a year. Use the continuous compound interest formula $A=Pe^{rt}$, where A is the amount in the account, P is the principal, r is the interest rate, and t is the time in years.

385. You buy a car that costs \$23,495, and it loses \$4,500 in value when you drive it off the car lot. The car then loses 10.785% in value each year (exponential depreciation). Rounded to the nearest cent, what is the value of the car after 5 years? Use the exponential equation $y=Ce^{kt}$, where y is the resulting value, C is the initial amount, k is the growth or decay rate, and t is the time in years.

386. A normal conversation is 60 decibels (dB). What is the intensity of the sound of the conversation (in watts per square meter)? Use the following formula, where L is decibels and I is the intensity of the sound: $L=10\log\left(\frac{I}{10^{-12}}\right)$.

387. Two earthquakes occurred in different parts of the country. The first earthquake had a magnitude of 8.0, and the second had a magnitude of 4.0. Compare the intensity of the earthquakes using the equation $M=\log\left(\frac{I}{I_0}\right)$, where M is the magnitude, I is the intensity measured from the epicenter, and I_0 is a constant.

388. A town had a population of 6,250 in 1975 and a population of 8,125 in 2010. If the rate of growth is exponential, what will the population be in 2040? Use the exponential model $y=Ce^{kt}$, where y is the resulting population, C is the initial amount, k is the growth or decay rate, and t is the time in years.

389. In 2000, a house was appraised at \$179,900. In 2013, it was reappraised at \$138,000. If the decline rate is exponential, what will the value of the house be in 2020? Round the value to the nearest dollar. Use the exponential model $y=Ce^{kt}$, where y is the resulting value, C is the initial amount, k is the growth or decay rate, and t is the time in years.

390. One student with a contagious flu virus goes to school when he's sick. The school district includes 7,500 students and 1,000 staff members. The growth of the virus is modeled by the following equation, where y is the number infected and t is time in days: $y=\frac{8{,}500}{1+999e^{-0.6t}}$. When 45% of the students and staff members fall ill, the schools will close. How many days before the schools are closed? Round to the nearest whole day.

Chapter 7

Trigonometry Basics

Trigonometric functions are special in several ways. The first characteristic that separates them from all the other types of functions is that input values are always angle measures. You input an angle measure, and the output is some real number. The angle measures can be in degrees or radians — a *degree* being one-360th of a slice of a circle, and a *radian* being about one-sixth of a circle. Each type has its place and use in the study of trigonometry.

Another special feature of trig functions is their *periodicity;* the function values repeat over and over and over, infinitely. This predictability works well with many types of physical phenomena, so trigonometric functions serve as models for many naturally occurring observations.

The Problems You'll Work On

In this chapter, you'll work with trigonometric functions and their properties in the following ways:

- Defining the basic trig functions using the sides of a right triangle
- Expanding the input values of trig functions by using the unit circle
- Exploring the right triangle and trig functions to solve practical problems
- Working with special right triangles and their unique ratios
- Changing angle measures from degrees to radians and vice versa
- Determining arc length of pieces of circles
- Using inverse trig functions to solve for angle measures
- Solving equations involving trig functions

What to Watch Out For

Don't let common mistakes trip you up; keep in mind that when working with trigonometric functions, some challenges will include the following:

- Setting up ratios for the basic trig functions correctly
- Recognizing the corresponding sides of right triangles when doing applications

- Remembering the counterclockwise rotation in the standard position of angles
- Measuring from the terminal side to the x-axis when determining reference angles
- Keeping the trig functions and their inverses straight from the functions and their reciprocals

Using Right Triangles to Determine Trig Functions

391–396 *Use the triangle to find the trig ratio.*

391.

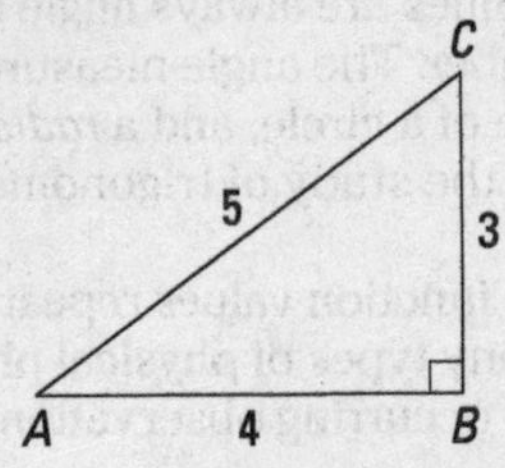

sin A

Illustration by Thomson Digital

392.

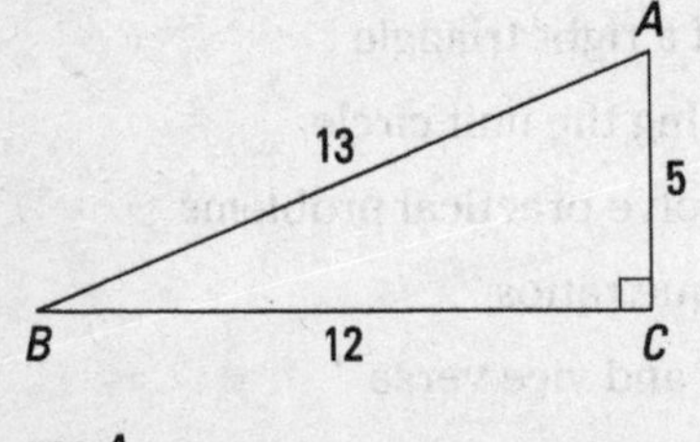

cos A

Illustration by Thomson Digital

393.

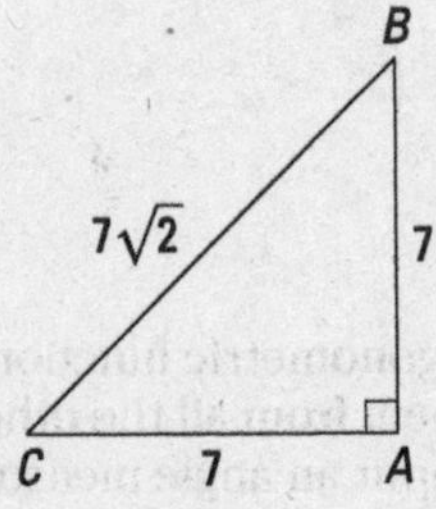

tan C

Illustration by Thomson Digital

394.

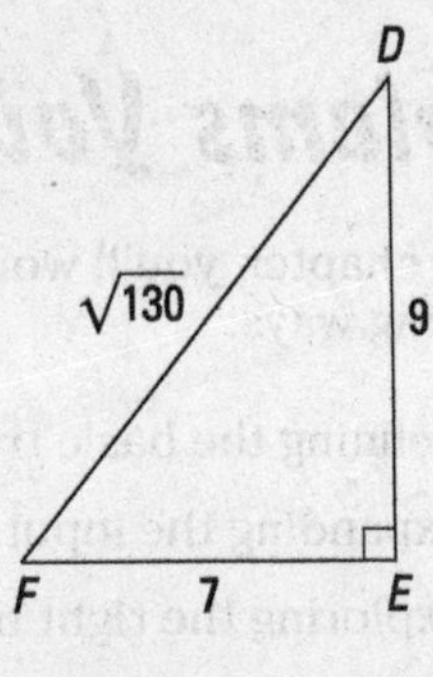

cot D

Illustration by Thomson Digital

395.

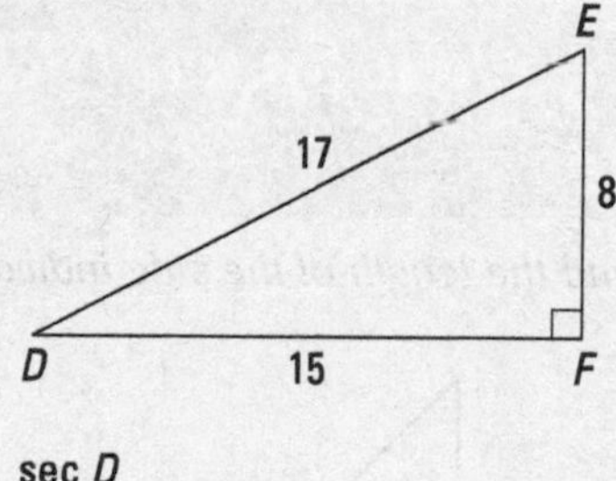

sec D

Illustration by Thomson Digital

396.

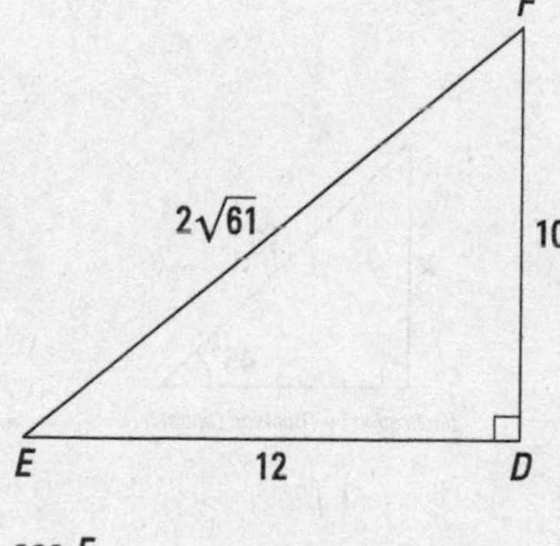

csc F

Illustration by Thomson Digital

Solving Problems by Using Right Triangles and Their Functions

397–406 *Solve the problem. Round your answer to the nearest tenth unless otherwise indicated.*

397. A ladder leaning against a house makes an angle of 30° with the ground. The foot of the ladder is 7 feet from the foot of the house. How long is the ladder?

398. A boy flying a kite lets out 300 feet of string, which makes an angle of 35° with the ground. Assuming that the string is straight, how high above the ground is the kite?

399. An airplane climbs at an angle of 9° with the ground. Find the ground distance it has traveled when it reaches an altitude of 400 feet.

400. A car is traveling up a grade with an angle of elevation of 4°. After traveling 1 mile, what is the vertical change in feet? (1 mile = 5,280 feet)

401. Jase is in a hot air balloon that is 600 feet above the ground, where he can see his brother Willie. The angle of depression from Jase's line of sight to Willie is 25°. How far is Willie from the point on the ground directly below the hot air balloon?

402. A bird is flying at a height of 36 feet and spots a windowsill 8 feet off the ground on which to perch. If the windowsill is at a 28° angle of depression from the bird, how far must the bird fly before it can land?

403. A submarine traveling 8 mph is descending at an angle of depression of 5°. How many minutes does the submarine take to reach a depth of 90 feet?

404. You're a block away from a building that is 820 feet tall. Your friend is between you and the building. The angle of elevation from your position to the top of the building is 42°. The angle of elevation from your friend's position to the top of the building is 71°. To the nearest foot, how far are you from your friend?

405. A lighthouse is located at the top of a hill 100 feet tall. From a point *P* on the ground, the angle of elevation to the top of the hill is 25°. From the same point *P*, the angle of elevation to the top of the lighthouse is 50°. How tall is the lighthouse?

406. A surveyor needs to find the distance *BC* across a lake as part of a project to build a bridge. The distance from point *A* to point *B* is 250 feet. The measurement of angle A is 40°, and the measurement of angle *B* is 112°. What is the distance *BC* across the lake to the nearest foot?

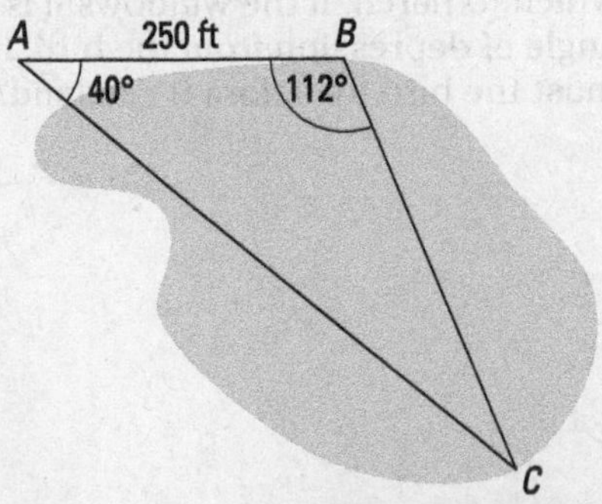

Illustration by Thomson Digital

Working with Special Right Triangles

407–410 *Find the length of the side indicated by x.*

407.

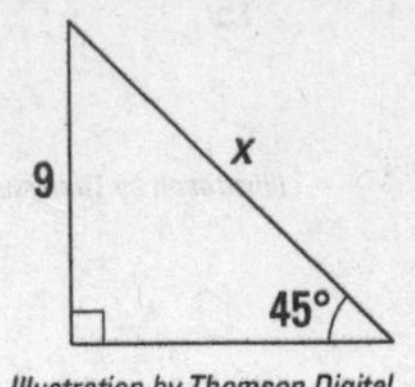

Illustration by Thomson Digital

408.

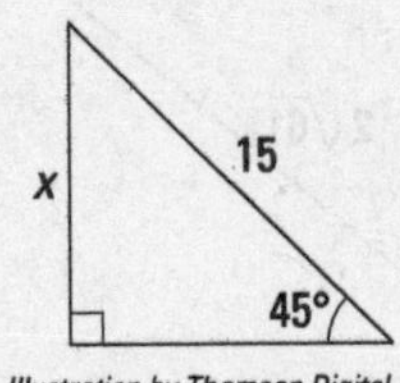

Illustration by Thomson Digital

409.

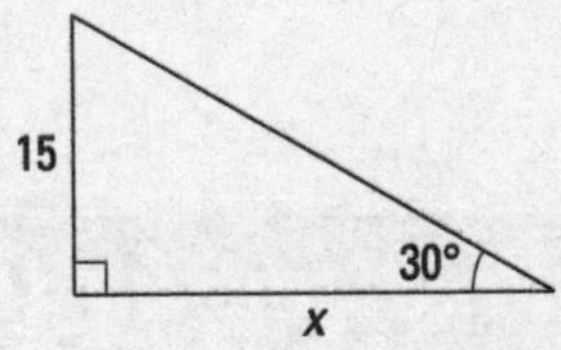

Illustration by Thomson Digital

410.

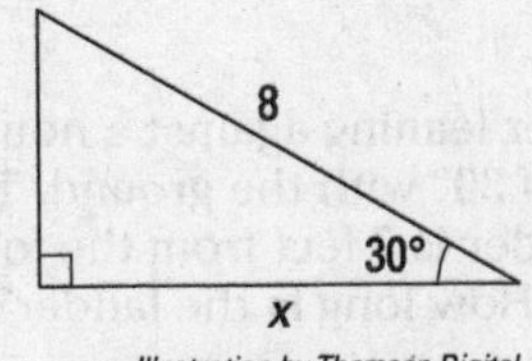

Illustration by Thomson Digital

Changing Radians to Degrees

***411–415** Convert the radian measure to degrees. Round any decimals to two places.*

411. $\frac{2\pi}{3}$

412. 12π

413. $-\frac{11\pi}{6}$

414. 3

415. 0.93

Changing Degrees to Radians

***416–420** Convert the degree measure to radians. Give the exact measure in terms of π.*

416. 225°

417. 60°

418. –405°

419. 36°

420. 167°

Finding Angle Measures (in Degrees) in Standard Position

***421–422** The terminal side of an angle θ in standard position on the unit circle contains the given point. Give the degree measure of θ.*

421. $\left(\frac{1}{2}, \frac{\sqrt{3}}{2}\right)$

422. $\left(-\frac{\sqrt{3}}{2}, -\frac{1}{2}\right)$

Determining Angle Measures (in Radians) in Standard Position

423–425 *The terminal side of an angle θ in standard position on the unit circle contains the given point. Give the radian measure of θ.*

423. $\left(-\frac{1}{2}, \frac{\sqrt{3}}{2}\right)$

424. $\left(-\frac{\sqrt{2}}{2}, -\frac{\sqrt{2}}{2}\right)$

425. $\left(-\frac{\sqrt{3}}{2}, \frac{1}{2}\right)$

Identifying Reference Angles

426–430 *Find the reference angle for the angle measure. (Recall that the quadrants in standard position are numbered counterclockwise, starting in the upper right-hand corner.)*

426. 167°

427. 342°

428. 265°

429. 792°

430. –748°

Determining Trig Functions by Using the Unit Circle

431–434 *Find the exact value of the trigonometric function. If any are not defined, write* undefined.

431. sin 225°

432. tan 330°

433. cos 405°

434. $\sec \frac{3\pi}{2}$

Calculating Trig Functions by Using Other Functions and Terminal Side Positions

435–440 Use the given information to find the exact value of the trigonometric function.

435. Given: $\cos\theta = \frac{5}{13}$; θ is in quadrant I

Find: $\tan\theta$

436. Given: $\sin\theta = \frac{1}{3}$; θ is in quadrant II

Find: $\cos\theta$

437. Given: $\tan\theta = -\frac{2}{3}$; θ is in quadrant IV

Find: $\sin\theta$

438. Given: $\sec\theta = -\frac{9}{8}$; θ is in quadrant III.

Find: $\csc\theta$

439. Given: $\csc\theta = 2$; $\sec\theta < 0$

Find: $\tan\theta$

440. Given: $\cot\theta = -\frac{8}{15}$, $\sin\theta < 0$

Find: $\cos\theta$

Using the Arc Length Formula

441–445 Solve the problem using the arc length formula $s = r\theta$, where r is the radius of the circle and θ is in radians.

441. The central angle in a circle with a radius of 20 cm is $\frac{5\pi}{4}$. Find the exact length of the intercepted arc.

442. The central angle in a circle of radius 6 cm is 85°. Find the exact length of the intercepted arc.

443. Find the radian measure of the central angle that intercepts an arc with a length of $\frac{39\pi}{4}$ inches in a circle with a radius of 13 inches.

444. The second hand of a clock is 18 inches long. In 25 seconds, it sweeps through an angle of 150°. How far does the tip of the second hand travel in 25 seconds?

445. How far does the tip of a 15 cm long minute hand on a clock move in 10 minutes?

Evaluating Inverse Functions

446–450 *Find an exact value of y.*

446. $y=\sin^{-1}\left(\frac{\sqrt{3}}{2}\right)$; give y in radians.

447. $y=\cos^{-1}\left(-\frac{1}{2}\right)$; give y in radians.

448. $y=\tan^{-1}\left(-\sqrt{3}\right)$; give y in radians.

449. $y=\sin^{-1}(\sin\pi)$; give y in radians.

450. $y=\cos\left(\tan^{-1}(-1)\right)$

Solving Trig Equations for x in Degrees

451–455 *Find all solutions of the equation in the interval $[0, 360°)$. (Recall that the quadrants in standard position are numbered counterclockwise, starting in the upper right-hand corner.)*

451. $2\cos x+6=5$

452. $-1+\tan x=\frac{-3+\sqrt{3}}{3}$

453. $4\sin^2 x-1=0$

454. $2\sin^2 x-\sin x-1=0$

455. $2\cos^2 x-2=3\cos x$

Calculating Trig Equations for x in Radians

456–460 *Find all solutions of the equation in the interval $[0, 2\pi)$. (Recall that the quadrants in standard position are numbered counterclockwise, starting in the upper right-hand corner.)*

456. $5\cos x-\sqrt{3}=3\cos x$

457. $3\tan^2 x-1=0$

458. $2\cos^2 x-\cos x=1$

459. $4\sin^3 x+2\sin^2 x-2\sin x-1=0$

460. $2\sin x\cos x-\sin x=0$

Chapter 8

Graphing Trig Functions

The graphs of trigonometric functions are usually easily recognizable — after you become familiar with the basic graph for each function and the possibilities for transformations of the basic graphs.

Trig functions are *periodic.* That is, they repeat the same function values over and over, so their graphs repeat the same curve over and over. The trick is to recognize how often this curve repeats and where one of the basic graphs starts for a particular function.

An interesting feature of four of the trig functions is that they have asymptotes — those not-really-there lines used as guides to the shape of a curve. The sine and cosine functions don't have asymptotes, because their domains are all real numbers. The other four functions have vertical asymptotes to mark where their domains have gaps.

The Problems You'll Work On

In this chapter, you'll work with the graphs of trigonometric functions in the following ways:

- Marking any intercepts on the x-axis to help graph the curves
- Locating and drawing in vertical asymptotes for the tangent, cotangent, secant, and cosecant functions
- Computing the change in the period of a function based on some transformation
- Adjusting the amplitude of the sine or cosine when the basic curve has a multiplier
- Making sideways moves when transformations involve horizontal translations
- Moving trig functions upward or downward with vertical translations

What to Watch Out For

When graphing trigonometric functions, some challenges will include

- Not misreading the period of the trig function when a transformation involves a fraction
- Drawing enough full cycles of the curve to show its characteristics properly

- Marking the axes appropriately for the situation
- Making use of intercepts when they're helpful in a graph

Recognizing Basic Trig Graphs

***461–465** Determine which trig function equation matches the graph.*

461.

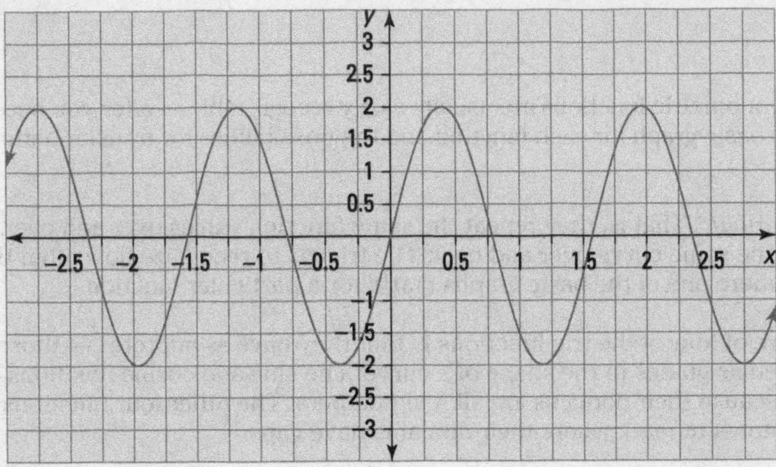

Illustration by Thomson Digital

(A) $f(x)=2\sin x$

(B) $f(x)=2\sin(2x)$

(C) $f(x)=2\sin(4x)$

(D) $f(x)=4\sin x$

(E) $f(x)=4\sin(2x)$

462.

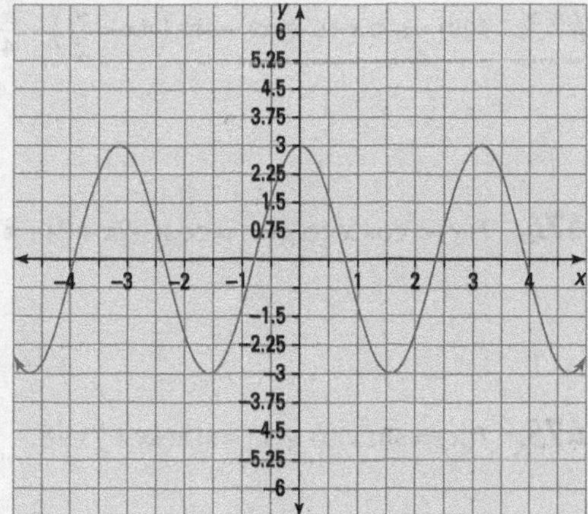

Illustration by Thomson Digital

(A) $f(x) = 3\cos x$

(B) $f(x) = -3\cos x$

(C) $f(x) = 3\cos(3x)$

(D) $f(x) = -3\cos(2x)$

(E) $f(x) = 3\cos(2x)$

463.

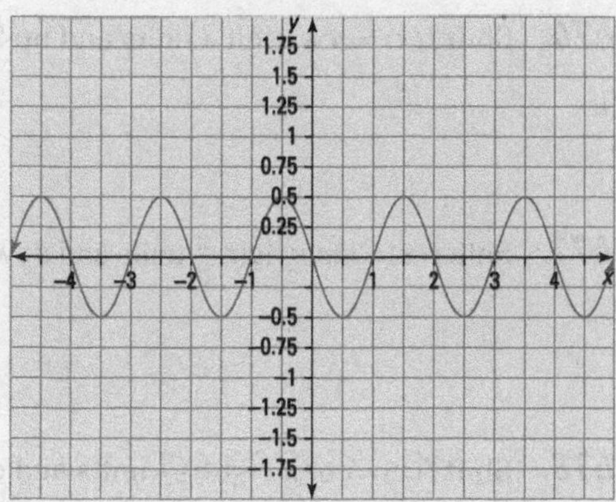

Illustration by Thomson Digital

(A) $f(x) = -\frac{1}{2}\sin(\pi x)$

(B) $f(x) = \frac{1}{2}\sin(2\pi x)$

(C) $f(x) = -\frac{1}{2}\sin(2x)$

(D) $f(x) = -\sin x$

(E) $f(x) = -\sin(2\pi x)$

464.

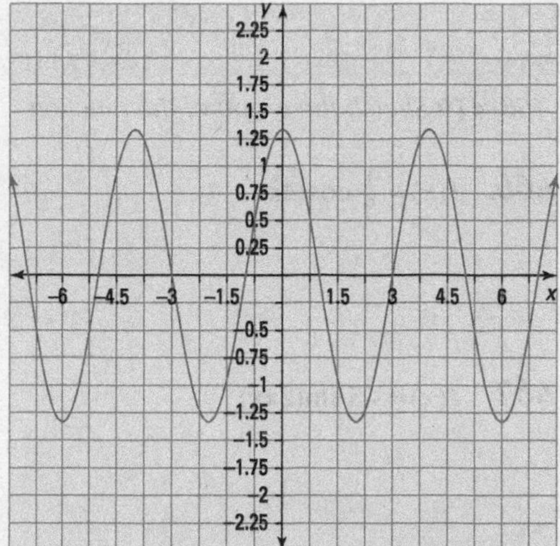

Illustration by Thomson Digital

(A) $f(x) = \frac{4}{3}\cos x$

(B) $f(x) = -\frac{4}{3}\cos(2\pi x)$

(C) $f(x) = \frac{4}{3}\cos\left(\frac{\pi}{2}x\right)$

(D) $f(x) = -\frac{4}{3}\cos\left(\frac{\pi}{2}x\right)$

(E) $f(x) = \frac{4}{3}\cos(\pi x)$

465.

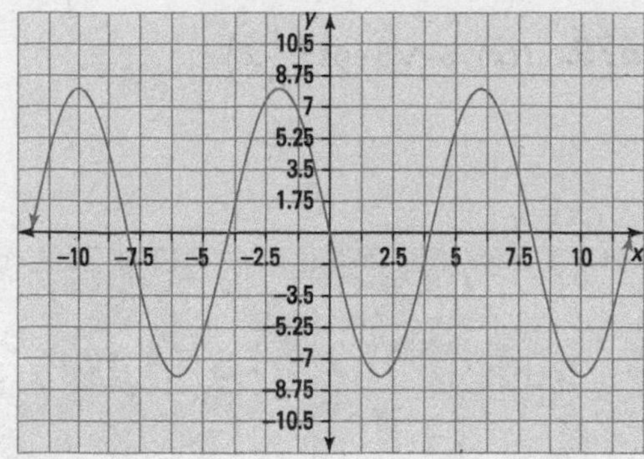

Illustration by Thomson Digital

(A) $f(x) = -8\sin\left(\frac{\pi}{4}x\right)$

(B) $f(x) = 8\sin(\pi x)$

(C) $f(x) = 8\sin\left(\frac{\pi}{4}x\right)$

(D) $f(x) = -8\sin(\pi x)$

(E) $f(x) = -8\sin(4\pi x)$

Graphing Sine and Cosine

***466–470** Sketch the graph of the function.*

466. $f(x)=\frac{1}{3}\cos(4\pi x)$

467. $f(x)=-5\sin(2\pi x)$

468. $f(x)=-\frac{7}{4}\cos(6x)$

469. $f(x)=\frac{\pi}{8}\sin\left(-\frac{2}{5}x\right)$

470. $f(x)=-\sqrt{5}\cos\left(\frac{1}{3}x\right)$

Applying Function Transformations to Graphs of Trig Functions

***471–475** Describe the transformation from function f to function g.*

471. $f(x)=\sin x$ to $g(x)=\sin\left(x-\frac{\pi}{3}\right)+3$

472. $f(x)=\cos x$ to $g(x)=\cos\left(x+\frac{\pi}{4}\right)-2$

473. $f(x)=\sin x$ to $g(x)=\sin\left(4x-\frac{\pi}{2}\right)-\frac{\pi}{4}$

474. $f(x)=\cos x$ to $g(x)=\cos(-2x+5)+\pi$

475. $f(x)=\sin x$ to $g(x)=\sin(4x+7)+8$

Writing New Trig Functions Using Transformations

***476–480** Given the description of the transformation from function f to function g, write the function equation for g.*

476. Shift $f(x)=\cos x$ left π units and up 5 units.

477. Shift $f(x)=\sin x$ right $\frac{\pi}{8}$ units and down π units.

478. Shift $f(x)=\cos x$ right $\frac{2\pi}{3}$ units and down 2π units.

479. Shift $f(x)=\sin x$ left $\frac{7}{2}$ units and down 3 units.

480. Shift $f(x)=\cos x$ right $\frac{4}{5}$ units and up $\frac{\pi}{6}$ units.

Graphing Tangent and Cotangent

481–484 *Sketch the graph of the function.*

481. $f(x) = \tan(4x)$

482. $f(x) = \cot(2x)$

483. $f(x) = -2\tan(\pi x)$

484. $f(x) = 3\cot\left(-\frac{\pi}{2}x\right)$

Interpreting Transformations of Trig Functions

485–495 *Describe the transformations from function f to function g.*

485. $f(x) = \tan x$ to $g(x) = -\frac{1}{2}\tan(6x)$

486. $f(x) = \tan x$ to $g(x) = 2\tan(3\pi x)$

487. $f(x) = \tan x$ to $g(x) = \sqrt{3}\tan\left(\frac{\pi}{3}x\right)$

488. $f(x) = \tan x$ to $g(x) = \frac{\pi}{4}\tan\left(-\frac{\pi}{6}x\right)$

489. $f(x) = \tan x$ to $g(x) = \tan(4x - \pi) + 3$

490. $f(x) = \cot x$ to $g(x) = \cot\left(3x + \frac{\pi}{2}\right) - 1$

491. $f(x) = \tan x$ to $g(x) = \tan(-2x + 2) + \pi$

492. $f(x) = \cot x$ to $g(x) = \cot(-3x - 4) + \frac{\pi}{2}$

493. $f(x) = \tan x$ to $g(x) = -3\tan(2x + 5) - \frac{\pi}{4}$

494. $f(x) = \cot x$ to $g(x) = -\frac{1}{2}\cot(5x - 2\pi) + 4$

495. $f(x) = \tan x$ to $g(x) = \frac{3}{4}\tan\left(\frac{1}{2}x - \frac{\pi}{3}\right) - \pi$

Graphing Secant and Cosecant

496–503 *Give a rule for the equations of the asymptotes. Then sketch the graph of the function.*

496. $f(x)=\sec(\pi x)$

497. $f(x)=\csc(2\pi x)$

498. $f(x)=-\sec(-3x)$

499. $f(x)=4\sec(5x)$

500. $f(x)=\frac{1}{3}\sec(4\pi x)$

501. $f(x)=-2\csc(-8x)$

502. $f(x)=-\sqrt{6}\sec\left(\frac{\pi}{6}x\right)$

503. $f(x)=\frac{\pi}{3}\csc\left(-\frac{\pi}{4}x\right)$

Interpreting Transformations from Function Rules

504–510 *Describe the transformation from function f to function g.*

504. $f(x)=\sec x$ to $g(x)=\sec(3x-4)+5$

505. $f(x)=\csc x$ to $g(x)=\csc(2x+\pi)+3$

506. $f(x)=\sec x$ to $g(x)=\sec(x-8)-\pi$

507. $f(x)=\csc x$ to $g(x)=\csc(-x+2\pi)+\frac{\pi}{4}$

508. $f(x)=\sec x$ to $g(x)=-2\sec\left(-2x+\frac{\pi}{2}\right)-\frac{3}{2}$

509. $f(x)=\csc x$ to $g(x)=\frac{1}{2}\csc(\pi x+3)-5$

510. $f(x)=\sec x$ to $g(x)=\frac{2}{3}\sec\left(\frac{1}{2}x+\frac{3}{2}\right)-\frac{\pi}{2}$

Chapter 9

Getting Started with Trig Identities

Don't have an identity crisis! In this chapter, you become more familiar with the possibilities for rewriting trigonometric expressions. A trig *identity* is really an equivalent expression or form of a function that you can use in place of the original. The equivalent format may make factoring easier, solving an application possible, and (later) performing an operation in calculus more manageable.

The trigonometric identities are divided into many different classifications. These groupings help you remember the identities and make determining which identity to use in a particular substitution easier. In a classic trig identity problem, you try to make one side of the equation match the other side. The best way to do so is to work on just one side — the left or the right — but sometimes you need to work on both sides to see just how to work the problem to the end.

The Problems You'll Work On

In this chapter, you'll work with the basic trigonometric identities in the following ways:

- Determining which trig functions are reciprocals of one another
- Creating Pythagorean identities from a right triangle whose hypotenuse measures 1 unit
- Determining the sign of identities whose angle measure is negated
- Matching up trig functions and their co-functions
- Using the periods of functions in identities
- Making the most of selected substitutions into identities
- Working on only one side of the identity
- Figuring out where to go with an identity by working both sides at once

What to Watch Out For

Don't let common mistakes trip you up; keep in mind that when working on trigonometric identities, some challenges will include the following:

- Keeping track of where the *1* goes in the Pythagorean identities
- Remembering the middle term when squaring binomials involving trig functions

- Correctly rewriting Pythagorean identities when solving for a squared term
- Recognizing the exponent notation

Proving Basic Trig Identities

***511–535** Prove the trig identity. Indicate your first identity substitution.*

511. $\sin\theta\csc\theta+\cot\theta\tan\theta=2$

512. $\frac{\sin\theta}{\csc\theta}+\frac{\cos\theta}{\sec\theta}=1$

513. $\frac{\sin(-\theta)}{\cos(-\theta)}=\tan(-\theta)$

514. $\sin^2\left(\frac{\pi}{2}-x\right)+\sin^2 x=1$

515. $\sin(2\pi-x)=-\sin x$

516. $\frac{\sec\theta}{\cos\theta}=\tan^2\theta+1$

517. $\frac{\sin^2 x+\cos^2 x}{\cos^2 x}=\sec^2 x$

518. $\sin^2(-\theta)+\cos^2(\theta)=1$

519. $\cos^2\left(\frac{\pi}{2}-x\right)+\sin x=\sin x(\sin x+1)$

520. $\frac{\sin x}{\cos x}(\tan x+\cot x)=\sec^2 x$

521. $\sin\theta\cot\theta=\frac{1}{\sec\theta}$

522. $\cos^2 x=(1-\sin x)(1+\sin x)$

523. $\frac{\csc(-\theta)}{\sec(-\theta)}=-\cot(\theta)$

524. $\tan\left(\frac{\pi}{2}-x\right)=-\tan\left(x-\frac{\pi}{2}\right)$

525. $\tan x(\cot x+\cot^3 x)=\csc^2 x$

526. $\tan\theta(\cos\theta - \cot\theta) = \sin\theta - 1$

527. $(\cot x - \csc x)(\cot x + \csc x) + 1 = 0$

528. $\tan\theta[\cos(-\theta) + \cos\theta] = 2\sin\theta$

529. $\cot^2 x + \tan^2\left(\frac{\pi}{2} - x\right) = 2(\csc^2 x - 1)$

530. $\sin x(\cot x - \csc x) = \cos x - 1$

531. $\frac{\tan\theta}{\csc\theta} \cdot \frac{1}{\csc^2\theta\sec\theta} = \sin^4\theta$

532. $\sin^2 x\left(\frac{1}{\cos^2 x} + \frac{1}{\sin^2 x}\right) = \sec^2 x$

533. $\sin\left(\frac{\pi}{2} - x\right)\tan x = \sin x$

534. $\cot(-\theta)\tan(-\theta) + \sec(-\theta)\cos(\theta) = 2$

535. $\sec^2 x\csc^2 x = \csc^2 x + \sec^2 x$

Returning to Basic Sine and Cosine to Solve Identities

536–540 *Determine the missing term or factor in the identity by changing all functions to those using sine or cosine.*

536. $\csc^2 x\tan^2 x + \sec^2 x\cot^2 x = \frac{1}{\cos^2 x} + \boxed{}$

537. $\frac{\tan x}{\sec x} + \frac{\cot x}{\csc x} = \sin x + \boxed{}$

538. $\tan^2 x(\csc^2 x - 1) = \boxed{}$

539. $\cot^2 x(\sec x - \tan x) = \frac{\cos x\boxed{}}{\sin^2 x}$

540. $\cot x(\sec x + \csc x) = \csc x\boxed{}$

Using Multiplication by a Conjugate to Solve Identities

541–545 *Determine the missing term or factor in the identity by multiplying by a conjugate.*

541. $\dfrac{\cos x}{1+\sec x}=\dfrac{\square}{\tan^2 x}$

542. $\dfrac{\tan x}{\cot x-1}=\dfrac{\square}{\csc^2 x-2}$

543. $\dfrac{1}{\sin x+\cos x}=\dfrac{\square}{1-2\cos^2 x}$

544. $\dfrac{1}{1-\cos x}+\dfrac{1}{1+\cos x}=2\square$

545. $\dfrac{\sin x}{1-\cos x}+\dfrac{\cos x}{1-\sin x}=\dfrac{1+\cos x}{\sin x}+\square$

Solving Identities After Raising a Binomial to a Power

546–550 *Determine the missing term or terms in the identity after computing the power of the binomial.*

546. $(1+\sin x)^2=2+2\sin x-\square$

547. $(\cot x+\tan x)^2=\csc^2 x+\square$

548. $(\sin x+\cos x)^2=2\sin x\cos x+\square$

549. $(\sin x+\csc x)^3=\sin^3 x+3\sin x+\square+\csc^3 x$

550. $(\sin x-\cos x)^4=1-4\cos^4 x+4\cos^2 x-\square$

Solving Identities After Factoring out a Common Function

551–555 *Determine the missing term or factor in the identity after factoring.*

551. $\sin x\tan^2 x+\sin x=\sec x\square$

552. $\tan^2 x\sin^2 x+\tan^2 x\cos^2 x=\square$

553. $\dfrac{\sin^2 x+2\sin x\tan x+\tan^2 x}{\sin x+\tan x}=\sin x+\square$

554. $\dfrac{\sin^2 x - \cos^2 x}{\sin x + \cos x} = \sin x - \boxed{}$

555. $\sin^2 x \tan x + \sin^2 x + \cos^2 x \tan x + \cos^2 x$
$= \tan x + \boxed{}$

Solving Identities After Combining Fractions

556–560 *Determine the missing term or factor in the identity after adding the fractions.*

556. $\dfrac{\tan x}{\sin x} + \dfrac{\cot x}{\cos x} = \sec x + \boxed{}$

557. $\dfrac{\sin x}{\cos x} + \dfrac{\cos x}{\sin x} = \csc x \boxed{}$

558. $\dfrac{\sin x}{\sec x} + \dfrac{\cos x}{\csc x} = 2\sin x \boxed{}$

559. $\dfrac{\sin x + \csc x}{\cos x} + \dfrac{\cos x + \sec x}{\sin x} = \dfrac{\boxed{}}{\sin x \cos x}$

560. $\dfrac{\sin x \sec x}{\tan^2 x} + \dfrac{\cos x \csc x}{\cot^2 x} = \cot x + \boxed{}$

Performing Algebraic Processes to Make Identities More Solvable

561–570 *Give the rewritten identity after performing the action. Then complete the solution of the identity.*

561. Distribute on the right and move all terms to the left.

$\sec x \tan x + \sec x \tan^2 x - \sec^3 x = \sec x(\tan x - 1)$

562. Split up the first fraction on the left and distribute on the right.

$\dfrac{1 + \sec x}{\tan x} - \dfrac{\tan x}{\sec x} = \cot x\,(1 + \cos x)$

563. Cross-multiply.

$\dfrac{\sin x - \cos x + 1}{\sin x + \cos x - 1} = \dfrac{\sin x + 1}{\cos x}$

564. Cross-multiply.

$\dfrac{\sin x - \cos x}{\cos^2 x} = \dfrac{\tan^2 x - 1}{\sin x + \cos x}$

565. Factor the numerator on the left and reduce the fraction.

$\dfrac{\sin^2 x + 4\sin x + 3}{\sin x + 3} = \dfrac{\cos^2 x}{1 - \sin x}$

566. Cross-multiply.

$$\frac{\tan x-1}{\tan x+1}=\frac{1-\cot x}{1+\cot x}$$

567. Add the middle fraction to both sides, and then add the two fractions on the right. Simplify by using reciprocal identities.

$$\frac{\tan x\cos x}{1+\sin x}-\frac{1-\sin x}{\csc x}=\frac{\sin^2 x}{\csc x+1}$$

568. Combine the two terms on the right.

$$\frac{\sin x}{1+\cos x}=\csc x-\cot x$$

569. Subtract the two terms on the left.

$$\frac{\tan x}{\sec x-1}-\frac{\tan x}{\sec x+1}=2\cot x$$

570. Square both sides of the equation.

$$\frac{1-\cot x}{\csc x}=\sqrt{1-2\sin x\cos x}$$

Chapter 10

Continuing with Trig Identities

The basic trig identities (see Chapter 9) will get you through most problems and applications involving trigonometry. But if you're going to broaden your horizons and study more and more mathematics, you'll find the additional identities in this chapter crucial to your success. Also, in some of the sciences, especially physics, these specialized identities come up in the most unlikely (and likely) places.

These trigonometric identities are broken into groups, depending on whether you're trying to combine angles or split them apart, increase exponents or reduce them, and so on. The groupings can help you to decide which identity to use in which situation. Keep a list of these identities handy, because you'll want to refer to them as you work through the problems.

The Problems You'll Work On

You'll work with the more-advanced trig identities in the following ways:

- Using the function values of two angles to determine the function value of the sum of the angles
- Applying the identities for the difference between two angles
- Making use of the half-angle identities
- Working from product-to-sum and sum-to-product identities
- Using the periods of functions in identities
- Applying power-reducing identities
- Deciding which identity to use first

What to Watch Out For

When you're working on these particular trig identities, some challenges will include the following:

- Applying the identities using the correct order of operations
- Simplifying the radicals correctly in half-angle identities
- Making the correct choices between positive and negative identities

Using Identities That Add or Subtract Angle Measures

571–573 *Use a sum or difference identity to determine the missing term in the identity.*

571. $\sin(45^\circ+\theta)=\frac{\sqrt{2}}{2}\left(\sin\theta+\boxed{}\right)$

572. $\tan(45^\circ+\theta)=\frac{1+\boxed{}}{1-\tan\theta}$

573. $\cos(60^\circ-\theta)=\frac{1}{2}\left(\cos\theta+\boxed{}\right)$

Confirming Double-Angle Identities

574–575 *Use a sum identity to determine the missing factor or term in the double-angle identity.*

574. $\sin 2\theta=2\boxed{}$

575. $\tan 2\theta=\frac{2\tan\theta}{1-\boxed{}}$

Using Identities That Double the Size of the Angle

576–578 *Use a double-angle identity to determine the missing term in the identity.*

576. $\sin 2\theta+2\sin^2\theta=2\sin\theta\left(\sin\theta+\boxed{}\right)$

577. $\cos^2\theta-\cos 2\theta=\sin^2\theta+\boxed{}$

578. $\cos^2 2\theta+4\sin^2\theta\cos^2\theta=\boxed{}$

Confirming the Statements of Multiple-Angle Identities

579–580 *Use sum and double-angle identities to determine the missing term or factor in the multiple-angle identity.*

579. $\cos(4\theta)=8\cos^4\theta-8\cos^2\theta+\boxed{}$

580. $\sin(4\theta)=4\sin\theta\cos\theta\boxed{}$

Creating Half-Angle Identities from Double-Angle Identities

581–582 Determine the missing term in the half-angle identity by starting with an identity for $\cos 2\theta$.

581. $\sin\left(\frac{\theta}{2}\right)=\pm\sqrt{\frac{1-\square}{2}}$

582. $\cos\left(\frac{\theta}{2}\right)=\pm\sqrt{\frac{1+\square}{2}}$

Creating a Half-Angle Identity for Tangent

583 Determine the missing factor in the half-angle identity by using the ratio identity for tangent and the half-angle identities for sine and cosine.

583. $\tan\left(\frac{\theta}{2}\right)=\frac{1-\cos\theta}{\square}=\frac{\square}{1+\cos\theta}$

Using Half-Angle Identities to Simplify Expressions

584–585 Determine the missing term in the identity.

584. $\dfrac{1-\tan^2\left(\frac{\theta}{2}\right)}{1+\tan^2\left(\frac{\theta}{2}\right)}=\square$

585. $\dfrac{\cos\left(\frac{\theta}{2}\right)+\sin\left(\frac{\theta}{2}\right)}{\cos\left(\frac{\theta}{2}\right)-\sin\left(\frac{\theta}{2}\right)}=\sec\theta+\square$

Creating Products of Trig Functions from Sums and Differences

586–588 Determine the missing factor in the product-to-sum identity.

586. $\sin a\,\square=\frac{1}{2}\left[\sin(a+b)+\sin(a-b)\right]$

587. $\cos a\,\square=\frac{1}{2}\left[\cos(a+b)+\cos(a-b)\right]$

588. $\sin a\,\square=\frac{1}{2}\left[\cos(a-b)-\cos(a+b)\right]$

Using Product-to-Sum Identities to Evaluate Expressions

589–590 Write the product as a sum using a product-to-sum identity. Then evaluate.

589. $\sin(75°)\cos(15°)$

590. $\cos(120°)\cos(60°)$

Using Sum-to-Product Identities to Evaluate Expressions

591–595 *Write the sum as a product using a sum-to-product identity. Then evaluate.*

591. $\sin 75° + \sin 15°$

592. $\sin 150° - \sin 90°$

593. $\cos 195° + \cos 75°$

594. $\cos 75° - \cos 15°$

595. $\cos 120° + \cos 90°$

Applying Power-Reducing Identities

596–600 *Use a power-reducing identity to determine the missing term in the identity.*

596. $\sin^2\left(\frac{\theta}{2}\right) = \frac{1-\square}{2}$

597. $\cos^2 4\theta - \sin^2 4\theta = \square$

598. $\tan^2(2\theta) = \sec^2 2\theta - \square$

599. $\sec^2\theta - \tan^2\theta = \square$

600. $\cot^2\theta - \cos^2\theta = \frac{\left(1+\square\right)^2}{2-2\cos(2\theta)}$

Using Identities to Determine Values of Functions at Various Angles

601–610 *Evaluate the angle's function using an appropriate identity.*

601. $\sin 75°$

602. $\cos 15°$

603. $\tan 165°$

604. $\sin \frac{\pi}{12}$

605. $\cos \frac{7\pi}{12}$

606. $\tan \frac{\pi}{12}$

607. $\cot 75°$

608. $\sec 165°$

609. $\csc \frac{5\pi}{24}$

610. $\tan \frac{\pi}{24}$

Working through Identities Using Multiple Methods

611–630 *Determine the missing term or factor in the identity.*

611. $\frac{1-\tan^2\theta}{1+\tan^2\theta} = \square$

612. $\frac{\sin\theta}{1-\cos\theta} + \frac{1-\cos\theta}{\sin\theta} = \square$

613. $(1+\cos\theta)(\csc\theta - \cot\theta) = \square$

614. $\cot^4\theta + 2\cot^2\theta + 1 = \square$

615. $\sin\frac{\theta}{2} = 2\sin\frac{\theta}{4}\cos\frac{\theta}{4} + \square$

616. $\frac{1-\cot\theta}{\tan\theta - 1} = \square$

617. $\frac{1}{\cos\theta - \cos^2\theta} = \frac{\square}{1-\cos\theta}$

618. $\cot^2\theta - \cos^2\theta\cot^2\theta + 1 - \cos^2\theta = \boxed{\quad}$

619. $\dfrac{\cot\theta + 4\sec\theta}{\cot\theta\sin\theta} = 4\sec^2\theta + \boxed{\quad}$

620. $\dfrac{\cot\theta}{\csc\theta - \sin\theta} = \dfrac{1}{\boxed{\quad}}$

621. $\cos 3\theta = 4\cos^3\theta - \boxed{\quad}$

622. $\cos 20\theta = 8\sin^4 5\theta - 8\sin^2 5\theta + \boxed{\quad}$

623. $\dfrac{\sin^3\theta + \cos^3\theta}{\sin\theta + \cos\theta} = \boxed{\quad} - \dfrac{1}{2}\sin 2\theta$

624. $\dfrac{\cos\theta}{\cos\theta - \sin\theta} - \dfrac{\sin\theta}{\cos\theta + \sin\theta} = \sec 2\theta + \boxed{\quad}$

625. $(\sec\theta + 2\cos\theta)^2 - (\sec\theta - 2\cos\theta)^2 = \boxed{\quad}$

626. $\dfrac{\sin 2\theta + \cos 2\theta}{\sin\theta\cos\theta} = \cot\theta - \tan\theta + \boxed{\quad}$

627. $\dfrac{1}{2}\sin^2 2\theta + \cos 2\theta + 2\sin^4\theta = \boxed{\quad}$

628. $\dfrac{2\sin 2\theta}{(\cos 2\theta - 1)^2} = \csc^2\theta\cot\theta + \boxed{\quad}$

629. $\sqrt{\dfrac{1+\cos\theta}{1-\cos\theta}} = \csc\theta + \boxed{\quad}$

630. $(\sin\theta - \sec\theta)^2 - \sin^2\theta = \left(\tan\theta - \boxed{\quad}\right)^2$

Chapter 11

Working with Triangles and Trigonometry

You may think that you're "not in Kansas anymore" when you leave the familiar world of right triangles and Mr. Pythagoras to enter this new world of oblique triangles. Trigonometry allows for some calculations that aren't possible with the geometric formulas and other types of measurement. The Law of Sines and Law of Cosines are relationships between the sides and angles of triangles that aren't right triangles.

The applications you can solve using these new laws are many and varied. Whole new worlds are opened to you now that you don't have triangle-type restrictions. The biggest challenge is in deciding which law to use, but even that is pretty straightforward.

The Problems You'll Work On

In this chapter, you'll work on solving for parts of triangles in the following ways:

- Using the Law of Sines and choosing the pair of ratios
- Applying the Law of Cosines and determining which version works
- Working with the ambiguous case and deciding which angle applies
- Finding the missing values of sides and angles of a triangle
- Computing areas of triangles by using a formula involving a trig function
- Using Heron's formula for the area of a triangle
- Practicing with practical applications

What to Watch Out For

Don't let common mistakes trip you up; keep in mind that when working with these formulas for triangles, some challenges will include

- Choosing the correct pair of ratios when using the Law of Sines
- Performing the order of operations correctly when applying the Law of Cosines

- Identifying the correct parts of the triangle when finding the area
- Writing the correct relationships between triangle parts when working on applications

Applying the Law of Sines to Find Sides

631–633 *Use the Law of Sines to find the indicated side. Round your answer to the nearest tenth.*

631.

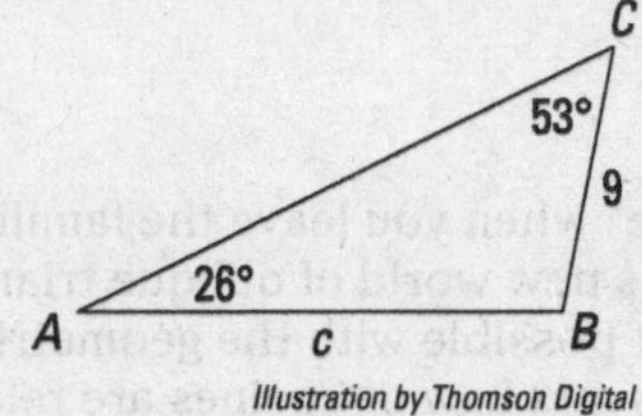

Illustration by Thomson Digital

Find c.

632.

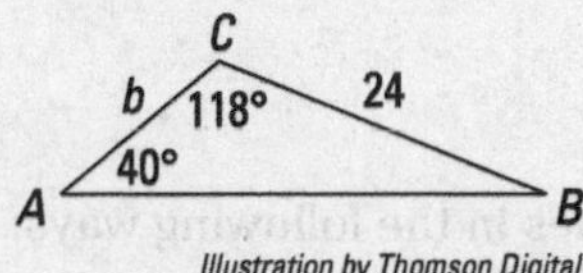

Illustration by Thomson Digital

Find b.

633.

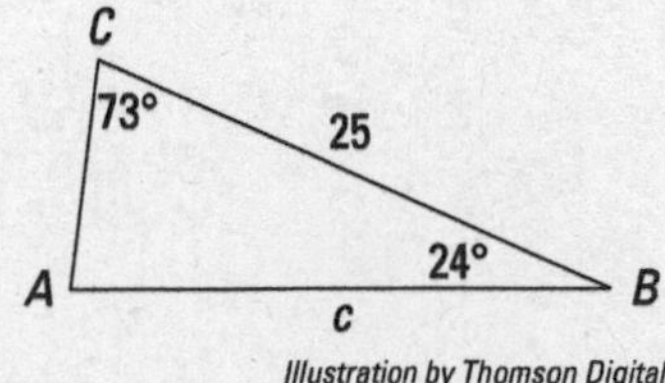

Illustration by Thomson Digital

Find c.

Utilizing the Law of Sines to Find Angles

634–636 *Use the Law of Sines to find the indicated angle. Round your answer to the nearest tenth of a degree.*

634.

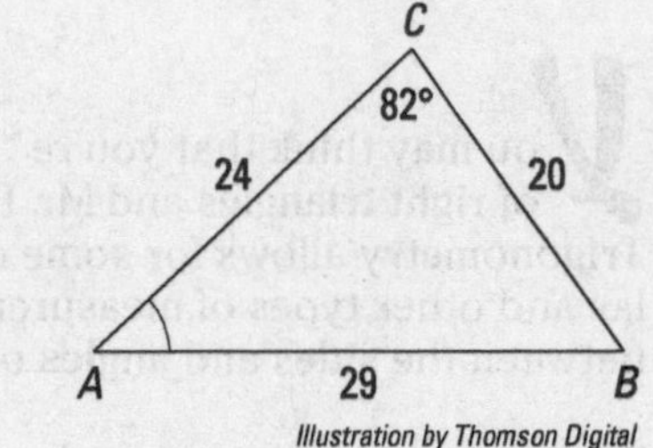

Illustration by Thomson Digital

Find $m\angle A$.

635.

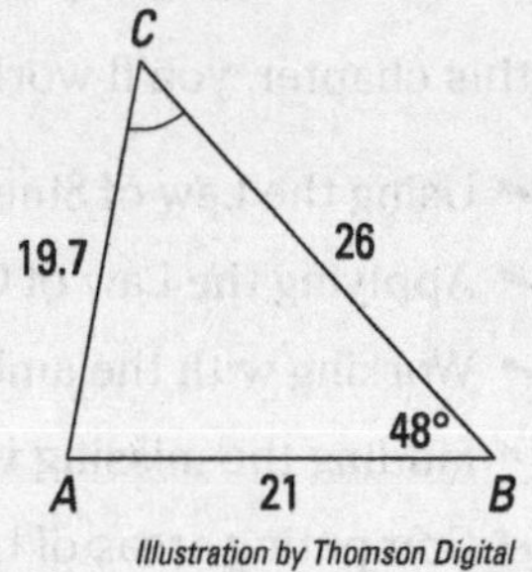

Illustration by Thomson Digital

Find $m\angle C$.

636.

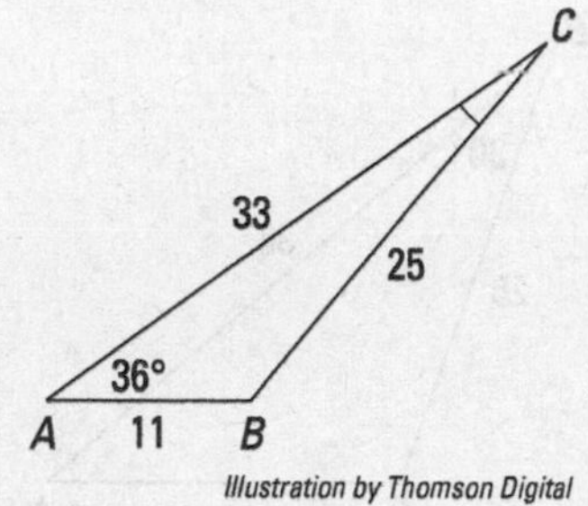

Illustration by Thomson Digital

Find $m\angle C$.

Using the Law of Sines for Practical Applications

***637–640** Use the Law of Sines to solve the problem.*

637. Points A and B are on one side of a river, 100 feet apart, with C on the opposite side. The angles A and B measure 70° and 60°, respectively. What is the distance from point A to point C, to the nearest foot?

638. A 14-foot pole is leaning. A wire attached to the top of the pole is anchored in the ground. The wire is 14.5 feet long and makes a 74.2° angle with the ground. What angle does the pole make with the ground? Round your answer to the nearest tenth of a degree.

639. Jake is taking a walk along a straight road. He decides to take a detour through some woods, so he walks on a path that makes an angle of 50° with the road. After walking for 1,450 feet, he turns 105° and heads back to the road. How far does Jake have to walk to get back to the road? Round your answer to the nearest foot.

640. Carlos measures the angle of elevation of the top of a tree as 75°. Elena (who is 20 feet farther from the tree on a straight, level path) measures the angle of elevation as 45°. How tall is the tree to the nearest foot?

Investigating the Ambiguous Case of the Law of Sines

***641–643** Determine the number of triangles that can be formed using the given measurements. If the answer is unknown, write, "Cannot be determined."*

641. $m\angle A = 95°, a = 33, b = 16$

642. $m\angle C = 83°, a = 18, c = 10$

643. $m\angle A = 25°, a = 8, b = 14$

Determining All Angles and Sides of a Triangle

644–645 *Use the Law of Sines to solve the triangle. Round your answers to the nearest tenth. If no triangle exists, write, "No solution."*

644. $m\angle A = 70°$, $a = 25$, $c = 26$

645. $m\angle A = 46°$, $a = 8$, $b = 13$

Finding Side Measures by Using the Law of Cosines

646–648 *Use the Law of Cosines to find the length of the missing side. Round your answer to the nearest tenth.*

646.

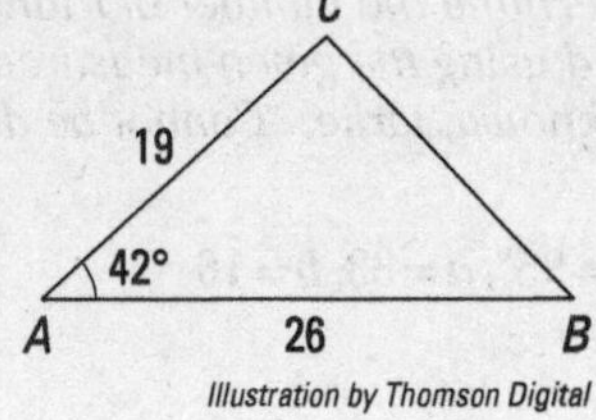

Illustration by Thomson Digital

647.

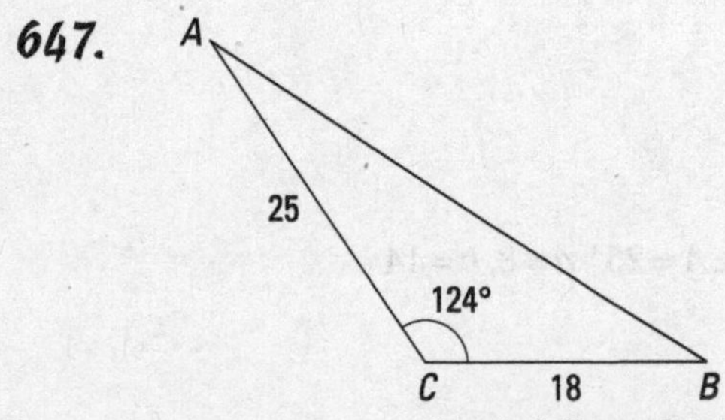

Illustration by Thomson Digital

648.

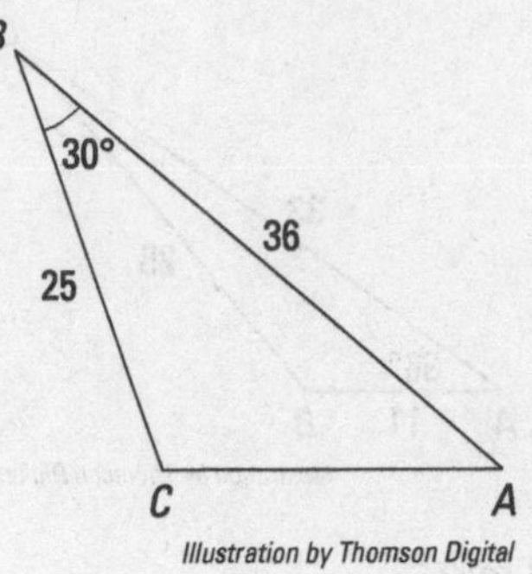

Illustration by Thomson Digital

Using the Law of Cosines to Determine an Angle

649–651 *Use the Law of Cosines to find the indicated angle. Round your answer to the nearest tenth of a degree.*

649.

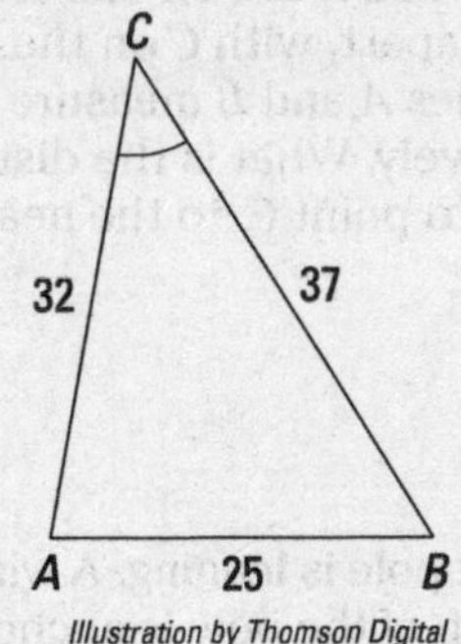

Illustration by Thomson Digital

Find $m\angle C$.

650.

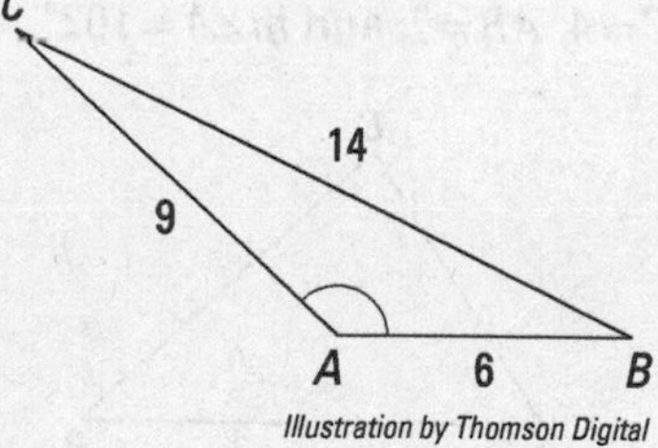

Illustration by Thomson Digital

Find $m\angle A$.

651.

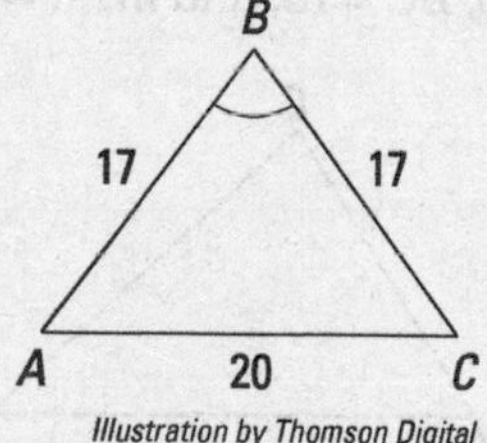

Illustration by Thomson Digital

Find $m\angle B$.

Applying the Law of Cosines to Real-World Situations

652–655 *Use the Law of Cosines to solve the problem.*

652. To approximate the length of a lake, a surveyor starts at one end of the lake and walks northeast 145 yards. He then turns 110° to the right and walks 270 yards until he arrives at the other end of the lake. What is the length of the lake to the nearest yard?

653. In a rhombus with a side length of 24, the longer diagonal has a length of 35. Find the larger angle of the rhombus. Round your answer to the nearest degree.

654. A baseball player in center field is playing 350 feet from a television camera that's behind home plate. The batter hits a fly ball that travels a horizontal distance of 420 feet from the camera. About how many feet does the center fielder run to catch the ball if the camera turns 8° while following the play?

655. A bicycle race follows a triangular course. The three legs of the race, in order, are 10.4 kilometers, 6.3 kilometers, and 8.3 kilometers. Find the angle between the first two legs of the race to the nearest degree.

Finding Areas of Triangles by Using the Sine

656–658 *Find the area of the triangle. Round your answer to the nearest tenth.*

656.

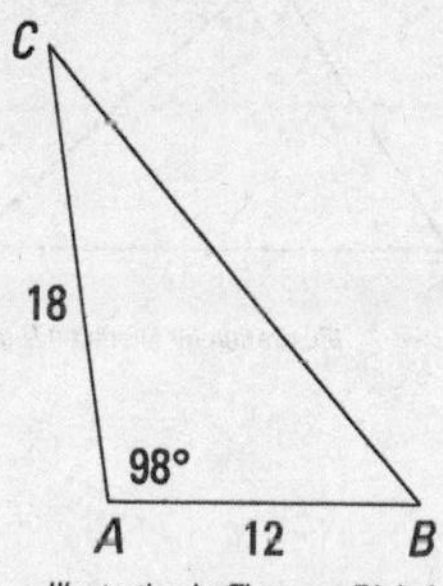

Illustration by Thomson Digital

657.

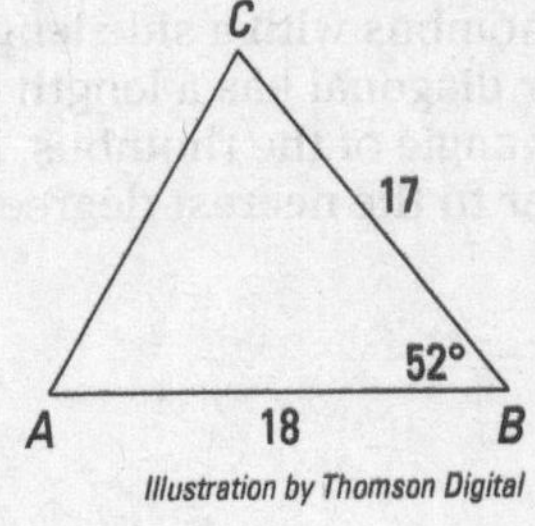

Illustration by Thomson Digital

658.

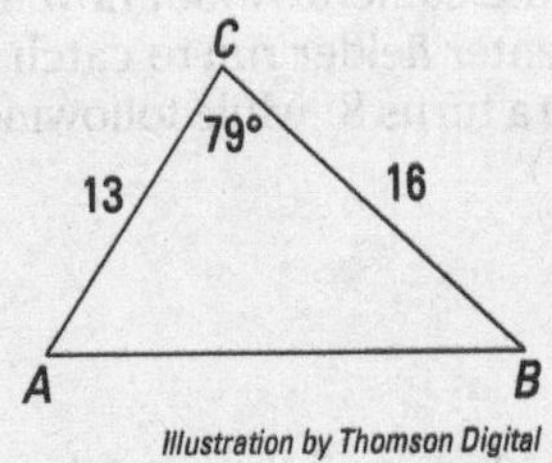

Illustration by Thomson Digital

660. $AC = 4$, $AB = 7$, and $m\angle A = 102°$

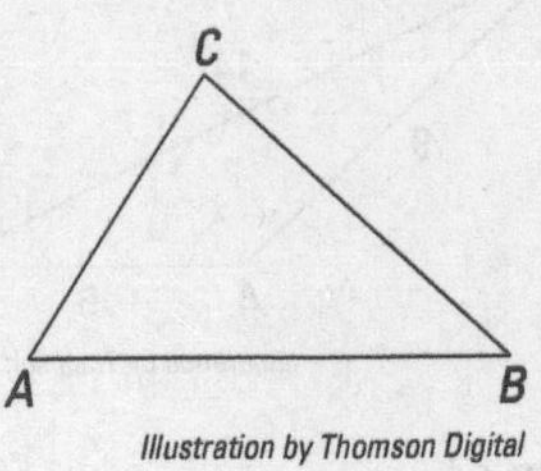

Illustration by Thomson Digital

661. $AC = 20$, $BC = 18$, and $m\angle C = 75°$

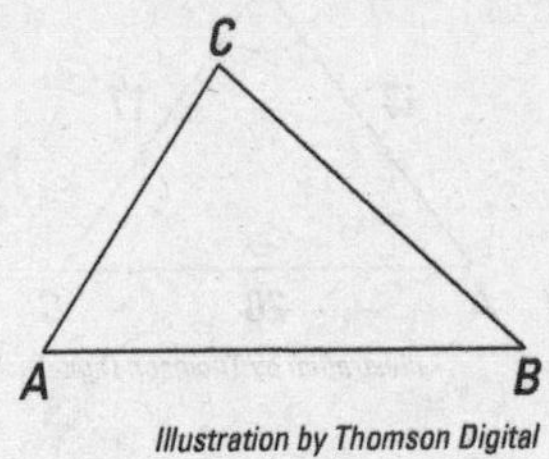

Illustration by Thomson Digital

Applying the Trig Formula for Area of a Triangle

***659–661** Use the figure and the given information to find the area of the triangle. (The triangle isn't necessarily drawn to scale.) Round your answer to the nearest tenth.*

659. $AB = 6$, $BC = 8$, and $m\angle B = 87°$

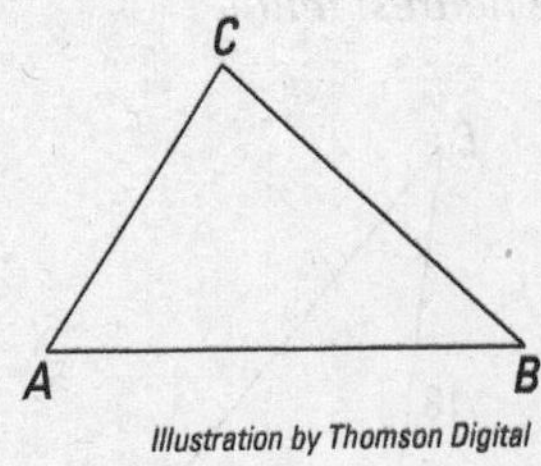

Illustration by Thomson Digital

Using the Trig Formula for Area in Various Situations

***662–663** Solve the problem.*

662. In an isosceles triangle, the two equal sides each measure 20 meters; they include an angle of 30°. Find the exact area of the isosceles triangle.

663. In a rhombus, each side is 15 inches, and one angle is 96°. Find the area of the rhombus to the nearest square inch.

Solving Area Problems Needing Additional Computations

664–669 *Find the area of the triangle. Round your answer to the nearest tenth.*

664.

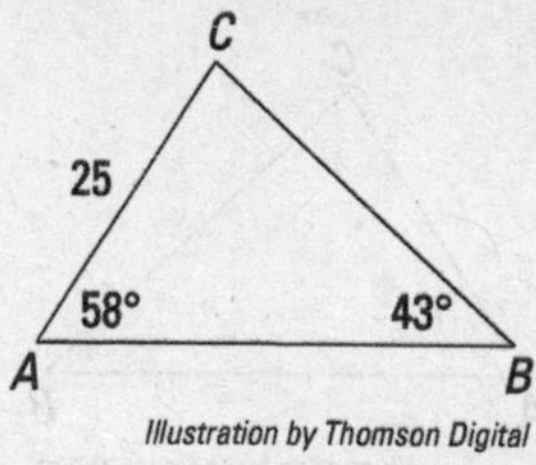

Illustration by Thomson Digital

665.

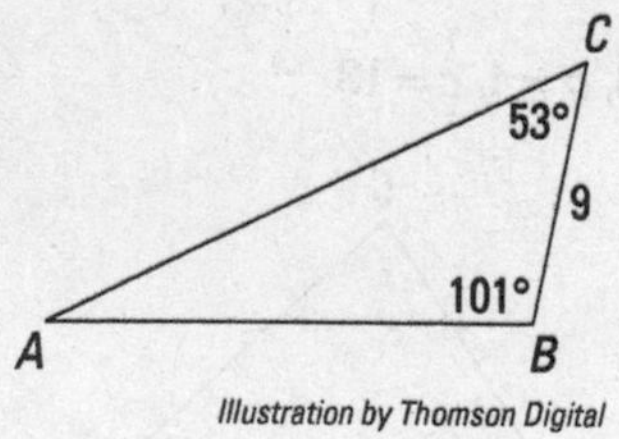

Illustration by Thomson Digital

666.

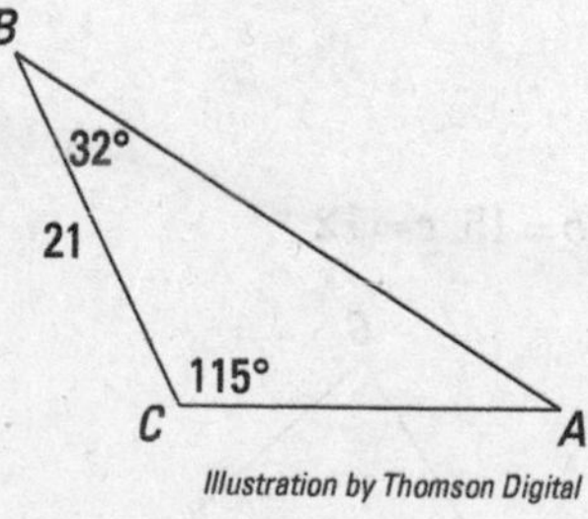

Illustration by Thomson Digital

667.

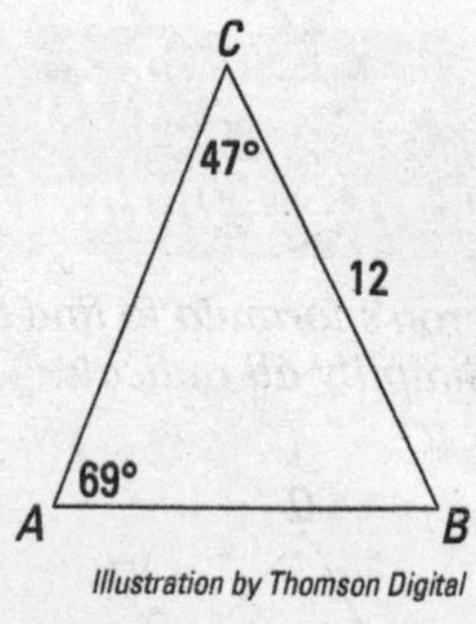

Illustration by Thomson Digital

668.

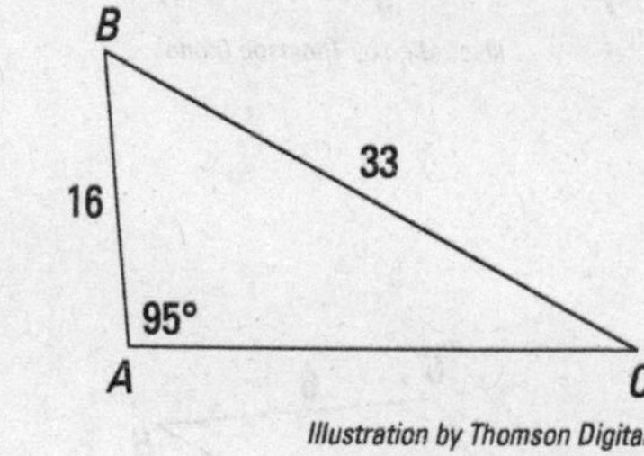

Illustration by Thomson Digital

669.

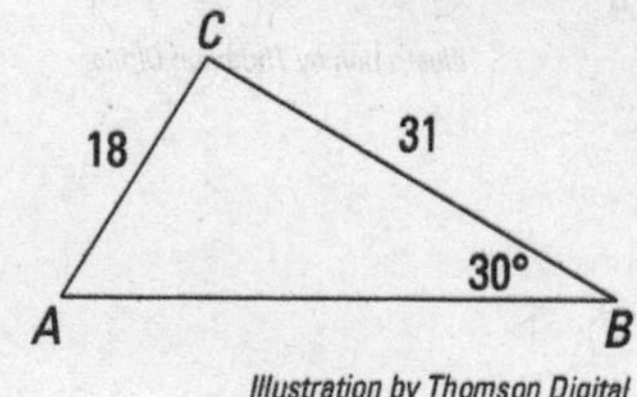

Illustration by Thomson Digital

670. Find the area of triangle *BCD*.

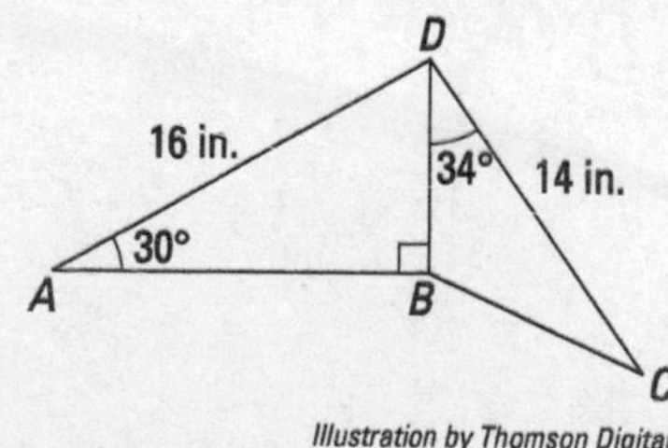

Illustration by Thomson Digital

Finding Areas of Triangles by Using Heron's Formula

671–673 *Use Heron's formula to find the exact area of the triangle. Simplify all radicals.*

671.

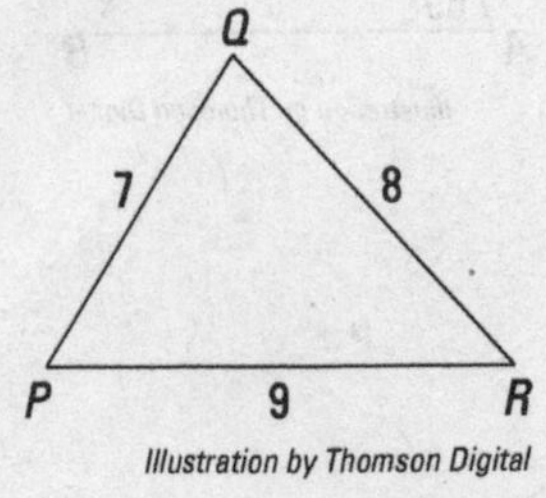

Illustration by Thomson Digital

672.

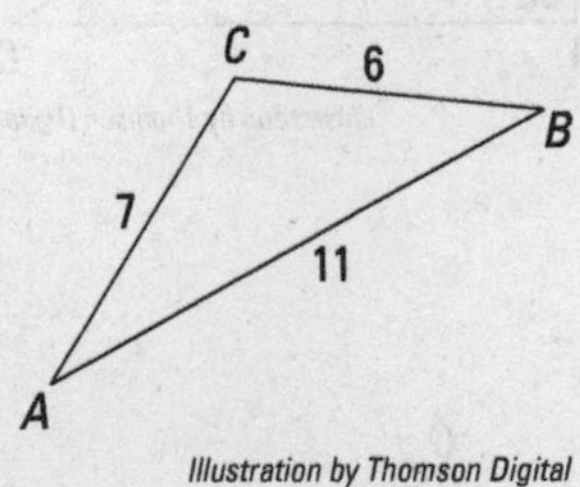

Illustration by Thomson Digital

673.

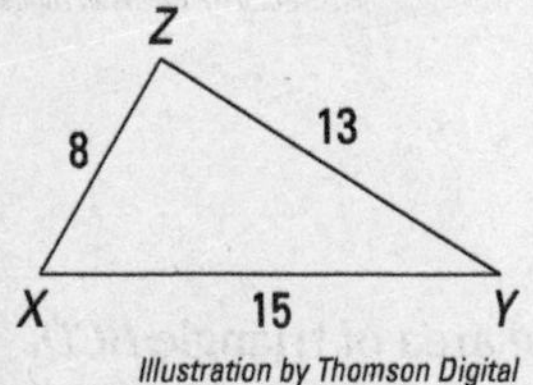

Illustration by Thomson Digital

Applying Heron's Formula

674–676 *Use the figure and the given information to find the area of the triangle with Heron's formula. (The triangle isn't necessarily drawn to scale.) Round your answer to the nearest tenth.*

674. $a=6, b=8, c=12$

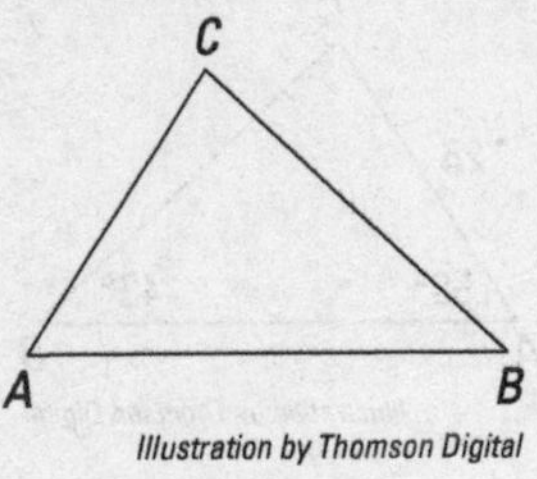

Illustration by Thomson Digital

675. $a=13, b=4, c=13$

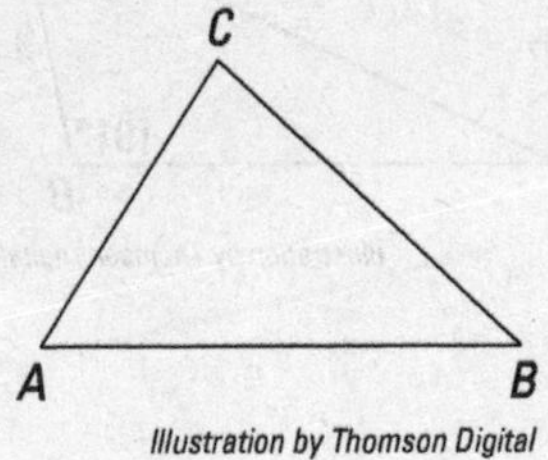

Illustration by Thomson Digital

676. $a=5, b=15, c=12$

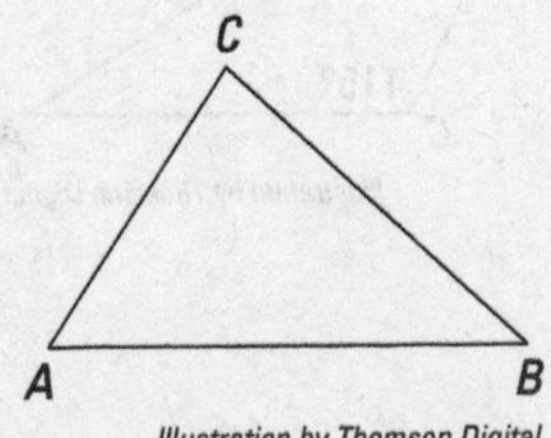

Illustration by Thomson Digital

Practical Applications Using Heron's Formula

***677–680** Use Heron's formula to solve the problem.*

677. If a triangular plot of land has sides that measure 375 feet, 250 feet, and 300 feet, what is the area of the plot? Round your answer to the nearest whole number.

678. The Bermuda Triangle is an imaginary area located off the southeastern Atlantic coast of the United States of America; it's noted for a supposedly high incidence of unexplained disappearances of ships and aircraft. The vertices of the triangle are generally believed to be Bermuda; Miami, Florida; and San Juan, Puerto Rico. The distance from Bermuda to Miami is about 1,040 miles. The distance from Miami to San Juan is about 1,000 miles, and the distance from San Juan to Bermuda is about 960 miles. Find the area of the Bermuda Triangle to the nearest one hundred square miles.

679. Find the area of triangle *BCD* to the nearest whole number.

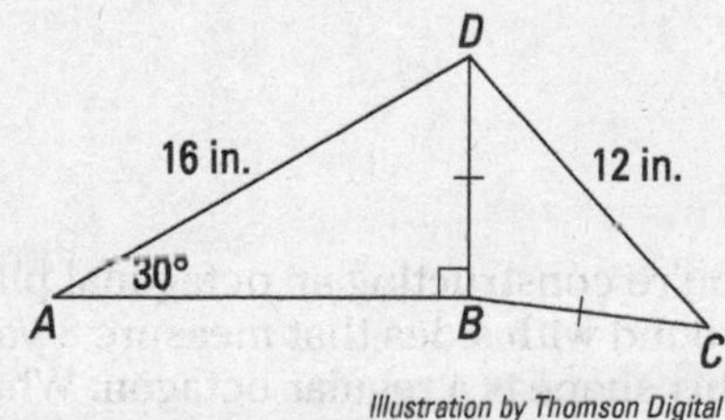

Illustration by Thomson Digital

680. Find the area of the triangle on the coordinate plane with vertices at A (–3, 1), B (1, 3), and C (3, –2).

Tackling Practical Applications by Using Triangular Formulas

***681–690** Solve the problem using the Law of Sines, the Law of Cosines, or an area formula.*

681. A man and a woman are walking toward each other on an east-west path through the park. A hot air balloon is directly above the path between them. The woman sees the balloon while looking east at an angle of elevation of 46°. The man sees the balloon while looking west at an angle of elevation of 58°. If the walkers are 75 yards apart, how far is the balloon from the man to the nearest tenth of a yard?

682. A surveyor has to stake the lot markers for a new public park. The diagram shows some of the dimensions of the park. How much fencing is necessary to enclose the entire park? Round your answer to the nearest meter.

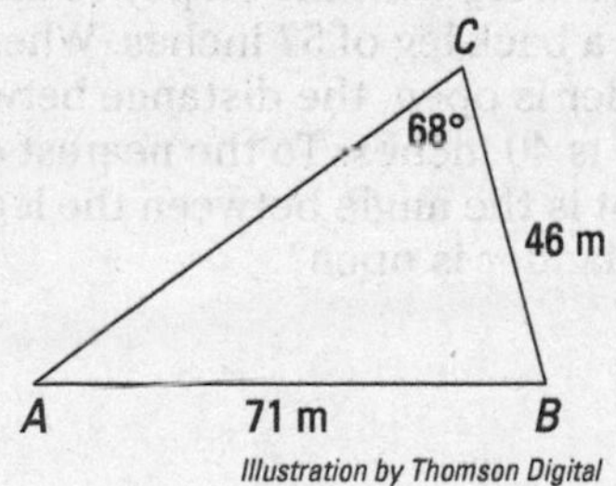

Illustration by Thomson Digital

683. Two tracking stations, *A* and *B*, are 20 miles apart. At noon, each station measures the angle of elevation of a rocket. The rocket isn't between the two stations. If the angle of elevation from station *A* is 45° and the angle of elevation from station *B* is 72°, find the altitude of the rocket to the nearest tenth of a mile. Assume that *D*, the point directly beneath the rocket, is on the same line as *A* and *B*.

684. A boat leaves port and heads due east for 25 miles. It then makes a turn 35° southeast and travels 15 miles. How far is the boat from port? Round your answer to the nearest tenth.

685. The lengths of the diagonals of a parallelogram are 8 centimeters and 12 centimeters. Find the lengths of the short sides of the parallelogram if the diagonals intersect at an angle of 36°. Round your answer to the nearest tenth.

686. The legs of a stepladder usually have different lengths. One stepladder has a front leg (the leg with the steps) 60 inches long and a back leg of 57 inches. When the ladder is open, the distance between the legs is 40 inches. To the nearest degree, what is the angle between the legs when the ladder is open?

687. A *perfect triangle* is a triangle whose side lengths are integers and whose area is numerically equal to its perimeter. If a particular perfect triangle has two sides measuring 9 and 17, what is the measure of the third side?

688. The diagram shows a field. Find the area of the field to the nearest square meter.

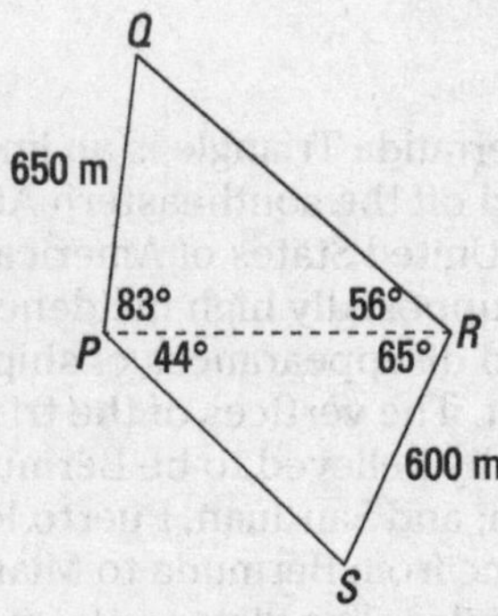

Illustration by Thomson Digital

689. Two hikers, Don and Marie, leave the same point at the same time. Marie walks due north at a rate of 3 miles per hour, and Don walks 62° northeast at a rate of 4 miles per hour. How far apart are they after three hours? Round your answer to the nearest tenth.

690. You're constructing an octagonal playground with sides that measure 8 yards. This shape is a regular octagon. What is the total area of the playground, rounded to the nearest tenth?

Chapter 12

Complex Numbers and Polar Coordinates

Complex numbers are unreal. Yes, that's the truth. A complex number has a term with a multiple of *i*, and *i* is the imaginary number equal to $\sqrt{-1}$. Many of the algebraic rules that apply to real numbers also apply to complex numbers, but you have to be careful because many rules are different for these numbers. The problems in this chapter will help you understand their particular properties.

You'll also work on graphing complex numbers. Polar coordinates are quite different from the usual (x, y) points on the Cartesian coordinate system. Polar coordinates bring together both angle measures and distances, all in one neat package. With the polar coordinate system, you can graph curves that resemble flowers and hearts and other elegant shapes.

The Problems You'll Work On

In this chapter, you'll work on complex numbers and polar coordinates in the following ways:

- Simplifying powers of *i* into one of four values
- Adding and subtracting complex numbers by combining like parts
- Multiplying complex numbers and simplifying resulting powers of *i*
- Dividing complex numbers by multiplying by a conjugate
- Interpreting graphs of basic polar coordinates
- Graphing polar equations such as cardioids and lemniscates

What to Watch Out For

When working with complex numbers and polar coordinates, some challenges will include the following:

- Multiplying imaginary numbers correctly
- Choosing the correct conjugate and simplifying the difference of squares correctly when dividing complex numbers

- Moving in a counterclockwise direction when graphing polar coordinates
- Recognizing which ray to use when graphing negative and multiple angle measures

Writing Powers of i in Their Simplest Form

691–695 *Write the power of i in its simplest form.*

691. i^9

692. i^{42}

693. i^{100}

694. i^{301}

695. $i^{4,003}$

Adding and Subtracting Complex Numbers

696–705 *Perform the operations. Write your answer in the form $a+bi$.*

696. $(3+4i)+(-2-5i)$

697. $(-4-2i)-(5-6i)$

698. $(4+3i)+(5-2i)-(6+i)$

699. $8+(3-5i)+(4-3i)$

700. $(6+4i)-(-7+3i)-6i$

701. $\left(5+i\sqrt{3}\right)+\left(6-4i\sqrt{3}\right)$

702. $\left(6+i\sqrt{8}\right)-\left(4-3i\sqrt{2}\right)$

703. $\left(-5-i\sqrt{40}\right)+\left(-2+i\sqrt{90}\right)$

704. $\left(-5+\frac{3}{4}i\right)-\left(-8-\frac{5}{4}i\right)$

705. $\left(-6-\frac{12}{5}i\right)+\left(7+\frac{3}{4}i\right)$

Multiplying Complex Numbers

***706–715** Multiply. Write your answer in $a+bi$ form.*

706. $6(3-2i)$

707. $-5(-4+3i)$

708. $2i(6+8i)$

709. $-3i(2-5i)$

710. $(2-3i)(2+3i)$

711. $(-8+2i)(-8-2i)$

712. $(5-4i)(6+3i)$

713. $(3-2i)^2$

714. $(2+3i)^3$

715. $(-1-4i)^4$

Using Multiplication to Divide Complex Numbers

***716–720** Perform the division by multiplying by a conjugate. Write your answer in the form $a+bi$.*

716. $\frac{2-3i}{2+3i}$

717. $\frac{-4-i}{6+2i}$

718. $\frac{5}{4-2i}$

719. $\frac{6i}{-1-i}$

720. $\frac{4+7i}{i}$

Solving Quadratic Equations with Complex Solutions

721–725 *Solve the equation. Write your answer in the form $a+bi$.*

721. $x^2-3x+4=0$

722. $x^2+4x+8=0$

723. $2x^2-5x+25=0$

724. $4x^2+x+4=0$

725. $5x^2+2x+3=0$

Graphing Complex Numbers

726–730 *Identify the point on the complex coordinate plane. Give your answer in the form $a+bi$.*

726.

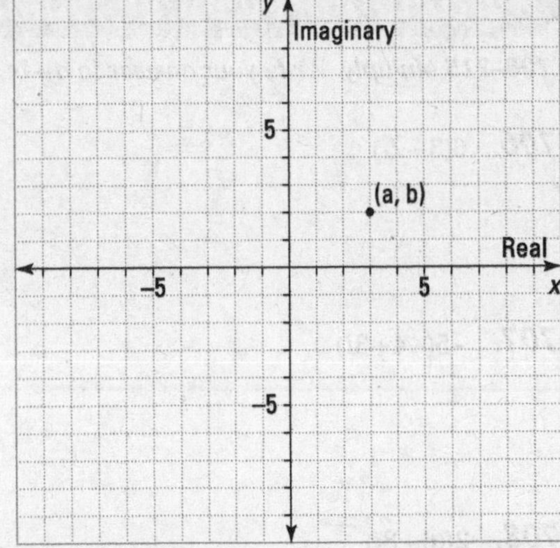

Illustration by Thomson Digital

727.

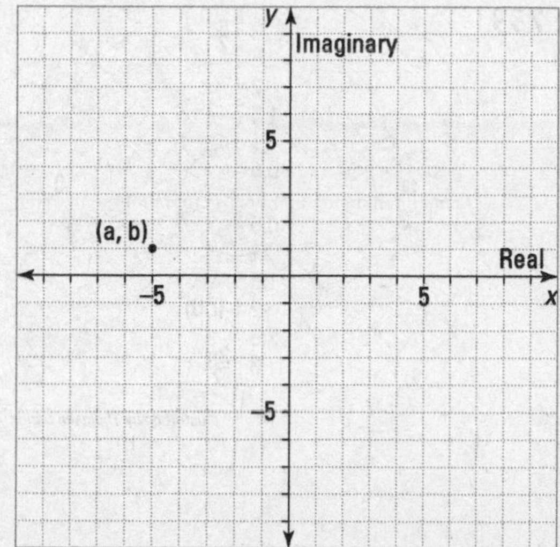

Illustration by Thomson Digital

728.

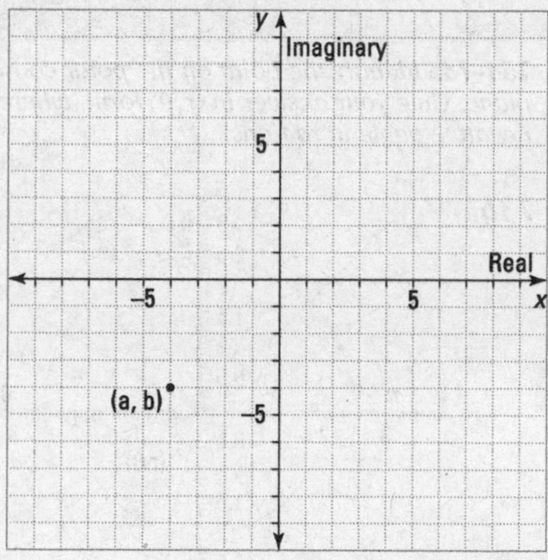

Illustration by Thomson Digital

729.

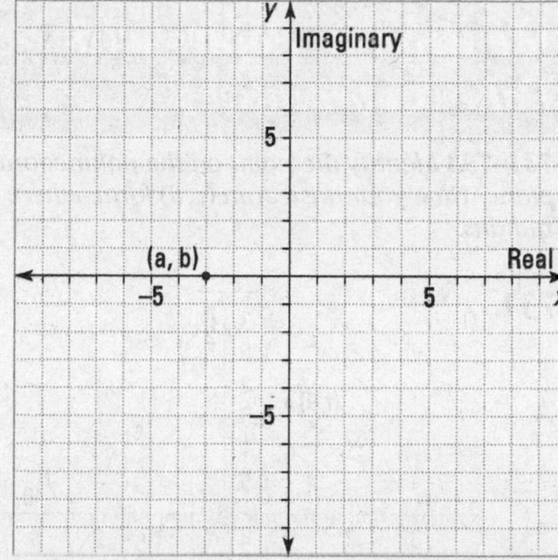

Illustration by Thomson Digital

730.

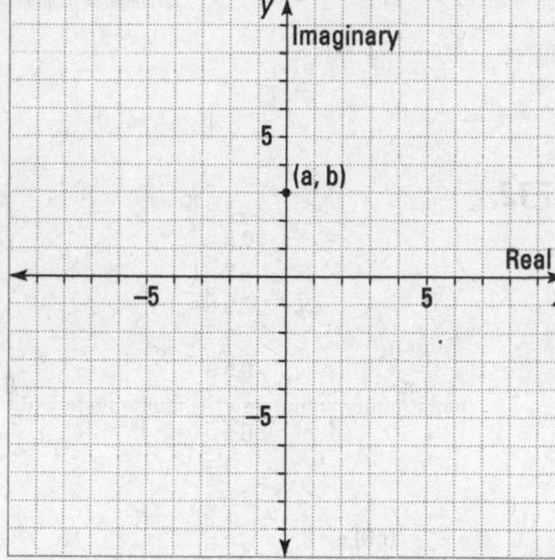

Illustration by Thomson Digital

Identifying Points with Polar Coordinates

731–733 *Identify the point on the polar coordinate plane. Give your answer in (r, θ) form, where θ is in radians.*

731.

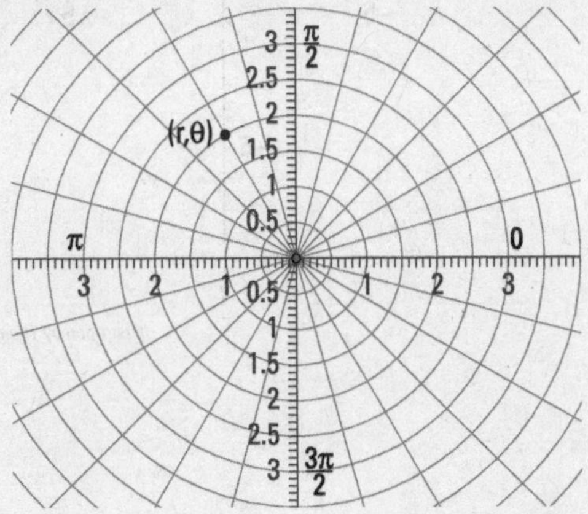

Illustration by Thomson Digital

732.

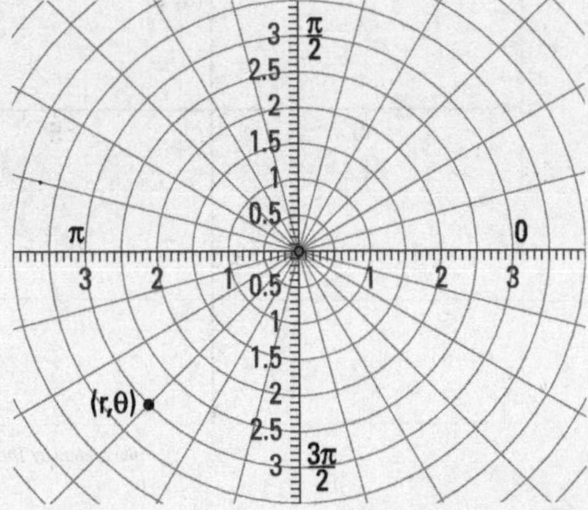

Illustration by Thomson Digital

733.

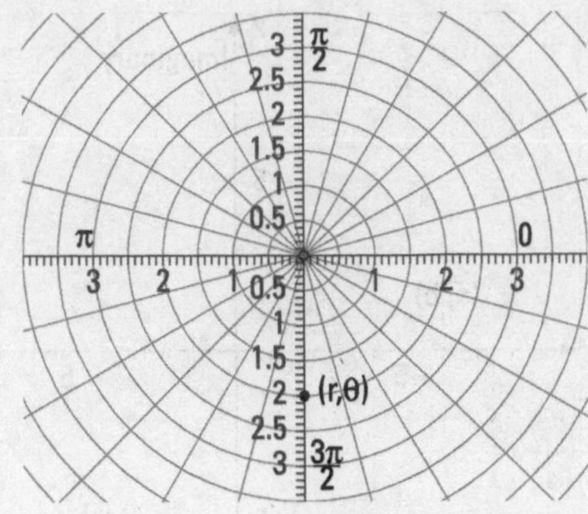

Illustration by Thomson Digital

Identifying Points Whose Angles Have Negative Measures

734–735 *Identify the point on the polar coordinate plane. Give your answer in (r, θ) form, where θ is a negative angle in radians.*

734.

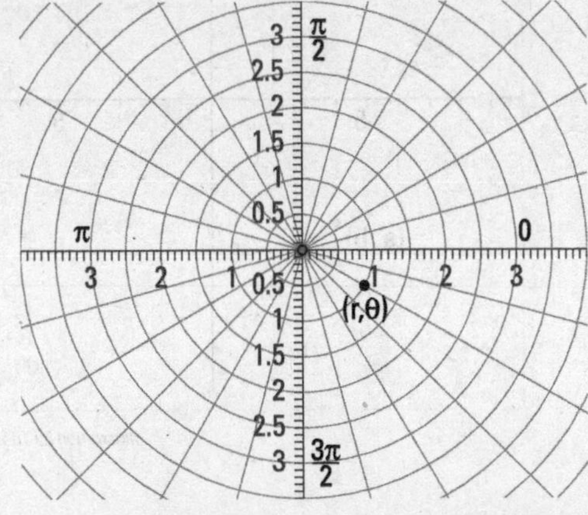

Illustration by Thomson Digital

735.

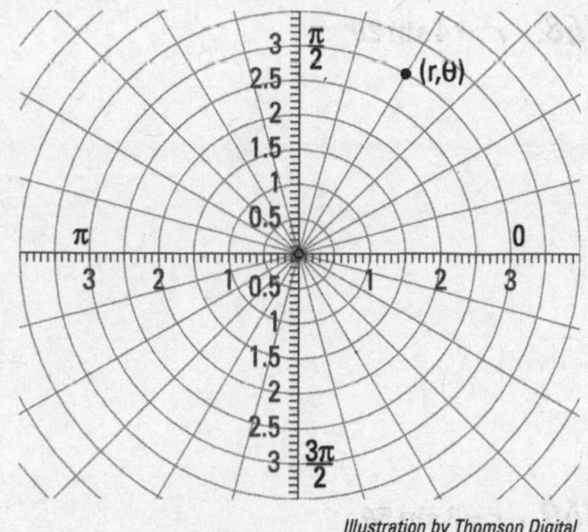

Illustration by Thomson Digital

Converting Polar to Rectangular Coordinates

736–740 *Change the polar coordinates to rectangular coordinates.*

736. $\left(2, \frac{3\pi}{2}\right)$

737. $\left(3, \frac{2\pi}{3}\right)$

738. $\left(4, \frac{5\pi}{6}\right)$

739. $\left(1, -\frac{5\pi}{3}\right)$

740. $(2, \pi)$

Converting Rectangular to Polar Coordinates

741–745 *Change the rectangular coordinates to polar coordinates.*

741. $\left(1, -\sqrt{3}\right)$

742. $(-1, 1)$

743. $\left(3, \sqrt{3}\right)$

744. $(3, 0)$

745. $\left(-\frac{7}{25}, -\frac{24}{25}\right)$

Note: Round your answer to three decimal places.

Recognizing Polar Curves

746–750 *Identify the polar curve and sketch its graph.*

746. $r = 2\theta$ for $\theta \geq 0$

747. $r = 2 + 2\sin\theta$

748. $r^2 = 4\sin 2\theta$

749. $r = 3\sin 5\theta$

750. $r = 2 + 3\cos\theta$

Chapter 13

Conic Sections

Conic sections can be described or illustrated with exactly what their name suggests: cones. Imagine an orange cone in the street, steering you in the right direction. Then picture some clever highway engineer placing one cone on top of the other, tip to tip. That engineer is trying to demonstrate how you can create conic sections (well, perhaps I stretch this a bit).

If you come along and slice one of those cones parallel to the ground, the cut edges form a circle. Slice the cone on an angle, and you have an ellipse. Slice the cone parallel to one of the sides, and you have a parabola. And, finally, slice through both cones together, perpendicular to the ground, and you have a hyperbola. If these descriptions just don't work for you, the problems in this chapter will certainly do the trick.

The Problems You'll Work On

In this chapter, you'll work on conic sections in the following ways:

- Recognizing which conic you have from the general equation
- Finding the centers of circles and ellipses
- Determining the foci of circles, ellipses, and parabolas
- Using the directrix of a parabola to complete the sketch
- Writing the equations of a hyperbola's asymptotes
- Changing basic conic section equations from parametric to rectangular

What to Watch Out For

When working with conic sections, some challenges will include the following:

- Determining the major axis of an ellipse
- Sketching the graph of a parabola in the correct direction
- Using the asymptotes of a hyperbola correctly in a graph
- Finding the square root in the equation of a circle when finding the radius

Identifying Conics from Their Equations

751–755 *Name the conic and its center.*

751. $\frac{(x-3)^2}{4}-\frac{(y+2)^2}{9}=1$

752. $\frac{(x+4)^2}{5}+\frac{(y-1)^2}{1}=1$

753. $(x-5)^2+y^2=16$

754. $9x^2+4(y+3)^2=36$

755. $6(y-8)^2-9(x+3)^2=54$

Rewriting Conic Equations in Standard Form

756–760 *Write the equation of the conic in standard form.*

756. $x^2+y^2-6x-7=0$

757. $25x^2+9y^2+100x-54y-44=0$

758. $25y^2-x^2-150y+200=0$

759. $4x^2+4y^2-32x+32y+127=0$

760. $4x^2+3y^2-8x-12y+13=0$

Writing Equations for Circles

761–770 *Write the equation of the circle described. Then graph the circle.*

761. center: (0, 0); radius: 4

762. center: (4, 3); radius: 5

763. center: (–2, 1); radius: 1

764. center: (0, –2); radius: $\frac{2}{3}$

765. center: (–5, 0); radius: $\sqrt{7}$

766. endpoints of diameter: (4, 7) and (–2, –1)

767. endpoints of diameter: (–3, –4) and (–1, 0)

768. tangent to both axes in third quadrant; radius: 3

769. tangent to $x=-1$ and $y=-2$ with center in third quadrant; radius: 7

770. tangent to both axes with center on the line $y=-x$; radius: 4

Determining Foci and Axes of Symmetry of Parabolas

***771–774** Identify the focus and axis of symmetry of the parabola.*

771. $y=(x-2)^2$

772. $y-2=-\frac{1}{4}(x+1)^2$

773. $x+1=\frac{1}{2}(y-4)^2$

774. $x-6=-2(y+1)^2$

Finding the Vertices and Directrixes of Parabolas

***775–777** Identify the vertex and directrix of the parabola.*

775. $y+2=\frac{1}{8}x^2$

776. $x=-\frac{1}{3}(y-3)^2$

777. $\frac{1}{2}x-1=\frac{1}{8}(y-2)^2$

Writing Equations of Parabolas

778–780 *Write the equation of the parabola described.*

778. vertex: (–3, 2); parabola goes through (0, 20); axis of symmetry: $x=-3$

779. vertex: (4, 1); parabola goes through (6, 3); axis of symmetry: $y=1$

780. vertex: (3, 0); axis of symmetry: $x=3$; directrix: $y=-1$

Determining Centers and Foci of Ellipses

781–787 *Identify the center and foci of the ellipse.*

781. $\frac{x^2}{2}+\frac{y^2}{1}=1$

782. $\frac{x^2}{36}+\frac{y^2}{100}=1$

783. $\frac{(x-1)^2}{9}+\frac{(y+2)^2}{25}=1$

784. $\frac{(x+3)^2}{25}+\frac{y^2}{169}=1$

785. $\frac{x^2}{100}+\frac{(y-1)^2}{64}=1$

786. $\frac{(x-7)^2}{5}+\frac{(y+7)^2}{4}=1$

787. $\frac{(x-2)^2}{16}+\frac{(y-2)^2}{20}=1$

Writing Equations of Ellipses

788–790 *Use the given information to write the equation of the ellipse. Then sketch its graph.*

788. center: (–3, 2); vertex: (2, 2); co-vertex: (–3, 3)

789. vertices: (1, 1), (1, –9); co-vertices: (–3, –4), (5, –4)

790. vertices: (–6, 5), (0, 5); foci: (–5, 5), (–1, 5)

Determining Asymptotes of Hyperbolas

791–797 *Identify the asymptotes of the hyperbola.*

791. $x^2 - y^2 = 1$

792. $\frac{x^2}{16} - \frac{y^2}{9} = 1$

793. $\frac{y^2}{49} - \frac{x^2}{576} = 1$

794. $\frac{(y-4)^2}{8} - \frac{(x-1)^2}{8} = 1$

795. $\frac{y^2}{20} - \frac{(x+5)^2}{5} = 1$

796. $\frac{(x+3)^2}{36} - \frac{(y-3)^2}{9} = 1$

797. $\frac{(y-4)^2}{18} - \frac{(x-4)^2}{6} = 1$

Writing Equations of Hyperbolas

798–800 *Write the standard equation of the hyperbola given the description. Then sketch its graph.*

798. center: (3, –2); foci: (–1, –2), (7, –2); $b = 3$

799. center: (0, 7); foci: (0, 1), (0, 13); $b = 5$

800. foci: $\left(-\sqrt{5}, 0\right), \left(\sqrt{5}, 0\right)$; asymptotes: $y = \frac{1}{2}x,\ y = -\frac{1}{2}x$

Changing Equation Format from Trig Functions to Algebraic

801–805 *Write the equation of the curve in rectangular form (algebraic terms) by eliminating the parameter θ.*

801. $x = 4\cos\theta,\ y = 4\sin\theta$

802. $x = 3 + 5\cos\theta,\ y = 2 + 3\sin\theta$

803. $x=-4+5\tan\theta,\ y=3\sec\theta$

804. $x=5+3\cos\theta,\ y=-2+3\sin\theta$

805. $x=3+\frac{1}{2}\tan\theta,\ y=6-\frac{1}{4}\sec\theta$

Changing Equation Format from Algebraic to Trig

806–810 *Change the equation of the conic from rectangular form (algebraic terms) to parametric form (using trig functions).*

806. $4x^2+25y^2=100$

807. $x^2+y^2-6x+14y+49=0$

808. $7x^2-11y^2-14x-44y=114$

809. $9x^2+16y^2+36x-64y=44$

810. $4y^2-8x^2-40y+16x+60=0$

Chapter 14

Systems of Equations and Inequalities

A *system of equations* is a collection of two or more equations involving two or more variables. If the number of equations is equal to the number of different variables, then you may be able to find a unique solution that's common to all the equations. Having the correct number of variables isn't a guarantee that you'll have that solution, and it's not terrible if a unique solution doesn't exist; sometimes you just write a rule to represent the many solutions shared by the equations in the collection.

A system of inequalities has an infinite number of solutions (unless it has none at all). You solve these systems using graphs of the separate statements.

The techniques for solving these systems involve algebraic manipulation and/or matrices and matrix mathematics. Lots of practice using these techniques is available here. When you're solving your own systems, the approach you use is up to you.

The Problems You'll Work On

In this chapter, you'll work on solving systems of equations and inequalities in the following ways:

- Using substitution to solve linear and nonlinear systems of equations
- Applying the elimination method when solving systems of linear equations
- Writing a rule for multiple solutions of systems of equations
- Graphing systems of inequalities
- Creating partial fractions using fraction decomposition
- Writing coefficient matrices and constant matrices to use in matrix solutions of systems
- Determining matrix inverses to use in solving systems of linear equations

What to Watch Out For

When you're working with systems of equations and inequalities, some challenges will include

- Recognizing that the answer may be *no solution*
- Distributing correctly when using substitution for solving systems
- Performing matrix operations correctly when doing row reductions and eliminating terms
- Writing solutions from resulting variable matrices

Using Substitution to Solve Systems of Linear Equations with Two Variables

***811–815** Solve the system by using substitution. Write your answer as an ordered pair, (x, y).*

811. $2x+y=8,\ 3x+2y=17$

812. $3x+5y=22,\ x-3y=-2$

813. $x+11y=4,\ 8y-3x=29$

814. $4x-y=6,\ 2y=8x-9$

815. $5x-y=9,\ 2y=10x-18$

Using Elimination to Solve Systems of Linear Equations with Two Variables

***816–820** Solve the system using elimination. Write your answer as an ordered pair, (x, y).*

816. $4x-3y=17,\ 2x+y=1$

817. $3x+5y=20,\ 3x+4y=16$

818. $6x+7y=9,\ 3x+8y=18$

819. $5x-3y=8,\ 6y=10x-3$

820. $3x+4y=6,\ 8y=12-6x$

Solving Systems of Equations Involving Nonlinear Functions

821–830 *Solve the system of equations. Write your answers as order pairs, (x,y).*

821. $y = 4x^2 - 3x - 2,\ y = 13x - 2$

822. $y = 3 - 4x^2,\ y = 8x - 9$

823. $y = \sqrt{x},\ y = x$

824. $y = x^3 - 3x^2 + 2x - 4,\ y = 2x - 4$

825. $y = x^4 - 9x^2 + 8,\ y = 4x - 4$

826. $y = x^3 - 3x^2 - 5x + 4,\ y = -x^2 - 2$

827. $y = -x^3 + 2x^2 - 3x + 7,\ y = -x^2 - 16x + 22$

828. $y = 2x^3 - 4x^2 + 10x - 5,\ y = 2x^2 + 82x + 59$

829. $y = \dfrac{x-3}{x+4},\ y = -7x - 27$

830. $y = \dfrac{x-6}{x^2-2},\ y = -3x + 4$

Solving Systems of Linear Equations

831–838 *Solve each system of equations. Write the solution as an ordered triple, (x,y,z).*

831. $\begin{cases} x + 3y - 2z = -7 \\ 2y + z = 1 \\ y - 4z = -13 \end{cases}$

832. $\begin{cases} 4x + 3z = 6 \\ x - y - z = -6 \\ y + 2z = 8 \end{cases}$

833. $\begin{cases} x - y + 2z = 1 \\ 3x + y - z = 13 \\ 2x - 4y + 3z = 6 \end{cases}$

834. $\begin{cases} 5x + 3y + z = -13 \\ 2x - 4y - z = -8 \\ -3x + y + 2z = 13 \end{cases}$

835. $\begin{cases} 2x-3y = 4 \\ 4x + 5z = 42 \\ 5y-3z = 14 \end{cases}$

836. $\begin{cases} x+y+z = -\frac{1}{2} \\ -x - z = 0 \\ y+z = -1 \end{cases}$

837. $\begin{cases} x+3y-2z = 19 \\ 4x-5y+z = -24 \\ 6x+y-3z = 14 \end{cases}$

838. $\begin{cases} 2x+y = -11 \\ x-5y+z = -4 \\ -3x-7y+z = 18 \end{cases}$

Solving Systems of Linear Equations with Four Variables

839–840 *Solve the system of equations. Write the solution as (x, y, z, w).*

839. $\begin{cases} x+2y-3z+w = -21 \\ 2x-3y+z-2w = 15 \\ x-4y-2z+3w = -3 \\ 3x + 2z - w = 7 \end{cases}$

840. $\begin{cases} 5x + 3z = 5 \\ 2x-3y + w = -3 \\ 4y+z-2w = 8 \\ x-y + w = -2 \end{cases}$

Graphing Systems of Inequalities

841–845 *Graph the system of inequalities.*

841. $\begin{cases} x+y \le 10 \\ 2x-y > 4 \end{cases}$

842. $\begin{cases} 2x+y \ge 8 \\ x \le 3 \\ y \le 7 \end{cases}$

843. $\begin{cases} x+3y \le 12 \\ 3x+y \le 6 \\ x \ge 0,\ y \ge 0 \end{cases}$

844. $\begin{cases} 2x+4y \ge 12 \\ 5x+2y \ge 20 \\ x \ge 0,\ y \ge 0 \end{cases}$

845. $\begin{cases} x + \ \ y \geq 3 \\ x + 2y \leq 16 \\ 4x + \ \ y \leq 24 \\ x \geq 0, y \geq 0 \end{cases}$

Decomposition of Fractions

***846–850** Write the fraction as the sum or difference of fractions using decomposition.*

846. $\dfrac{3x+7}{x^2-1}$

847. $\dfrac{11x+8}{x^2+x-6}$

848. $\dfrac{28-3x}{x^2+7x}$

849. $\dfrac{6x^2+11x-30}{x^3-3x^2-10x}$

850. $\dfrac{3x^2-12x+64}{x^3-16x}$

Operating on Matrices

***851–855** Perform the matrix operation.*

851. Given $A=\begin{bmatrix} 1 & 2 \\ -3 & 4 \end{bmatrix}$ and $B=\begin{bmatrix} 0 & -3 \\ 4 & -7 \end{bmatrix}$, find $A+B$.

852. Given $A=\begin{bmatrix} 1 & 2 \\ -3 & 4 \end{bmatrix}$ and $B=\begin{bmatrix} 0 & -3 \\ 4 & -7 \end{bmatrix}$, find $B-A$.

853. Given $A=\begin{bmatrix} 1 & 2 \\ -3 & 4 \end{bmatrix}$ and $B=\begin{bmatrix} 0 & -3 \\ 4 & -7 \end{bmatrix}$, find $4A-2B$.

854. Given $A=\begin{bmatrix} 1 & 2 \\ -3 & 4 \end{bmatrix}$ and $B=\begin{bmatrix} 0 & -3 \\ 4 & -7 \end{bmatrix}$, find $A \cdot B$.

855. Given $B=\begin{bmatrix} 0 & -3 \\ 4 & -7 \end{bmatrix}$ and $C=\begin{bmatrix} 1 & 0 & 2 & 3 \\ 4 & 1 & -4 & 2 \end{bmatrix}$, find $B \cdot C$.

Changing Matrices to the Echelon Form

856–860 *Use matrix operations to rewrite the matrix in echelon form.*

856. $\begin{bmatrix} 1 & 3 & 2 \\ 4 & 0 & 3 \\ 2 & -1 & -3 \end{bmatrix}$

857. $\begin{bmatrix} 5 & -1 & 3 \\ 1 & 0 & 2 \\ 4 & 1 & -3 \end{bmatrix}$

858. $\begin{bmatrix} 2 & -4 & 4 \\ -1 & 6 & 3 \\ 0 & -8 & 6 \end{bmatrix}$

859. $\begin{bmatrix} 1 & 0 & 4 & -2 \\ 3 & 1 & -2 & 5 \\ 4 & 3 & 2 & -10 \\ 0 & 1 & -6 & -3 \end{bmatrix}$

860. $\begin{bmatrix} 3 & -8 & 1 & 3 \\ 1 & -3 & 4 & 2 \\ 5 & -13 & -6 & 1 \\ -4 & 13 & -2 & 1 \end{bmatrix}$

Solving Systems of Equations Using Augmented Matrices

861–870 *Solve the system of equations using an augmented matrix.*

861. $\begin{cases} 2x+y=1 \\ x-2y=8 \end{cases}$

862. $\begin{cases} 5x-3y=7 \\ x-2y=7 \end{cases}$

863. $\begin{cases} 2x-y=1 \\ 3x+y=9 \end{cases}$

864. $\begin{cases} 2x-3y=10 \\ 4x+y=-1 \end{cases}$

865. $\begin{cases} x+3y-2z=-13 \\ 2x-y+3z=9 \\ x+8y+6z=1 \end{cases}$

866. $\begin{cases} 4x-3y-z=11 \\ 2x+y+3z=3 \\ x+2y+5z=0 \end{cases}$

867. $\begin{cases} x+3y-4z=-25 \\ 2x-y+11z=66 \\ 3x+4y+z=5 \end{cases}$

868. $\begin{cases} x+4z=15 \\ 2x-3y=23 \\ 5y-3z=-21 \end{cases}$

869. $\begin{cases} x+2y+z-w=2 \\ 2x-3y+2w=3 \\ 4x+y-z=1 \\ 2y+4z+w=5 \end{cases}$

870. $\begin{cases} 5x+3y-z=12 \\ x-2y+3w=0 \\ 2x+4z-3w=-18 \\ 4y+3z-w=1 \end{cases}$

Solving Systems of Equations Using the Inverse of the Coefficient Matrix

***871–875** Solve the system of equations using the inverse of the coefficient matrix.*

871. $\begin{cases} 2x+3y=7 \\ 3x+4y=8 \end{cases}$

872. $\begin{cases} 5x-2y=16 \\ 12x-5y=39 \end{cases}$

873. $\begin{cases} 3x+5y=12 \\ 7x+12y=29 \end{cases}$

874. $\begin{cases} x+2y+3z=-2 \\ x-2y+4z=-3 \\ x+y+3z=-2 \end{cases}$

875. $\begin{cases} -7x+2y+z=7 \\ 2x-3y+5z=-28 \\ 4x-3z=8 \end{cases}$

Applying Cramer's Rule to Solve Systems of Equations

876–880 *Use Cramer's Rule to solve the system of equations.*

876. $\begin{cases} 4x+5y=12 \\ 5x+7y=18 \end{cases}$

877. $\begin{cases} 9x-4y=21 \\ 6x+7y=72 \end{cases}$

878. $\begin{cases} 3x+11y=-29 \\ 12x-5y=31 \end{cases}$

879. $\begin{cases} 5x+8y=11 \\ 13x+4y=79 \end{cases}$

880. $\begin{cases} x+3y-2z=8 \\ 4x+3y-5z=23 \\ 2x+5y+4z=38 \end{cases}$

Chapter 15

Sequences and Series

A *sequence* is a list of items; in mathematics, a sequence usually consists of numbers such as 1, 2, 4, 8, . . . or 1, 1, 2, 3, 5, 8, 13, . . . See the patterns in these two sequences? That's part of what this chapter is about: finding patterns in lists of numbers so you can write the rest of the numbers in the list.

A *series* is the sum of a list of numbers, such as $1 + 2 + 4 + 8$. Many times, you can find a formula to help you add up the numbers in a series. Formulas are especially helpful when you have a lot of numbers to add or if they're fractions or alternating negative and positive.

You can classify many sequences and series by type, which helps you determine particular terms or sums. Recognize the type, and you're pretty much home free when doing your calculations.

The Problems You'll Work On

In this chapter, you'll work on sequences and series in the following ways:

- Finding a particular term in a sequence when given the general formula
- Determining a formula or standard expression for all the terms in a sequence
- Writing terms in a recursively formed sequence
- Becoming familiar with summation notation
- Computing the sum of terms in arithmetic and geometric series
- Generating terms by using the binomial theorem

What to Watch Out For

Don't let common mistakes trip you up; keep in mind that when working with sequences and series, some challenges will include

- Correctly formulating terms in a sequence by using the order of operations
- Recognizing when a sequence or series is arithmetic or geometric
- Using the infinite series sum only when the ratio of consecutive terms has an absolute value less than 1
- Assigning the correct signs in the terms of an alternating sequence

Finding Terms of Sequences

881–885 *Given the general term of the sequence, find the fifth term.*

881. $a_n = 2^n - 1$

882. $a_n = \dfrac{4n+1}{2n-1}$

883. $a_n = n^2 - 3n$

884. $a_n = \dfrac{(-1)^n}{n^2+1}$

885. $a_n = 3\sin\left(\dfrac{n\pi}{2}\right)$

Determining Rules for Sequences

886–890 *Write an expression for the general term of the sequence.*

886. 1, 3, 9, 27, 81, ...

887. 1, 2, 6, 24, 120, ...

888. $2e, 4e^2, 6e^3, 8e^4, 10e^5, \ldots$

889. 2, 5, 10, 17, 26, ...

890. $4, \dfrac{5}{4}, \dfrac{2}{3}, \dfrac{7}{16}, \dfrac{8}{25}, \ldots$

Working with Recursively Defined Sequences

891–895 *Find the third and fourth terms of the sequence.*

891. $a_1 = 1, a_2 = 1, a_n = 2a_{n-2} - 3a_{n-1}$

892. $a_1 = 1, a_2 = 2, a_n = \dfrac{a_{n-2} + a_{n-1}}{a_{n-1}}$

893. $a_1 = 0, a_2 = 1, a_n = 2^{a_{n-2}} + 2^{a_{n-1}}$

894. $a_1 = e, a_2 = \dfrac{1}{e}, a_n = (a_{n-2})(a_{n-1})$

895. $a_1 = 1, a_2 = 1, a_n = a_{n-2} + a_{n-1}$

Adding Terms in an Arithmetic Series

***896–900** Find the sum of the arithmetic series.*

896. $\sum_{i=1}^{10}(4+3i)$

897. $\sum_{i=1}^{8}(6-2i)$

898. $\sum_{i=8}^{11}(5i)$

899. $\sum_{i=3}^{9}\left(\frac{i}{2}+1\right)$

900. $\sum_{i=0}^{6}\left(\frac{1}{3}i+\frac{2}{3}\right)$

Summing Terms of a Series

***901–905** Find the sum of the series.*

901. $\sum_{i=1}^{4}\left(3+\frac{12}{i}\right)$

902. $\sum_{i=1}^{5}(-1)^{i+1}2^{i-1}$

903. $\sum_{i=0}^{5}(\cos i\pi)$

904. $\sum_{i=1}^{10}\left(\frac{1}{i}-\frac{1}{i+1}\right)$

905. $\sum_{i=0}^{5}\left(\frac{1}{3}\right)^{i}$

Finding Rules and Summing Terms of a Series

***906–910** Find the sum of the first ten terms of the series.*

906. 1, 3, 5, 7, ...

907. –8, –3, 2, 7, ...

908. $1, \frac{4}{3}, \frac{5}{3}, 2, \ldots$

909. 12, 9, 6, 3, ...

910. –2, –10, –18, –26, ...

Calculating the Sum of a Geometric Series

911–915 *Find the sum of the geometric series.*

911. $\sum_{i=0}^{5} 2^i$

912. $\sum_{i=0}^{6} 4(-3)^i$

913. $\sum_{i=0}^{7} 2\left(\frac{2}{3}\right)^i$

914. $\sum_{i=0}^{8} (2)^{i-1}$

915. $\sum_{i=1}^{6} \left(\frac{1}{3}\right)^i$

Determining Formulas and Finding Sums

916–920 *Find the sum of the infinite series.*

916. $16+8+4+2+\ldots$

917. $27+18+12+8+\ldots$

918. $\frac{64}{27}+\frac{16}{9}+\frac{4}{3}+1+\ldots$

919. $64-16+4-1+\ldots$

920. $e^3+e^2+e+1+\ldots$

Counting Items by Using Combinations

921–925 *Compute the value of the combination.*

921. ${}_6C_2$

922. ${}_9C_8$

923. ${}_4C_4$

924. ${}_5C_0$

925. ${}_{54}C_6$

Constructing Pascal's Triangle

***926–930** Construct Pascal's triangle to include the coefficients of the expansion of the binomial.*

926. $(a+b)^2$

927. $(a+b)^4$

928. $(a+b)^5$

929. $(a+b)^6$

930. $(a+b)^1$

Applying Pascal's Triangle

***931–935** Expand the binomial by using Pascal's triangle.*

931. $(x+y)^3$

932. $(x+2)^4$

933. $(y-3)^5$

934. $(2x-5)^3$

935. $(4x-3y)^4$

Utilizing the Binomial Theorem

***936–940** Use the binomial theorem to find the coefficient of the term in the expansion of the binomial.*

936. x^7y^2 term in $(x+y)^9$

937. x^3y^5 term in $(x-y)^8$

938. x^3 term in $(x+4)^6$

939. x^4y^3 term in $(3x-y)^7$

940. x^3y^2 term in $(2x+5y)^5$

Chapter 16

Introducing Limits and Continuity

"What's the limit on this road?" "The sky's the limit." "We need to introduce term limits." "Have you reached your limit?" People talk about limits in many ways. In mathematics, a *limit* suggests that you're approaching some value. Some functions, such as a rational function with a horizontal asymptote, have a limit as the x values move toward positive or negative infinity — that is, as the value of x gets very small or very large. Limits are another way of describing the characteristics of particular functions.

Although limits are often demonstrated graphically (a picture is worth a thousand words?), you can describe limits more precisely using algebra. In this chapter, you find examples of both techniques.

Coupled with limits is the concept of *continuity* — whether a function is defined for all real numbers or not. Investigating a function's continuity (or lack thereof) is part of this chapter.

The Problems You'll Work On

In this chapter, you'll work on limits and continuity in the following ways:

- Looking at graphs for information on a function's limits
- Using analytic techniques to investigate limits
- Performing algebraic operations to solve for a function's limits
- Determining where a function is continuous
- Searching for any removable discontinuities

What to Watch Out For

When you're working with limits and continuity, some challenges include the following:

- Recognizing a function's behavior at negative infinity or positive infinity
- Using the correct technique for an analytic look at limits
- Factoring correctly when investigating limits algebraically
- Using the correct conjugates in algebraic procedures
- Forgetting that the "removable" part of a removable discontinuity doesn't really change a function's continuity; a function with a removable discontinuity is *not* continuous

Determining Limits from Graphs

941–945 Given the graph of $f(x)$, find $\lim_{x \to 2} f(x)$.

941.

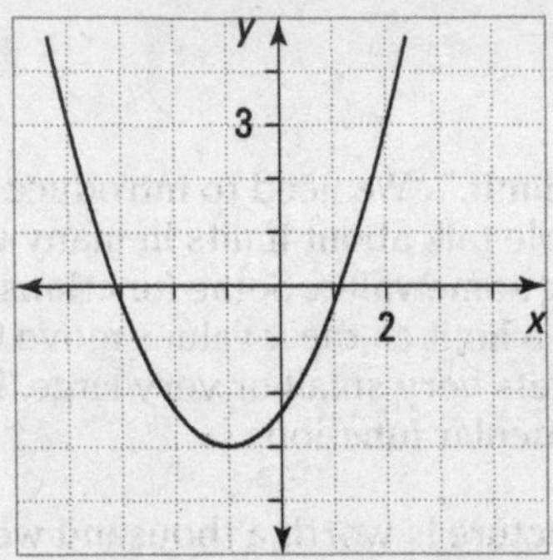

Illustration by Thomson Digital

942.

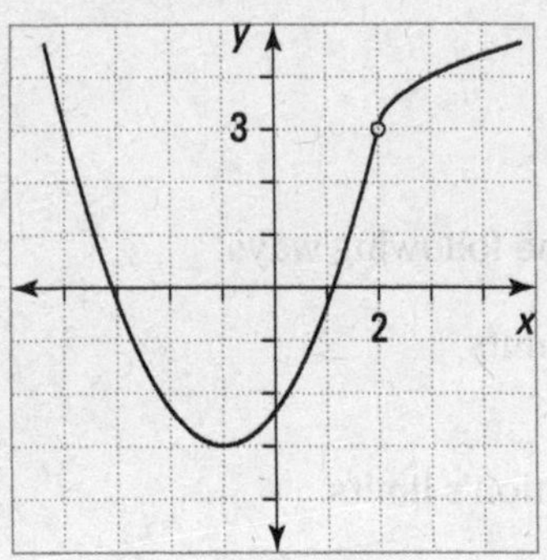

Illustration by Thomson Digital

943.

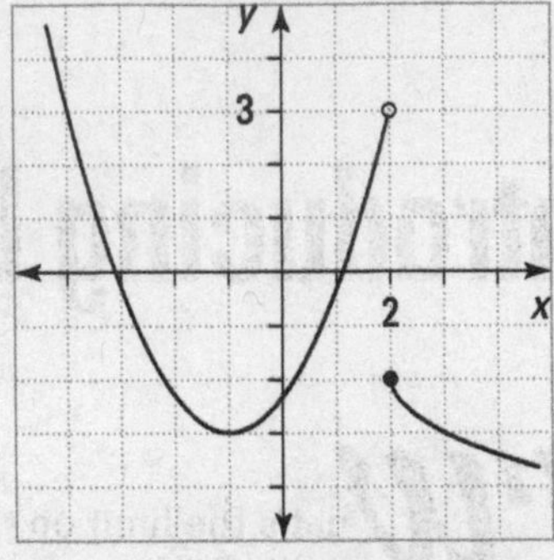

Illustration by Thomson Digital

944.

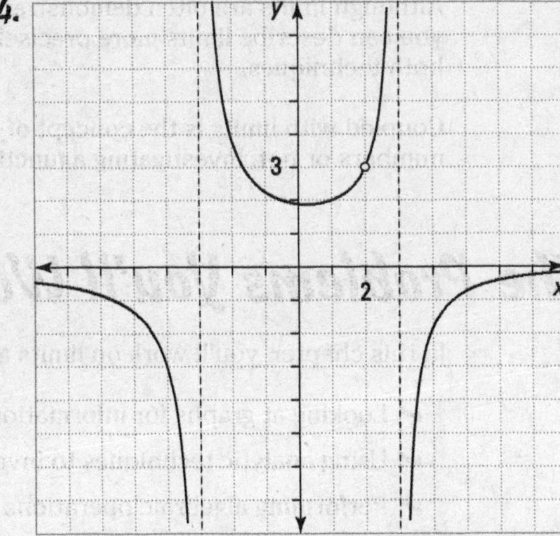

Illustration by Thomson Digital

945.

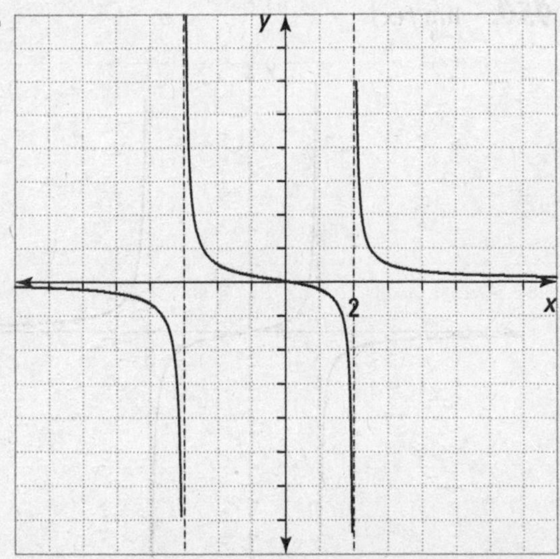

Illustration by Thomson Digital

Determining One-Sided Limits

946–950 *Given the graph of f(x), find the one-sided limit.*

946. $\lim_{x \to -1^+} f(x)$

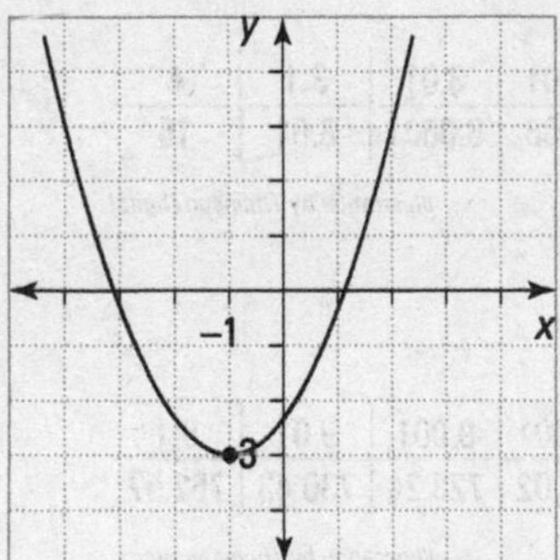

Illustration by Thomson Digital

947. $\lim_{x \to -3^+} f(x)$

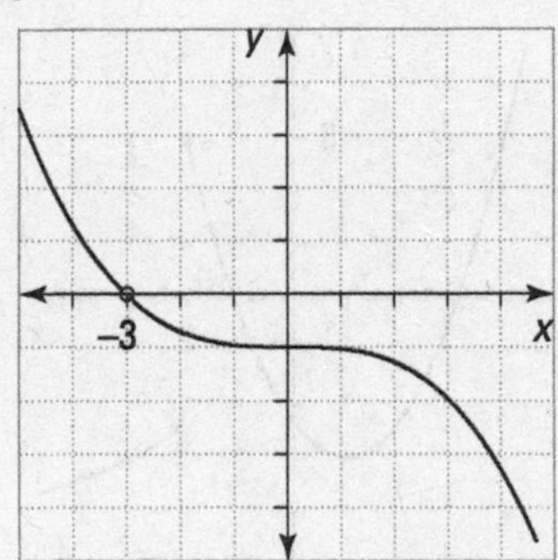

Illustration by Thomson Digital

948. $\lim_{x \to -3^-} f(x)$

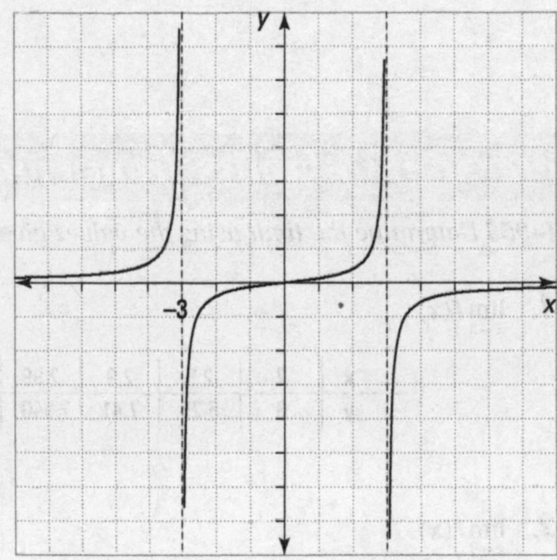

Illustration by Thomson Digital

949. $\lim_{x \to 2^+} f(x)$

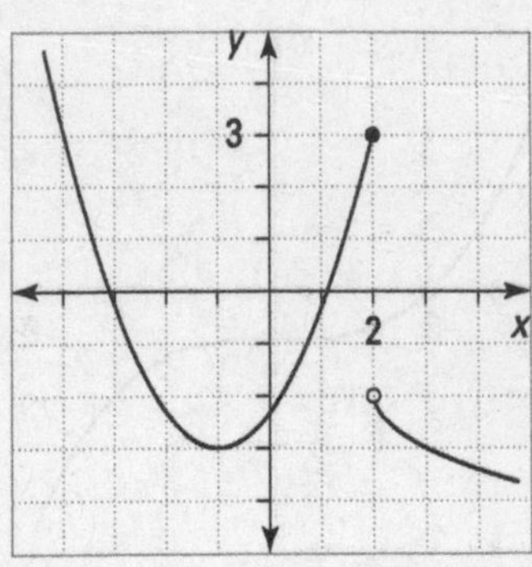

Illustration by Thomson Digital

950. $\lim_{x \to 3^-} f(x)$

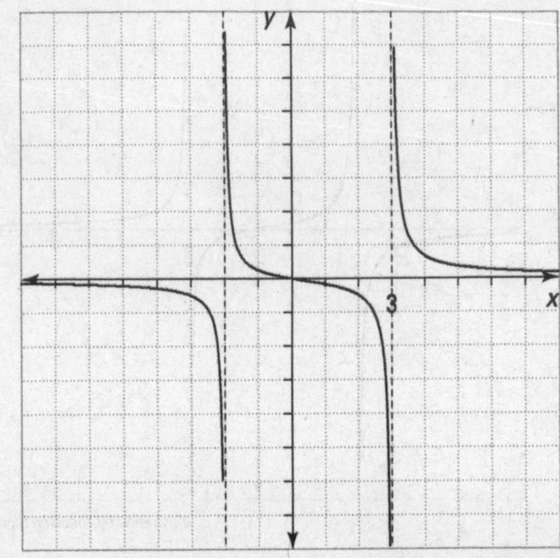

Illustration by Thomson Digital

Determining Limits from Function Values

951–955 *Determine the limit using the values given in the chart.*

951. $\lim_{x \to 3} f(x)$

x	2	2.5	2.9	2.99	2.999	3	3.001	3.01	3.1	4
y	3	5.25	7.41	7.9401	7.994	?	8.006	8.0601	8.61	15

Illustration by Thomson Digital

952. $\lim_{x \to 9} f(x)$

x	8	8.9	8.99	8.999	8.9999	9	9.0001	9.001	9.01	9.1
y	511	703.97	725.57	727.76	727.98	?	728.02	728.24	730.43	752.57

Illustration by Thomson Digital

953. $\lim_{x \to -2} f(x)$

x	–3	–2.999	–2.99	–2.01	–2.001	–2	–1.999	–1.99	–1.9	–1.5
y	–28	–27.97	–27.73	–9.121	–9.012	?	–8.988	–8.881	–7.859	–4.375

Illustration by Thomson Digital

954. $\lim_{x \to 3} f(x)$

x	2.5	2.9	2.99	2.999	2.9999	3	3.0001	3.001	3.01	3.1
y	–7	–39	–399	–3999	–39999	?	40001	4001	401	41

Illustration by Thomson Digital

955. $\lim_{x \to 5} f(x)$

x	4.5	4.9	4.99	4.999	4.9999	5	5.0001	5.001	5.01	5.1
y	22	590	59,900	5,999,000	599,990,000	?	600,010,000	6,001,000	60,100	610

Illustration by Thomson Digital

Determining Limits from Function Rules

956–980 *Find the limit.*

956. $\lim_{x \to 3} \frac{x+5}{x^2-x+2}$

957. $\lim_{x \to -2} \frac{x^2-4}{x+3}$

958. $\lim_{x \to 1} \frac{x^2-5x+4}{x^2-1}$

959. $\lim_{x \to -3} \frac{2x^2+x-15}{x^2-9}$

960. $\lim_{x \to 5} \frac{x^2+5x}{x^2-25}$

961. $\lim_{x \to 5} \frac{3x^2+16x+5}{x^2-3x-10}$

962. $\lim_{x \to 3} \frac{x^3-27}{x^4-81}$

963. $\lim_{x \to -2} \frac{x^3+8}{x^2-4}$

964. $\lim_{x \to -3} \frac{x^3+5x^2-9x-45}{x^3+3x^2-4x-12}$

965. $\lim_{x \to -2} \frac{x^3+x^2-25x-25}{4x^3+8x^2-9x-18}$

966. $\lim_{x \to \frac{\pi}{4}} \frac{\sin x}{\cos x}$

967. $\lim_{x\to\frac{\pi}{2}} \frac{\tan x+1}{\cos x}$

968. $\lim_{x\to\frac{\pi}{3}} \frac{\sin^2 x-1}{\cos x}$

969. $\lim_{x\to 0} \frac{1-\sec^2 x}{\sec x-1}$

970. $\lim_{x\to -\frac{\pi}{6}} \frac{\cos x}{1+2\sin x}$

971. $\lim_{x\to 9} \frac{\sqrt{x}-3}{x-9}$

972. $\lim_{x\to 4} \frac{5-x}{2-\sqrt{x}}$

973. $\lim_{x\to 3} \frac{\sqrt{x-2}-1}{x-3}$

974. $\lim_{x\to 0} \frac{\frac{1}{x-5}+\frac{1}{5}}{x}$

975. $\lim_{x\to\frac{\pi}{4}} \frac{\sin^2 x-\frac{1}{2}}{\tan x-1}$

976. $\lim_{x\to 0} f(x)$ where $f(x)=\begin{cases} x^2+1, & x<0 \\ 2x+1, & x\geq 0 \end{cases}$

977. $\lim_{x\to -2} g(x)$ where $g(x)=\begin{cases} 2x, & x<-2 \\ x^2-8, & x\geq -2 \end{cases}$

978. $\lim_{x\to 0} h(x)$ where $h(x)=\begin{cases} x-1, & x<-1 \\ x^2-3, & -1\leq x<0 \\ 4x-3, & x\geq 0 \end{cases}$

979. $\lim_{x\to -1} k(x)$ where $k(x)=\begin{cases} \frac{1}{x-1}, & x<-1 \\ \sqrt{x+1}, & -1\leq x<3 \\ \frac{4}{\sqrt{x+1}}, & x\geq 3 \end{cases}$

980. $\lim_{x\to 3} k(x)$ where $k(x)=\begin{cases} \frac{1}{x-1}, & x<-1 \\ \sqrt{x+1}, & -1\leq x<3 \\ \frac{4}{\sqrt{x+1}}, & x\geq 3 \end{cases}$

Applying Laws of Limits

***981–990** Use the limit laws to evaluate the expression.*

981. If $\lim_{x\to 1} f(x)=4$ and $\lim_{x\to 1} g(x)=9$, then $\lim_{x\to 1} \left[f(x)+g(x)\right]=$

982. If $\lim_{x\to 0} f(x) = 12$ and $\lim_{x\to 0} g(x) = -12$, then $\lim_{x\to 0}\left[f(x)\cdot g(x)\right] =$

983. If $\lim_{x\to -2} f(x) = 24$ and $\lim_{x\to -2} g(x) = 0$, then $\lim_{x\to -2}\left[\frac{f(x)}{g(x)}\right] =$

984. If $\lim_{x\to -4} f(x) = -\frac{1}{3}$, then $\lim_{x\to -4}\left[f(x)\right]^{-3} =$

985. If $\lim_{x\to -2} g(x) = -1$, then $\lim_{x\to -2}\left[5\cdot g(x)\right] =$

986. If $\lim_{x\to 5} f(x) = -6$ and $\lim_{x\to 5} g(x) = -3$, then $\lim_{x\to 5}\left[2f(x) - \left(g(x)\right)^2\right] =$

987. If $\lim_{x\to -3} f(x) = -8$ and $\lim_{x\to -3} g(x) = 25$, then $\lim_{x\to -3}\left[\frac{3f(x) - 4g(x)}{\sqrt{g(x)}}\right] =$

988. If $\lim_{x\to 6} f(x) = 4$ and $\lim_{x\to 6} g(x) = 3$, then $\lim_{x\to 6}\left[\frac{\left(f(x)\right)^2 + 3\left(g(x)\right)^3}{2g(x) - 1}\right] =$

989. If $\lim_{x\to -2} f(x) = 5$ and $\lim_{x\to -2} g(x) = -4$, then $\lim_{x\to -2}\left[\frac{\left(f(x)\right)^2 - \left(g(x)\right)^2}{g(x) + 1}\right] =$

990. If $\lim_{x\to 0} f(x) = 16$ and $\lim_{x\to 0} g(x) = 81$, then $\lim_{x\to 0}\left[\frac{4\sqrt{f(x)} + \sqrt{g(x)}}{\sqrt{f(x)} - \sqrt{g(x)}}\right] =$

Investigating Continuity

***991–1,001** Determine for which values of x the function is not continuous.*

991. $f(x) = \frac{x-4}{x-3}$

992. $f(x) = \frac{x^2 - 16}{x-4}$

993. $f(x) = \sqrt{16 - x^2}$

994. $f(x) = \frac{x+3}{x^2 - 9}$

995. $f(x)=\dfrac{\sin x}{\cos x+1}$ on the interval $0\le x\le 2\pi$

996. $f(x)=\tan x$ on the interval $-\pi\le x\le 0$

997. $f(x)=\dfrac{\sin 2x}{1+\cos 2x}$ on the interval $0\le x\le 2\pi$

998. $f(x)=\dfrac{e^x+1}{e^x-1}$

999. $f(x)=\dfrac{2e^{x-1}+2}{2e^{x-1}-2}$

1,000. $f(x)=\begin{cases}\sin x, & x<0\\ \cos x, & x\ge 0\end{cases}$

1,001. $f(x)=\begin{cases}1+x^2, & x\le 0\\ |x-1|, & 0<x\le 1\\ e^x-1, & x>1\end{cases}$

Part II
The Answers

To access the free Cheat Sheet created specifically for this book, go to www.dummies.com/cheatsheet/1001precalculus.

In this part . . .

You find answers and explanations for all 1,001 problems. You already know that you can often solve a particular mathematics problem in more than one way, and you have several more resources that you can refer to for those options (all published by Wiley):

- *Algebra I For Dummies* (Mary Jane Sterling)
- *Algebra II For Dummies* (Mary Jane Sterling)
- *Trigonometry For Dummies* (Mary Jane Sterling)
- *Pre-Calculus For Dummies* (Yang Kuang, PhD, and Elleyne Kase)

And after you've mastered pre-calculus and are ready to move on, you find

- *Calculus For Dummies* (Mark Ryan)
- *Math Word Problems For Dummies* (Mary Jane Sterling)
- *Linear Algebra For Dummies* (Mary Jane Sterling)

Visit `www.dummies.com` for more information.

Chapter 17

Answers

1. $\frac{\sqrt{3}}{7}$

You can write a rational number as a fraction with an integer in both the numerator and denominator. You can't write the square root of 3 as an integer.

2. $3i+1$

You can write a rational number as a fraction with an integer in both the numerator and denominator. The number $3i+1$ is an imaginary number with $i=\sqrt{-1}$, which you can't write as an integer.

3. 0

A natural number (also referred to as a *counting* number) is a positive integer. The number 0 isn't positive.

4. 1.9

A natural number (also referred to as a *counting* number) is a positive integer. The number 1.9 is actually a fraction (rational number) and isn't a natural number.

5. $\sqrt{20}$

An integer is a positive or negative counting number or 0. The number $\sqrt{20}$ is irrational.

6. $-\frac{100}{6}$

An integer is a positive or negative counting number or 0. You can reduce the number $-\frac{100}{6}$ to $-\frac{50}{3}$, but that's still a rational number, which you can't write as an integer.

7. $\sqrt{25}$

You can write the square root of 25 as 5. It is rational and not irrational.

8. $\sqrt{0}$

The square root of 0 is 0. It isn't irrational.

9. i^2

Because $i^2 = -1$, you can write it as an integer.

10. $2 + i^4$

Because $i^4 = 1$, you can write the expression as $2 + 1 = 3$.

11. commutative

The commutative property of addition says that changing the order of the terms in addition doesn't change the result.

12. associative

The associative property of addition says that changing the grouping of the terms in addition doesn't change the result.

13. associative

The associative property of multiplication says that changing the grouping of the factors in multiplication doesn't change the result.

14. identity

When you add the additive identity, 0, to a term, it doesn't change that term.

15. distributive

The distributive property of multiplication over addition says that multiplying each term in the parentheses gives the same result as the factored form.

16. inverse

Multiplying a number by its multiplicative inverse results in the multiplicative identity, 1.

17. distributive

The distributive property of multiplication over addition says that multiplying each term in the parentheses gives the same result as the factored form.

18. commutative

The commutative property of multiplication says that changing the order of the factors in a multiplication problem doesn't change the result.

19. transitive

The transitive property says that if one value is equal to both a second value and a third, then the second and third values are equal to one another.

20. inverse

Adding a number to its additive inverse results in 0.

21. 36

The order of operations says to perform powers (exponents) before multiplication:

$$4(3)^2 = 4(9) = 36$$

22. 4

The order of operations says to first perform the addition in the numerator of the fraction because the fraction bar acts like a grouping symbol:

$$\frac{6+18}{6} = \frac{24}{6} = 4$$

An alternate way to simplify is to factor the terms in the numerator and then divide:

$$\frac{6+18}{6} = \frac{\cancel{6}(1+3)}{\cancel{6}} = 1+3 = 4$$

23. $\frac{21}{68}$

First, find common denominators for the fractions in the numerator and denominator; then add the fractions and reduce the results:

$$\frac{\frac{1}{3}+\frac{1}{6}}{\frac{5}{7}+\frac{19}{21}} = \frac{\frac{2}{6}+\frac{1}{6}}{\frac{15}{21}+\frac{19}{21}} = \frac{\frac{3}{6}}{\frac{34}{21}} = \frac{\frac{1}{2}}{\frac{34}{21}}$$

Multiply the numerator by the reciprocal of the denominator:

$$\frac{\frac{1}{2}}{\frac{34}{21}} = \frac{1}{2} \cdot \frac{21}{34} = \frac{21}{68}$$

Answers 1–100

24. $\frac{10}{3}$

First, find common denominators for the fractions in the numerator and denominator; then subtract and add the fractions and reduce the results:

$$\frac{\frac{13}{6}-\frac{1}{2}}{\frac{1}{6}+\frac{1}{3}}=\frac{\frac{13}{6}-\frac{3}{6}}{\frac{1}{6}+\frac{2}{6}}=\frac{\frac{10}{6}}{\frac{3}{6}}=\frac{\frac{5}{3}}{\frac{1}{2}}$$

Multiply the numerator by the reciprocal of the denominator:

$$\frac{\frac{5}{3}}{\frac{1}{2}}=\frac{5}{3}\cdot\frac{2}{1}=\frac{10}{3}$$

25. $\frac{3}{2}$

Using the order of operations, you first simplify what's under the radical. Applying the order of operations there, first multiply the three numbers together and then subtract the product from the 4:

$$\frac{-4+\sqrt{4-4(2)(-12)}}{4}=\frac{-4+\sqrt{4-(-96)}}{4}=\frac{-4+\sqrt{100}}{4}$$

Find the square root of 100 and then add the two numbers in the numerator together before reducing the fraction:

$$\frac{-4+10}{4}=\frac{6}{4}=\frac{3}{2}$$

26. 3

Using the order of operations, you first simplify what's under the radical. Applying the order of operations there, first multiply the three numbers together and then subtract the product from the 400:

$$\frac{-20-\sqrt{400-4(-5)(-15)}}{-10}=\frac{-20-\sqrt{400-300}}{-10}=\frac{-20-\sqrt{100}}{-10}$$

Find the square root of 100 and then subtract the two numbers in the numerator before reducing the fraction:

$$\frac{-20-10}{-10}=\frac{-30}{-10}=3$$

27. 5

Using the order of operations, first square the 18 and then multiply the last two numbers under the radical. Add their results. In the denominator, add the two numbers in the absolute value.

$$\frac{\sqrt{18^2+4(19)}}{|-7+3|}=\frac{\sqrt{324+76}}{|-4|}$$

Add the numbers under the radical, find the square root, and then apply the absolute value before reducing the fraction:

$$\frac{\sqrt{400}}{4}=\frac{20}{4}=5$$

28. $\frac{9}{11}$

Perform the operations under each of the radicals, using the order of operations:

$$\frac{\sqrt{25-16}+\sqrt{9(6-2)}}{\sqrt{61^2-60^2}}=\frac{\sqrt{9}+\sqrt{9(4)}}{\sqrt{3{,}721-3{,}600}}=\frac{\sqrt{9}+\sqrt{36}}{\sqrt{121}}$$

Find the roots and then add the terms in the numerator:

$$\frac{3+6}{11}=\frac{9}{11}$$

29. $\frac{1}{24}$

Complete the computations within the absolute value symbols before applying that operation. In the denominator, square the 3 and subtract.

$$\frac{|-4(3)+2(7)|}{6(17-3^2)}=\frac{|-12+14|}{6(17-9)}$$

Find the absolute value in the numerator and multiply in the denominator; then reduce the fraction:

$$\frac{|2|}{6(8)}=\frac{2}{48}=\frac{1}{24}$$

30. $-\frac{2}{121}$

Square each of the terms in the absolute value and parentheses. Then find their differences.

$$\frac{\left|3^2-4^2\right|-\left|4^2-5^2\right|}{\left(5^2-6^2\right)^2}=\frac{|9-16|-|16-25|}{(25-36)^2}=\frac{|-7|-|-9|}{(-11)^2}$$

Find the absolute values in the numerator and the square in the denominator; then find the difference in the numerator:

$$\frac{7-9}{121}=\frac{-2}{121}$$

31.

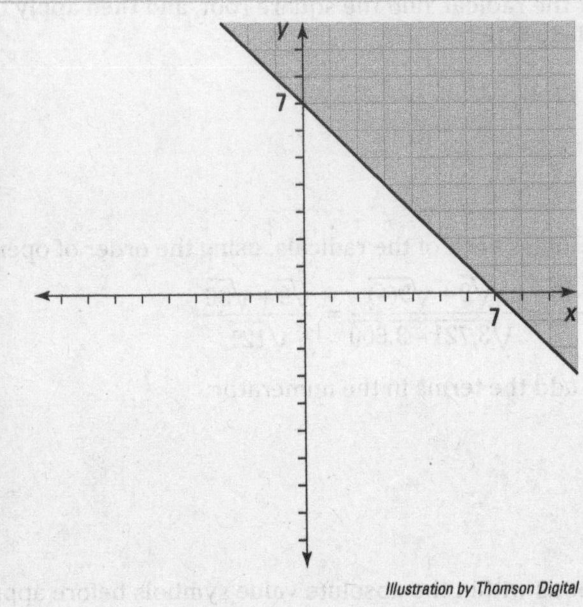

Illustration by Thomson Digital

First, find the intercepts of the line $x+y=7$ and draw the line through those two intercepts. You find the x-intercept by letting $y=0$ in the equation $x+y=7$, giving you (7, 0). You find the y-intercept by letting $x=0$ in the same equation, giving you (0, 7).

Next, use the test point (0, 0) to see whether it's part of the solution. Because $0+0$ isn't greater than 7, the origin isn't in the solution, so shade on the other side of the line.

32.

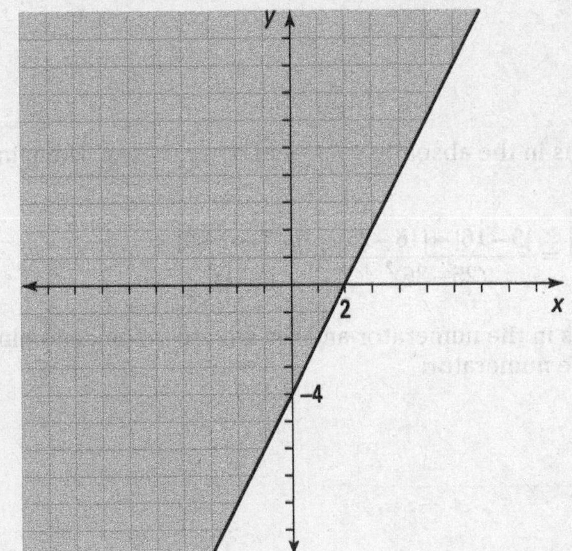

Illustration by Thomson Digital

First, find the intercepts of the line $2x-y=4$ and draw the line through those two intercepts. You find the x-intercept by letting $y=0$ in the equation $2x-y=4$, giving you (2, 0). You find the y-intercept by letting $x=0$ in the same equation, giving you (0, –4).

Next, use the test point (0, 0) to see whether it's part of the solution. Because 0 – 0 is less than 4, the origin is part of the solution, so shade on that side of the line.

33.

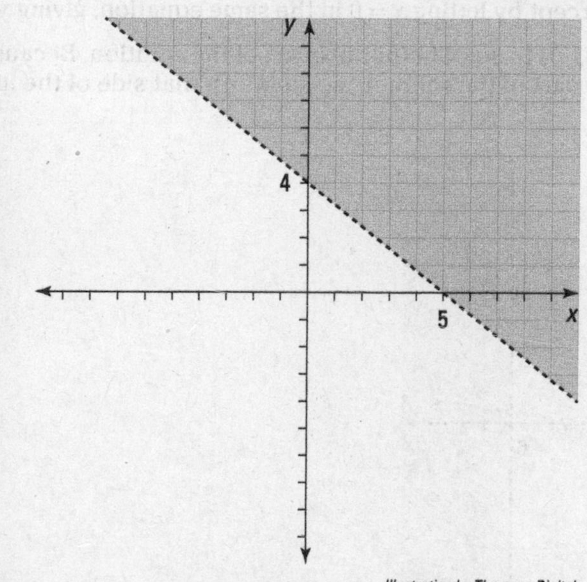

Illustration by Thomson Digital

First, find the intercepts of the line $4x+5y=20$ and draw the line through those two intercepts; use a dashed line because the points on the line aren't included in the solution. You find the x-intercept by letting $y = 0$ in the equation $4x+5y=20$, giving you (5, 0). You find the y-intercept by letting $x = 0$ in the same equation, giving you (0, 4).

Next, use the test point (0, 0) to see whether it's part of the solution. Because 0 + 0 isn't greater than 20, the origin isn't part of the solution, so shade on the other side of the line.

34.

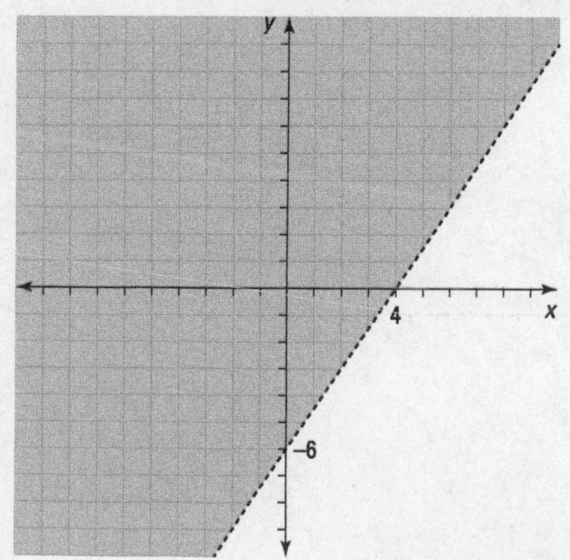

Illustration by Thomson Digital

First, find the intercepts of the line $3x - 2y = 12$ and draw the line through those two intercepts; use a dashed line because the points on the line aren't included in the solution. You find the x-intercept by letting $y = 0$ in the equation $3x - 2y = 12$, giving you (4, 0). You find the y-intercept by letting $x = 0$ in the same equation, giving you (0, –6).

Next, use the test point (0, 0) to see whether it's part of the solution. Because 0 – 0 is less than 12, the origin is part of the solution, so shade on that side of the line.

35.

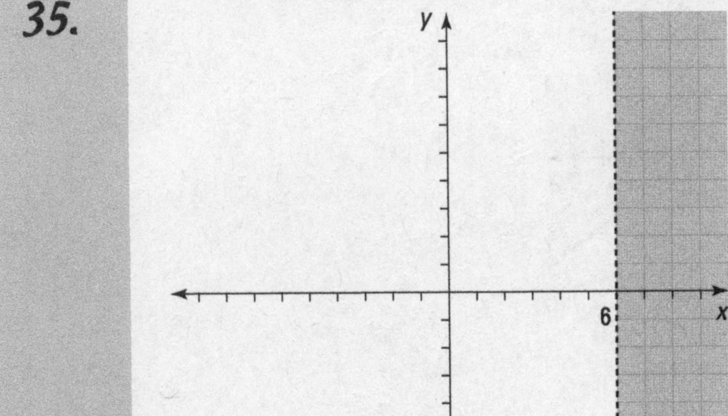

Illustration by Thomson Digital

First, draw the vertical line through the intercept on the x-axis, (6, 0); make the line dashed because the points on the line aren't part of the solution. Because all the x values must be greater than 6, shade to the right of the vertical line.

36.

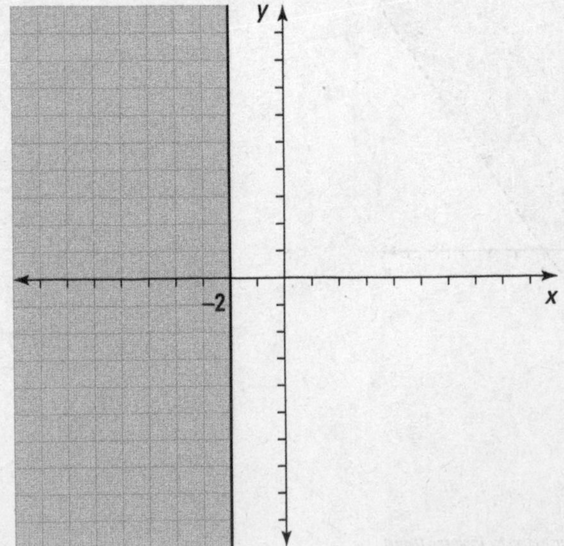

Illustration by Thomson Digital

First, draw the vertical line through the intercept on the *x*-axis, (–2, 0). Because all the *x* values must be less than –2, shade to the left of the vertical line.

37.

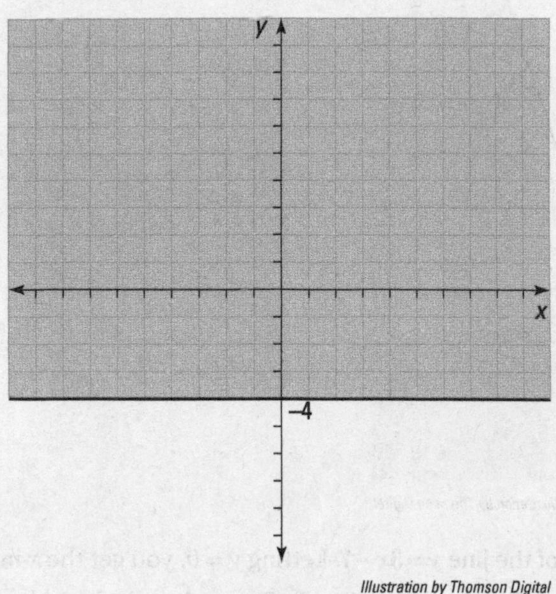

Illustration by Thomson Digital

First, draw the horizontal line through the intercept on the *y*-axis, (0, –4). Because all the *y* values must be greater than –4, shade above the horizontal line.

38.

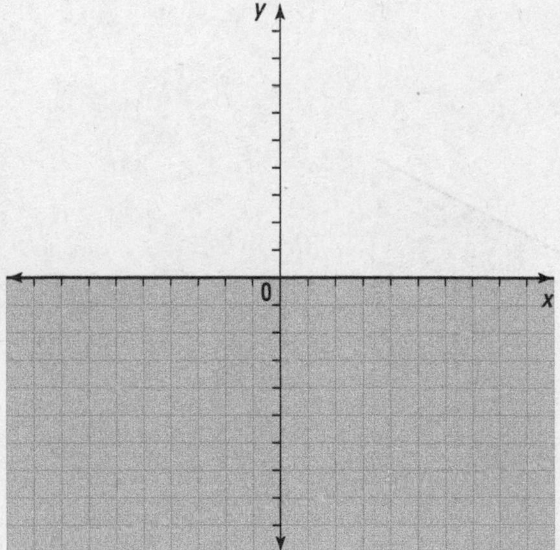

Illustration by Thomson Digital

First, draw the horizontal line through the origin, along the *x*-axis. Because all the *y* values must be less than 0, shade below the axis.

39.

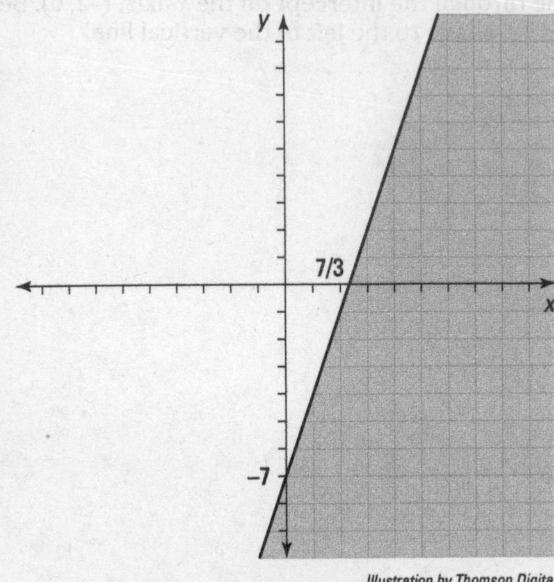

Illustration by Thomson Digital

First, find the intercepts of the line $y = 3x - 7$. Letting $y = 0$, you get the x-intercept $\left(\frac{7}{3}, 0\right)$. Then, letting $x = 0$, you get the y-intercept $(0, -7)$. Draw a line through those two points. Use the test point $(0, 0)$ to see whether it's a solution. Because 0 isn't less than $0 - 7$, the origin isn't part of the solution, so you shade on the other side of the line.

40.

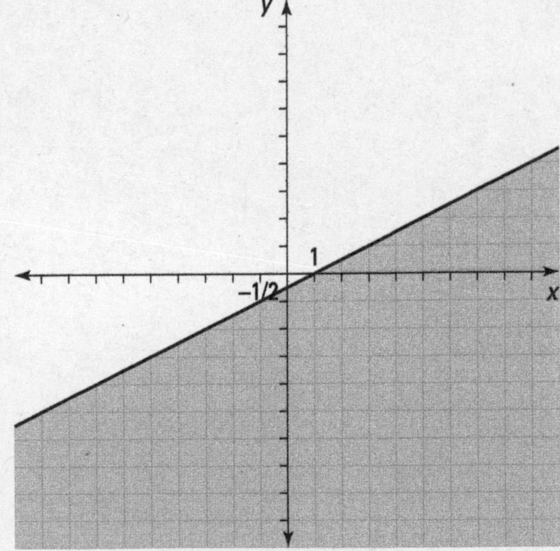

Illustration by Thomson Digital

First, find the intercepts of the line $x = 2y + 1$. Letting $y = 0$, you get the x-intercept $(1, 0)$; and letting $x = 0$, you get the y-intercept $\left(0, -\frac{1}{2}\right)$. Draw a line through those two points. Use the test point $(0, 0)$ to see whether it's a solution. Because 0 isn't greater than $0 + 1$, the origin isn't part of the solution, so you shade on the other side of the line.

41. 1

Use the slope formula, $m = \frac{y_2 - y_1}{x_2 - x_1}$:

$$m = \frac{9-3}{4-(-2)} = \frac{6}{6} = 1$$

42. $-\frac{5}{2}$

Use the slope formula, $m = \frac{y_2 - y_1}{x_2 - x_1}$:

$$m = \frac{2-(-3)}{-6-(-4)} = \frac{5}{-2} = -\frac{5}{2}$$

43. No slope

Use the slope formula, $m = \frac{y_2 - y_1}{x_2 - x_1}$:

$$m = \frac{-7-(-3)}{4-4} = \frac{-4}{0}$$

You can't divide by 0, so the line has no slope. The line is vertical.

44. 0

Use the slope formula, $m = \frac{y_2 - y_1}{x_2 - x_1}$:

$$m = \frac{-9-(-9)}{2-(-2)} = \frac{0}{4} = 0$$

The line is horizontal.

45. 10 units

Use the distance formula, $d = \sqrt{(x_2 - x_1)^2 + (y_2 - y_1)^2}$:

$$d = \sqrt{(-8-(-2))^2 + (-1-7)^2} = \sqrt{(-6)^2 + (-8)^2} = \sqrt{36+64} = \sqrt{100} = 10$$

46. 25 units

Use the distance formula, $d = \sqrt{(x_2 - x_1)^2 + (y_2 - y_1)^2}$:

$$d = \sqrt{(0-7)^2 + (16-(-8))^2} = \sqrt{(-7)^2 + (24)^2} = \sqrt{49+576} = \sqrt{625} = 25$$

Answers 1–100

47. $2\sqrt{41}$ units

Use the distance formula, $d=\sqrt{(x_2-x_1)^2+(y_2-y_1)^2}$:

$$d=\sqrt{(6-(-4))^2+(-5-3)^2}=\sqrt{(10)^2+(-8)^2}=\sqrt{100+64}=\sqrt{164}$$

You can simplify the radical:

$$\sqrt{164}=\sqrt{4\cdot 41}=2\sqrt{41}$$

48. (1, –3)

Use the midpoint formula, $M=\left(\frac{x_1+x_2}{2},\frac{y_1+y_2}{2}\right)$:

$$M=\left(\frac{-5+7}{2},\frac{2+(-8)}{2}\right)=\left(\frac{2}{2},\frac{-6}{2}\right)=(1,-3)$$

49. $\left(1,-\frac{1}{2}\right)$

Use the midpoint formula, $M=\left(\frac{x_1+x_2}{2},\frac{y_1+y_2}{2}\right)$:

$$M=\left(\frac{6+(-4)}{2},\frac{3+(-4)}{2}\right)=\left(\frac{2}{2},\frac{-1}{2}\right)=\left(1,-\frac{1}{2}\right)$$

50. $\left(\frac{7}{12},-\frac{3}{8}\right)$

Use the midpoint formula, $M=\left(\frac{x_1+x_2}{2},\frac{y_1+y_2}{2}\right)$:

$$M=\left(\frac{\frac{1}{3}+\frac{5}{6}}{2},\frac{-\frac{1}{2}+\left(-\frac{1}{4}\right)}{2}\right)=\left(\frac{\frac{2}{6}+\frac{5}{6}}{2},\frac{-\frac{2}{4}+\left(-\frac{1}{4}\right)}{2}\right)$$

$$=\left(\frac{\frac{7}{6}}{2},\frac{-\frac{3}{4}}{2}\right)=\left(\frac{7}{12},-\frac{3}{8}\right)$$

51. 12 units

Find the lengths of the sides of the triangle by using the distance formula, $d=\sqrt{(x_2-x_1)^2+(y_2-y_1)^2}$, for each side:

$$d_{AB}=\sqrt{(1-1)^2+(1-4)^2}=\sqrt{0+9}=3$$

$$d_{AC}=\sqrt{(1-5)^2+(1-1)^2}=\sqrt{16+0}=4$$

$$d_{BC}=\sqrt{(1-5)^2+(4-1)^2}=\sqrt{16+9}=\sqrt{25}=5$$

The perimeter of a triangle is the sum of the lengths of the sides, so $P=3+4+5=12$.

52. $10\sqrt{10}$ units

The opposite sides of a parallelogram are congruent, and you find the perimeter of a parallelogram with $P=2(L+W)$, so you need to find the lengths of two adjacent sides by using the distance formula, $d=\sqrt{(x_2-x_1)^2+(y_2-y_1)^2}$, and put those lengths in the formula:

$$d_{DE}=\sqrt{(0-9)^2+(10-13)^2}=\sqrt{81+9}=\sqrt{90}=3\sqrt{10}$$

$$d_{EF}=\sqrt{(9-11)^2+(13-7)^2}=\sqrt{4+36}=\sqrt{40}=2\sqrt{10}$$

So $P=2\left(3\sqrt{10}+2\sqrt{10}\right)=2\left(5\sqrt{10}\right)=10\sqrt{10}$.

53. (4, 3)

The center of a rhombus is at the intersection of the diagonals, which is also the midpoint of each diagonal. Find the midpoint of either diagonal by using $M=\left(\frac{x_1+x_2}{2},\frac{y_1+y_2}{2}\right)$ to find the midpoint of the rhombus. Using the diagonal HK, the midpoint is

$$M=\left(\frac{0+8}{2},\frac{3+3}{2}\right)=\left(\frac{8}{2},\frac{6}{2}\right)=(4,3)$$

54. right and scalene

Find the lengths of the three sides by using the distance formula, $d=\sqrt{(x_2-x_1)^2+(y_2-y_1)^2}$:

$$d_{AB}=\sqrt{(1-4)^2+(1-5)^2}=\sqrt{9+16}=\sqrt{25}=5$$

$$d_{AC}=\sqrt{(1-9)^2+(1-(-5))^2}=\sqrt{64+36}=\sqrt{100}=10$$

$$d_{BC}=\sqrt{(4-9)^2+(5-(-5))^2}=\sqrt{25+100}=\sqrt{125}$$

The triangle isn't isosceles or equilateral. Use the Pythagorean theorem $(a^2+b^2=c^2)$ to see whether it's a right triangle:

$$5^2+10^2\stackrel{?}{=}\left(\sqrt{125}\right)^2$$

$$25+100=125$$

This triangle is a right triangle; because the measures of the sides are all different, it's also scalene.

55. equilateral

Find the lengths of the three sides by using the distance formula, $d=\sqrt{(x_2-x_1)^2+(y_2-y_1)^2}$:

$$d_{AB}=\sqrt{(0-0)^2+(0-12)^2}=\sqrt{0+144}=12$$

$$d_{AC} = \sqrt{\left(0-6\sqrt{3}\right)^2 + (0-6)^2} = \sqrt{108+36} = \sqrt{144} = 12$$

$$d_{BC} = \sqrt{\left(0-6\sqrt{3}\right)^2 + (12-6)^2} = \sqrt{108+36} = \sqrt{144} = 12$$

The three lengths are the same, so the triangle is equilateral.

56. 12 units

The altitude to side AC is drawn from vertex B perpendicular to AC. Because AC lies along the *x*-axis, the altitude is a vertical line through point B and intersects the *x*-axis at the point (5, 0).

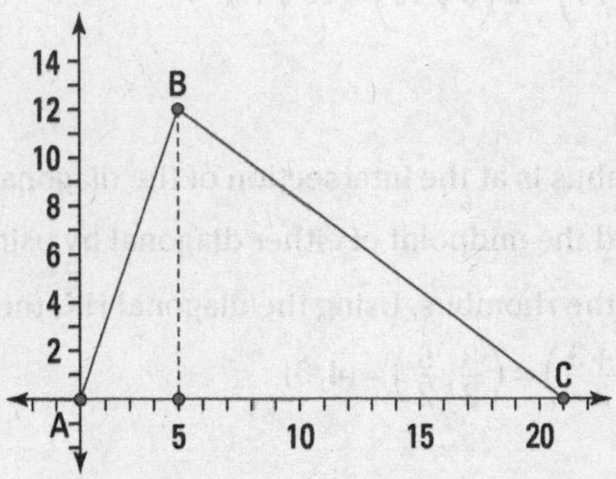

Illustration by Thomson Digital

The distance from (5, 0) to (5, 12) is 12 units.

57. 9 units

The triangle is a right triangle because sides DE and DF are parallel to the *y*-axis and *x*-axis, respectively, making DE perpendicular to DF.

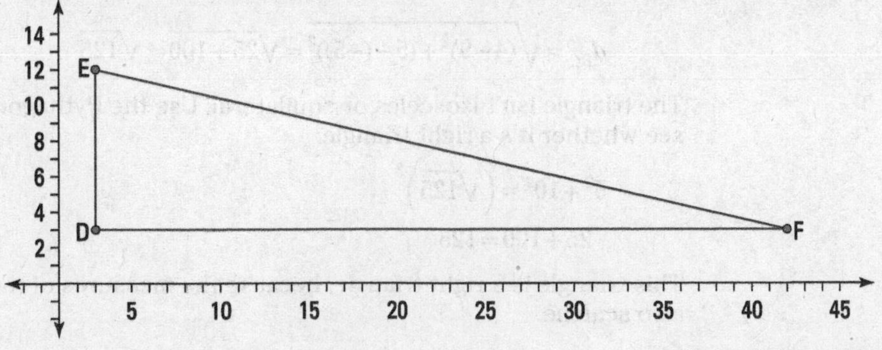

Illustration by Thomson Digital

So the altitude to DF is side DE, which has a length of $12-3=9$ units.

58. 40 square units

You find the area of a parallelogram by multiplying the length of one side by the distance between that side and the other side parallel to it; another form for area is $A=bh$.

The sides QR and PS are both parallel to the x-axis, and the distance between them is $12 - 7 = 5$.

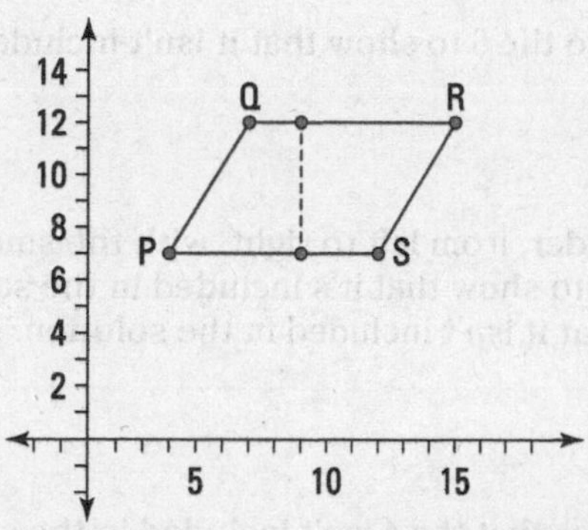

Illustration by Thomson Digital

The lengths of QR and PS are 8 units. So the area of the parallelogram is $A = 8(5) = 40$ square units.

59. 65π square units

The radius of a circle is half its diameter, so find the length of the diameter by using $d = \sqrt{(x_2 - x_1)^2 + (y_2 - y_1)^2}$ and halve that amount.

$$d = \sqrt{(6-(-8))^2 + (13-21)^2} = \sqrt{196+64} = \sqrt{260} = 2\sqrt{65}$$

Half of $2\sqrt{65}$ is $\sqrt{65}$, which is the length of the radius.

You find the area of a circle with the formula $A = \pi r^2$, so this circle has area $A = \pi\left(\sqrt{65}\right)^2 = 65\pi$ square units.

60. 84 square units

The triangle has its base along the x-axis, so the altitude drawn to the base from point B has a length of 12 units.

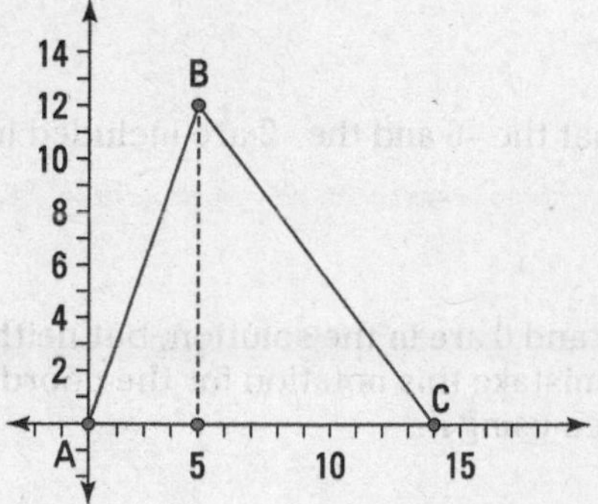

Illustration by Thomson Digital

You find the area of a triangle with the formula $A = \frac{1}{2}bh$. The base of the triangle, AC, has a length of $14 - 0 = 14$ units, so the area of the triangle is $A = \frac{1}{2}(14)(12) = 84$ square units.

61. $(-\infty, 6)$

Use a parenthesis next to the 6 to show that it isn't included in the solution.

62. $[-5, 8)$

Write the numbers in order, from left to right, with the smaller number to the left. Use a bracket next to the –5 to show that it's included in the solution. Use a parenthesis next to the 8 to show that it isn't included in the solution.

63. $4 < x \leq 16$

The parenthesis indicates that the 4 isn't included in the solution; the bracket indicates that the 16 is included in the solution.

64. $x \geq -8$

The x indicates all numbers starting with –8 and larger. You use the greater-than-or-equal-to symbol to include the –8.

65. $3 \leq x < 7$

The 3 is included in the solution, but the 7 is not.

66. $0 < x < 11$

Neither the 0 nor the 11 is a part of the solution.

67. $-5 < x \leq 7$ or $x > 11$

The numbers between –5 and 7, including the 7, are in the solution. Also, the numbers greater than 11 are in the solution.

68. $[-6, -2]$

Use brackets to show that the –6 and the –2 are included in the solution.

69. $(-10, 0)$

The points between –10 and 0 are in the solution, but neither the –10 nor the 0 is included. Many people mistake this notation for the coordinates of a point, so make your intention clear when using it.

70. $(-\infty, 3) \cup [5, \infty)$

All the numbers smaller than 3 are in the solution, but the number 3 isn't. Also, the number 5 and all the numbers bigger than 5 are in the solution.

71. $x > 3$

Add 3 to each side of the inequality:

$$4x - 3 > 9$$
$$4x > 12$$

Divide each side by 4:

$$x > 3$$

In interval notation, you write the solution as $(3, \infty)$.

72. $x \geq -7$

Subtract $7x$ from each side of the inequality:

$$5x - 6 \leq 7x + 8$$
$$-2x - 6 \leq 8$$

Add 6 to each side:

$$-2x \leq 14$$

Divide each side by –2, reversing the direction of the inequality:

$$x \geq -7$$

In interval notation, you write the solution as $[-7, \infty)$.

73. $x > 2$

Add $-4x$ to each side of the inequality:

$$16 - 3x < 4x + 2$$
$$16 - 7x < 2$$

Subtract 16 from each side:

$$-7x < -14$$

Divide each side by –7, reversing the direction of the inequality.

$$x > 2$$

In interval notation, you write the solution as $(2, \infty)$.

74. $x \leq -3$

Multiply each side of the inequality by –3, reversing the direction of the inequality:

$$-\frac{5}{3}x - 3 \geq 1 - \frac{1}{3}x$$
$$5x + 9 \leq -3 + x$$

Subtract x from each side:

$$4x + 9 \leq -3$$

Subtract 9 from each side:

$$4x \leq -12$$

Divide each side by 4:

$$x \leq -3$$

In interval notation, you write the solution as $(-\infty, -3]$.

75. $x \leq -1$ or $x \geq 8$

First factor the quadratic and set it equal to 0 to find the critical numbers:

$$x^2 - 7x - 8 = 0$$
$$(x-8)(x+1) = 0$$
$$x = 8, -1$$

Use a sign line to find the sign of each factor in the intervals determined by the critical numbers:

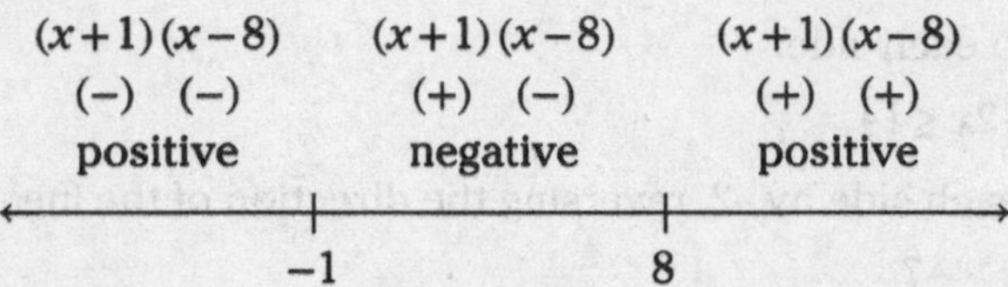

You want the product to be positive or zero, so all x less than –1 or greater than 8 will give you that product, as will –1 and 8:

$$x \leq -1 \text{ or } x \geq 8$$

In interval notation, you write the solution as $(-\infty, -1] \cup [8, \infty)$.

76. $x < 4$ or $x > 6$

First factor the quadratic and set it equal to 0 to find the critical numbers:

$$24 - 10x + x^2 = 0$$
$$(6-x)(4-x) = 0$$
$$x = 6, 4$$

Use a sign line to find the sign of each factor in the intervals determined by the critical numbers:

$(4-x)(6-x)$	$(4-x)(6-x)$	$(4-x)(6-x)$
(+) (+)	(–) (+)	(–) (–)
positive	negative	positive

4   6

You want the product to be positive, so all x less than 4 or greater than 6 will give you that product:

$$x < 4 \text{ or } x > 6$$

In interval notation, you write the solution as $(-\infty, 4) \cup (6, \infty)$.

77. $x \leq -3$ or $0 \leq x \leq 3$

First factor the polynomial and set it equal to 0 to find the critical numbers:

$$x^3 - 9x = 0$$
$$x(x-3)(x+3) = 0$$
$$x = 0, 3, -3$$

Use a sign line to find the sign of each factor in the intervals determined by the critical numbers:

$(x)\ (x^2-9)$	$(x)\ (x^2-9)$	$(x)\ (x^2-9)$	$(x)\ (x^2-9)$
(–) (+)	(–) (–)	(+) (–)	(+) (+)
negative	positive	negative	positive

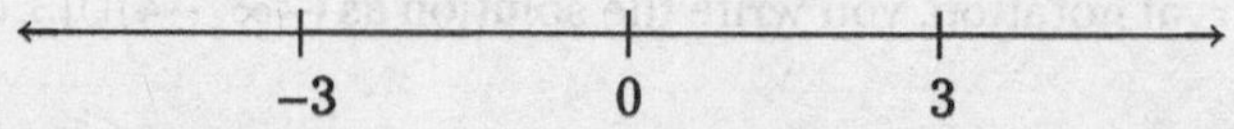

You want the product to be negative or zero, so all x less than –3 or between 0 and 3 will give you that product, as will –3, 0, and 3:

$$x \leq -3 \text{ or } 0 \leq x \leq 3$$

In interval notation, you write the solution as $(-\infty, -3] \cup [0, 3]$.

78. $x < -1$ or $1 < x < 6$

First use grouping to factor the polynomial. Set the function equal to 0 to find the critical numbers:

$$x^3 - 6x^2 - x + 6 = 0$$
$$x^2(x-6) - 1(x-6) = 0$$
$$(x-6)(x^2-1) = 0$$
$$(x-6)(x-1)(x+1) = 0$$
$$x = 6, 1, -1$$

Use a sign line to find the sign of each factor in the intervals determined by the critical numbers:

$(x-6)\ (x^2-1)$	$(x-6)\ (x^2-1)$	$(x-6)\ (x^2-1)$	$(x-6)\ (x^2-1)$
(–) (+)	(–) (–)	(–) (+)	(+) (+)
negative	positive	negative	positive

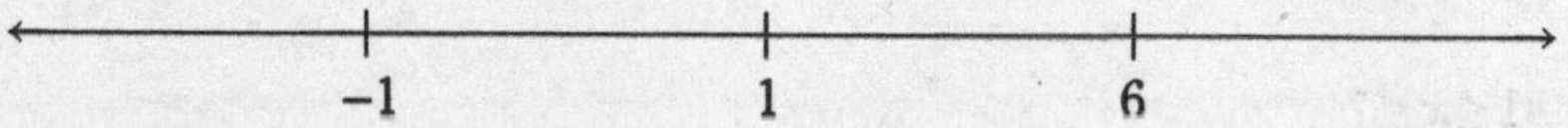

You want the product to be negative, so all x less than –1 or between 1 and 6 will give you that product:

$$x < -1 \text{ or } 1 < x < 6$$

In interval notation, you write the solution as $(-\infty, -1) \cup (1, 6)$.

79. $x < -4$ or $x \geq 3$

First determine the critical numbers by setting the numerator and denominator equal to 0:

$$x - 3 = 0 \quad \text{and} \quad x + 4 = 0$$
$$x = 3 \qquad\qquad x = -4$$

Use a sign line to find the sign of each factor in the intervals determined by the critical numbers:

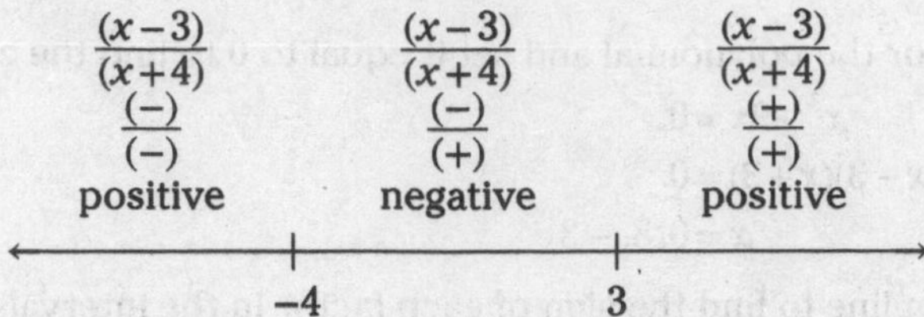

You want the quotient to be positive or zero, so all x less than –4 or greater than 3 will give you that quotient; also, 3 will give you a 0.

$x<-4$ or $x\geq 3$

In interval notation, you write the solution as $(-\infty,-4)\cup[3,\infty)$.

80. $x<-7$ or $-5<x<5$

First determine the critical numbers by setting the numerator and denominator equal to 0:

$$x+7=0 \qquad\qquad x^2-25=0$$
$$x=-7 \quad \text{and} \quad (x-5)(x+5)=0$$
$$x=5,\ -5$$

Use a sign line to find the sign of each factor in the different intervals determined by the critical numbers:

$\frac{(x+7)}{(x^2-25)}$	$\frac{(x+7)}{(x^2-25)}$	$\frac{(x+7)}{(x^2-25)}$	$\frac{(x+7)}{(x^2-25)}$
$\frac{(-)}{(+)}$	$\frac{(+)}{(+)}$	$\frac{(+)}{(-)}$	$\frac{(+)}{(+)}$
negative	positive	negative	positive

Critical numbers on the line: -7, -5, 5

You want the quotient to be negative, so all x less than –7 or between –5 and 5 will give you that quotient:

$x<-7$ or $-5<x<5$

In interval notation, you write the solution as $(-\infty,-7)\cup(-5,5)$.

81. $1\leq x\leq 7$

Rewrite the absolute value inequality as

$-3\leq x-4\leq 3$

Then add 4 to each expression:

$1\leq x\leq 7$

In interval notation, you write the solution as $[1,7]$.

82. $x<-1$ or $x>4$

Rewrite the absolute value inequality as

$$2x-3<-5 \quad \text{or} \quad 2x-3>5$$

Then add 3 to each side of each expression:

$$2x<-2 \quad \text{or} \quad 2x>8$$

Divide each side by 2:

$$x<-1 \quad \text{or} \quad x>4$$

In interval notation, you write the solution as $(-\infty, -1)\cup(4, \infty)$.

83. $-3<x<2$

First divide each side of the inequality by 4:

$$|2x+1|<5$$

Now rewrite the absolute value inequality as

$$-5<2x+1<5$$

Subtract 1 from each expression:

$$-6<2x<4$$

Divide each expression by 2:

$$-3<x<2$$

In interval notation, you write the solution as $(-3, 2)$.

84. $-7\le x\le -3$

First subtract 3 from each side of the inequality:

$$-|x+5|\ge -2$$

Next, multiply each side by –1, reversing the direction of the inequality sign:

$$|x+5|\le 2$$

Now rewrite the absolute value inequality as

$$-2\le x+5\le 2$$

Subtract 5 from each expression:

$$-7\le x\le -3$$

In interval notation, you write the solution as $[-7, -3]$.

85. no solution

First subtract 5 from each side of the inequality:

$$|3x-1|\le -3$$

You can stop right there. The absolute value expression is either a positive number or 0. It can't be negative, so there's no solution.

86. $x^{4/3}$

The root outside the radical, 3, goes in the denominator of the fraction. The power, 4, goes in the numerator.

87. $x^{-1/2}$

Remember that when you don't see a root indicated outside the radical, you assume that it's the square root. That means you can then write the fraction as $\frac{1}{x^{1/2}}$. When you bring the denominator up to the numerator, the sign of the exponent becomes negative: $x^{-1/2}$.

88. $\sqrt[3]{x^5}$

The denominator of the fraction, 3, becomes the index of the radical. The numerator of the fraction, 5, is the power of x under the radical.

89. $\frac{1}{\sqrt[4]{x^3}}$

First rewrite the expression as a fraction with the power of x in the denominator: $\frac{1}{x^{3/4}}$.

The 4 in the denominator of the exponent means that the root is 4; the 3 in the numerator of the exponent means that the x is raised to the third power.

90. x^2

When multiplying expressions with the same base, you add the exponents:

$$x^{2/3} \cdot x^{4/3} = x^{\frac{2}{3}+\frac{4}{3}} = x^{\frac{6}{3}} = x^2$$

91. x^{-1}

When multiplying expressions with the same base, you add the exponents:

$$x^{1/5} \cdot x^{-6/5} = x^{\frac{1}{5}+\left(-\frac{6}{5}\right)} = x^{-\frac{5}{5}} = x^{-1}$$

92. $x^{1/3}$

When dividing expressions with the same base, you subtract the exponents:

$$\frac{x^{2/3}}{x^{1/3}} = x^{\frac{2}{3}-\frac{1}{3}} = x^{1/3}$$

93. x

When dividing expressions with the same base, you subtract the exponents.

$$\frac{x^{1/2}}{x^{-1/2}} = x^{\frac{1}{2}-\left(-\frac{1}{2}\right)} = x^{\frac{1}{2}+\frac{1}{2}} = x^1 = x$$

94. $x^{-7/10}$

When dividing expressions with the same base, you subtract the exponents:

$$\frac{x^{-3/5}}{x^{1/10}} = x^{-\frac{3}{5}-\frac{1}{10}}$$

Find a common denominator before combining the fractions:

$$x^{-\frac{3}{5}-\frac{1}{10}} = x^{-\frac{6}{10}-\frac{1}{10}} = x^{-7/10}$$

95. $2x^{1/2}(1+2x)$

The GCF is $2x^{1/2}$. Factoring gives you

$$2x^{1/2}+4x^{3/2} = 2x^{1/2}(1+2x)$$

96. $y^{1/3}(-9y^{1/3}+4)$

The GCF is $y^{1/3}$. Factoring gives you

$$-9y^{2/3}+4y^{1/3} = y^{1/3}(-9y^{1/3}+4)$$

97. $2x^{-3/5}(3x-2)$

The GCF is $2x^{-3/5}$. Factoring gives you

$$6x^{2/5}-4x^{-3/5} = 2x^{-3/5}(3x-2)$$

Remember that when dividing, you subtract exponents:

$$\frac{6x^{2/5}}{2x^{-3/5}} = 3x^{\frac{2}{5}-\left(-\frac{3}{5}\right)} = 3x^{\frac{2}{5}+\frac{3}{5}} = 3x^{\frac{5}{5}} = 3x^1 = 3x$$

98. $5y^{-8/5}(3y+2)$

The GCF is $5y^{-8/5}$. Remember, the factor with the lowest power of the variable is the GCF. Factoring gives you

$$15y^{-3/5}+10y^{-8/5} = 5y^{-8/5}(3y+2)$$

99. $x^{1/3}(4+8x^{1/3}-5x^{2/3})$

The GCF is $x^{1/3}$. Dividing each term by that factor gives you

$$\frac{4x^{1/3}}{x^{1/3}}+\frac{8x^{2/3}}{x^{1/3}}-\frac{5x}{x^{1/3}} = 4+8x^{\frac{2}{3}-\frac{1}{3}}-5x^{1-\frac{1}{3}}$$

Therefore, $4x^{1/3}+8x^{2/3}-5x = x^{1/3}(4+8x^{1/3}-5x^{2/3})$.

100. $-16y^{-11/4}(y^2+2y-3)$

The GCF is $-16y^{-11/4}$. The GCF could also be $+16y^{-11/4}$ if you prefer a positive factor in front. The exponent $-\frac{11}{4}$ is the smallest of the three exponents on y.

Factoring gives you

$$\begin{aligned}&-16y^{-3/4}-32y^{-7/4}+48y^{-11/4}\\&=-16y^{-11/4}\left(y^{-\frac{3}{4}-\left(-\frac{11}{4}\right)}+2y^{-\frac{7}{4}-\left(-\frac{11}{4}\right)}-3\right)\\&=-16y^{-11/4}(y^{8/4}+2y^{4/4}-3)\\&=-16y^{-11/4}(y^2+2y-3)\end{aligned}$$

101. $x=5$

Square both sides of the equation:

$$\begin{aligned}\left(\sqrt{4x+5}\right)^2&=(5)^2\\4x+5&=25\end{aligned}$$

Now solve for x:

$$\begin{aligned}4x+5&=25\\4x&=20\\x&=5\end{aligned}$$

Check the solution in the original equation:

$$\begin{aligned}\sqrt{4(5)+5}&\stackrel{?}{=}5\\\sqrt{20+5}&=\sqrt{25}=5\end{aligned}$$

It checks.

102. $x=-3$

Square both sides of the equation:

$$\begin{aligned}\left(\sqrt{10-2x}\right)^2&=(4)^2\\10-2x&=16\end{aligned}$$

Now solve for x:

$$\begin{aligned}10-2x&=16\\-2x&=6\\x&=-3\end{aligned}$$

Check the solution in the original equation:

$$\begin{aligned}\sqrt{10-2(-3)}&\stackrel{?}{=}4\\\sqrt{10+6}&=\sqrt{16}=4\end{aligned}$$

It checks.

103. $x = 9$

First subtract 4 from both sides of the equation:

$$4+\sqrt{3x-2}=9$$
$$\sqrt{3x-2}=5$$

Next, square both sides of the equation:

$$\left(\sqrt{3x-2}\right)^2=(5)^2$$
$$3x-2=25$$

Solve for x:

$$3x=27$$
$$x=9$$

Check the solution in the original equation:

$$4+\sqrt{3(9)-2}\stackrel{?}{=}9$$
$$4+\sqrt{27-2}\stackrel{?}{=}9$$
$$4+\sqrt{25}\stackrel{?}{=}9$$
$$4+5=9$$

It checks.

104. $x = -3$

First add 3 to both sides of the equation:

$$\sqrt{4-7x}-3=2$$
$$\sqrt{4-7x}=5$$

Next, square both sides of the equation:

$$\left(\sqrt{4-7x}\right)^2=(5)^2$$
$$4-7x=25$$

Solve for x:

$$-7x=21$$
$$x=-3$$

Check the solution in the original equation:

$$\sqrt{4-7(-3)}-3\stackrel{?}{=}2$$
$$\sqrt{4+21}-3\stackrel{?}{=}2$$
$$\sqrt{25}-3\stackrel{?}{=}2$$
$$5-3=2$$

It checks.

105. $x = 5$

Square both sides of the equation. Be careful to square the binomial correctly.

$$\left(\sqrt{6x-5}\right)^2 = (2x-5)^2$$
$$6x-5 = 4x^2 - 20x + 25$$

Set the quadratic equation equal to 0 and solve for x:

$$0 = 4x^2 - 26x + 30$$
$$= 2\left(2x^2 - 13x + 15\right)$$
$$= 2(2x-3)(x-5)$$

The solutions of the quadratic equation are $x = \frac{3}{2}$ and $x = 5$.

Check $x = \frac{3}{2}$ in the original equation:

$$\sqrt{6\left(\frac{3}{2}\right) - 5} \stackrel{?}{=} 2\left(\frac{3}{2}\right) - 5$$
$$\sqrt{9-5} \stackrel{?}{=} 3-5$$
$$\sqrt{4} \neq -2$$

This solution is extraneous.

Now check $x = 5$ in the original equation:

$$\sqrt{6(5)-5} \stackrel{?}{=} 2(5) - 5$$
$$z\sqrt{30-5} \stackrel{?}{=} 10-5$$
$$\sqrt{25} = 5$$

This solution checks.

106. $x = -1$

Square both sides of the equation. Be careful to square the binomial correctly.

$$\left(\sqrt{7-2x}\right)^2 = (x+4)^2$$
$$7 - 2x = x^2 + 8x + 16$$

Set the quadratic equation equal to 0 and solve for x:

$$0 = x^2 + 10x + 9$$
$$= (x+1)(x+9)$$

The solutions of the quadratic equation are $x = -1$ and $x = -9$.

Check $x = -1$ in the original equation:

$$\sqrt{7-2(-1)} \stackrel{?}{=} -1+4$$
$$\sqrt{7+2} \stackrel{?}{=} 3$$
$$\sqrt{9} = 3$$

This solution works.

Now check $x = -9$ in the original equation:

$$\sqrt{7-2(-9)} \stackrel{?}{=} -9+4$$
$$\sqrt{7+18} \stackrel{?}{=} -5$$
$$\sqrt{25} \neq -5$$

This solution is extraneous.

107. $x=-\frac{11}{4}, x=-2$

First add 6 to each side of the equation:

$$\sqrt{5x+14}=2x+6$$

Now square both sides of the equation. Be careful to square the binomial correctly.

$$\left(\sqrt{5x+14}\right)^2=(2x+6)^2$$
$$5x+14=4x^2+24x+36$$

Set the quadratic equation equal to 0 and solve for x:

$$0=4x^2+19x+22$$
$$=(4x+11)(x+2)$$

The solutions of the quadratic equation are $x=-\frac{11}{4}$ and $x=-2$.

Check $x=-\frac{11}{4}$ in the original equation:

$$\sqrt{5\left(-\frac{11}{4}\right)+14}-6 \stackrel{?}{=} 2\left(-\frac{11}{4}\right)$$
$$\sqrt{-\frac{55}{4}+\frac{56}{4}}-6 \stackrel{?}{=} -\frac{11}{2}$$
$$\sqrt{\frac{1}{4}}-6 \stackrel{?}{=} -\frac{11}{2}$$
$$\frac{1}{2}-\frac{12}{2}=-\frac{11}{2}$$

This solution works.

Now check $x = -2$ in the original equation:

$$\sqrt{5(-2)+14}-6 \stackrel{?}{=} 2(-2)$$
$$\sqrt{-10+14}-6 \stackrel{?}{=} -4$$
$$\sqrt{4}-6 \stackrel{?}{=} -4$$
$$2-6=-4$$

This solution also works.

108. $x=-2, x=-3$

First add x to each side of the equation:

$$\sqrt{3x+10}=4+x$$

Now square both sides of the equation. Be careful to square the binomial correctly.

$$\left(\sqrt{3x+10}\right)^2 = (4+x)^2$$
$$3x+10 = 16+8x+x^2$$

Set the quadratic equation equal to 0 and solve for x:

$$0 = x^2+5x+6$$
$$= (x+3)(x+2)$$

The solutions of the quadratic equation are $x = -3$ and $x = -2$.

Check $x = -3$ in the original equation:

$$\sqrt{3(-3)+10}-(-3) \stackrel{?}{=} 4$$
$$\sqrt{-9+10}+3 \stackrel{?}{=} 4$$
$$\sqrt{1}+3 \stackrel{?}{=} 4$$
$$1+3 = 4$$

This solution works.

Now check $x = -2$ in the original equation:

$$\sqrt{3(-2)+10}-(-2) \stackrel{?}{=} 4$$
$$\sqrt{-6+10}+2 \stackrel{?}{=} 4$$
$$\sqrt{4}+2 \stackrel{?}{=} 4$$
$$2+2 = 4$$

This solution also works.

109. $x = 3, x = -1$

First add $\sqrt{x+1}$ to each side of the equation:

$$\sqrt{2x+3} = 1+\sqrt{x+1}$$

Now square both sides of the equation. Be careful to square the binomial with the radical correctly.

$$\left(\sqrt{2x+3}\right)^2 = \left(1+\sqrt{x+1}\right)^2$$
$$2x+3 = 1+2\sqrt{x+1}+x+1$$

Now subtract x and 2 from each side to isolate the radical on the right:

$$x+1 = 2\sqrt{x+1}$$

Square both sides of the equation. Then set the quadratic equation equal to 0 and solve for x:

$$(x+1)^2 = \left(2\sqrt{x+1}\right)^2$$
$$x^2+2x+1 = 4(x+1)$$
$$x^2+2x+1 = 4x+4$$
$$x^2-2x-3 = 0$$
$$(x-3)(x+1) = 0$$

The two solutions of the quadratic are $x = 3$ and $x = -1$.

Check $x = 3$ in the original equation:

$$\sqrt{2(3)+3}-\sqrt{3+1}\stackrel{?}{=}1$$
$$\sqrt{6+3}-\sqrt{4}\stackrel{?}{=}1$$
$$\sqrt{9}-2\stackrel{?}{=}1$$
$$3-2=1$$

This solution works.

Now check $x = -1$ in the original equation:

$$\sqrt{2(-1)+3}-\sqrt{-1+1}\stackrel{?}{=}1$$
$$\sqrt{-2+3}-\sqrt{0}\stackrel{?}{=}1$$
$$\sqrt{1}-0\stackrel{?}{=}1$$
$$1-0=1$$

This solution works.

110. $x=-5, x=-\frac{38}{9}$

First subtract $\sqrt{x+6}$ from each side of the equation:

$$\sqrt{11-5x}=7-\sqrt{x+6}$$

Now square both sides of the equation. Be careful to square the binomial with the radical correctly.

$$\left(\sqrt{11-5x}\right)^2=\left(7-\sqrt{x+6}\right)^2$$
$$11-5x=49-14\sqrt{x+6}+x+6$$

Now subtract x and 55 (that is, 49 + 6) from each side to isolate the radical on the right:

$$-44-6x=-14\sqrt{x+6}$$

Before squaring both sides of the equation, divide each term by 2. Then set the quadratic equation equal to 0 and solve for x:

$$-22-3x=-7\sqrt{x+6}$$
$$(-22-3x)^2=\left(-7\sqrt{x+6}\right)^2$$
$$484+132x+9x^2=49(x+6)$$
$$484+132x+9x^2=49x+294$$
$$9x^2+83x+190=0$$
$$(x+5)(9x+38)=0$$

The two solutions of the quadratic are $x = -5$ and $x=-\frac{38}{9}$.

Check $x = -5$ in the original equation:

$$\sqrt{11-5(-5)}+\sqrt{-5+6}\stackrel{?}{=}7$$
$$\sqrt{11+25}+\sqrt{1}\stackrel{?}{=}7$$
$$\sqrt{36}+1\stackrel{?}{=}1$$
$$6+1=7$$

This solution works.

Now check $x=-\frac{38}{9}$ in the original equation:

$$\sqrt{11-5\left(-\frac{38}{9}\right)}+\sqrt{-\frac{38}{9}+6}\stackrel{?}{=}7$$
$$\sqrt{\frac{99}{9}+\frac{190}{9}}+\sqrt{-\frac{38}{9}+\frac{54}{9}}\stackrel{?}{=}7$$
$$\sqrt{\frac{289}{9}}+\sqrt{\frac{16}{9}}\stackrel{?}{=}7$$
$$\frac{17}{3}+\frac{4}{3}\stackrel{?}{=}7$$
$$\frac{21}{3}=7$$

This solution also works.

111. $\frac{2\sqrt{6}}{3}$

Multiply both the numerator and the denominator by $\sqrt{6}$:

$$\frac{4}{\sqrt{6}}\cdot\frac{\sqrt{6}}{\sqrt{6}}=\frac{4\sqrt{6}}{6}$$

Reduce the fraction:

$$\frac{{}^{2}\cancel{4}\sqrt{6}}{{}_{3}\cancel{6}}=\frac{2\sqrt{6}}{3}$$

112. $2\sqrt{5}$

Multiply both the numerator and the denominator by $\sqrt{5}$:

$$\frac{10}{\sqrt{5}}\cdot\frac{\sqrt{5}}{\sqrt{5}}=\frac{10\sqrt{5}}{5}$$

Reduce the fraction:

$$\frac{{}^{2}\cancel{10}\sqrt{5}}{{}_{1}\cancel{5}}=2\sqrt{5}$$

113. $\frac{6\sqrt{3}+\sqrt{6}}{3}$

Multiply both the numerator and the denominator by $\sqrt{3}$:

$$\frac{6+\sqrt{2}}{\sqrt{3}}\cdot\frac{\sqrt{3}}{\sqrt{3}}=\frac{6\sqrt{3}+\sqrt{6}}{3}$$

114. $5\sqrt{2}-\sqrt{7}$

Multiply both the numerator and the denominator by $\sqrt{2}$:

$$\frac{10-\sqrt{14}}{\sqrt{2}}\cdot\frac{\sqrt{2}}{\sqrt{2}}=\frac{10\sqrt{2}-\sqrt{28}}{2}$$

In the numerator, simplify the radical on the right:

$$\frac{10\sqrt{2}-\sqrt{4}\sqrt{7}}{2}=\frac{10\sqrt{2}-2\sqrt{7}}{2}$$

Reduce the fraction:

$$\frac{{}^{5}\cancel{10}\sqrt{2}-\cancel{2}\sqrt{7}}{\cancel{2}}=5\sqrt{2}-\sqrt{7}$$

115. $5\left(\sqrt{2}-1\right)$

Multiply both the numerator and the denominator by the conjugate of the denominator:

$$\frac{5}{1+\sqrt{2}}\cdot\frac{1-\sqrt{2}}{1-\sqrt{2}}=\frac{5\left(1-\sqrt{2}\right)}{1-2}$$

Simplify the fraction:

$$\frac{5\left(1-\sqrt{2}\right)}{-1}=-5\left(1-\sqrt{2}\right)=5\left(\sqrt{2}-1\right)$$

116. $2\left(3+\sqrt{3}\right)$

Multiply both the numerator and the denominator by the conjugate of the denominator:

$$\frac{12}{3-\sqrt{3}}\cdot\frac{3+\sqrt{3}}{3+\sqrt{3}}=\frac{12\left(3+\sqrt{3}\right)}{9-3}$$

Simplify the fraction:

$$\frac{12\left(3+\sqrt{3}\right)}{6}=\frac{{}^{2}\cancel{12}\left(3+\sqrt{3}\right)}{\cancel{6}}=2\left(3+\sqrt{3}\right)$$

117. $2\left(3-\sqrt{5}\right)$

Multiply both the numerator and the denominator by the conjugate of the denominator:

$$\frac{8}{\sqrt{5}+3}\cdot\frac{\sqrt{5}-3}{\sqrt{5}-3}=\frac{8\left(\sqrt{5}-3\right)}{5-9}$$

Simplify the fraction:

$$\frac{8\left(\sqrt{5}-3\right)}{-4}=\frac{{}^{2}\not{8}\left(\sqrt{5}-3\right)}{-\not{4}}=-2\left(\sqrt{5}-3\right)=2\left(3-\sqrt{5}\right)$$

118. $\frac{2\sqrt{6}-\sqrt{15}}{3}$

Multiply both the numerator and the denominator by the conjugate of the denominator:

$$\frac{\sqrt{6}}{4+\sqrt{10}}\cdot\frac{4-\sqrt{10}}{4-\sqrt{10}}=\frac{4\sqrt{6}-\sqrt{60}}{16-10}$$

In the numerator, simplify the radical on the right. Then simplify and reduce the fraction:

$$\frac{4\sqrt{6}-\sqrt{4}\sqrt{15}}{16-10}=\frac{4\sqrt{6}-2\sqrt{15}}{6}=\frac{{}^{2}\not{4}\sqrt{6}-\not{2}\sqrt{15}}{{}^{3}\not{6}}=\frac{2\sqrt{6}-\sqrt{15}}{3}$$

119. $\frac{2\sqrt{3}-\sqrt{30}-2\sqrt{5}+5\sqrt{2}}{-3}$

Multiply both the numerator and the denominator by the conjugate of the denominator:

$$\frac{\sqrt{6}-\sqrt{10}}{\sqrt{2}+\sqrt{5}}\cdot\frac{\sqrt{2}-\sqrt{5}}{\sqrt{2}-\sqrt{5}}=\frac{\sqrt{12}-\sqrt{30}-\sqrt{20}+\sqrt{50}}{2-5}$$

Simplify the radicals in the numerator; then subtract in the denominator:

$$\frac{\sqrt{4}\sqrt{3}-\sqrt{30}-\sqrt{4}\sqrt{5}+\sqrt{25}\sqrt{2}}{2-5}=\frac{2\sqrt{3}-\sqrt{30}-2\sqrt{5}+5\sqrt{2}}{-3}$$

None of the radicals combine.

120. $\frac{7\sqrt{2}+\sqrt{42}-\sqrt{21}-3}{4}$

Multiply both numerator and denominator by the conjugate of the denominator:

$$\frac{\sqrt{14}-\sqrt{3}}{\sqrt{7}-\sqrt{3}}\cdot\frac{\sqrt{7}+\sqrt{3}}{\sqrt{7}+\sqrt{3}}=\frac{\sqrt{98}+\sqrt{42}-\sqrt{21}-3}{7-3}$$

Simplify the radical in the numerator; then subtract in the denominator:

$$\frac{\sqrt{49}\sqrt{2}+\sqrt{42}-\sqrt{21}-3}{7-3}=\frac{7\sqrt{2}+\sqrt{42}-\sqrt{21}-3}{4}$$

121. –79

Substitute –3 for n in the expression for $f(n)$:

$$\begin{aligned} f(-3) &= 4(-3)^3 + (-3)^2 - 6(-3) + 2 \\ &= 4(-27) + 9 + 18 + 2 \\ &= -108 + 29 = -79 \end{aligned}$$

122. 11

Substitute –6 for x in the expression for $g(x)$:

$$\begin{aligned} g(-6) &= |-6-3| + 2 \\ &= |-9| + 2 \\ &= 9 + 2 \\ &= 11 \end{aligned}$$

123. –256

Substitute 3 for y in the expression for $h(y)$:

$$h(3) = -2^{2(3)+2} = -2^8$$

Next, find the 8th power of 2. Using the order of operations, you perform the power before performing multiplication or division. The negative sign in front of the 2 acts as a –1 multiplier, so you perform it after the power.

$$-2^8 = -(2^8) = -(256) = -256$$

124. 27

This function is a piecewise function, so evaluate it by using the portion of the function defined for $x = -5$. In this case, because $-5 \leq -4$, use $g(x) = x^2 + 2$.

$$g(-5) = (-5)^2 + 2 = 25 + 2 = 27$$

125. $2x^2 + 4xh + 2h^2 - 3x - 3h$

Replace x with $x + h$ and then simplify the expression.

$$\begin{aligned} f(x+h) &= 2(x+h)^2 - 3(x+h) \\ &= 2\left(x^2 + 2xh + h^2\right) - 3x - 3h \\ &= 2x^2 + 4xh + 2h^2 - 3x - 3h \end{aligned}$$

126. domain: {–2, 0, 1, 2, 5}; range: {2, 3, 4, 6}

The domain is the set of x values, and the range is the set of y values. f is a function because each x value maps to only one y value.

127. domain: $-5 < x$; range: $0 < y \leq 5$ or $y < -1$

The domain is the set of x values, and the range is the set of y values for which the function is defined. In this case, x is defined for all real numbers greater than –5. You don't include –5 because there's an open dot on –5 in the graph. If $-5 < x \leq 1$, then the range is $0 < y \leq 5$. If $x > 1$, then the range is all real numbers less than –1. So the range is $0 < y \leq 5$ or $y < -1$.

128. domain: $(-\infty, \infty)$; range: $(-\infty, \infty)$

The domain and range of a linear function are all real numbers, written $(-\infty, \infty)$. The graph of the function makes this fact clear.

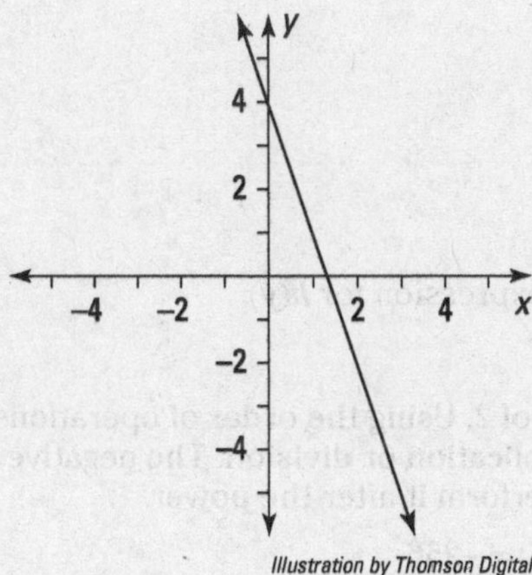

Illustration by Thomson Digital

129. domain: $(-\infty, \infty)$; range: $[1, \infty)$

The domain of any polynomial function is all real numbers. However, the value of the function can only be a positive real number because squaring any number gives a positive result. Letting $x = 0$ gives $f(x) = 1$, which is the minimum value of the function, so the range is the set of all real numbers greater than or equal to 1. The graph of the function makes this fact clear.

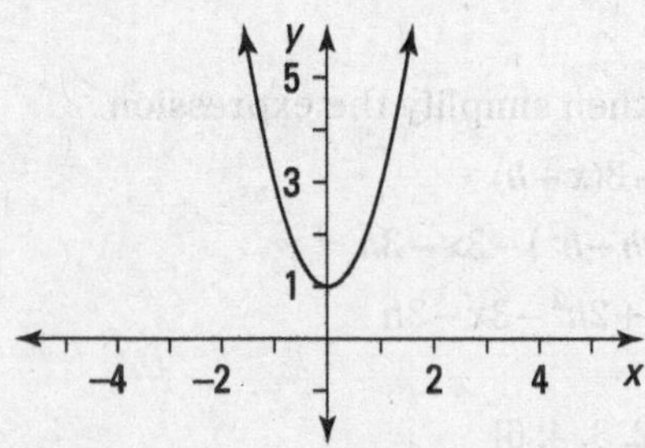

Illustration by Thomson Digital

130. domain: [0, 8]; range: [0, 256]

The definition of the function restricts the values of t to the real numbers between 0 and 8, inclusive. Look at a factored version of the function equation: $h(t) = 16t(8 - t)$.

Use the TABLE function of a graphing calculator to find some values of the function.

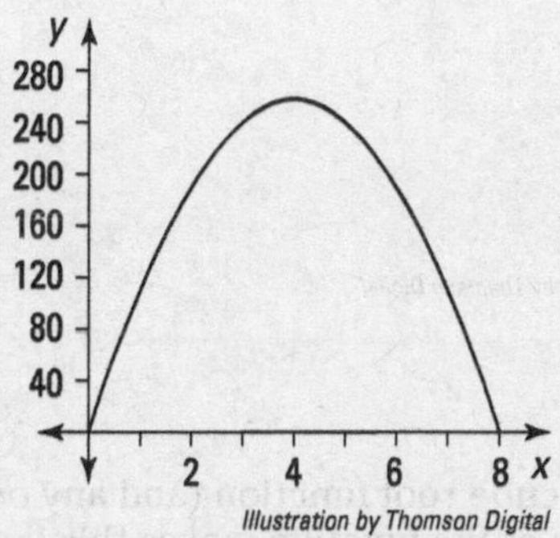

Illustration by Thomson Digital

The minimum value of the function is 0 and the maximum value of the function is 256, so the range is [0, 256]. The graph of the function makes this fact clear.

Illustration by Thomson Digital

131. domain: $(-\infty, \infty)$; range: $(-\infty, \infty)$

The domain of any polynomial function is all real numbers. The range of an odd degree polynomial is always all real numbers. The graph of the function makes this fact clear.

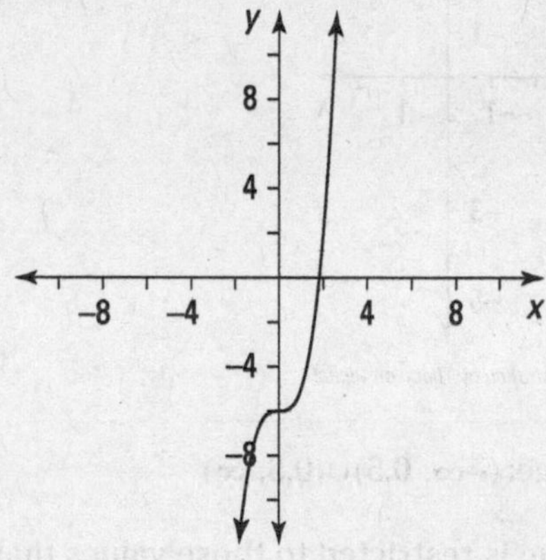

Illustration by Thomson Digital

132. domain: $(-\infty, 2.5]$; range: $[0, \infty)$

The argument of the square root must be greater than or equal to zero because the square root of a negative number is not defined. So solve $5-2x\geq 0$.

$$5-2x \geq 0$$
$$5 \geq 2x$$
$$2.5 \geq x$$

Therefore, the domain is all real numbers less than or equal to 2.5.

Square roots are always positive or zero. Therefore, the range is all real numbers greater than or equal to zero. The graph of the function makes this fact clear.

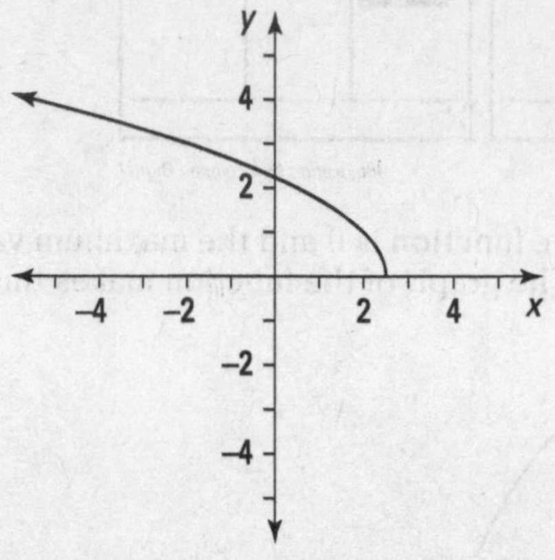

Illustration by Thomson Digital

133.

domain: $(-\infty, \infty)$; range: $(-\infty, \infty)$

The domain and range of the cube root function (and any odd-numbered integer root) is all real numbers. The graph of the function makes this fact clear.

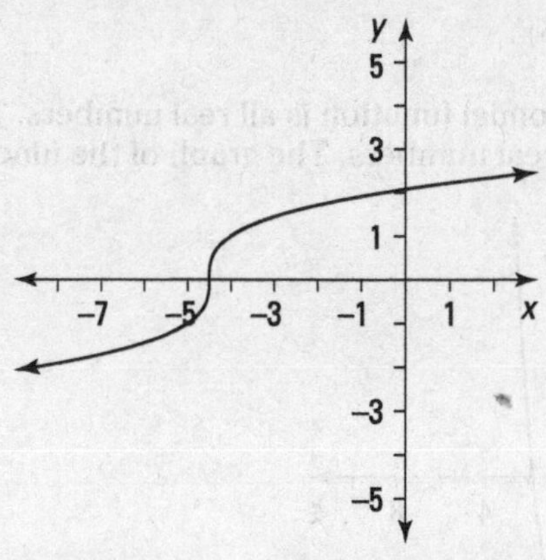

Illustration by Thomson Digital

134.

domain: $(-\infty, -3) \cup (-3, \infty)$; range: $(-\infty, 0.5) \cup (0.5, \infty)$

The domain of the function is restricted to those values that don't make the denominator equal zero.

$$\begin{aligned} 2x+6&=0 \\ 2x&=-6 \\ x&=-3 \end{aligned}$$

Thus, the domain is $(-\infty, -3) \cup (-3, \infty)$.

Use the TABLE function of a graphing calculator to find some values of the function. You see that the values of Y_1 approach 0.5 as x becomes very small or very large.

However, the value of the function will never equal 0.5. (Try to solve $\frac{x-5}{2x+6}=0.5$. You end up with $x-5=x+6$, which isn't true!) Therefore, 0.5 is not in the range of $g(x)$.

X	Y1
-1000	.50401
-500	.50805
-100	.54124
-50	.58511
-20	.73529
-5	2.5
-3	ERROR

X=-3

X	Y1
-1	-1.5
0	-.8333
4	-.0714
6	.05556
20	.32609
50	.42453
100	.46117

X=100

X	Y1
500	.49205
1000	.49601
10000	.4996

X=10000

Illustration by Thomson Digital

The graph of the function makes this fact clear.

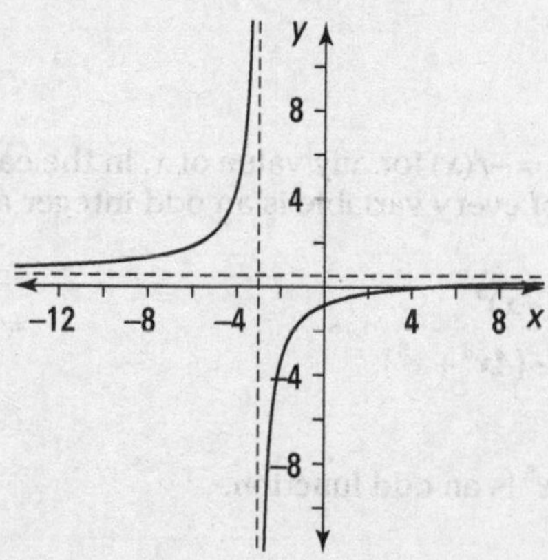

Illustration by Thomson Digital

135. domain: $(-\infty, \infty)$; range: $[3, \infty)$

The domain of an absolute value function is all real numbers. The absolute value part of the function will always be greater than or equal to zero. (In this case, you know the absolute value will equal zero if $x = 4$.) Because 3 is being added to the absolute value, the range is $[3, \infty)$. The graph of the function makes this fact clear.

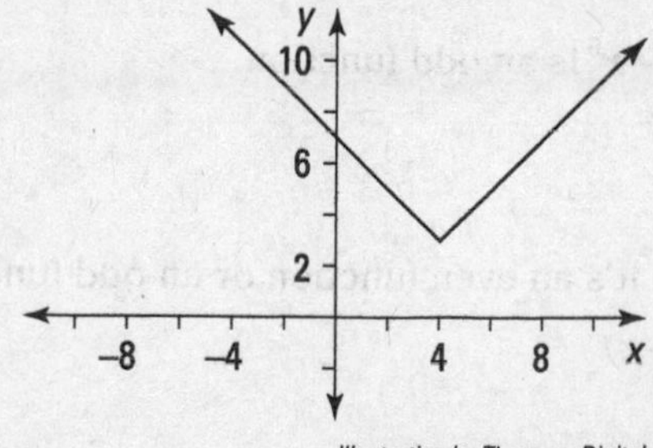

Illustration by Thomson Digital

136. $f(x)=-3x^2+4$

A function is *even* if $f(-x)=f(x)$ for any value of x. In the case of a polynomial function, it's *even* if the exponent of every variable is an even integer.

$$f(-x)=-3(-x)^2+4=-3x^2+4=f(x)$$

Therefore, $f(x)=-3x^2+4$ is an even function.

137. $g(x)=1-x^2-2x^4+x^6$

A function is *even* if $g(-x)=g(x)$ for any value of x. In the case of a polynomial function, it's *even* if the exponent of every variable is an even integer.

$$\begin{aligned} g(-x)&=1-(-x)^2-2(-x)^4+(-x)^6 \\ &=1-x^2-2x^4+x^6 \\ &=g(x) \end{aligned}$$

Therefore, $g(x)=1-x^2-2x^4+x^6$ is an even function.

138. $f(x)=-4x^3+x^5$

A function is *odd* if $f(-x)=-f(x)$ for any value of x. In the case of a polynomial function, it's *odd* if the exponent of every variable is an odd integer and there are no constant terms.

$$\begin{aligned} f(-x)&=-4(-x)^3+(-x)^5 \\ &=4x^3-x^5=-\left(4x^3+x^5\right) \\ &=-f(x) \end{aligned}$$

Therefore, $f(x)=-4x^3+x^5$ is an odd function.

139. $g(x)=x+2x^3-x^5$

A function is *odd* if $g(-x)=-g(x)$ for any value of x. In the case of a polynomial function, it's *odd* if the exponent of every variable is an odd integer and there are no constant terms.

$$\begin{aligned} g(-x)&=(-x)+2(-x)^3-(-x)^5 \\ &=-x-2x^3+x^5=-\left(x+2x^3-x^5\right) \\ &=-g(x) \end{aligned}$$

Therefore, $g(x)=x+2x^3-x^5$ is an odd function.

140. $h(x)=x^2-2x$

Test $h(x)$ to see whether it's an even function or an odd function:

$$\begin{aligned} h(-x)&=(-x)^2-2(-x) \\ &=x^2+2x \\ &\neq h(x) \end{aligned}$$

$$f(x+h)-f(x)=\frac{2}{(x+h)+1}-\frac{2}{x+1}$$
$$=\frac{2(x+1)-2(x+h+1)}{(x+h+1)(x+1)}$$
$$=\frac{2x+2-2x-2h-2}{(x+h+1)(x+1)}$$
$$=-\frac{2h}{(x+h+1)(x+1)}$$

Finally, plug this expression into the numerator of the difference quotient and simplify:

$$\frac{f(x+h)-f(x)}{h}=-\frac{2h}{(x+h+1)(x+1)}\cdot\frac{1}{h}$$
$$=-\frac{2}{(x+h+1)(x+1)}$$

181. $f^{-1}(x)=\frac{x+5}{3}$

To find the inverse, first rewrite the function as $y=3x-5$. Then exchange the y and x:

$$x=3y-5$$

Solve for y:

$$x+5=3y$$
$$\frac{x+5}{3}=y$$

Finally, rewrite the function, replacing the y with $f^{-1}(x)$:

$$\frac{x+5}{3}=f^{-1}(x)$$

Here are the graphs of $f(x)$ and $f^{-1}(x)$:

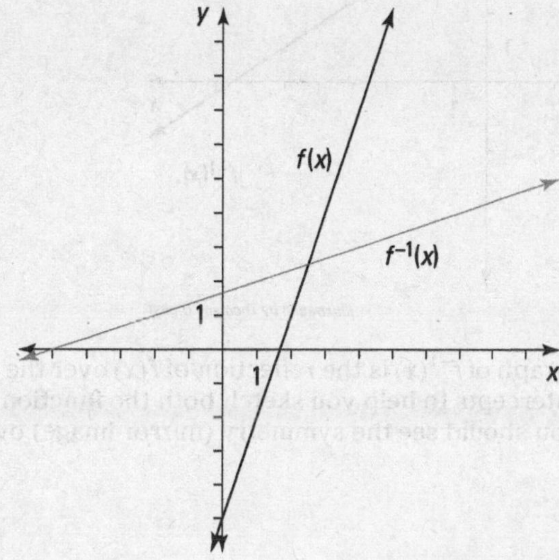

Illustration by Thomson Digital

Note that the graph of $f^{-1}(x)$ is the reflection of $f(x)$ over the line $y=x$. After using the slopes and y-intercepts to help you sketch both the function and its inverse on the same graph, you should see the symmetry (mirror image) over the diagonal $y=x$.

182. $f^{-1}(x)=-\frac{3}{2}x+\frac{15}{2}$

To find the inverse, first rewrite the function as $y=-\frac{2}{3}x+5$. Then exchange the y and x:

$$x=-\frac{2}{3}y+5$$

Solve for y:

$$x-5=-\frac{2}{3}y$$

$$-\frac{3}{2}(x-5)=y$$

Finally, rewrite the function, replacing the y with $f^{-1}(x)$:

$$-\frac{3}{2}(x-5)=f^{-1}(x)$$

In slope-intercept form, that's $f^{-1}(x)=-\frac{3}{2}x+\frac{15}{2}$.

Here are the graphs of $f(x)$ and $f^{-1}(x)$:

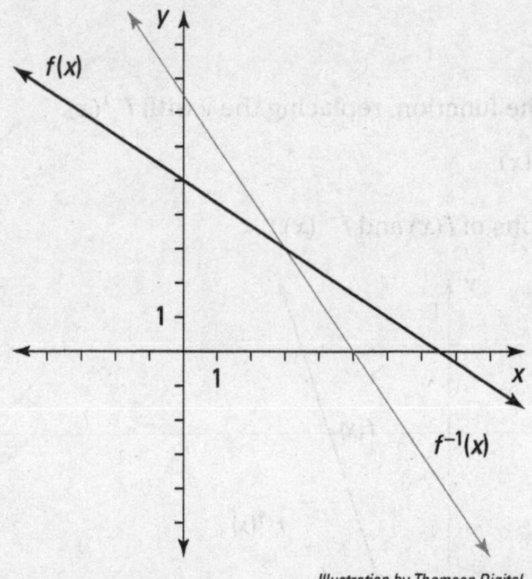

Illustration by Thomson Digital

Note that the graph of $f^{-1}(x)$ is the reflection of $f(x)$ over the line $y=x$. After using the slopes and y-intercepts to help you sketch both the function and its inverse on the same graph, you should see the symmetry (mirror image) over the diagonal $y=x$.

$$h(-x) = (-x)^2 - 2(-x)$$
$$= x^2 + 2x$$
$$\neq -h(x)$$

Therefore, $h(x) = x^2 - 2x$ is neither even nor odd. Note that the exponents of the variables are a mixture of even and odd integers.

141. $\{(1, a), (2, d), (3, b)\}$

This relation is a one-to-one function because each *x* value maps to just one *y* value and each *y* value maps to just one *x* value.

142. $\{(2, 3), (4, 2), (1, 5), (3, 4)\}$

This relation is a one-to-one function because each *x* value maps to just one *y* value and each *y* value maps to just one *x* value.

143.

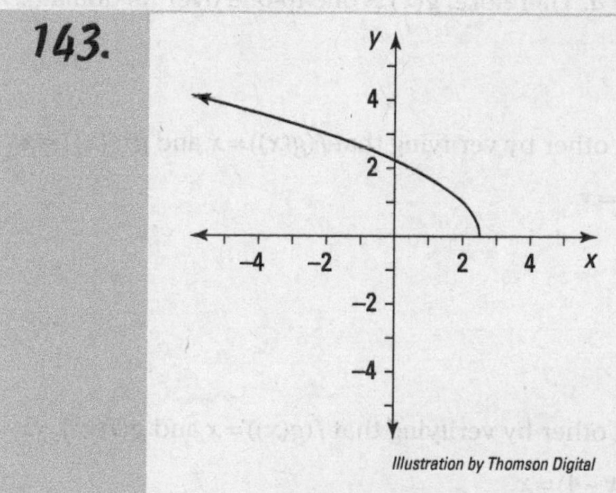

Illustration by Thomson Digital

You can use two tests to determine whether a graph is a one-to-one function. The *horizontal line test* states that if a horizontal line intersects a function's graph more than once, then the function is not one-to-one. The *vertical line test* states that a relation is a function if there are no vertical lines that intersect the graph at more than one point. This graph passes both the horizontal line test and the vertical line test, so it's a one-to-one function.

144. $f(x) = -x^3$

Suppose that *a* and *b* are real numbers and that $f(a) = f(b)$. To prove that $f(x)$ is one-to-one, show that $a = b$:

$$f(a) = -a^3$$
$$f(b) = -b^3$$

Let $f(a) = f(b)$. Then, substituting, you get $-a^3 = -b^3$ and $a^3 = b^3$.

Taking the cube root of each side, $a = b$, so the function is one-to-one over its domain.

145. $g(x)=\frac{2x+1}{x}$

Suppose that a and b are real numbers and that $g(a)=g(b)$. To prove that $g(x)$ is one-to-one, show that $a=b$:

$$g(a)=\frac{2a+1}{a}$$
$$g(b)=\frac{2b+1}{b}$$

Let $g(a)=g(b)$. Then, substituting, you get

$$\frac{2a+1}{a}=\frac{2b+1}{b}$$

Cross-multiply and distribute:

$$b(2a+1)=a(2b+1)$$
$$2ab+b=2ab+a$$

Subtract $2ab$ from each side to get $b=a$. Therefore, $g(x)$ is one-to-one over its domain.

146. $f(x)=3x-1,\ g(x)=\frac{x+1}{3}$

Show that f and g are inverses of each other by verifying that $f(g(x))=x$ and $g(f(x))=x$:

$$f(g(x))=3\left(\frac{x+1}{3}\right)-1=x+1-1=x$$
$$g(f(x))=\frac{(3x-1)+1}{3}=\frac{3x}{3}=x$$

147. $f(x)=4+\frac{3}{2}x,\ g(x)=\frac{2}{3}(x-4)$

Show that f and g are inverses of each other by verifying that $f(g(x))=x$ and $g(f(x))=x$:

$$f(g(x))=4+\frac{3}{2}\left(\frac{2}{3}(x-4)\right)=4+(x-4)=x$$
$$g(f(x))=\frac{2}{3}\left(\left(4+\frac{3}{2}x\right)-4\right)=\frac{8}{3}+x-\frac{8}{3}=x$$

148. $f(x)=-\frac{2}{x}-1,\ g(x)=-\frac{2}{x+1}$

Show that f and g are inverses of each other by verifying that $f(g(x))=x$ and $g(f(x))=x$:

$$f(g(x))=-\frac{2}{\left(-\frac{2}{x+1}\right)}-1=-2\left(-\frac{x+1}{2}\right)-1=x+1-1=x$$
$$g(f(x))=-\frac{2}{\left(-\frac{2}{x}-1\right)+1}=-\frac{2}{-\frac{2}{x}}=-2\left(-\frac{x}{2}\right)=x$$

149. $f(x)=\frac{x-2}{x+2},\ g(x)=\frac{-2x-2}{x-1}$

Show that f and g are inverses of each other by verifying that $f(g(x))=x$ and $g\left(f(x)\right)=x$:

$$\begin{aligned}
f(g(x))&=\frac{\left(\frac{-2x-2}{x-1}\right)-2}{\left(\frac{-2x-2}{x-1}\right)+2}=\frac{\left(\frac{-2x-2}{x-1}\right)-2}{\left(\frac{-2x-2}{x-1}\right)+2}\cdot\frac{x-1}{x-1}\\
&=\frac{-2x-2-2(x-1)}{-2x-2+2(x-1)}\\
&=\frac{-2x-2-2x+2}{-2x-2+2x-2}=\frac{-4x}{-4}=x
\end{aligned}$$

$$\begin{aligned}
g\left(f(x)\right)&=\frac{-2\left(\frac{x-2}{x+2}\right)-2}{\frac{x-2}{x+2}-1}=\frac{-2\left(\frac{x-2}{x+2}\right)-2}{\frac{x-2}{x+2}-1}\cdot\frac{x+2}{x+2}\\
&=\frac{-2(x-2)-2(x+2)}{x-2-(x+2)}=\frac{-2x+4-2x-4}{x-2-x-2}\\
&=\frac{-4x}{-4}=x
\end{aligned}$$

150. $f(x)=-(x-2)^3,g(x)=2-\sqrt[3]{x}$

Show that f and g are inverses of each other by verifying that $f(g(x))=x$ and $g\left(f(x)\right)=x$:

$$\begin{aligned}
f(g(x))&=-\left[\left(2-\sqrt[3]{x}\right)-2\right]^3\\
&=-\left(-\sqrt[3]{x}\right)^3\\
&=-(-x)\\
&=x
\end{aligned}$$

$$\begin{aligned}
g\left(f(x)\right)&=2-\sqrt[3]{-(x-2)^3}\\
&=2+\sqrt[3]{(x-2)^3}\\
&=2+x-2\\
&=x
\end{aligned}$$

151. $f^{-1}=\{(1,\ -4),\ (2,\ -3),\ (0,\ 0),\ (10,\ 1),\ (3,\ 2)\}$

Find the inverse by interchanging the x and y values:

$$f^{-1}=\{(1,\ -4),\ (2,\ -3),\ (0,\ 0),\ (10,\ 1),\ (3,\ 2)\}$$

152. $f^{-1}(x)=-\frac{x-3}{2}$

Change $f(x)$ to y:

$$y=-2x+3$$

Interchange x and y:

$$x=-2y+3$$

Now solve for *y*:

$$x=-2y+3$$
$$x-3=-2y$$
$$-\frac{x-3}{2}=y$$

Rename *y* as $f^{-1}(x)$:

$$f^{-1}(x)=-\frac{x-3}{2}$$

153. $f^{-1}(x)=\frac{-5x+10}{3}$

Change $f(x)$ to *y*:

$$y=-\frac{3}{5}x+2$$

Interchange *x* and *y*:

$$x=-\frac{3}{5}y+2$$

Now solve for *y*:

$$x=-\frac{3}{5}y+2$$
$$x-2=-\frac{3}{5}y$$
$$-\frac{5}{3}(x-2)=y$$
$$\frac{-5x+10}{3}=y$$

Rename *y* as $f^{-1}(x)$:

$$f^{-1}(x)=\frac{-5x+10}{3}$$

154. $f^{-1}(x)=-\frac{4}{x-1}$

Change $f(x)$ to *y*:

$$y=1-\frac{4}{x}$$

Interchange *x* and *y*:

$$x=1-\frac{4}{y}$$

Now solve for *y* by first multiplying each term in the equation by *y*:

$$x=1-\frac{4}{y}$$
$$xy=y-4$$
$$xy-y=-4$$
$$y(x-1)=-4$$
$$y=-\frac{4}{x-1}$$

Rename y as $f^{-1}(x)$:

$$f^{-1}(x) = -\frac{4}{x-1}$$

155. $f^{-1}(x) = x^3 + 5$

Change $f(x)$ to y:

$$y = \sqrt[3]{x-5}$$

Interchange x and y:

$$x = \sqrt[3]{y-5}$$

Now solve for y:

$$x = \sqrt[3]{y-5}$$
$$x^3 = y-5$$
$$x^3 + 5 = y$$

Rename y as $f^{-1}(x)$:

$$f^{-1}(x) = x^3 + 5$$

156. $f^{-1}(x) = \frac{1-3x}{x}$

Change $f(x)$ to y:

$$y = \frac{1}{x+3}$$

Interchange x and y:

$$x = \frac{1}{y+3}$$

Now solve for y:

$$x = \frac{1}{y+3}$$
$$x(y+3) = 1$$
$$y+3 = \frac{1}{x}$$
$$y = \frac{1}{x} - 3$$
$$y = \frac{1-3x}{x}$$

Rename y as $f^{-1}(x)$:

$$f^{-1}(x) = \frac{1-3x}{x}$$

157. $f^{-1}(x) = -\frac{4}{x-2} - 2$

Change $f(x)$ to y:

$$y = \frac{4}{-x-2} + 2$$

Interchange x and y:

$$x=\frac{4}{-y-2}+2$$

Now solve for y:

$$x=\frac{4}{-y-2}+2$$
$$x-2=\frac{4}{-y-2}$$
$$-y-2=\frac{4}{x-2}$$
$$-y=\frac{4}{x-2}+2$$
$$y=-\frac{4}{x-2}-2$$

Rename y as $f^{-1}(x)$:

$$f^{-1}(x)=-\frac{4}{x-2}-2$$

158. $f^{-1}(x)=\sqrt[3]{\frac{2x+3}{x}}$

Change $f(x)$ to y:

$$y=\frac{3}{x^3-2}$$

Interchange x and y:

$$x=\frac{3}{y^3-2}$$

Now solve for y:

$$x=\frac{3}{y^3-2}$$
$$y^3-2=\frac{3}{x}$$
$$y^3=\frac{3}{x}+2$$
$$y^3=\frac{2x+3}{x}$$
$$y=\sqrt[3]{\frac{2x+3}{x}}$$

Rename y as $f^{-1}(x)$:

$$f^{-1}(x)=\sqrt[3]{\frac{2x+3}{x}}$$

159. $f^{-1}(x)=\frac{-5x-7}{4x-1}$

Change $f(x)$ to y:

$$y=\frac{x-7}{4x+5}$$

Interchange x and y:

$$x=\frac{y-7}{4y+5}$$

Now solve for y:

$$\begin{aligned} x&=\frac{y-7}{4y+5}\\ 4xy+5x&=y-7\\ 4xy-y&=-5x-7\\ y(4x-1)&=-5x-7\\ y&=\frac{-5x-7}{4x-1} \end{aligned}$$

Rename y as $f^{-1}(x)$:

$$f^{-1}(x)=\frac{-5x-7}{4x-1}$$

160. $f^{-1}(x)=\sqrt{1-x^2},\ 0\leq x\leq 1$

Note that the domain of f is restricted to those values of x between 0 and 1, inclusive. The domain is restricted so that f is one-to-one and an inverse exists. Also note that the range of f is restricted to those values of y between 0 and 1, inclusive. The range of f becomes the domain of f^{-1}.

Change $f(x)$ to y:

$$y=\sqrt{1-x^2}$$

Interchange x and y:

$$x=\sqrt{1-y^2}$$

Now solve for y:

$$\begin{aligned} x&=\sqrt{1-y^2}\\ x^2&=1-y^2\\ x^2-1&=-y^2\\ y^2&=1-x^2\\ y&=\sqrt{1-x^2} \end{aligned}$$

Rename y as $f^{-1}(x)$:

$$f^{-1}(x)=\sqrt{1-x^2},\quad 0\leq x\leq 1$$

The domain of f^{-1} the same as the range of f, so it's restricted to those values of y between 0 and 1, inclusive. This function, on its restricted domain, is its own inverse.

161. x^2+x-4

Find $(f+g)(x)$ by adding the expressions for $f(x)$ and $g(x)$ and simplifying:

$$\begin{aligned}(f+g)(x)&=f(x)+g(x)\\&=(x^2+2x+1)+(-x-5)\\&=x^2+x-4\end{aligned}$$

162. n^2+3n+8

Find $(k-h)(n)$ by subtracting the expression for $h(n)$ from the expression for $k(n)$ and simplifying:

$$\begin{aligned}(k-h)(n)&=k(n)-h(n)\\&=(n^2+5)-(-3n-3)\\&=n^2+5+3n+3\\&=n^2+3n+8\end{aligned}$$

163. $3a^2+19a-14$

Find $(h\cdot g)(a)$ by multiplying the expressions for $g(a)$ and $h(a)$. Use the FOIL method to multiply the binomials and then simplify:

$$\begin{aligned}(h\cdot g)(a)&=h(a)\cdot g(a)\\&=(3a-2)(a+7)\\&=3a^2+21a-2a-14\\&=3a^2+19a-14\end{aligned}$$

164. $x-4, x\neq -1$

Find $\left(\frac{f}{g}\right)(x)$ by dividing the expression for $f(x)$ by the expression for $g(x)$. Simplify the quotient if possible.

$$\begin{aligned}\left(\frac{f}{g}\right)(x)&=\frac{f(x)}{g(x)}\\&=\frac{x^2-3x-4}{x+1}\\&=\frac{(x-4)(x+1)}{x+1}\\&=x-4\end{aligned}$$

The value of x cannot be -1.

165. 112

First find $f(-2)$ and $g(-2)$ by substituting -2 for x in the expressions for $f(x)$ and $g(x)$. Then multiply the results.

$$f(-2)=8-3(-2)=8+6=14$$
$$g(-2)=(-2)^2-2(-2)=4+4=8$$
$$(g\cdot f)(-2)=8(14)=112$$

Alternatively, find $(g\cdot f)(x)$ by multiplying the expressions for $g(x)$ and $f(x)$. Then substitute -2 for x in the resulting expression to find $(g\cdot f)(-2)$.

$$\begin{aligned}(g\cdot f)(x)&=(8-3x)\left(x^2-2x\right)\\&=8x^2-16x-3x^3+6x^2\\&=-3x^3+14x^2-16x\\(g\cdot f)(-2)&=-3(-2)^3+14(-2)^2-16(-2)\\&=-3(-8)+14(4)+32\\&=24+56+32\\&=112\end{aligned}$$

166. $3x^2+1$

The notation $(f\circ g)(x)$ means the same thing as $f(g(x))$. So substitute x^2 for x in $f(x)$:

$$\begin{aligned}(f\circ g)(x)&=f(g(x))\\&=f\left(x^2\right)\\&=3x^2+1\end{aligned}$$

167. $4x-5$

The notation $(f\circ g)(x)$ means the same thing as $f(g(x))$. So substitute $x-2$ for x in $f(x)$. Then simplify the resulting expression.

$$\begin{aligned}(f\circ g)(x)&=f(g(x))\\&=f(x-2)\\&=4(x-2)+3\\&=4x-8+3\\&=4x-5\end{aligned}$$

168. x^2+4x

The notation $(f\circ g)(x)$ means the same thing as $f(g(x))$. So substitute $x+1$ for each x in $f(x)$. Then simplify the resulting expression:

$$\begin{aligned}(f\circ g)(x)&=f(g(x))\\&=f(x+1)\\&=(x+1)^2+2(x+1)-3\\&=x^2+2x+1+2x+2-3\\&=x^2+4x\end{aligned}$$

169. x

The notation $(f \circ g)(x)$ means the same thing as $f(g(x))$. So substitute $\sqrt{x+3}$ for x in $f(x)$. Then simplify the resulting expression:

$$\begin{aligned}(f \circ g)(x) &= f(g(x))\\ &= \left(\sqrt{x+3}\right)^2 - 3\\ &= x+3-3\\ &= x\end{aligned}$$

170. 2^{3x}

The notation $(g \circ f)(x)$ means the same thing as $g(f(x))$. So substitute $3x$ for x in $g(x)$:

$$(g \circ f)(x) = g(f(x)) = 2^{3x}$$

171. $-3x^2 - 1$

The notation $(g \circ f)(x)$ means the same thing as $g(f(x))$. So substitute $-x^2 - 2$ for x in $g(x)$. Then simplify the resulting expression.

$$\begin{aligned}(g \circ f)(x) &= g(f(x))\\ &= g\left(-x^2-2\right)\\ &= 3\left(-x^2-2\right)+5\\ &= -3x^2-6+5\\ &= -3x^2-1\end{aligned}$$

172. $\frac{10x-7}{15x-7}$

The notation $(g \circ f)(x)$ means the same thing as $g(f(x))$. So substitute $5x-3$ for each x in $g(x)$. Then simplify the resulting expression.

$$\begin{aligned}(g \circ f)(x) &= g(f(x))\\ &= \frac{2(5x-3)-1}{3(5x-3)+2}\\ &= \frac{10x-6-1}{15x-9+2}\\ &= \frac{10x-7}{15x-7}\end{aligned}$$

173. $x^4 - 4x^3 + 2x^2 + 4x$

The notation $(f \circ f)(x)$ means the same thing as $f(f(x))$. So substitute $x^2 - 2x$ for each x in $f(x)$. Then simplify the resulting expression.

$$\begin{aligned}(f \circ f)(x) &= f(f(x))\\ &= \left(x^2-2x\right)^2 - 2\left(x^2-2x\right)\\ &= x^4-4x^3+4x^2-2x^2+4x\\ &= x^4-4x^3+2x^2+4x\end{aligned}$$

174. −159

The notation $(f \circ g)(12)$ means the same thing as $f(g(12))$. First, find $g(12)$ by substituting 12 for x in the expression for $g(x)$. Next, find $f(g(12))$ by substituting the value of $g(12)$ for x in the expression for $f(x)$. Finally, simplify the resulting expression.

$$(f \circ g)(12) = f(g(12))$$

$$g(12) = 3(12) - 9 = 36 - 9 = 27$$

$$f(27) = -6(27) + 3 = -162 + 3 = -159$$

175. 4

The notation $(f \circ g)(-3)$ means the same thing as $f(g(-3))$. First, find $g(-3)$ by substituting −3 for x in the expression for $g(x)$. Next, find $f(g(-3))$ by substituting the value of $g(-3)$ for x in the expression for $f(x)$. Finally, simplify the resulting expression.

$$(f \circ g)(-3) = f(g(-3))$$

$$g(-3) = (-3)^2 = 9$$

$$f(9) = \sqrt{9+7} = \sqrt{16} = 4$$

176. $4x^2 + 8x - 4$

The notation $(h \circ g \circ f)(x)$ means the same thing as $h(g(f(x)))$. Work from the inside out to evaluate the function. First, substitute $x+1$, the expression for $f(x)$, for x in the expression for $g(x)$:

$$(h \circ g \circ f)(x) = h(g(f(x))) = h(g(x+1))$$

Next, evaluate $g(x+1)$:

$$\begin{aligned} g(x+1) &= (x+1)^2 - 2 \\ &= x^2 + 2x + 1 - 2 \\ &= x^2 + 2x - 1 \end{aligned}$$

Now evaluate $h\left(x^2 + 2x - 1\right)$ by substituting $\left(x^2 + 2x - 1\right)$ for x in the expression for $h(x)$:

$$\begin{aligned} h\left(x^2 + 2x - 1\right) &= 4\left(x^2 + 2x - 1\right) \\ &= 4x^2 + 8x - 4 \end{aligned}$$

177. 4

To find $f(x+h)$, substitute $x+h$ for x in the expression for $f(x)$. Then simplify the resulting expression.

$$\begin{aligned} f(x+h) &= 4(x+h) + 3 \\ &= 4x + 4h + 3 \end{aligned}$$

Now plug this result and the expression for $f(x)$ into the difference quotient and simplify:

$$\frac{f(x+h) - f(x)}{h} = \frac{(4x+4h+3) - (4x+3)}{h} = \frac{4h}{h} = 4$$

178. $2x+h$

To find $f(x+h)$, substitute $x+h$ for x in the expression for $f(x)$. Then simplify the resulting expression.

$$\begin{aligned} f(x+h) &= (x+h)^2 - 4 \\ &= x^2 + 2xh + h^2 - 4 \end{aligned}$$

Now plug this result and the expression for $f(x)$ into the difference quotient and simplify:

$$\begin{aligned} \frac{f(x+h)-f(x)}{h} &= \frac{(x^2+2xh+h^2-4)-(x^2-4)}{h} \\ &= \frac{2xh+h^2}{h} = \frac{h(2x+h)}{h} = 2x+h \end{aligned}$$

179. $4x+5+2h$

To find $f(x+h)$, substitute $x+h$ for x in the expression for $f(x)$. Then simplify the resulting expression.

$$\begin{aligned} f(x+h) &= 2(x+h)^2 + 5(x+h) - 3 \\ &= 2\left(x^2+2xh+h^2\right) + 5x + 5h - 3 \\ &= 2x^2 + 4xh + 2h^2 + 5x + 5h - 3 \end{aligned}$$

To make the computation a little easier, work on the numerator of the difference quotient next. Find $f(x+h)-f(x)$, the numerator of the difference quotient, by using the result you found for $f(x+h)$ and subtracting the expression for $f(x)$ from it. Be sure to simplify the result.

$$\begin{aligned} f(x+h)-f(x) &= \left(2x^2+4xh+2h^2+5x+5h-3\right) - \left(2x^2+5x-3\right) \\ &= 4xh + 2h^2 + 5h \end{aligned}$$

Finally, plug this expression into the numerator of the difference quotient and simplify:

$$\begin{aligned} \frac{f(x+h)-f(x)}{h} &= \frac{4xh+2h^2+5h}{h} \\ &= \frac{h(4x+2h+5)}{h} \\ &= 4x+5+2h \end{aligned}$$

180. $-\frac{2}{(x+h+1)(x+1)}$

To find $f(x+h)$, substitute $x+h$ for x in the expression for $f(x)$.

$$f(x+h) = \frac{2}{(x+h)+1}$$

To make the computation a little easier, work on the numerator of the difference quotient next. Find $f(x+h)-f(x)$, the numerator of the difference quotient, by using the result you found for $f(x+h)$ and subtracting the expression for $f(x)$ from it. Be sure to simplify the result; the lowest common denominator for the two fractions in the second step is $(x+h+1)(x+1)$.

183. $f^{-1}(x)=\sqrt[3]{x}$

To find the inverse, first rewrite the function as $y=x^3$. Then exchange the y and x:

$$x=y^3$$

Solve for y:

$$x=y^3$$
$$\sqrt[3]{x}=y$$

Finally, rewrite the function, replacing the y with $f^{-1}(x)$:

$$\sqrt[3]{x}=f^{-1}(x)$$

Here are the graphs of $f(x)$ and $f^{-1}(x)$, which have a common intercept at (0, 0) and shared points at (1, 1) and (–1, –1):

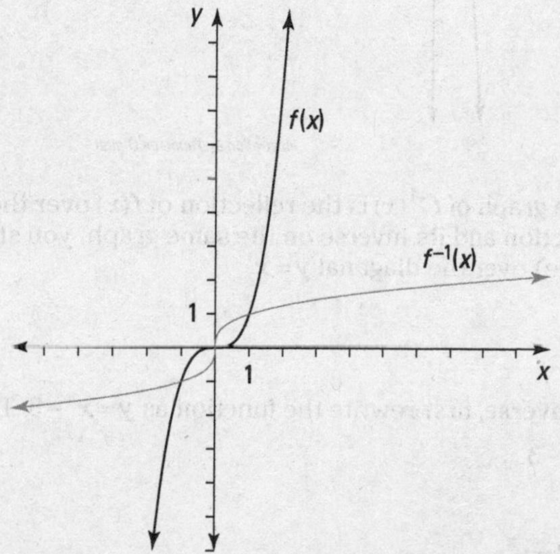

Illustration by Thomson Digital

Note that the graph of $f^{-1}(x)$ is the reflection of $f(x)$ over the line $y=x$. After sketching both the function and its inverse on the same graph, you should see the symmetry (mirror image) over the diagonal $y=x$.

184. $f^{-1}(x)=\sqrt[5]{x}$

To find the inverse, first rewrite the function as $y=x^5$. Then exchange the y and x:

$$x=y^5$$

Solve for y:

$$x=y^5$$
$$\sqrt[5]{x}=y$$

Finally, rewrite the function, replacing the y with $f^{-1}(x)$:

$$\sqrt[5]{x}=f^{-1}(x)$$

Here are the graphs of $f(x)$ and $f^{-1}(x)$, which have a common intercept at (0, 0) and shared points at (1, 1) and (–1, –1):

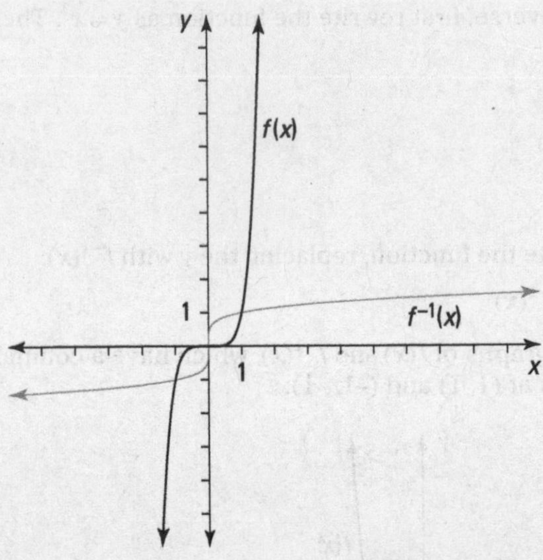

Illustration by Thomson Digital

Note that the graph of $f^{-1}(x)$ is the reflection of $f(x)$ over the line $y=x$. After sketching both the function and its inverse on the same graph, you should see the symmetry (mirror image) over the diagonal $y=x$.

185. $f^{-1}(x)=\sqrt[3]{x+3}$

To find the inverse, first rewrite the function as $y=x^3-3$. Then exchange the y and x:

$$x=y^3-3$$

Solve for y:

$$x+3=y^3$$
$$\sqrt[3]{x+3}=y$$

Finally, rewrite the function, replacing the y with $f^{-1}(x)$:

$$\sqrt[3]{x+3}=f^{-1}(x)$$

Here are the graphs of $f(x)$ and $f^{-1}(x)$:

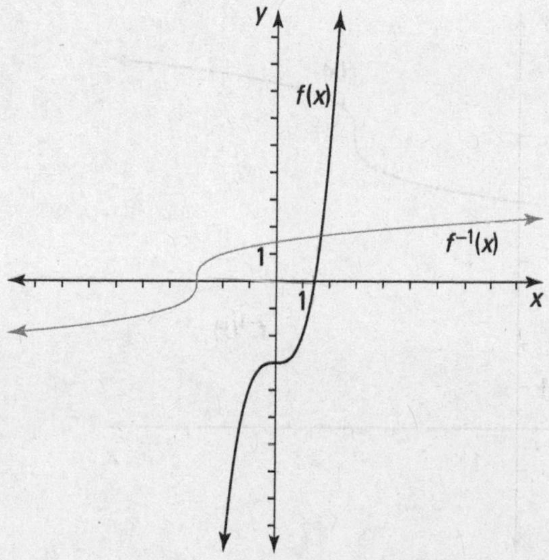

Illustration by Thomson Digital

Note that the graph of $f^{-1}(x)$ is the reflection of $f(x)$ over the line $y = x$. After sketching both the function and its inverse on the same graph, you should see the symmetry (mirror image) over the diagonal $y = x$.

Also note that the functions give you the "opposite operations in the opposite order": In $f(x)$, you cube x and then add 3; in $f^{-1}(x)$, you subtract 3 and then take the cube root of the result.

186. $f^{-1}(x) = (x-7)^3 + 4$

To find the inverse, first rewrite the function as $y = \sqrt[3]{x-4} + 7$. Then exchange the y and x:

$$x = \sqrt[3]{y-4} + 7$$

Solve for y:

$$\begin{aligned} x - 7 &= \sqrt[3]{y-4} \\ (x-7)^3 &= \left(\sqrt[3]{y-4}\right)^3 \\ (x-7)^3 &= y - 4 \\ (x-7)^3 + 4 &= y \end{aligned}$$

Finally, rewrite the function, replacing the y with $f^{-1}(x)$:

$$(x-7)^3 + 4 = f^{-1}(x)$$

Here are the graphs of $f(x)$ and $f^{-1}(x)$:

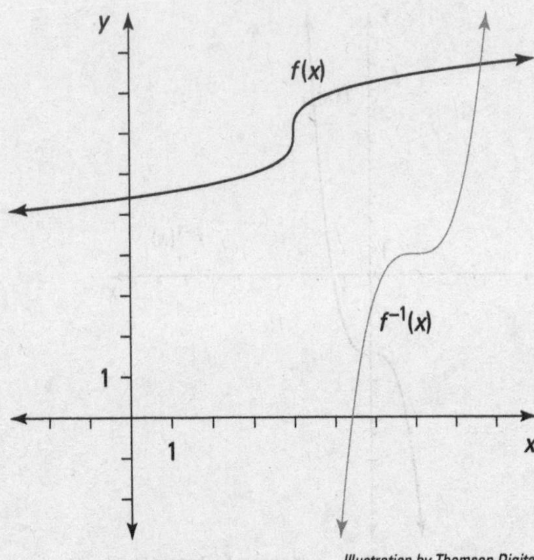

Illustration by Thomson Digital

Note that the graph of $f^{-1}(x)$ is the reflection of $f(x)$ over the line $y=x$. After sketching both the function and its inverse on the same graph, you should see the symmetry (mirror image) over the diagonal $y=x$.

Also note that the functions give you the "opposite operations in the opposite order": In $f(x)$, you subtract 4, then take a root, and then add 7; in $f^{-1}(x)$, you subtract 7, and then cube the result, and then add 4.

187. $f^{-1}(x)=\dfrac{1+2x}{x}$

To find the inverse, first rewrite the function as $y=\dfrac{1}{x-2}$. Then exchange the y and x:

$$x=\frac{1}{y-2}$$

Solve for y:

$$\begin{aligned}(y-2)x &= \frac{1}{\cancel{y-2}}\cdot\frac{\cancel{y-2}}{1}\\ x(y-2) &= 1\\ xy-2x &= 1\\ xy &= 1+2x\\ y &= \frac{1+2x}{x}\end{aligned}$$

Finally, rewrite the function, replacing the y with $f^{-1}(x)$:

$$f^{-1}(x)=\frac{1+2x}{x}$$

Here are the graphs of $f(x)$ and $f^{-1}(x)$:

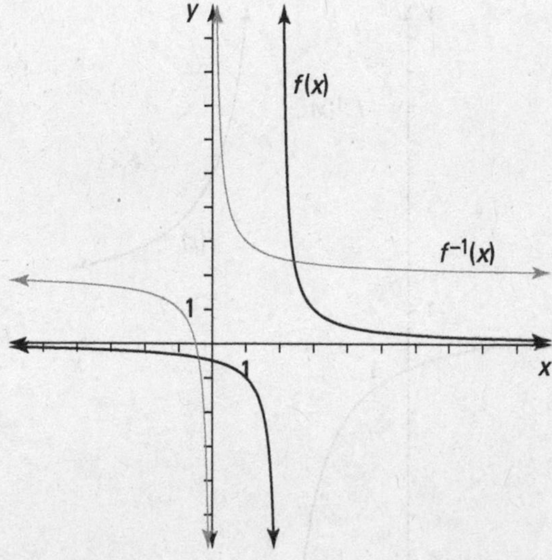

Illustration by Thomson Digital

Note that the graph of $f^{-1}(x)$ is the reflection of $f(x)$ over the line $y=x$. After using the intercepts and asymptotes to help you sketch both the function and its inverse on the same graph, you should see the symmetry (mirror image) over the diagonal $y=x$. When the function and its inverse intersect, they do so at a point on the line $y=x$.

188. $f^{-1}(x)=\dfrac{4x+3}{x-1}$

To find the inverse, first rewrite the function as $y=\dfrac{x+3}{x-4}$. Then exchange the y and x's:

$$x=\frac{y+3}{y-4}$$

Solve for y:

$$(y-4)x=\frac{y+3}{\cancel{y-4}}\cdot\frac{\cancel{y-4}}{1}$$

$$(y-4)x=y+3$$

$$xy-4x=y+3$$

$$xy-y=3+4x$$

$$y(x-1)=3+4x$$

$$y=\frac{3+4x}{x-1}$$

Finally, rewrite the function, replacing the y with $f^{-1}(x)$:

$$f^{-1}(x)=\frac{3+4x}{x-1}$$

Here are the graphs of $f(x)$ and $f^{-1}(x)$:

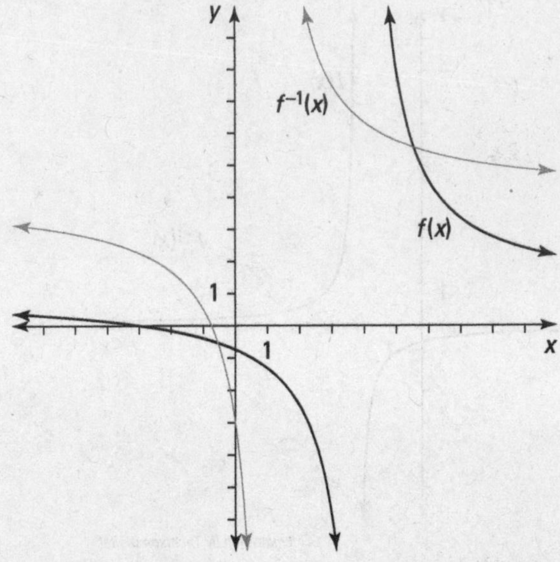

Illustration by Thomson Digital

Note that the graph of $f^{-1}(x)$ is the reflection of $f(x)$ over the line $y=x$. After using the intercepts and asymptotes to help you sketch both the function and its inverse on the same graph, you should see the symmetry (mirror image) over the diagonal $y=x$. When the function and its inverse intersect, they do so at a point on the line $y=x$.

189. $f^{-1}(x)=\frac{8}{x^3}+3$

To find the inverse, first rewrite the function as $y=\frac{2}{\sqrt[3]{x-3}}$. Then exchange the y and x:

$$x=\frac{2}{\sqrt[3]{y-3}}$$

Solve for y, starting by multiplying each side by the denominator on the right:

$$x\sqrt[3]{y-3}=\frac{2}{\cancel{\sqrt[3]{y-3}}}\cdot\frac{\cancel{\sqrt[3]{y-3}}}{1}$$

$$x\sqrt[3]{y-3}=2$$

$$\sqrt[3]{y-3}=\frac{2}{x}$$

$$y-3=\left(\frac{2}{x}\right)^3$$

$$y=\frac{8}{x^3}+3$$

Finally, rewrite the function, replacing the y with $f^{-1}(x)$:

$$f^{-1}(x)=\frac{8}{x^3}+3$$

Here are the graphs of $f(x)$ and $f^{-1}(x)$:

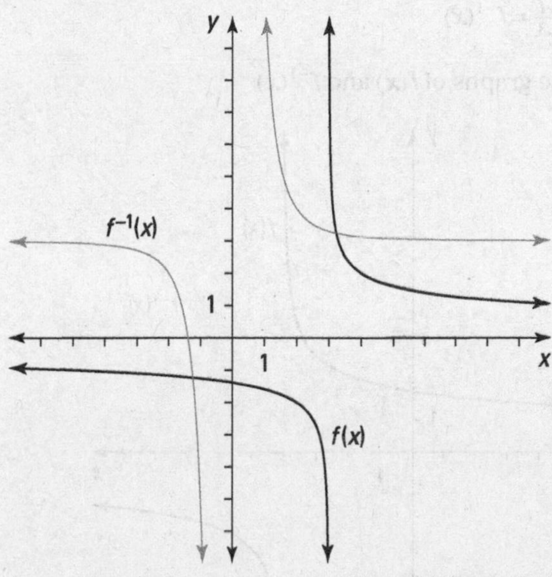

Illustration by Thomson Digital

Note that the graph of $f^{-1}(x)$ is the reflection of $f(x)$ over the line $y=x$. After sketching both the function and its inverse on the same graph, you should see the symmetry (mirror image) over the diagonal $y=x$. When the function and its inverse intersect, they do so at a point on the line $y=x$.

190. $f^{-1}(x)=4-\frac{27}{x^3}$

To find the inverse, first rewrite the function as $y=\frac{3}{\sqrt[3]{4-x}}$. Then exchange the y and x:

$$x=\frac{3}{\sqrt[3]{4-y}}$$

Solve for y, starting by multiplying each side by the denominator on the right:

$$\begin{aligned} x\sqrt[3]{4-y}&=\frac{3}{\cancel{\sqrt[3]{4-y}}}\cdot\frac{\cancel{\sqrt[3]{4-y}}}{1}\\ x\sqrt[3]{4-y}&=3\\ \sqrt[3]{4-y}&=\frac{3}{x}\\ 4-y&=\left(\frac{3}{x}\right)^3\\ 4-\left(\frac{3}{x}\right)^3&=y\\ 4-\frac{27}{x^3}&=y \end{aligned}$$

Finally, rewrite the function, replacing the y with $f^{-1}(x)$:

$$4-\frac{27}{x^3}=f^{-1}(x)$$

Here are the graphs of $f(x)$ and $f^{-1}(x)$:

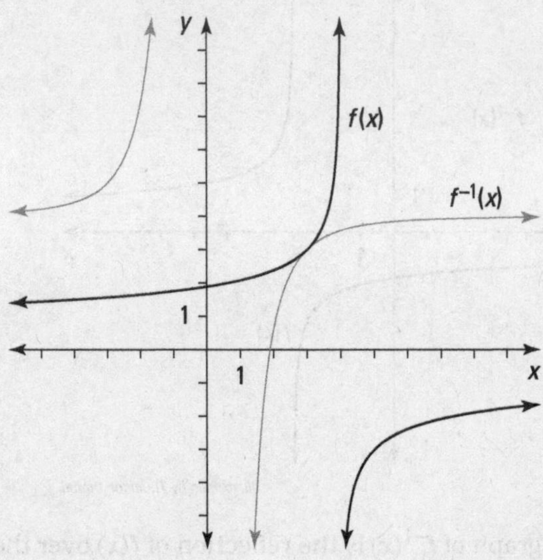

Illustration by Thomson Digital

Note that the graph of $f^{-1}(x)$ is the reflection of $f(x)$ over the line $y=x$. After sketching both the function and its inverse on the same graph, you should see the symmetry (mirror image) over the diagonal $y=x$. When the function and its inverse intersect, they do so at a point on the line $y=x$.

191.

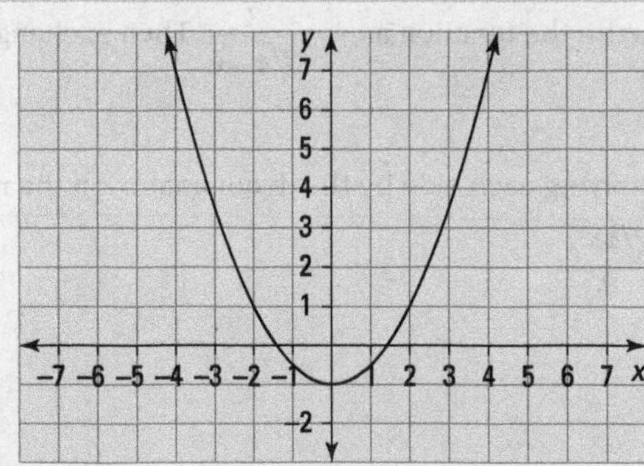

Illustration by Thomson Digital

Rewrite the equation in vertex form, $y=a(x-h)^2+k$:

$$y=\frac{1}{2}(x-0)^2-1$$

The vertex is (h, k), so in this parabola, you see the vertex at (0, –1).

The parabola opens upward if a is positive and downward if a is negative, and if $|a| > 1$, then the curve is steep. Therefore, this parabola opens upward and is flattened rather than steep.

The y-intercept is (0, –1), the vertex. Find the x-intercepts by setting $y = 0$ and solving for x:

$$0=\frac{1}{2}x^2-1$$
$$1=\frac{1}{2}x^2$$
$$2=x^2$$
$$x=\pm\sqrt{2}$$

So the x-intercepts are $\left(\sqrt{2},0\right)$ and $\left(-\sqrt{2},0\right)$.

192.

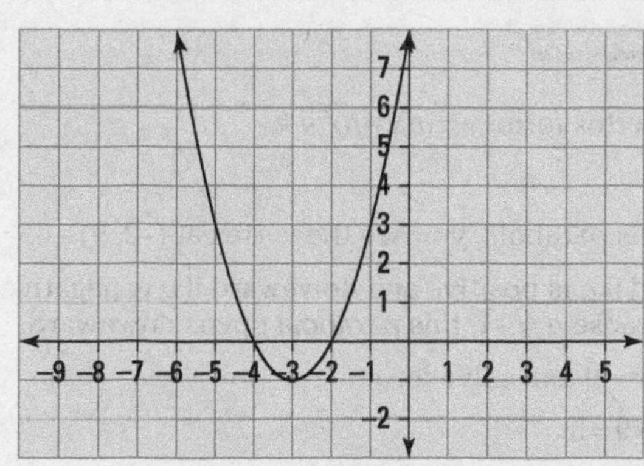

Illustration by Thomson Digital

Rewrite the equation in vertex form, $y=a(x-h)^2+k$:

$$y=\left(x^2+6x+9\right)+8-9=(x+3)^2-1$$

The vertex is (h, k), so in this parabola, you see the vertex at (–3, –1).

The parabola opens upward if a is positive and downward if a is negative, and if $|a| > 1$, then the curve is steep. Because $a = 1$, this parabola opens upward.

To find the y-intercept, let $x = 0$ and solve for y; the y-intercept is (0, 8). Find the x-intercepts by setting $y = 0$ and solving for x:

$$0=x^2+6x+8$$
$$0=(x+2)(x+4)$$
$$x=-2,-4$$

Therefore, the x-intercepts are (–2, 0) and (–4, 0).

193.

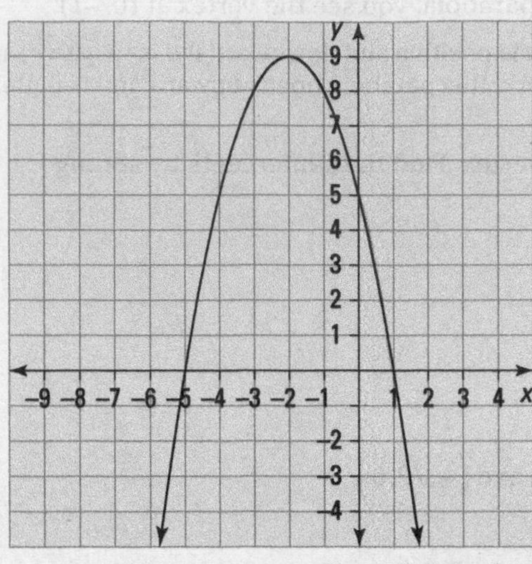

Illustration by Thomson Digital

Start with the equation in vertex form, $y=a(x-h)^2+k$:

$$y=-(x+2)^2+9$$

The vertex is (h, k), so in this parabola, you see the vertex at (–2, 9).

The parabola opens upward if a is positive and downward if a is negative, and if $|a|>1$, then the curve is steep. Because $a=-1$, this parabola opens downward.

To find the y-intercept, let $x=0$ and solve for y:

$$y=-(0+2)^2+9=-4+9=5$$

So the y-intercept is (0, 5). Find the x-intercepts by setting $y=0$ and solving for x:

$$0=-(x+2)^2+9$$
$$0=-\left(x^2+4x+4\right)+9$$
$$0=-x^2-4x-4+9$$
$$0=-x^2-4x+5$$
$$0=-\left(x^2+4x-5\right)$$
$$0=-(x+5)(x-1)$$
$$x=-5, 1$$

Therefore, the x-intercepts are (–5, 0) and (1, 0).

194.

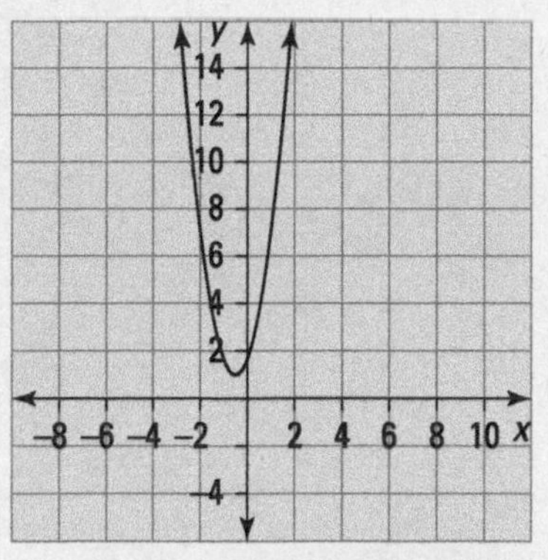

Illustration by Thomson Digital

Start with the equation in vertex form, $y = a(x-h)^2 + k$:

$$y = 4\left(x + \frac{1}{2}\right)^2 + 1$$

The vertex is (h, k), so in this parabola, you see the vertex at $\left(-\frac{1}{2}, 1\right)$.

The parabola opens upward if a is positive and downward if a is negative, and if $|a| > 1$, then the curve is steep. Because $a = 4$, this parabola is steep and opens upward.

To find the y-intercept, let $x = 0$ and solve for y:

$$y = 4\left(0 + \frac{1}{2}\right)^2 + 1 = 4\left(\frac{1}{4}\right) + 1 = 2$$

So the y-intercept is (0, 2). Find the x-intercepts by setting $y = 0$ and solving for x:

$$0 = 4\left(x + \frac{1}{2}\right)^2 + 1$$

$$0 = 4\left(x^2 + x + \frac{1}{4}\right) + 1$$

$$0 = 4x^2 + 4x + 1 + 1$$

$$0 = 4x^2 + 4x + 2$$

Use the quadratic formula to solve for x:

$$x = \frac{-4 \pm \sqrt{4^2 - 4(4)(2)}}{2(4)} = \frac{-4 \pm \sqrt{16 - 32}}{8} = \frac{-4 \pm \sqrt{-16}}{8}$$

There are no real roots, so there are no x-intercepts.

195.

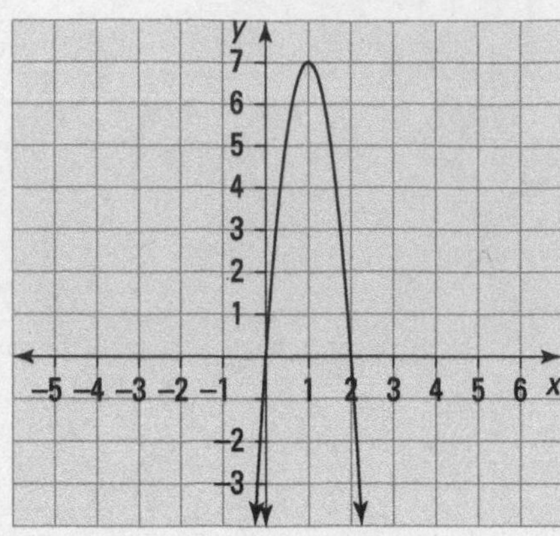

Illustration by Thomson Digital

Rewrite the equation in vertex form, $y = a(x-h)^2 + k$:

$$y = -7\left(x^2 - 2x\right) = -7\left(x^2 - 2x + 1\right) + 7 = -7(x-1)^2 + 7$$

The vertex is (h, k), so in this parabola, you see the vertex at (1, 7).

The parabola opens upward if a is positive and downward if a is negative, and if $|a| > 1$, then the curve is steep. Because $a = -7$, this parabola is steep and opens downward.

To find the y-intercept, let $x = 0$ and solve for y:

$$y = -7\left(0^2\right) + 14(0) = 0$$

So the y-intercept is (0, 0). Find the x-intercepts by setting $y = 0$ and solving for x:

$$\begin{aligned} 0 &= -7x^2 + 14x \\ 0 &= -7x(x-2) \\ x &= 0, 2 \end{aligned}$$

Therefore, the x-intercepts are (0, 0) and (2, 0).

196. $y = -x^2 + 9$

The vertex is (0, 9), and the graph opens downward. Using the vertex form of a quadratic equation, $y = a(x-h)^2 + k$, these characteristics are represented by $y = a(x-0)^2 + 9$, where a is a negative number.

The x-intercepts are (–3, 0) and (3, 0). Substitute (3, 0) into the equation and solve for a:

$$\begin{aligned} 0 &= a(3-0)^2 + 9 \\ -9 &= a(9) \\ a &= -1 \end{aligned}$$

So the equation of the parabola is $y = -1(x-0)^2 + 9$.

197. $y = (x+1)^2 - 4$

The vertex is (–1, –4), and the graph opens upward. Using the vertex form of a quadratic equation, $y = a(x-h)^2 + k$, these characteristics are represented by $y = a(x+1)^2 - 4$, where a is a positive number.

The x-intercepts are (–3, 0) and (1, 0). Choosing one of the intercepts, (1, 0), substitute the values into the equation and solve for a:

$$0=a(1+1)^2-4$$
$$4=a(4)$$
$$a=1$$

So the equation of the parabola is $y=(x+1)^2-4$.

198. $y=-2(x+3)^2+8$

The vertex is (–3, 8), and the graph opens downward. Using the vertex form of a quadratic equation, $y=a(x-h)^2+k$, these characteristics are represented by $y=a(x+3)^2+8$, where a is a negative number.

The x-intercepts are (–5, 0) and (–1, 0). Substitute (–1, 0) into the equation and solve for a:

$$0=a(-1+3)^2+8$$
$$-8=a(4)$$
$$a=-2$$

So the equation of the parabola is $y=-2(x+3)^2+8$.

199. $y=-\frac{3}{2}(x-4)^2+\frac{27}{2}$

It's hard to tell exactly where the vertex is, because it isn't at an intersection on the grid. But the x-intercepts are easy to determine; they're at (1, 0) and (7, 0). Also, the curve goes through the point (3, 12).

An equation representing the intercepts is $y=a(x-1)(x-7)$, where a is a negative number. Substituting 3 for x and 12 for y from the point on the curve, you can solve for a:

$$12=a(3-1)(3-7)$$
$$12=a(2)(-4)$$
$$12=a(-8)$$
$$a=-\frac{12}{8}=-\frac{3}{2}$$

Now replace the a in the equation that uses the intercepts:

$$y=-\frac{3}{2}(x-1)(x-7)$$

To put this equation in vertex form, first multiply the 7 in the parentheses by $-\frac{3}{2}$ and move the product outside the parentheses. Then complete the square using the two remaining terms in the parentheses.

Recall that to complete the square of a quadratic ax^2+bx, add and subtract the square of half b to the variable terms $ax^2+bx+\frac{b^2}{4}-\frac{b^2}{4}$. The first three terms should be a perfect square trinomial. Multiply the last term by the $-\frac{3}{2}$ and move it outside the parentheses.

Here are the calculations:

$$y=-\frac{3}{2}\left(x^2-8x+7\right)$$

$$y=-\frac{3}{2}\left(x^2-8x\right)+\left(-\frac{3}{2}\cdot 7\right)=-\frac{3}{2}\left(x^2-8x\right)-\frac{21}{2}$$

$$y=-\frac{3}{2}\left(x^2-8x+16-16\right)-\frac{21}{2}$$

$$y=-\frac{3}{2}\left(x^2-8x+16\right)-\frac{21}{2}+\left(-\frac{3}{2}\cdot(-16)\right)$$

$$y=-\frac{3}{2}\left(x^2-8x+16\right)-\frac{21}{2}+\frac{48}{2}$$

$$y=-\frac{3}{2}(x-4)^2+\frac{27}{2}$$

This equation, which is in vertex form, tells you that the vertex is $\left(4, \frac{27}{2}\right)$.

200. $y=\frac{1}{8}(x-1)^2-\frac{49}{8}$

It's hard to tell exactly where the vertex is, because it isn't at an intersection on the grid. But the *x*-intercepts are easy to determine; they're at (–6, 0) and (8, 0). Also, the curve has a *y*-intercept of (0, –6).

An equation representing the *x*-intercepts is $y=a(x+6)(x-8)$, where *a* is a positive number. Substituting 0 for *x* and –6 for *y* from the point on the curve, you can solve for *a:*

$$-6=a(0+6)(0-8)$$

$$-6=a(6)(-8)$$

$$-6=a(-48)$$

$$a=\frac{6}{48}=\frac{1}{8}$$

Now replace the *a* in the equation using the intercepts:

$$y=\frac{1}{8}(x+6)(x-8)$$

To put this equation in the standard form, multiply the binomials, and complete the square.

Recall that to complete the square of a quadratic ax^2+bx, add and subtract the square of half *b* to the variable terms $ax^2+bx+\frac{b^2}{4}-\frac{b^2}{4}$. The first three terms should be a perfect square trinomial. Multiply the last term by $\frac{1}{8}$ and move it outside the parentheses. Here are the calculations:

$$y=\frac{1}{8}\left(x^2-2x-48\right)$$

$$y=\frac{1}{8}\left(x^2-2x\right)+\left(\frac{1}{8}\cdot(-48)\right)=\frac{1}{8}\left(x^2-2x\right)-6$$

$$y=\frac{1}{8}\left(x^2-2x+1-1\right)-6$$

$$y=\frac{1}{8}(x-1)^2-6+\frac{1}{8}(-1)$$

$$y=\frac{1}{8}(x-1)^2-\frac{49}{8}$$

This equation, which is in vertex form, tells you that the vertex is $\left(1, -\frac{49}{8}\right)$.

201. domain: $[-4, \infty)$; intercepts: $(-4, 0)$, $(0, 2)$

The domain begins at $x = -4$ with the x-intercept $(-4, 0)$ and the curve rises as x goes to positive infinity. The y-intercept is $(0, 2)$.

The graph of $f(x) = \sqrt{x+4}$ looks like half of a parabola:

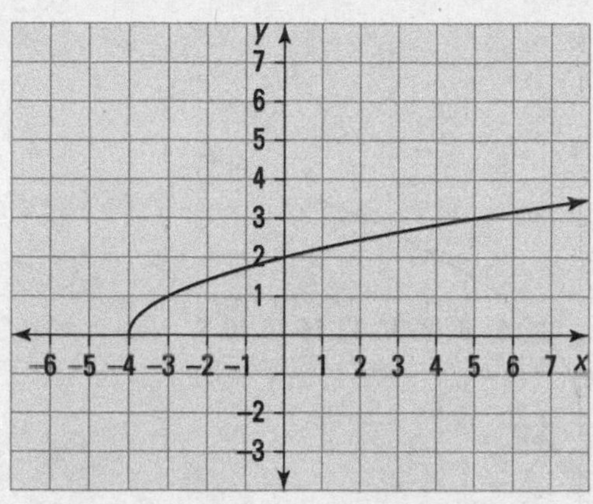

Illustration by Thomson Digital

202. domain: $\left[\frac{3}{2}, \infty\right)$; intercept: $(2, 0)$

The domain begins at $\frac{3}{2}$ with the point $\left(\frac{3}{2}, 1\right)$ and the curve falls as x goes toward positive infinity. The function has an x-intercept at $(2, 0)$ and no y-intercept.

The graph of $f(x) = -\sqrt{2x-3}+1$ looks like half of a parabola:

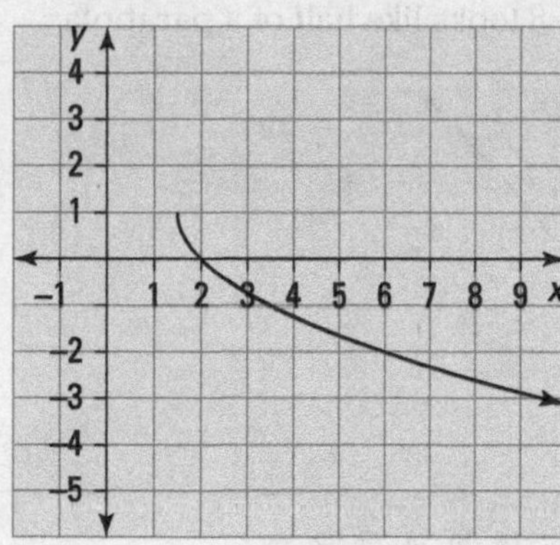

Illustration by Thomson Digital

203. domain: $[0, \infty)$; intercepts: $(0, -6)$, $(4, 0)$

The domain begins at 0 with $(0, -6)$, which is the *y*-intercept, and the curve rises as *x* goes to positive infinity. The *x*-intercept is $(4, 0)$.

The graph of $f(x) = 3\sqrt{x} - 6$ looks like half of a parabola:

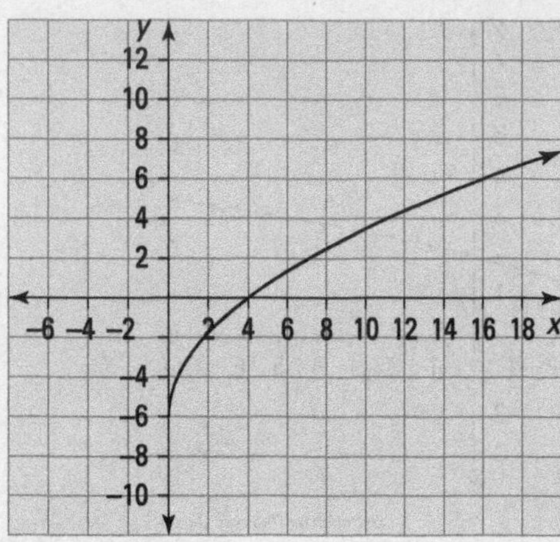

Illustration by Thomson Digital

204. domain: $[0, \infty)$; intercepts: $\left(\frac{4}{5}, 0\right)$, $(0, -8)$

The domain begins at 0 with $(0, -8)$, which is the *y*-intercept, and the curve rises as *x* goes to positive infinity. The *x*-intercept is $\left(\frac{4}{5}, 0\right)$.

The graph of $f(x) = 4\sqrt{5x} - 8$ looks like half of a parabola:

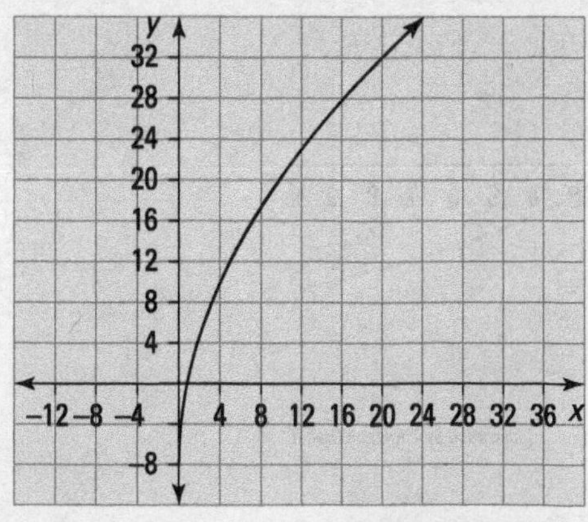

Illustration by Thomson Digital

205. domain: $\left[\frac{7}{3},\infty\right)$; intercept: $\left(\frac{71}{3},0\right)$

The domain begins at $\frac{7}{3}$ with $\left(\frac{7}{3},4\right)$ and the curve falls as x goes to positive infinity. There is no y-intercept. The x-intercept is $\left(\frac{71}{3},0\right)$. The graph looks like half a parabola.

The graph of $f(x)=-\frac{1}{2}\sqrt{3x-7}+4$ looks like half of a parabola:

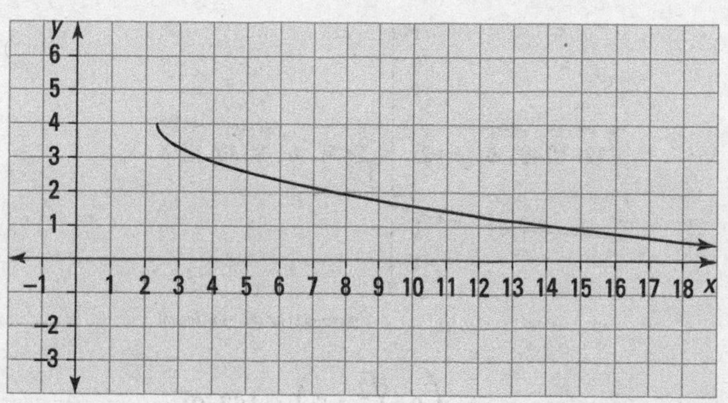

Illustration by Thomson Digital

206. domain: $(-\infty,\infty)$; intercepts: $(-3, 0)$, $\left(0,-\sqrt[3]{3}\right)$

The domain is all real numbers. The curve falls moving from left to right. The x-intercept is $(-3, 0)$, and the y-intercept is $\left(0,-\sqrt[3]{3}\right)$.

The graph of $f(x)=-\sqrt[3]{x+3}$ looks like two halves of a parabola, with one half rotated around 180 degrees from the other:

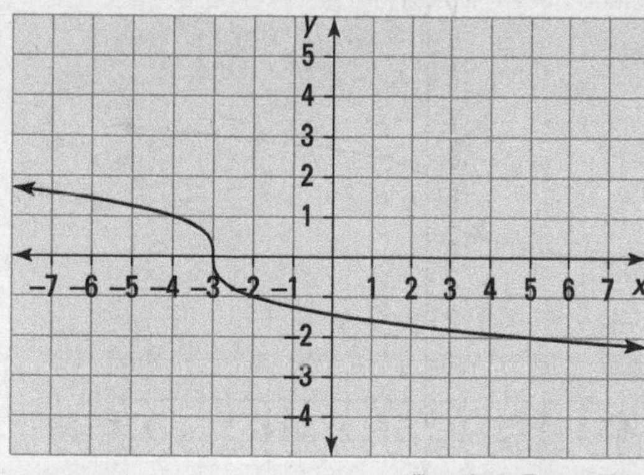

Illustration by Thomson Digital

207. domain: $(-\infty, \infty)$; intercepts: $\left(\frac{31}{3}, 0\right), \left(0, \sqrt[3]{-4}-3\right)$

The domain is all real numbers. The curve rises moving from left to right. The x-intercept is $\left(\frac{31}{3}, 0\right)$, and the y-intercept is $\left(0, \sqrt[3]{-4}-3\right)$ which is about (0, –4.6).

The graph of $f(x)=\sqrt[3]{3x-4}-3$ looks like two halves of a parabola, with one half rotated around 180 degrees from the other:

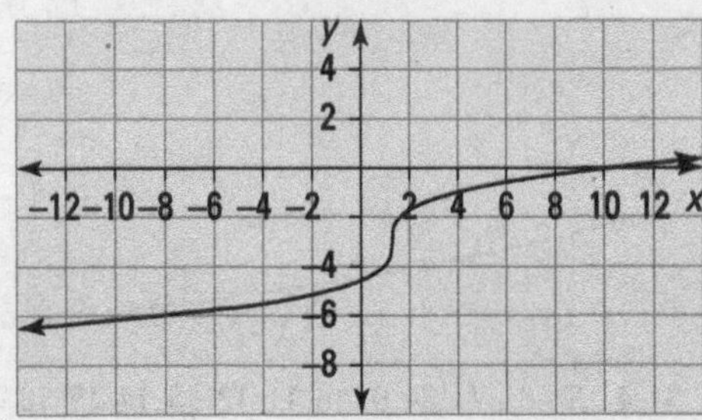

Illustration by Thomson Digital

208. domain: $(-\infty, \infty)$; intercepts: $\left(0, \frac{\sqrt[3]{2}}{2}+5\right)$, (–167, 0)

The domain is all real numbers. The curve rises moving from left to right. The y-intercept is $\left(0, \frac{\sqrt[3]{2}}{2}+5\right)$, and the x-intercept is (–167, 0).

The graph of $f(x)=\frac{1}{2}\sqrt[3]{6x+2}+5$ looks like two halves of a parabola, with one half rotated around 180 degrees from the other:

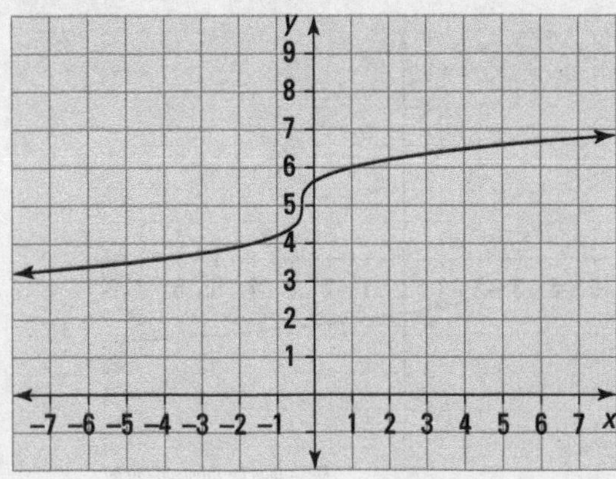

Illustration by Thomson Digital

209. domain: $(-\infty, \infty)$; intercepts: $(0, -7)$, $\left(\frac{343}{27}, 0\right)$

The domain is all real numbers. The curve rises moving from left to right. The y-intercept is $(0, -7)$, and the x-intercept is $\left(\frac{343}{27}, 0\right)$.

The graph of $f(x) = 3\sqrt[3]{x} - 7$ looks like two halves of a parabola, with one half rotated around 180 degrees from the other:

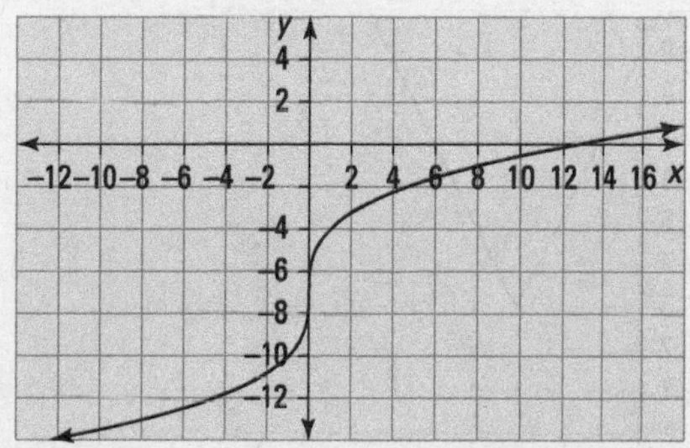

Illustration by Thomson Digital

210. domain: $(-\infty, \infty)$; intercepts: $\left(0, 2\sqrt[3]{2} + 1\right)$, $\left(\frac{17}{8}, 0\right)$

The domain is all real numbers. The curve falls moving from left to right. The y-intercept is $\left(0, 2\sqrt[3]{2} + 1\right)$, and the x-intercept is $\left(\frac{17}{8}, 0\right)$.

The graph of $f(x) = -2\sqrt[3]{x - 2} + 1$ looks like two halves of a parabola, with one half rotated around 180 degrees from the other:

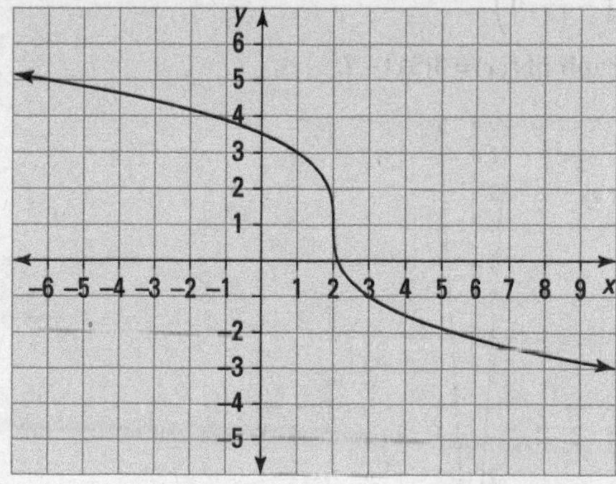

Illustration by Thomson Digital

211. range: $[3, \infty)$

The domain of the function is all real numbers, and the range is from 3 to positive infinity (including 3). The graph has the characteristic *V* shape with the lowest point at (3, 3). There is no *x*-intercept; the *y*-intercept is (0, 9).

Here's the graph of $f(x) = |2x - 6| + 3$:

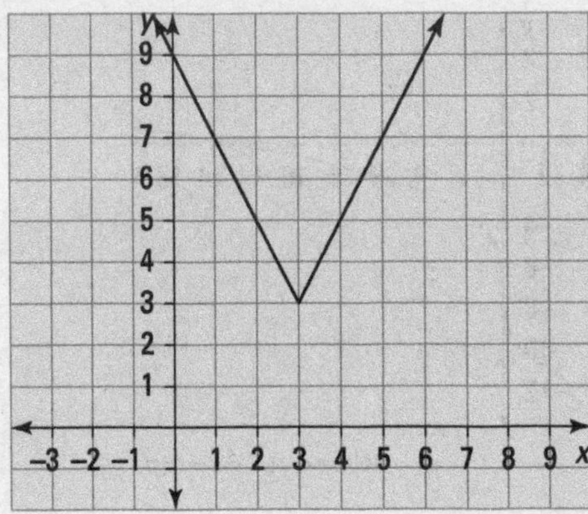

Illustration by Thomson Digital

212. range: $[-7, \infty)$

The domain of the function is all real numbers, and the range is from −7 to positive infinity (including −7). The graph has the characteristic *V* shape, made more steep with the multiplier of 3; the lowest point is (0, −7), the *y*-intercept. The *x*-intercepts are $\left(\frac{7}{15}, 0\right)$ and $\left(-\frac{7}{15}, 0\right)$.

Here's the graph of $f(x) = 3|5x| - 7$:

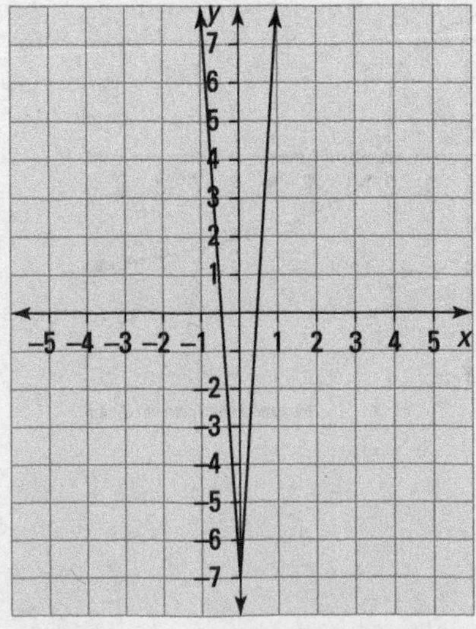

Illustration by Thomson Digital

213. range: $[0, \infty)$

The domain is all real numbers, and the range goes from 0 to positive infinity. You can draw the curve starting with the graph of the parabola $y = x^2 - 4$ with the portion between $x = -2$ and $x = 2$ reflected over the x-axis. The y-intercept is (0, 4), and the x-intercepts are (–2, 0) and (2, 0).

Here's the graph of $f(x) = |x^2 - 4|$:

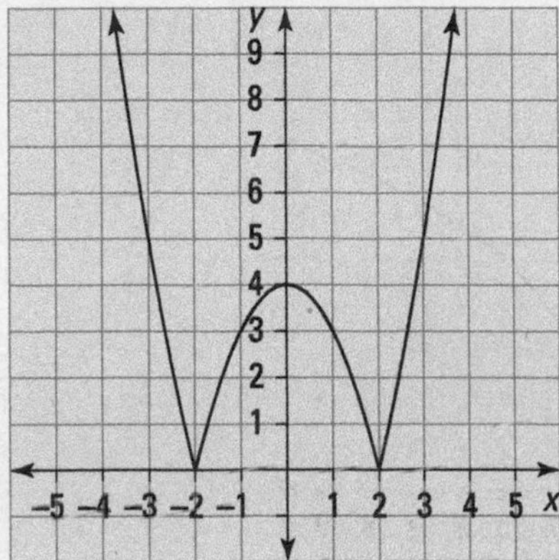

Illustration by Thomson Digital

214. range: $[0, \infty)$

The domain is all real numbers, and the range goes from 0 to positive infinity. You can draw the curve starting with the graph of the cubic $y = x^3 + 1$ with the portion from $-\infty$ to –1 reflected over the x-axis. The y-intercept is (0, 1), and the x-intercept is (–1, 0).

Here's the graph of $f(x) = |x^3 + 1|$:

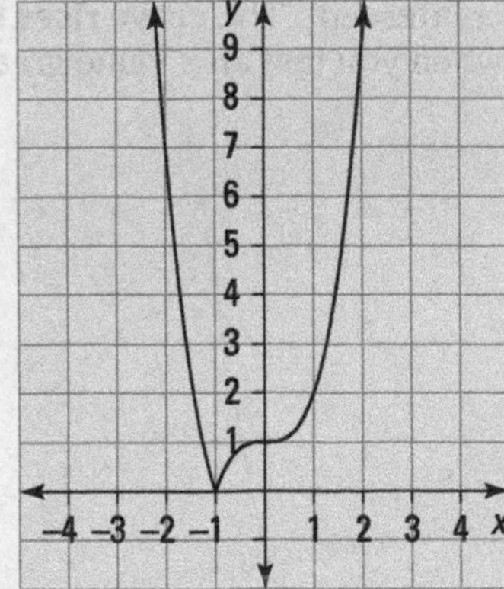

Illustration by Thomson Digital

215. range: $[0, \infty)$

The domain is all real numbers, and the range goes from 0 to positive infinity. You can draw the curve starting with the graph of the cube root $y = 4\sqrt[3]{x-3}$ with the portion from $-\infty$ to 3 reflected over the x-axis. The curve comes down to the point (3, 0) with a vertical tangent. The y-intercept is $\left(0, 4\sqrt[3]{3}\right)$, and the x-intercept is (3, 0).

Here's the graph of $f(x) = \left|4\sqrt[3]{x-3}\right|$:

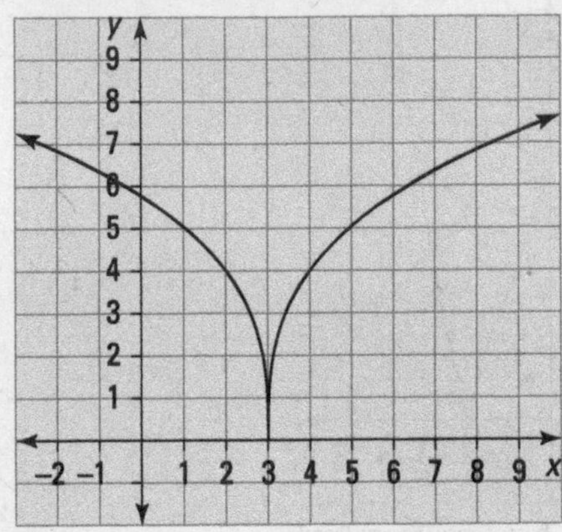

Illustration by Thomson Digital

216. intercepts: (0, 0), (4, 0), (–5, 0)

Find the x-intercepts by letting $y = 0$ and solving for x. The x-intercepts of $y = x(x-4)(x+5)$ are (0, 0), (4, 0), and (–5, 0).

Find the y-intercept by letting $x = 0$ and solving for y. The y-intercept is (0, 0).

To sketch the graph, note that the exponents on the factors are odd numbers, so the curve crosses the x-axis at each x-intercept. The curve rises to the right as x approaches positive infinity, as determined when you test an x value greater than the right-most intercept. Here's the graph:

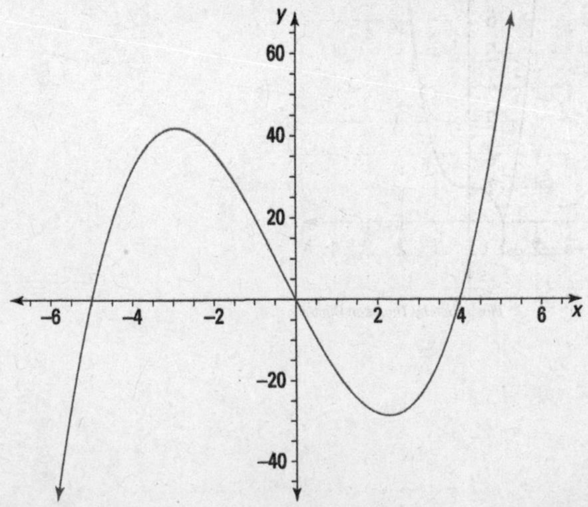

Illustration by Thomson Digital

217. intercepts: (3,0), (–2,0), (0, 36)

Find the x-intercepts by letting $y = 0$ and solving for x. The x-intercepts of $y = (x-3)^2(x+2)^2$ are (3, 0) and (–2, 0).

Find the y-intercept by letting $x = 0$ and solving for y. The y-intercept is (0, 36).

To sketch the graph, note that the exponents on the factors are even numbers, so the curve just touches the x-axis at each x-intercept. The curve rises to the right as x approaches positive infinity, as determined when you test an x value greater than the right-most intercept. Here's the graph:

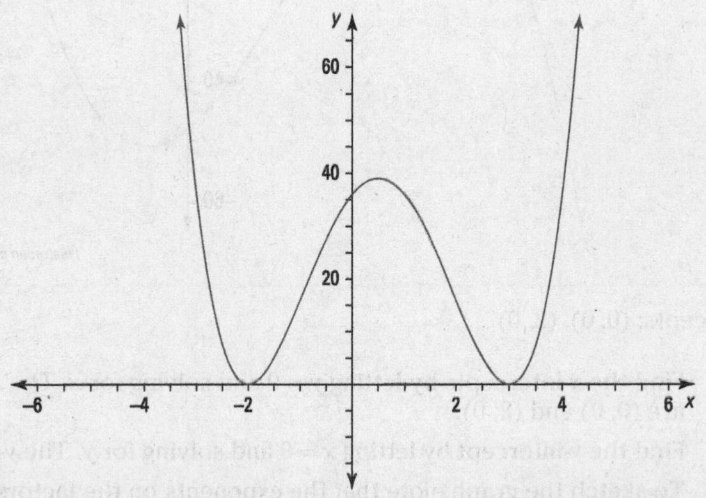

Illustration by Thomson Digital

218. intercepts: (–4, 0), (3, 0), (0, –48)

Find the x-intercepts by letting $y = 0$ and solving for x. The x-intercepts of $y = (x+4)^2(x-3)$ are (–4, 0) and (3, 0).

Find the y-intercept by letting $x = 0$ and solving for y. The y-intercept is (0, –48).

To sketch the graph, note that the curve crosses the x-axis at the x-intercept (3, 0) because the exponent on $(x-3)$ is odd, and the curve just touches the x-axis at (–4, 0) because the exponent on $(x+4)$ is even. The curve rises to the right as x approaches positive infinity, as determined when you test an x value greater than the right-most intercept. Here's the graph:

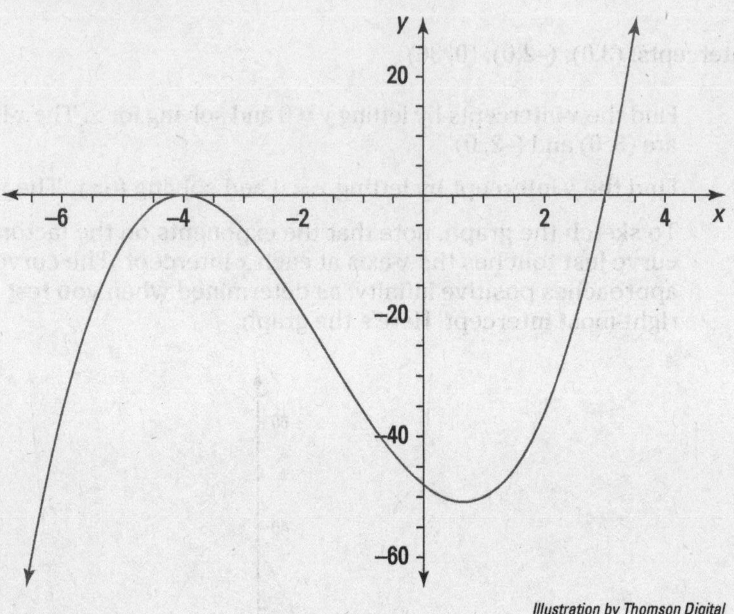

Illustration by Thomson Digital

219. intercepts: (0, 0), (2, 0)

Find the x-intercepts by letting $y = 0$ and solving for x. The x-intercepts of $y = x^3(x-2)^3$ are (0, 0) and (2, 0).

Find the y-intercept by letting $x = 0$ and solving for y. The y-intercept is (0, 0).

To sketch the graph, note that the exponents on the factors are odd numbers, so the curve crosses the x-axis at both x-intercepts. The curve also flattens out at each x-intercept, with a low point between the intercepts at $x = 1$. The curve rises to the right as x approaches positive infinity, as determined when you test an x value greater than the right-most intercept. Here's the graph:

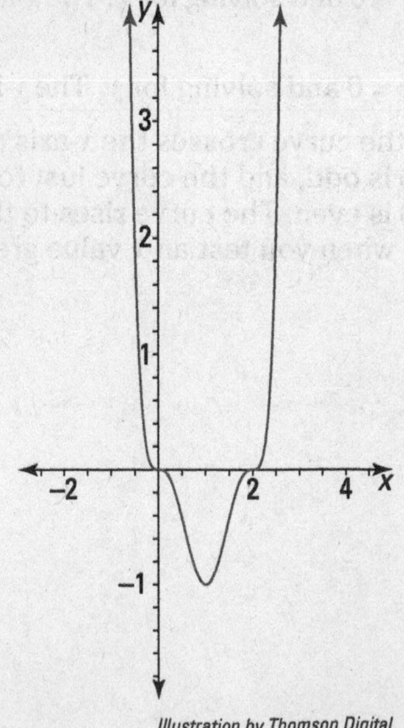

Illustration by Thomson Digital

220. intercepts: (–5, 0), (–2, 0), (2, 0), (5, 0), (0, 100)

Find the x-intercepts by letting $y = 0$ and solving for x. The x-intercepts of $y = (x^2 - 4)(x^2 - 25)$ are (–2, 0), (2, 0), (–5, 0), and (5, 0).

Find the y-intercept by letting $x = 0$ and solving for y. The y-intercept is (0, 100).

To sketch the graph, note that the exponent on each factor is odd, so the curve crosses the x-axis at each intercept. The function is symmetric about the y-axis, because replacing x with $-x$ doesn't change the equation. The curve rises to the right as x approaches positive infinity, as determined when you test an x value greater than the right-most intercept. Here's the graph:

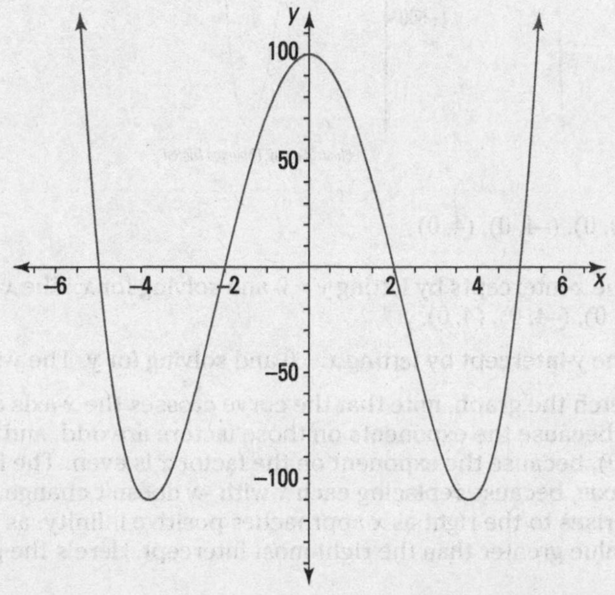

Illustration by Thomson Digital

221. intercepts: (0, 0), (–3, 0), (3, 0), (–6, 0), (6, 0)

Find the x-intercepts by letting $y = 0$ and solving for x. The x-intercepts of $y = x(x^2 - 9)(x^2 - 36)$ are (0, 0), (–3, 0), (3, 0), (–6, 0), and (6, 0).

Find the y-intercept by letting $x = 0$ and solving for y. The y-intercept is (0, 0).

To sketch the graph, note that the exponents on the factors are all odd, so the curve crosses the x-axis at each intercept. The function is symmetric about the origin, because replacing each x with $-x$ changes $f(x)$ to $-f(x)$. The curve rises to the right as x approaches positive infinity, as determined when you test an x value greater than the right-most intercept. Here's the graph:

Answers 201–300

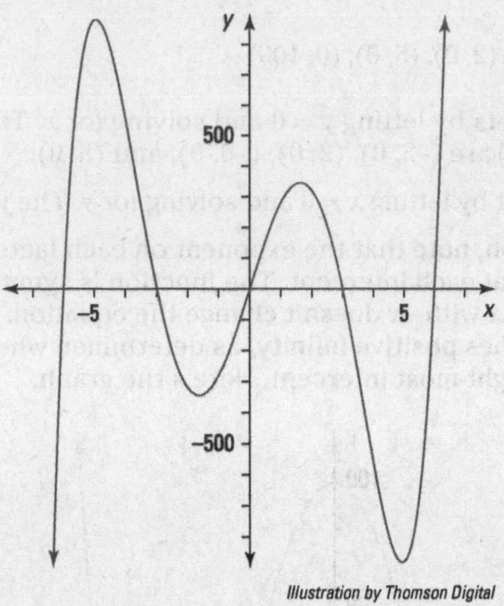

Illustration by Thomson Digital

222. intercepts: (0, 0), (–4, 0), (4, 0)

Find the x-intercepts by letting $y = 0$ and solving for x. The x-intercepts of $y = x^2(x^2 - 16)$ are (0, 0), (–4, 0), (4, 0).

Find the y-intercept by letting $x = 0$ and solving for y. The y-intercept is (0, 0).

To sketch the graph, note that the curve crosses the x-axis at the intercepts (–4, 0) and (4, 0), because the exponents on those factors are odd, and the curve touches the axis at (0, 0), because the exponent on the factor x is even. The function is symmetric about the y-axis, because replacing each x with $-x$ doesn't change the function equation. The curve rises to the right as x approaches positive infinity, as determined when you test an x-value greater than the right-most intercept. Here's the graph:

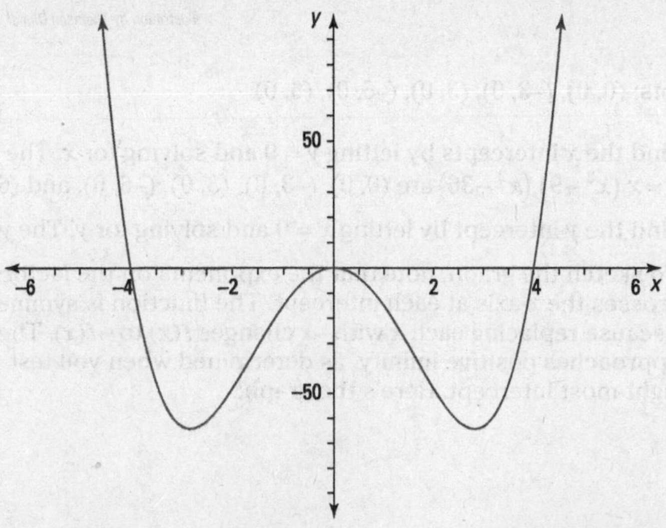

Illustration by Thomson Digital

223. intercepts: (–5, 0), (5, 0), (–1, 0), (0, –25)

Find the *x*-intercepts by letting $y = 0$ and solving for *x*. The *x*-intercepts of $y = (x^2 - 25)(x+1)^2$ are (–5, 0), (5, 0), and (–1, 0).

Find the *y*-intercept by letting $x = 0$ and solving for *y*. The *y*-intercept is (0, –25).

To sketch the graph, note that the curve crosses the *x*-axis at the intercepts (–5, 0) and (5, 0), because the exponents on those factors are odd, and the curve touches the axis at (–1, 0), because the exponent on that factor is even. The curve rises to the right as *x* approaches positive infinity. Here's the graph:

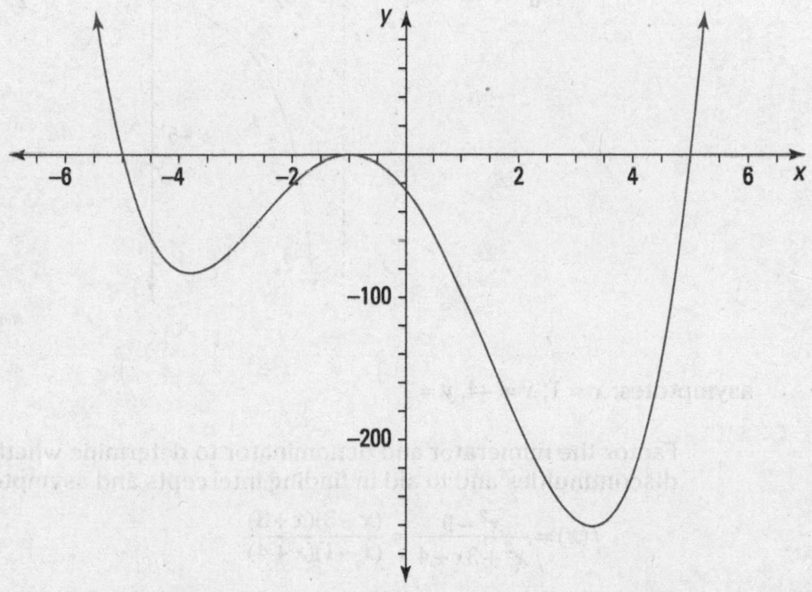

Illustration by Thomson Digital

224. asymptotes: $x = -3$, $y = 1$

Find the vertical asymptote of $y = \frac{x-2}{x+3}$ by setting the denominator equal to 0 and solving for *x*. The vertical asymptote is $x = -3$.

The horizontal asymptote is $y = 1$, because the highest power in both the numerator and denominator is 1, and the ratio formed by the two variables raised to the first power is $\frac{1}{1} = 1$.

For the graph, you also need to identify the intercepts. Find the *x*-intercepts by letting $y = 0$ and solving for *x*. The *x*-intercept of $y = \frac{x-2}{x+3x}$ is (2, 0). Find the *y*-intercept by letting $x = 0$ and solving for *y*. The *y*-intercept is $\left(0, -\frac{2}{3}\right)$. Here's the graph:

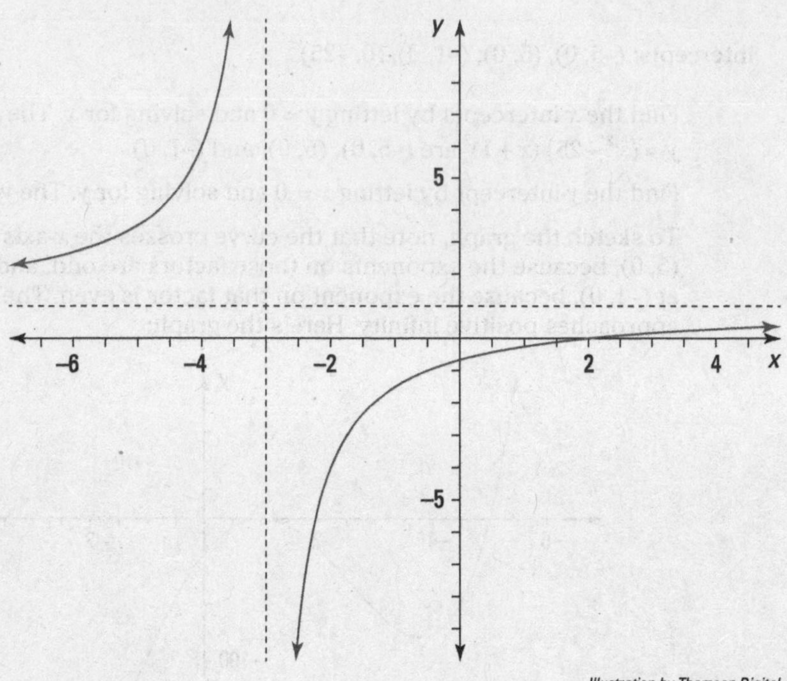

Illustration by Thomson Digital

225. asymptotes: $x = 1, x = -4, y = 1$

Factor the numerator and denominator to determine whether there are any removable discontinuities and to aid in finding intercepts and asymptotes:

$$f(x) = \frac{x^2 - 9}{x^2 + 3x - 4} = \frac{(x-3)(x+3)}{(x-1)(x+4)}$$

No common factors appear in the numerator and denominator, so there's no removable discontinuity.

Find the vertical asymptotes by setting the denominator equal to 0 and solving for x. The vertical asymptotes are $x = 1$ and $x = -4$.

The horizontal asymptote is $y = 1$, because the highest power in both the numerator and denominator is 2, and the ratio formed by the two variables raised to the second power is $\frac{1}{1} = 1$.

For the graph, you also need to identify the intercepts. Find the x-intercepts by letting $y = 0$ and solving for x. The x-intercepts are $(-3, 0)$ and $(3, 0)$. Find the y-intercept by letting $x = 0$ and solving for y. The y-intercept is $\left(0, \frac{9}{4}\right)$. Here's the graph:

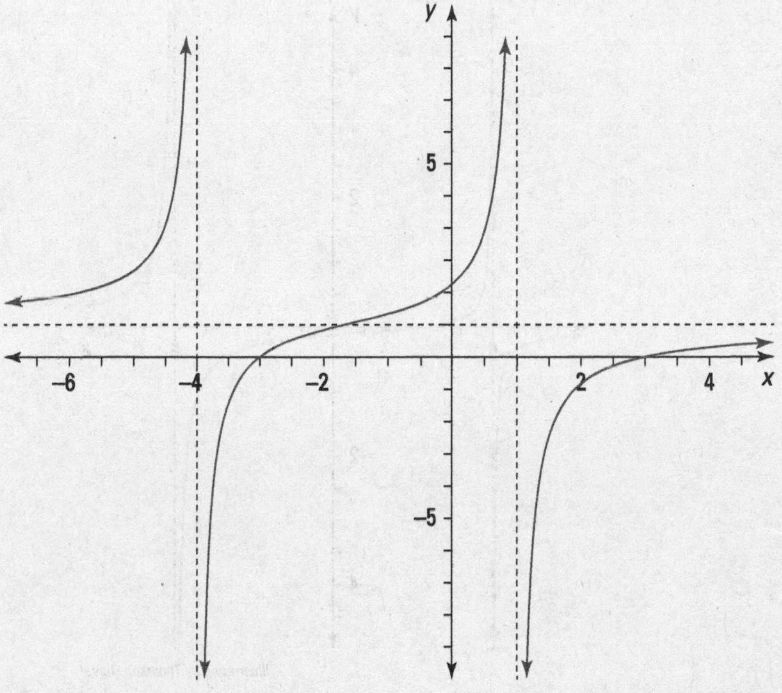

Illustration by Thomson Digital

226. asymptotes: $x = 5$, $x = -5$, $y = 0$

Factor the denominator to determine whether there are any removable discontinuities and to aid in finding intercepts and asymptotes:

$$f(x) = \frac{x+3}{x^2-25} = \frac{x+3}{(x-5)(x+5)}$$

No common factors appear in the numerator and denominator, so there's no removable discontinuity.

Find the vertical asymptotes by setting the denominator equal to 0 and solving for x. The vertical asymptotes are $x = 5$ and $x = -5$.

The horizontal asymptote is $y = 0$, because the highest power in the numerator is smaller than the highest power in the denominator.

For the graph, you also need to identify the intercepts. Find the x-intercepts by letting $y = 0$ and solving for x. The x-intercept is $(-3, 0)$. Find the y-intercept by letting $x = 0$ and solving for y. The y-intercept is $\left(0, -\frac{3}{25}\right)$. Here's the graph:

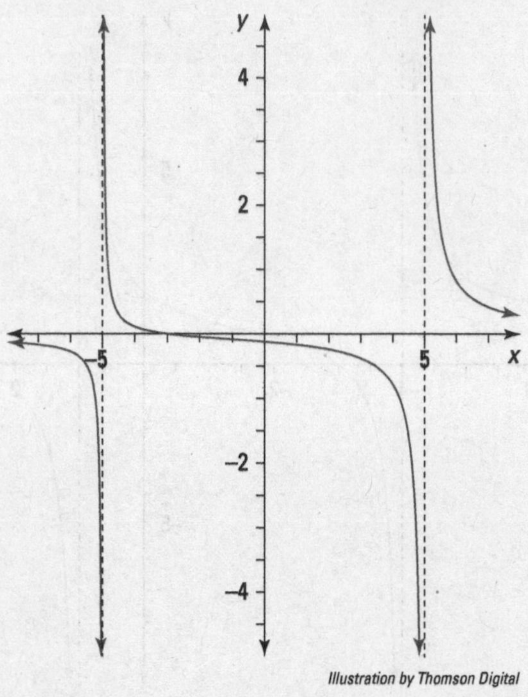

Illustration by Thomson Digital

227. asymptotes: $x = -5$, $y = 1$

Factor the numerator and the denominator to determine whether there are any removable discontinuities and to aid in finding intercepts and asymptotes:

$$f(x) = \frac{x^2 - 5x}{x^2 - 25} = \frac{x\cancel{(x-5)}}{\cancel{(x-5)}(x+5)}$$

The common factor of $(x - 5)$ in the numerator and denominator produces a removable discontinuity at $x = 5$.

After factoring out the common factor, find the asymptotes. The vertical asymptote is $x = -5$; you find that asymptote by setting the denominator equal to 0 and solving for x.

The horizontal asymptote is $y = 1$, because the highest power in both the numerator and denominator of the reduced fraction is 1, and the ratio formed by the two variables raised to the first power is $\frac{1}{1} = 1$.

For the graph, you also need to identify the intercepts. Find the x-intercepts by letting $y = 0$ and solving for x. The x-intercept is (0, 0). Find the y-intercept by letting $x = 0$ in the factored version and solving for y. The y-intercept is also (0, 0). Here's the graph:

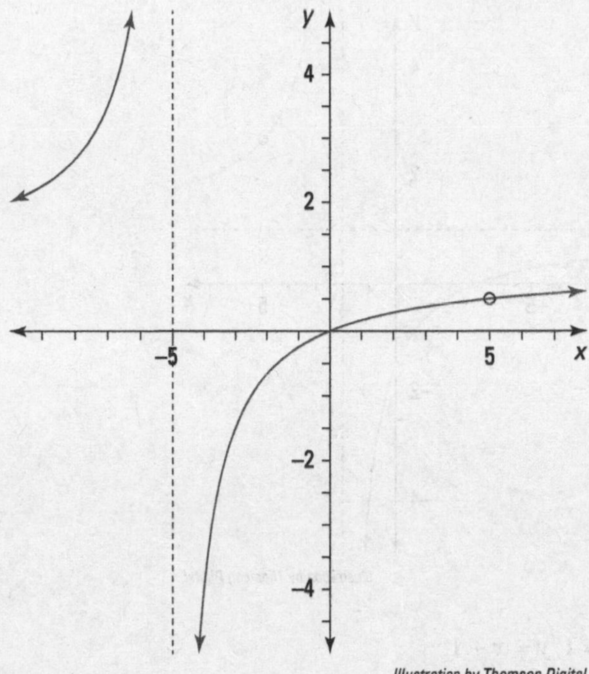

Illustration by Thomson Digital

228. asymptotes: $x = 2$, $y = 1$

Factor the numerator and the denominator to determine whether there are any removable discontinuities and to aid in finding intercepts and asymptotes:

$$f(x) = \frac{x^2 - 2x - 15}{x^2 - 7x + 10} = \frac{(x+3)\cancel{(x-5)}}{\cancel{(x-5)}(x-2)}$$

The common factor of $(x - 5)$ in the numerator and denominator produces a removable discontinuity at $x = 5$.

After factoring out the common factor, find the asymptotes. The vertical asymptote is $x = 2$; you find that asymptote by setting the denominator equal to 0 and solving for x.

The horizontal asymptote is $y = 1$, because the highest power in both the numerator and denominator of the reduced fraction is 1, and the ratio formed by the two variables raised to the first power is $\frac{1}{1} = 1$.

For the graph, you also need to identify the intercepts. Find the x-intercepts by letting $y = 0$ and solving for x. The x-intercept is $(-3, 0)$. Find the y-intercept by letting $x = 0$ in the factored version and solving for y. The y-intercept is $\left(0, -\frac{3}{2}\right)$. Here's the graph:

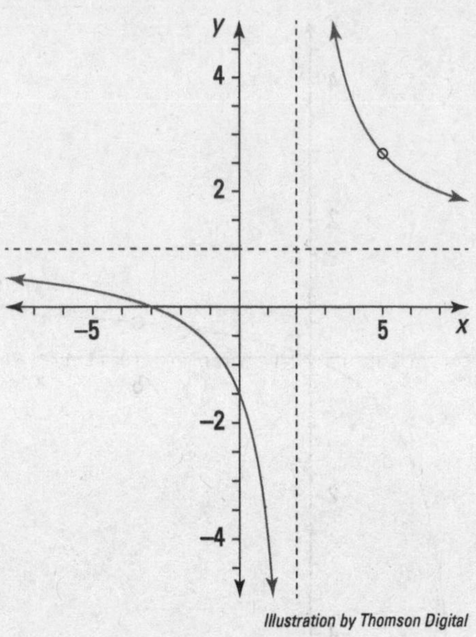

Illustration by Thomson Digital

229. asymptotes: $x = 1$, $y = x + 1$

The vertical asymptote of $f(x)=\frac{x^2-4}{x-1}$ is $x = 1$; you find that asymptote by setting the denominator equal to 0 and solving for x.

Find the slant/oblique asymptote by dividing the numerator by the denominator and writing an equation using the quotient without the remainder. The slant/oblique asymptote is $y=x+1$.

For the graph, you also need to identify the intercepts. Find the x-intercepts by letting $y = 0$ and solving for x. The x-intercepts of are (–2, 0) and (2, 0). Find the y-intercept by letting $x = 0$ and solving for y. The y-intercept is (0, 4). Here's the graph:

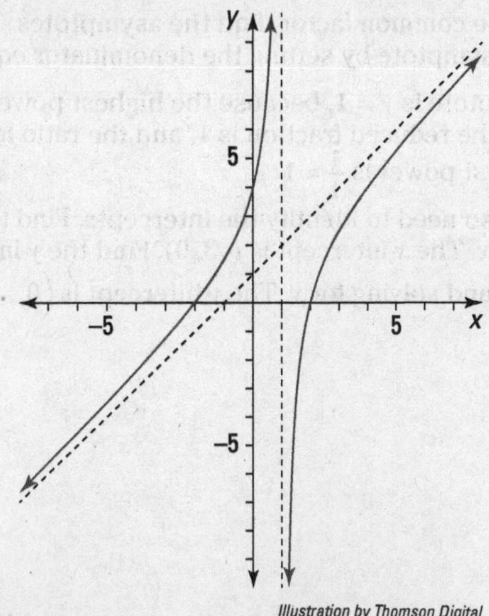

Illustration by Thomson Digital

230. asymptotes: $x=1$, $x=-1$, $y=x-2$

Find the vertical asymptotes of $y=\frac{x^3-2x^2+x-2}{x^2-1}$ by setting the denominator equal to 0 and solving for x. The vertical asymptotes are $x=1$ and $x=-1$.

Find the slant/oblique asymptote by dividing the numerator by the denominator and writing an equation using the quotient without the remainder. The slant/oblique asymptote is $y=x-2$.

For the graph, you also need to identify the intercepts. Find the x-intercepts by letting $y=0$ and solving for x. The x-intercept is (2, 0). Find the y-intercept by letting $x=0$ and solving for y. The y-intercept is (0, 2). Here's the graph:

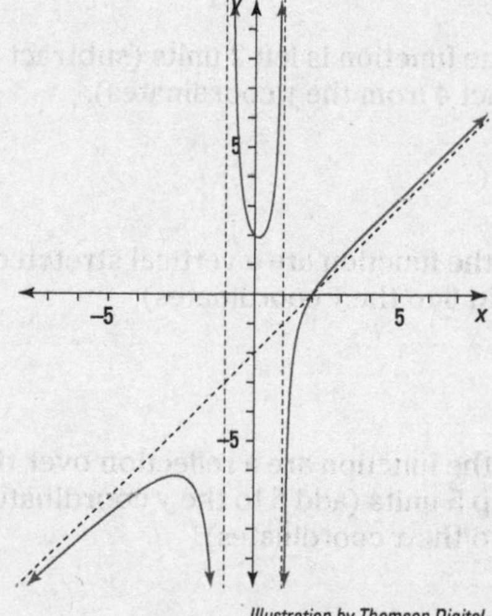

Illustration by Thomson Digital

231. reflection over the x-axis, right 5 units, down 7 units

The negative sign in front of the parentheses in the g function represents a reflection over the x-axis. The –5 inside the parentheses represents a shift right 5 units, and the –7 on the outside represents a shift down 7 units.

232. shrink of $\frac{1}{2}$, left 4 units

The $\frac{1}{2}$ in front of the parentheses in the g function represents a shrink of $\frac{1}{2}$ (the graph gets flatter). The +4 inside the parentheses represents a shift left 4 units.

233. reflection over the x-axis, left 3 units, up 2 units

The negative sign in front of the absolute value in the g function represents a reflection over the x-axis. The +3 inside the absolute value represents a shift left 3 units, and the +2 on the outside represents a shift up 2 units.

234. stretch of 3, down 8 units

The 3 in front of the x term in the g function represents a stretch of 3 (the graph gets steeper). The –8 after the x term represents a shift down 8 units.

235. stretch of 4, reflection over the y-axis, up 3 units

The 4 in front of the square root sign in the g function represents a stretch of 4; the graph gets steeper. The negative sign under the square root sign represents a reflection over the y-axis. The +3 after the square root sign represents a shift up 3 units.

236. (1, 0), (–4, 2), (3, –5)

The transformation for the function is left 2 units (subtract 2 from the x coordinates) and down 4 units (subtract 4 from the y coordinates).

237. (3, 26), (–2, 36), (5, 1)

The transformations for the function are a vertical stretch of 5 (5 times the y coordinates) and up 6 units (add 6 to the y coordinates).

238. (6, 1), (1, –1), (8, 6)

The transformations for the function are a reflection over the x-axis (the signs of the y coordinates change), up 5 units (add 5 to the y coordinates after the sign change), and right 3 units (add 3 to the x coordinates).

239. (–2, –6), (–7, –5), (0, –8.5)

The transformations for the function are a shrink of $\frac{1}{2}$ ($\frac{1}{2}$ times the y coordinates), left 5 units (subtract 5 from the x coordinates), and down 8 units (subtract 8 from the y coordinates after finding half).

240. (–3, 1), (2, 3), (–5, –4)

The transformations for the function are a reflection over the y-axis (the signs of the x coordinates change) and down 3 units (subtract 3 from the y coordinates).

241. vertex: (0, –3)

The parabola is shifted 3 units down from the basic graph's vertex of (0, 0). It's steeper because of the 2 multiplier. Here's the graph:

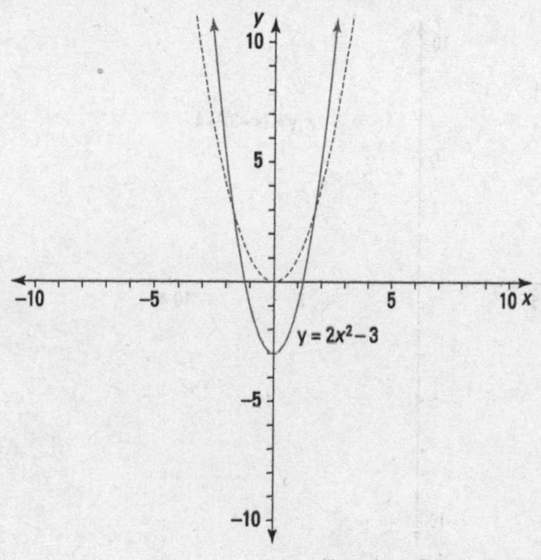

Illustration by Thomson Digital

242. vertex: (2, 0)

The parabola is shifted 2 units right from the basic graph's vertex of (0, 0). It's flatter because of the $\frac{1}{3}$ multiplier. Here's the graph:

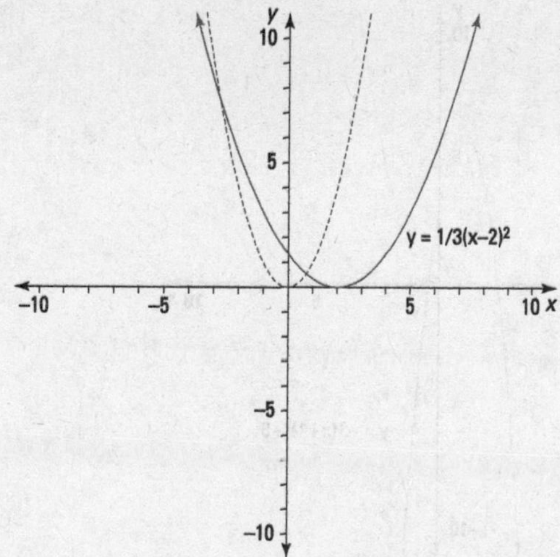

Illustration by Thomson Digital

243. vertex: (3, 4)

The parabola is shifted 3 units right and 4 units up from the basic graph's vertex of (0, 0). Here's the graph:

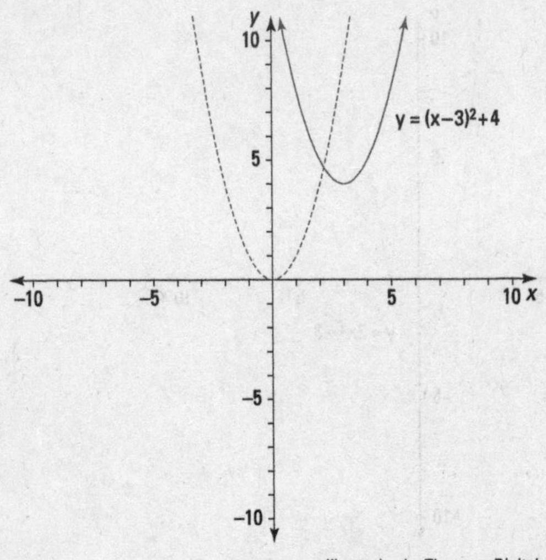

Illustration by Thomson Digital

244. vertex: (–1, 6)

The parabola is shifted 1 unit left and 6 units up from the basic graph's vertex of (0, 0). It's steeper and opens downward because of the –3 multiplier. Here's the graph:

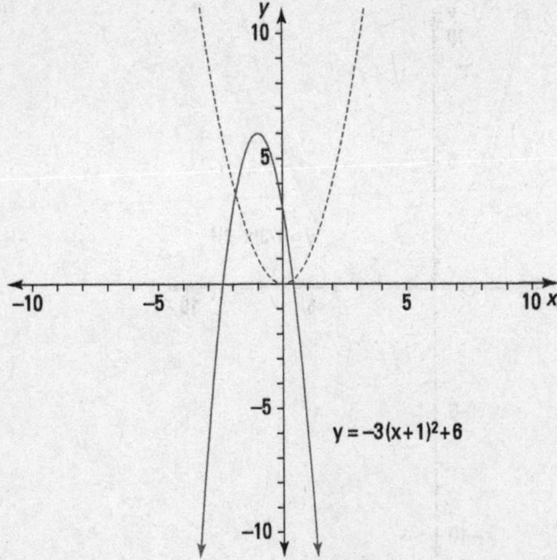

Illustration by Thomson Digital

245. vertex: (4, 2)

The parabola is shifted 4 units right and 2 units up from the basic graph's vertex of (0, 0). It's flatter and opens downward because of the $-\frac{1}{4}$ multiplier.

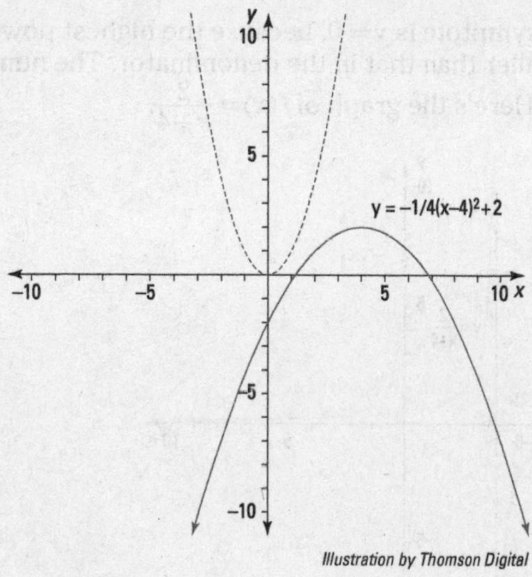

Illustration by Thomson Digital

246. vertical asymptote: $x = 3$

From the basic graph's intersection of vertical and horizontal asymptotes at (0, 0), the curve is shifted to 3 units the right. The vertical asymptote, which you find by setting the denominator equal to 0 and solving for x, is $x = 3$.

The horizontal asymptote is $y = 0$, because the highest power of the variable in the numerator is smaller than that in the denominator. Here's the graph of $f(x) = \frac{1}{x-3}$:

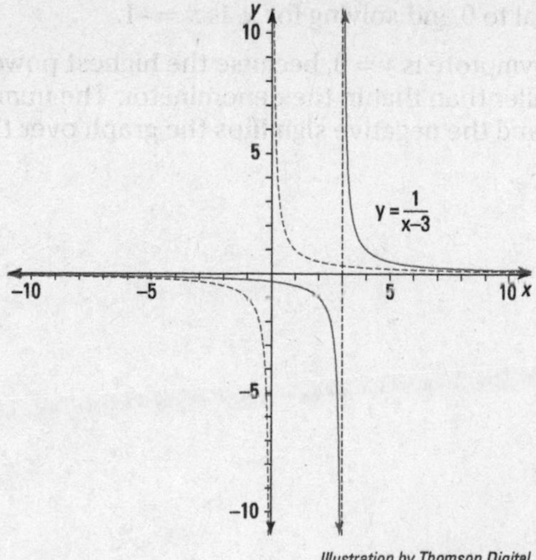

Illustration by Thomson Digital

247. vertical asymptote: $x = -4$

From the basic graph's intersection of vertical and horizontal asymptotes at (0, 0), the curve is shifted 4 units to the left. The vertical asymptote, which you find by setting the denominator equal to 0 and solving for x, is $x = -4$.

The horizontal asymptote is $y = 0$, because the highest power of the variable in the numerator is smaller than that in the denominator. The numerator of 2 stretches the graph vertically. Here's the graph of $f(x) = \frac{2}{x+4}$:

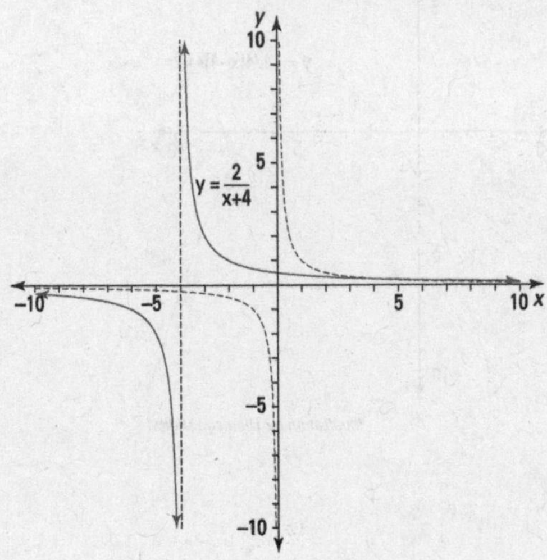

Illustration by Thomson Digital

248. vertical asymptote: $x = -1$

From the basic graph's intersection of vertical and horizontal asymptotes at (0, 0), the curve is shifted 1 unit to the left. The vertical asymptote, which you find by setting the denominator equal to 0 and solving for x, is $x = -1$.

The horizontal asymptote is $y = 0$, because the highest power of the variable in the numerator is smaller than that in the denominator. The numerator of 3 stretches the graph vertically, and the negative sign flips the graph over the x-axis. Here's the graph of $f(x) = -\frac{3}{x+1}$:

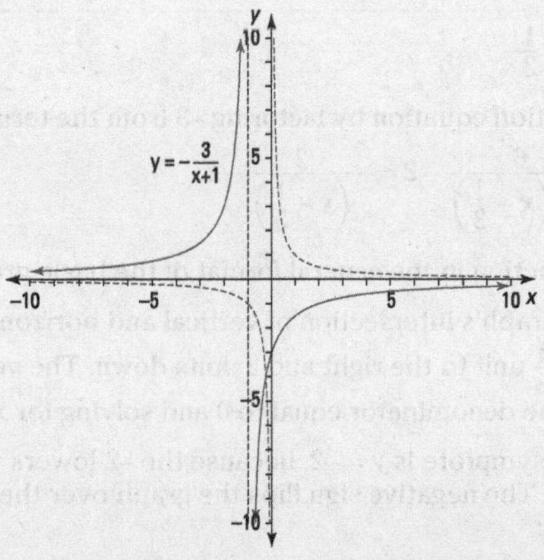

Illustration by Thomson Digital

249. vertical asymptote: $x = 5$

Rewrite the function equation by factoring –1 from the terms in the denominator:

$$f(x) = -\frac{1}{x-5} + 1$$

This puts the function in the general format of the basic graph, $f(x) = \frac{1}{x}$.

From the basic graph's intersection of vertical and horizontal asymptotes at (0, 0), the curve is shifted 5 units to the right and 1 unit up. The vertical asymptote, which you find by setting the denominator equal to 0 and solving for x, is $x = 5$.

The horizontal asymptote is $y = 1$, because the +1 raises the basic graph 1 unit up off the x-axis. The negative sign flips the graph over the horizontal asymptote. Here's the graph:

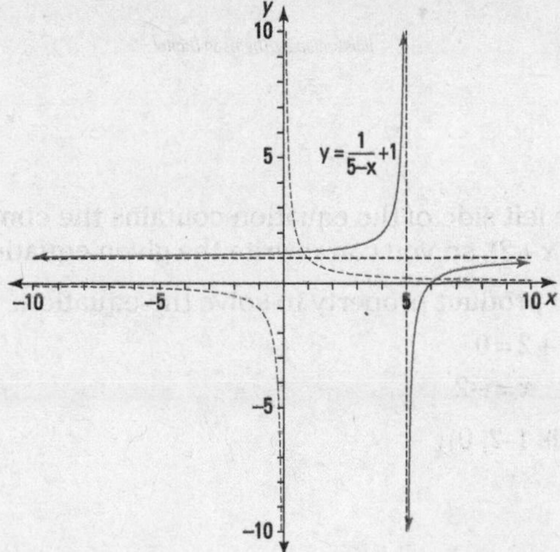

Illustration by Thomson Digital

250. vertical asymptote: $x = \frac{1}{3}$

Rewrite the function equation by factoring –3 from the terms in the denominator:

$$f(x) = \frac{6}{-3\left(x - \frac{1}{3}\right)} - 2 = -\frac{2}{\left(x - \frac{1}{3}\right)} - 2$$

This puts the function in the general format of the basic graph, $f(x) = \frac{1}{x}$.

From the basic graph's intersection of vertical and horizontal asymptotes at (0, 0), the curve is shifted $\frac{1}{3}$ unit to the right and 2 units down. The vertical asymptote, which you find by setting the denominator equal to 0 and solving for x, is $x = \frac{1}{3}$.

The horizontal asymptote is $y = -2$, because the –2 lowers the basic graph 2 units below the x-axis. The negative sign flips the graph over the horizontal asymptote. Here's the graph:

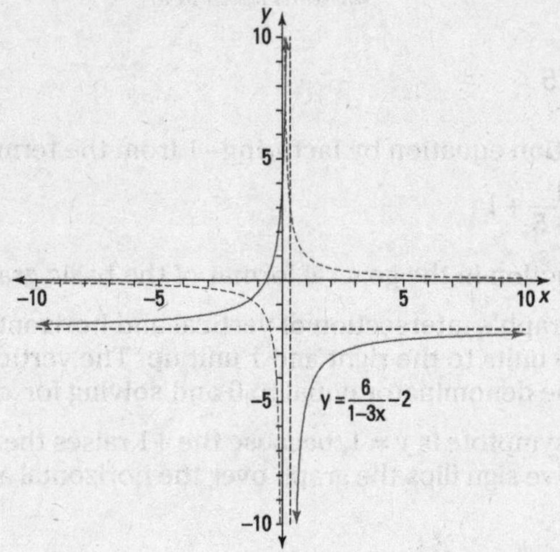

Illustration by Thomson Digital

251. {–2, 0}

Each term on the left side of the equation contains the common factor x, so factor it out: $x^2 + 2x = x(x+2)$, so you can rewrite the given equation as $x(x+2) = 0$.

Now use the zero product property to solve the equation:

$$x = 0 \qquad x + 2 = 0$$
$$x = -2$$

The solution set is {–2, 0}.

252. {–3, 4}

Subtract 9 from each side to set the equation equal to 0:

$$x^2 - x - 3 = 9$$
$$x^2 - x - 12 = 0$$

Now factor the left side:

$$x^2 - x - 12 = 0$$
$$(x+3)(x-4) = 0$$

Then use the zero product property to solve the equation:

$$x + 3 = 0 \qquad x - 4 = 0$$
$$x = -3 \qquad x = 4$$

The solution set is {–3, 4}.

253. $\left\{-\frac{5}{2}, \frac{5}{2}\right\}$

The expression $4x^2 - 25$ is a difference of squares, so you can factor it as $4x^2 - 25 = (2x-5)(2x+5)$. Rewrite the given equation as $(2x-5)(2x+5) = 0$ and use the zero product property to solve for x:

$$2x + 5 = 0 \qquad 2x - 5 = 0$$
$$2x = -5 \qquad 2x = 5$$
$$x = -\frac{5}{2} \qquad x = \frac{5}{2}$$

The solution set is $\left\{-\frac{5}{2}, \frac{5}{2}\right\}$.

254. $\left\{\frac{5}{7}\right\}$

The expression $49x^2 - 70x + 25$ is a perfect square trinomial, so there's just one binomial factor and one solution:

$$49x^2 - 70x + 25 = 0$$
$$(7x-5)(7x-5) = 0$$
$$(7x-5)^2 = 0$$
$$7x - 5 = 0$$
$$7x = 5$$
$$x = \frac{5}{7}$$

The solution set is $\left\{\frac{5}{7}\right\}$.

255. {–4, 7}

First, subtract 24 from each side to set the equation equal to zero:

$$3x^2 - 9x - 60 = 24$$
$$3x^2 - 9x - 84 = 0$$

All three terms on the left side of the equation have a common factor of 3, so factor it out:

$$3x^2 - 9x - 84 = 0$$
$$3(x^2 - 3x - 28) = 0$$

Now divide both sides by 3:

$$3(x^2 - 3x - 28) = 0$$
$$x^2 - 3x - 28 = 0$$

Factor the left side of the equation to rewrite it as $(x+4)(x-7)=0$ and then use the zero product property to solve for x:

$$x + 4 = 0 \qquad x - 7 = 0$$
$$x = -4 \qquad x = 7$$

The solution set is {–4, 7}.

256. $\left\{\frac{1}{7}, 2\right\}$

First, factor the left side of the equation:

$$7x^2 - 15x + 2 = 0$$
$$(7x - 1)(x - 2) = 0$$

Now use the zero product property to solve for x:

$$7x - 1 = 0 \qquad x - 2 = 0$$
$$7x = 1 \qquad x = 2$$
$$x = \frac{1}{7}$$

The solution set is $\left\{\frac{1}{7}, 2\right\}$.

257. $\left\{-\frac{2}{5}, 3\right\}$

First, factor the left side of the equation:

$$5x^2 - 13x - 6 = 0$$
$$(5x + 2)(x - 3) = 0$$

Now use the zero product property to solve for x:

$$5x+2=0 \qquad x-3=0$$
$$5x=-2 \qquad x=3$$
$$x=-\frac{2}{5}$$

The solution set is $\left\{-\frac{2}{5}, 3\right\}$.

258. $\left\{\frac{4}{7}, \frac{7}{5}\right\}$

First, factor the left side of the equation:

$$35x^2-69x+28=0$$
$$(7x-4)(5x-7)=0$$

Now use the zero product property to solve for x:

$$7x-4=0 \qquad 5x-7=0$$
$$7x=4 \qquad 5x=7$$
$$x=\frac{4}{7} \qquad x=\frac{7}{5}$$

The solution set is $\left\{\frac{4}{7}, \frac{7}{5}\right\}$.

259. $\{-3, 2\}$

You can write $x^6+19x^3-216=0$ as a quadratic equation by letting $u=x^3$ (because x^3 is a factor of both x^6 and x^3). Then $u^2=(x^3)^2=x^6$. When you substitute u for x^3, you get $u^2+19u-216=0$. Factor the left side of the equation:

$$u^2+19u-216=0$$
$$(u+27)(u-8)=0$$

Use the zero product property to solve for u:

$$u+27=0 \qquad u-8=0$$
$$u=-27 \qquad u=8$$

You're almost there, but you aren't looking for the value of u; what you need is the value of x. Substitute the u values into the equation $x^3=u$ to solve for x:

$$x^3=-27 \qquad x^3=8$$
$$x=\sqrt[3]{-27} \qquad x=\sqrt[3]{8}$$
$$x=-3 \qquad x=2$$

The solution set is $\{-3, 2\}$.

Answers 201–300

260. $\{\pm\sqrt{3}, \pm\sqrt{7}\}$

You can write $x^4 - 10x^2 + 21 = 0$ as a quadratic equation by letting $u = x^2$ (because x^2 is a factor of both x^4 and x^2). Then $u^2 = (x^2)^2 = x^4$. When you substitute u for x^2, you get $u^2 - 12u + 21 = 0$. Factor the left side of the equation:

$$u^2 - 10u + 21 = 0$$
$$(u-3)(u-7) = 0$$

Use the zero product property to solve for u:

$$u - 3 = 0 \qquad u - 7 = 0$$
$$u = 3 \qquad u = 7$$

You're almost there, but you aren't looking for the value of u; what you need is the value of x. Substitute the u values into the equation $x^2 = u$ to solve for x:

$$x^2 = 3 \qquad x^2 = 7$$
$$x = \pm\sqrt{3} \qquad x = \pm\sqrt{7}$$

The solution set is $\left\{\pm\sqrt{3}, \pm\sqrt{7}\right\}$.

261. $\left\{-\frac{5}{4}, 3\right\}$

The given equation is a quadratic equation already written in standard form $(ax^2 + bx + c = 0)$, so you know that $a = 4$, $b = -7$, and $c = -15$. Plug the values of a, b, and c into the quadratic formula, $x = \frac{-b \pm \sqrt{b^2 - 4ac}}{2a}$, and solve:

$$\begin{aligned} x &= \frac{-(-7) \pm \sqrt{(-7)^2 - 4(4)(-15)}}{2(4)} \\ &= \frac{7 \pm \sqrt{49 + 240}}{8} \\ &= \frac{7 \pm \sqrt{289}}{8} \\ &= \frac{7 \pm 17}{8} \end{aligned}$$

Find both x values:

$$x = \frac{7-17}{8} = -\frac{10}{8} = -\frac{5}{4}$$
$$x = \frac{7+17}{8} = \frac{24}{8} = 3$$

The solution set is $\left\{-\frac{5}{4}, 3\right\}$.

262. $\left\{\frac{1-\sqrt{17}}{2}, \frac{1+\sqrt{17}}{2}\right\}$

First write the equation in standard form $(ax^2+bx+c=0)$ in order to determine a, b, and c:

$$x^2=x+4$$
$$x^2-x-4=0$$

So $a = 1$, $b = -1$, and $c = -4$.

Then plug the values of a, b, and c into the quadratic formula, $x=\frac{-b\pm\sqrt{b^2-4ac}}{2a}$, and solve:

$$\begin{aligned} x&=\frac{-(-1)\pm\sqrt{(-1)^2-4(1)(-4)}}{2(1)}\\ &=\frac{1\pm\sqrt{1+16}}{2}\\ &=\frac{1\pm\sqrt{17}}{2} \end{aligned}$$

The solution set is $\left\{\frac{1-\sqrt{17}}{2}, \frac{1+\sqrt{17}}{2}\right\}$.

263. $\left\{\frac{-2-\sqrt{5}}{2}, \frac{-2+\sqrt{5}}{2}\right\}$

First write the equation in standard form $(ax^2+bx+c=0)$ in order to determine a, b, and c:

$$4x^2+8x=1$$
$$4x^2+8x-1=0$$

So $a = 4$, $b = 8$, and $c = -1$. Plug the values of a, b, and c into the quadratic formula, $x=\frac{-b\pm\sqrt{b^2-4ac}}{2a}$, and solve:

$$\begin{aligned} x&=\frac{-8\pm\sqrt{8^2-4(4)(-1)}}{2(4)}\\ &=\frac{-8\pm\sqrt{64+16}}{8}\\ &=\frac{-8\pm\sqrt{80}}{8}\\ &=\frac{-8\pm4\sqrt{5}}{8}\\ &=\frac{-2\pm\sqrt{5}}{2} \end{aligned}$$

The solution set is $\left\{\frac{-2-\sqrt{5}}{2}, \frac{-2+\sqrt{5}}{2}\right\}$.

264. $\left\{\frac{5-\sqrt{65}}{4}, \frac{5+\sqrt{65}}{4}\right\}$

The equation is already written in standard form $(ax^2+bx+c=0)$, so you know that $a=2$, $b=-5$, and $c=-5$. Plug the values of a, b, and c into the quadratic formula, $x=\frac{-b\pm\sqrt{b^2-4ac}}{2a}$, and solve:

$$\begin{aligned} x &= \frac{-(-5)\pm\sqrt{(-5)^2-4(2)(-5)}}{2(2)} \\ &= \frac{5\pm\sqrt{25+40}}{4} \\ &= \frac{5\pm\sqrt{65}}{4} \end{aligned}$$

The solution set is $\left\{\frac{5-\sqrt{65}}{4}, \frac{5+\sqrt{65}}{4}\right\}$.

265. $\left\{4-\sqrt{11}, 4+\sqrt{11}\right\}$

First, expand the left side of the equation by multiplying $(x-4)(x-4)$, and then write the equation in standard form $(ax^2+bx+c=0)$ in order to determine a, b, and c:

$$\begin{aligned} (x-4)^2 &= 11 \\ x^2-8x+16 &= 11 \\ x^2-8x+5 &= 0 \end{aligned}$$

So $a=1$, $b=-8$, and $c=5$. Plug the values of a, b, and c into the quadratic formula, $x=\frac{-b\pm\sqrt{b^2-4ac}}{2a}$, and solve:

$$\begin{aligned} x &= \frac{-(-8)\pm\sqrt{(-8)^2-4(1)(5)}}{2(1)} \\ &= \frac{8\pm\sqrt{64-20}}{2} \\ &= \frac{8\pm\sqrt{44}}{2} \\ &= \frac{8\pm2\sqrt{11}}{2} \\ &= 4\pm\sqrt{11} \end{aligned}$$

The solution set is $\left\{4-\sqrt{11}, 4+\sqrt{11}\right\}$.

266. {–4, 2}

First, add the constant term, 8, to both sides of the equation so that you have only terms with the variable on the left side of the equation:

$$x^2+2x-8=0$$
$$x^2+2x=8$$

Next, complete the square by taking half of the linear coefficient (the *b* value from the quadratic equation — in this case, 2) and squaring the result: $(2\div 2)^2=1^2$. Add that value to both sides of the equation:

$$x^2+2x=8$$
$$x^2+2x+1^2=8+1^2$$

Now simplify the equation, factor the left side, and take the square root of both sides:

$$x^2+2x+1=9$$
$$(x+1)^2=9$$
$$x+1=\pm 3$$

Isolate the variable to solve the equation:

$$x+1=\pm 3$$
$$x=-1\pm 3$$

The expression -1 ± 3 equals –4 or 2, so the solution set is {–4, 2}.

267. {–9, –2}

In this equation, terms containing the variable are only on the left side. So complete the square by taking half of the linear coefficient (what would be the *b* value if you made this a quadratic equation — in this case, 11) and squaring the result. Add that value, $\left(\frac{11}{2}\right)^2$, to both sides of the equation:

$$x^2+11x=-18$$
$$x^2+11x+\left(\frac{11}{2}\right)^2=-18+\left(\frac{11}{2}\right)^2$$

Next, simplify the equation, factor the left side, and take the square root of both sides:

$$x^2+11x+\left(\frac{11}{2}\right)^2=-18+\frac{121}{4}$$
$$x^2+11x+\left(\frac{11}{2}\right)^2=-\frac{72}{4}+\frac{121}{4}$$
$$\left(x+\frac{11}{2}\right)^2=\frac{49}{4}$$
$$x+\frac{11}{2}=\pm\frac{7}{2}$$

Isolate the variable to solve the equation:

$$x+\frac{11}{2}=\pm\frac{7}{2}$$
$$x=-\frac{11}{2}\pm\frac{7}{2}$$

The expression $-\frac{11}{2}\pm\frac{7}{2}$ equals $-\frac{18}{2}$ or $-\frac{4}{2}$, so the solution set is {–9, –2}.

Answers 201–300

268. $\left\{-2-\sqrt{7},\ -2+\sqrt{7}\right\}$

In this equation, terms containing the variable are only on the left side. So complete the square by taking half of the linear coefficient (what would be the *b* value if you made this a quadratic equation — in this case, 4) and squaring the result: $(4\div 2)^2=2^2$. Add that value to both sides of the equation:

$$x^2+4x=3$$
$$x^2+4x+2^2=3+2^2$$

Next, simplify the equation, factor the left side, and take the square root of both sides:

$$x^2+4x+4=7$$
$$(x+2)^2=7$$
$$x+2=\pm\sqrt{7}$$

Isolate the variable to solve the equation:

$$x+2=\pm\sqrt{7}$$
$$x=-2\pm\sqrt{7}$$

The solution set is $\left\{-2-\sqrt{7},-2+\sqrt{7}\right\}$.

269. $\left\{-1,\frac{2}{5}\right\}$

First, divide every term by the leading coefficient, 5, so that $a=1$:

$$5x^2+3x=2$$
$$x^2+\frac{3}{5}x=\frac{2}{5}$$

Now complete the square by taking half of the linear coefficient (what would be the *b* value if you made this a quadratic equation — in this case, $\frac{3}{5}$) and squaring the result. Add that value to both sides of the equation:

$$x^2+\frac{3}{5}x=\frac{2}{5}$$
$$x^2+\frac{3}{5}x+\left(\frac{3}{10}\right)^2=\frac{2}{5}+\left(\frac{3}{10}\right)^2$$

Simplify the equation, factor the left side, and then take the square root of both sides:

$$x^2+\frac{3}{5}x+\left(\frac{3}{10}\right)^2=\frac{2}{5}+\frac{9}{100}$$
$$x^2+\frac{3}{5}x+\left(\frac{3}{10}\right)^2=\frac{40}{100}+\frac{9}{100}$$
$$\left(x+\frac{3}{10}\right)^2=\frac{49}{100}$$
$$x+\frac{3}{10}=\pm\sqrt{\frac{49}{100}}$$
$$x+\frac{3}{10}=\pm\frac{7}{10}$$

Isolate the variable to solve the equation:

$$x+\frac{3}{10}=\pm\frac{7}{10}$$
$$x=-\frac{3}{10}\pm\frac{7}{10}$$

The expression $-\frac{3}{10}\pm\frac{7}{10}$ equals –1 or $\frac{2}{5}$, so the solution set is $\left\{-1,\frac{2}{5}\right\}$.

270. $\left\{\frac{1}{3},5\right\}$

First you have to get only terms with the variable on the left side of the equation, so add the constant term, 2, to both sides of the equation:

$$3x^2-16x-2=-7$$
$$3x^2-16x=-5$$

Next, divide every term by the leading coefficient, 3, so that $a = 1$:

$$3x^2-16x=-5$$
$$x^2-\frac{16}{3}x=-\frac{5}{3}$$

Now complete the square by taking half of the linear coefficient (the b value from the quadratic equation — in this case, $-\frac{16}{3}$) and squaring the result. Add that value to both sides of the equation:

$$x^2-\frac{16}{3}x+\left(-\frac{16}{6}\right)^2=-\frac{5}{3}+\left(-\frac{16}{6}\right)^2$$

Simplify the equation, factor the left side, and then take the square root of both sides:

$$x^2-\frac{16}{3}x+\left(-\frac{16}{6}\right)^2=-\frac{5}{3}+\left(-\frac{16}{6}\right)^2$$
$$x^2-\frac{16}{3}x+\left(-\frac{8}{3}\right)^2=-\frac{5}{3}+\left(-\frac{8}{3}\right)^2$$
$$x^2-\frac{16}{3}x+\left(-\frac{8}{3}\right)^2=-\frac{15}{9}+\frac{64}{9}$$
$$x^2-\frac{16}{3}x+\left(-\frac{8}{3}\right)^2=\frac{49}{9}$$
$$\left(x-\frac{8}{3}\right)^2=\frac{49}{9}$$
$$x-\frac{8}{3}=\pm\frac{7}{3}$$

Isolate the variable to solve the equation:

$$x-\frac{8}{3}=\pm\frac{7}{3}$$
$$x=\frac{8}{3}\pm\frac{7}{3}$$

The expression $\frac{8}{3}\pm\frac{7}{3}$ equals $\frac{1}{3}$ or 5, so the solution set is $\left\{\frac{1}{3},5\right\}$.

Answers 201–300

271. $0, \pm\sqrt{5}$

Both left-side terms have a common factor of $-3x^4$, so factor it out of each term:

$$-3x^6+15x^4=0$$
$$-3x^4\left(x^2-5\right)=0$$

Set each factor equal to zero and solve for x to find the x-intercepts:

$$-3x^4=0 \qquad x^2-5=0$$
$$x=0 \qquad x^2=5$$
$$x=\pm\sqrt{5}$$

272. $0, \frac{3}{2}$

All left-side terms have a common factor of $2x$, so factor it out of each term:

$$8x^3-24x^2+18x=0$$
$$2x\left(4x^2-12x+9\right)=0$$

Set each factor equal to zero and solve for x to find the x-intercepts:

$$2x=0 \qquad 4x^2-12x+9=0$$
$$x=0 \qquad (2x-3)^2=0$$
$$2x-3=0$$
$$2x=3$$
$$x=\frac{3}{2}$$

273. $-\frac{7}{5}, -\frac{1}{3}, 0$

All left-side terms have a common factor of x^2, so factor it out of each term:

$$15x^4+26x^3+7x^2=0$$
$$x^2\left(15x^2+26x+7\right)=0$$

Then factor the expression inside the parentheses to get

$$x^2(5x+7)(3x+1)=0$$

Set each factor equal to zero and solve for x to find the x-intercepts:

$$x^2=0 \qquad 5x+7=0 \qquad 3x+1=0$$
$$x=0 \qquad 5x=-7 \qquad 3x=-1$$
$$x=-\frac{7}{5} \qquad x=-\frac{1}{3}$$

274. $-2, 2$

This equation is in the form of the difference of squares. The formula for factoring the difference of squares is $a^2 - b^2 = (a+b)(a-b)$, so you can rewrite the given equation as follows:

$$x^4 - 16 = 0$$
$$(x^2+4)(x^2-4) = 0$$

Set each factor equal to zero and solve for x to find the x-intercepts. Start with the first factor, x^2+4:

$$x^2+4=0$$
$$x^2=-4$$
$$x=\pm 2i$$

Remember that $\sqrt{-1}=i$, so $\sqrt{-4}=\sqrt{-1}\sqrt{4}=i\cdot 2$. These aren't real roots or intercepts, so you disregard this solution when choosing your answer.

On to the second factor. Notice that x^2-4 is also the difference of squares. Factor it further to get

$$(x+2)(x-2)=0$$

Then set those factors equal to zero and solve for x:

$$x+2=0 \qquad x-2=0$$
$$x=-2 \qquad x=2$$

These are real roots, so the x-intercepts are -2 and 2.

275. $0, -5\pm\sqrt{22}$

All the left-side terms have a common factor of $2x^2$, so factor it out of each term:

$$4x^4+20x^3+6x^2=0$$
$$2x^2(x^2+10x+3)=0$$

Now set each factor equal to zero and solve for x to find the x-intercepts. The first factor is pretty straightforward:

$$2x^2=0$$
$$x^2=0$$
$$x=0$$

You need to use the quadratic formula, $x=\frac{-b\pm\sqrt{b^2-4ac}}{2a}$, to solve $x^2+10x+3=0$ because it isn't factorable. It is in standard form $(ax^2+bx+c=0)$, though, so you can easily determine that $a=1$, $b=10$, and $c=3$. Plug those values into the quadratic formula:

$$x=\frac{-10\pm\sqrt{10^2-4(1)(3)}}{2(1)}$$
$$=\frac{-10\pm\sqrt{100-12}}{2}$$
$$=\frac{-10\pm\sqrt{88}}{2}=\frac{-10\pm\sqrt{4}\sqrt{22}}{2}$$
$$=\frac{-10\pm2\sqrt{22}}{2}=\frac{-\cancel{10}^{5}\pm\cancel{2}\sqrt{22}}{\cancel{2}}$$
$$=-5\pm\sqrt{22}$$

276. $-\frac{2}{3}$

You can use the sum of cubes to solve this equation. The formula for factoring the sum of cubes is $a^3+b^3=(a+b)\left(a^2-ab+b^2\right)$. Plug in the values from the given equation, noting that both 27 and 8 are perfect cubes:

$$27x^3+8=0$$
$$\left(3^3\right)\left(x^3\right)+2^3=0$$
$$(3x+2)\left(3^2x^2-(2\cdot3x)+2^2\right)=0$$
$$(3x+2)\left(9x^2-6x+4\right)=0$$

Now set each factor equal to zero and solve for x to find the x-intercepts. The first factor is pretty straightforward:

$$3x+2=0$$
$$3x=-2$$
$$x=-\frac{2}{3}$$

So $-\frac{2}{3}$ is an x-intercept.

You need to use the quadratic formula, $x=\frac{-b\pm\sqrt{b^2-4ac}}{2a}$, to solve $9x^2-6x+4=0$ because it isn't factorable. It is in standard form $\left(ax^2+bx+c=0\right)$, though, so you can easily determine that $a=9$, $b=-6$, and $c=4$. Plug those values into the quadratic formula:

$$x=\frac{-b\pm\sqrt{b^2-4ac}}{2a}$$
$$x=\frac{-(-6)\pm\sqrt{(-6)^2-4(9)(4)}}{2(9)}$$
$$x=\frac{6\pm\sqrt{36-144}}{18}$$
$$x=\frac{6\pm\sqrt{-108}}{18}$$
$$x=\frac{6\pm6i\sqrt{3}}{18}$$
$$x=\frac{1\pm i\sqrt{3}}{3}$$

(Remember that $\sqrt{-1}=i$, so $\sqrt{-108}=\sqrt{-1}\sqrt{36}\sqrt{3}=i\cdot 6\sqrt{3}$.) These roots aren't real, so they aren't intercepts.

277. $1, \pm\sqrt{5}$

First, factor by grouping. Break up the polynomial into sets of two and then find the greatest common factor of each set and factor it out. Finally, factor again.

$$\begin{aligned} x^3-x^2-5x+5&=0\\ (x^3-x^2)-(5x-5)&=0\\ x^2(x-1)-5(x-1)&=0\\ (x^2-5)(x-1)&=0 \end{aligned}$$

Next, set each factor equal to zero and solve for x to find the x-intercepts:

$$\begin{aligned} x^2-5&=0 & x-1&=0\\ x^2&=5 & x&=1\\ x&=\pm\sqrt{5} \end{aligned}$$

278. $-\frac{2}{3}, \pm 1$

First, factor by grouping. Break up the polynomial into sets of two and then find the greatest common factor of each set and factor it out. Finally, factor again.

$$\begin{aligned} 3x^3+2x^2-3x-2&=0\\ (3x^3+2x^2)-(3x+2)&=0\\ x^2(3x+2)-(3x+2)&=0\\ (x^2-1)(3x+2)&=0\\ (x+1)(x-1)(3x+2)&=0 \end{aligned}$$

Next, set each factor equal to zero and solve for x to find the x-intercepts:

$$\begin{aligned} x+1&=0 & x-1&=0 & 3x+2&=0\\ x&=-1 & x&=1 & 3x&=-2\\ & & & & x&=-\frac{2}{3} \end{aligned}$$

279. $\pm 2, 9$

First, factor by grouping. Break up the polynomial into sets of two and then find the greatest common factor of each set and factor it out. Finally, factor again.

$$\begin{aligned} x^3-9x^2-4x+36&=0\\ (x^3-9x^2)-(4x-36)&=0\\ x^2(x-9)-4(x-9)&=0\\ (x^2-4)(x-9)&=0\\ (x+2)(x-2)(x-9)&=0 \end{aligned}$$

Next, set each factor equal to zero and solve for x to find the x-intercepts:

$$x+2=0 \qquad x-2=0 \qquad x-9=0$$
$$x=-2 \qquad x=2 \qquad x=9$$

280. $-\frac{1}{5}, \pm\sqrt{\frac{6}{5}}$

First, factor by grouping. Break up the polynomial into sets of two and then find the greatest common factor of each set and factor it out. Finally, factor again.

$$\begin{aligned} 25x^3+5x^2-30x-6&=0 \\ (25x^3+5x^2)-(30x+6)&=0 \\ 5x^2(5x+1)-6(5x+1)&=0 \\ (5x^2-6)(5x+1)&=0 \end{aligned}$$

Next, set each factor equal to zero and solve for x to find the x-intercepts:

$$5x^2-6=0 \qquad 5x+1=0$$
$$5x^2=6 \qquad 5x=-1$$
$$x^2=\frac{6}{5} \qquad x=-\frac{1}{5}$$
$$x=\pm\sqrt{\frac{6}{5}}$$

281. possible number of positive real roots: 2 or 0; possible number of negative real roots: 1

Descartes's rule of signs says that the number of positive roots is equal to the number of changes in sign in $f(x)$ or is less than that number by an even number (so you keep subtracting 2 until you get either 1 or 0).

In $f(x)=6x^3-5x^2-3x+1$, there's a change in sign between the first and second terms and between the third and fourth terms, so the possible number of positive real roots is 2 or 0. (Remember, $6x^3$ has a positive sign even though the plus sign isn't written.)

To find the possible number of negative real roots, find $f(-x)$ and count the number of sign changes from term to term. The number of negative roots is equal to the changes in sign for $f(-x)$ or must be less than that number by an even number.

$$\begin{aligned} f(-x)&=6(-x)^3-5(-x)^2-3(-x)+1 \\ &=-6x^3-5x^2+3x+1 \end{aligned}$$

There's one sign change between the second and third terms, so the possible number of negative real roots is 1.

282. possible number of positive real roots: 2 or 0; possible number of negative real roots: 2 or 0

Descartes's rule of signs says that the number of positive roots is equal to the number of changes in sign in $f(x)$ or is less than that number by an even number (so you keep subtracting 2 until you get either 1 or 0).

In $f(x)=18x^6-24x^4-25x^2+60$, there's a change in sign between the first and second terms and between the third and fourth terms, so the possible number of positive real roots is 2 or 0. (Remember, $18x^6$ has a positive sign even though the plus sign isn't written.)

To find the possible number of negative real roots, find $f(-x)$ and count the number of sign changes from term to term. The number of negative roots is equal to the changes in sign for $f(-x)$ or must be less than that number by an even number.

$$\begin{aligned} f(-x) &= 18(-x)^6-24(-x)^4-25(-x)^2+60 \\ &= 18x^6-24x^4-25x^2+60 \end{aligned}$$

There's a sign change between the first and second terms and between the third and fourth terms, so the possible number of negative real roots is 2 or 0.

283. possible number of positive real roots: 3 or 1; possible number of negative real roots: 1

Descartes's rule of signs says that the number of positive roots is equal to the number of changes in sign in $f(x)$ or is less than that number by an even number (so you keep subtracting 2 until you get either 1 or 0).

In $f(x)=7x^6-5x^4+2x^3-1$, there's a change in sign between the first and second terms, between the second and third terms, and between the third and fourth terms, so the possible number of positive real roots is 3 or 1. (Remember, $7x^6$ has a positive sign even though the plus sign isn't written.)

To find the possible number of negative real roots, find $f(-x)$ and count the number of sign changes from term to term. The number of negative roots is equal to the changes in sign for $f(-x)$ or must be less than that number by an even number.

$$\begin{aligned} f(-x) &= 7(-x)^6-5(-x)^4+2(-x)^3-1 \\ &= 7x^6-5x^4-2x^3-1 \end{aligned}$$

There's a sign change between the first and second terms, so the possible number of negative real roots is 1.

284. possible number of positive real roots: 4 or 2 or 0; possible number of negative real roots: 0

Descartes's rule of signs says that the number of positive roots is equal to the number of changes in sign in $f(x)$ or is less than that number by an even number (so you keep subtracting 2 until you get either 1 or 0).

In $f(x)=2x^4-3x^3+8x^2-7x+5$, there's a change in sign between the first and second terms, between the second and third terms, between the third and fourth terms, and between the fourth and fifth terms, so the possible number of positive real roots is 4 or 2 or 0. (Remember, $2x^4$ has a positive sign even though the plus sign isn't written.)

To find the possible number of negative real roots, find $f(-x)$ and count the number of sign changes from term to term. The number of negative roots is equal to the changes in sign for $f(-x)$ or must be less than that number by an even number.

$$\begin{aligned} f(-x) &= 2(-x)^4-3(-x)^3+8(-x)^2-7(-x)+5 \\ &= 2x^4+3x^3+8x^2+7x+5 \end{aligned}$$

There are no sign changes, so the possible number of negative real roots is 0.

285. possible number of positive real roots: 0; possible number of negative real roots: 5 or 3 or 1

Descartes's rule of signs says that the number of positive roots is equal to the number of changes in sign in $f(x)$ or is less than that number by an even number (so you keep subtracting 2 until you get either 1 or 0).

In $f(x)=x^5+x^4+x^3+x^2+x+1$, there are no changes in sign, so the possible number of positive real roots is 0. (Remember, x^5 has a positive sign even though the plus sign isn't written.)

To find the possible number of negative real roots, find $f(-x)$ and count the number of sign changes from term to term. The number of negative roots is equal to the changes in sign for $f(-x)$ or must be less than that number by an even number.

$$\begin{aligned} f(-x) &= (-x)^5+(-x)^4+(-x)^3+4(-x)^2+8(-x)+1 \\ &= -x^5+x^4-x^3+x^2-x+1 \end{aligned}$$

There's a sign change between each pair of terms, so there are five changes in sign. The possible number of negative real roots is 5 or 3 or 1.

286. $\pm1, \pm7, \pm\frac{1}{7}$

The rational root theorem says that if you take all the factors of the constant term in a polynomial and divide them by all the factors of the leading coefficient, you produce a list of all the possible rational roots of the polynomial.

In $f(x)=7x^3-18x^2+17x-7$, the constant term is -7, and its factors are ±1 and ±7. The leading coefficient is 7, and its factors are ±1 and ±7. Now you can divide. The possible rational roots are $\frac{\pm1}{\pm1}=\pm1$, $\frac{\pm7}{\pm1}=\pm7$, and $\frac{\pm1}{\pm7}=\pm\frac{1}{7}$.

287. $\pm1, \pm5, \pm\frac{1}{3}, \pm\frac{5}{3}$

The rational root theorem says that if you take all the factors of the constant term in a polynomial and divide them by all the factors of the leading coefficient, you produce a list of all the possible rational roots of the polynomial.

In $f(x)=3x^3+2x-5$, the constant term is -5, and its factors are ±1 and ±5. The leading coefficient is 3, and its factors are ±1 and ±3. Now you can divide. The possible rational roots are $\frac{\pm1}{\pm1}=\pm1$, $\frac{\pm5}{\pm1}=\pm5$, $\frac{\pm1}{\pm3}=\pm\frac{1}{3}$, and $\frac{\pm5}{\pm3}=\pm\frac{5}{3}$.

288. $\pm1, \pm\frac{1}{2}, \pm\frac{1}{4}$

The rational root theorem says that if you take all the factors of the constant term in a polynomial and divide them by all the factors of the leading coefficient, you produce a list of all the possible rational roots of the polynomial.

In $f(x)=4x^3-7x^2+4x-1$, the constant term is -1, and its factors are ±1. The leading coefficient is 4, and its factors are ±1, ±2, and ±4. Now you can divide. The possible rational roots are $\frac{\pm1}{\pm1}=\pm1$, $\frac{\pm1}{\pm2}=\pm\frac{1}{2}$, and $\frac{\pm1}{\pm4}=\pm\frac{1}{4}$.

289. $\pm 1, \pm 2, \pm 4, \pm 8, \pm\frac{1}{4}, \pm\frac{1}{2}$

The rational root theorem says that if you take all the factors of the constant term in a polynomial and divide them by all the factors of the leading coefficient, you produce a list of all the possible rational roots of the polynomial.

In $f(x)=4x^3-5x^2+3x+8$, the constant term is 8, and its factors are ± 1, ± 2, ± 4, and ± 8. The leading coefficient is 4, and its factors are ± 1, ± 2, and ± 4. Now you can divide:

$$\frac{\pm 1}{\pm 1}=\pm 1,\ \frac{\pm 1}{\pm 2}=\pm\frac{1}{2},\ \frac{\pm 1}{\pm 4}=\pm\frac{1}{4},\ \pm\frac{\pm 2}{\pm 1}=\pm 2,\ \frac{\pm 2}{\pm 2}=\pm 1,\ \frac{\pm 2}{\pm 4}=\pm\frac{1}{2},$$
$$\pm\frac{\pm 4}{\pm 1}=\pm 4,\ \frac{\pm 4}{\pm 2}=\pm 2,\ \frac{\pm 4}{\pm 4}=\pm 1,\ \pm\frac{\pm 8}{\pm 1}=\pm 8,\ \frac{\pm 8}{\pm 2}=\pm 4,\ \frac{\pm 8}{\pm 4}=\pm 2$$

Eliminating the duplicates, you have ± 1, ± 2, ± 4, ± 8, $\pm\frac{1}{4}$, and $\pm\frac{1}{2}$.

290. $\pm 1, \pm 2, \pm 4, \pm 8, \pm\frac{1}{5}, \pm\frac{2}{5}, \pm\frac{4}{5}, \pm\frac{8}{5}$

The rational root theorem says that if you take all the factors of the constant term in a polynomial and divide them by all the factors of the leading coefficient, you produce a list of all the possible rational roots of the polynomial.

In $f(x)=5x^3+2x^2-10x-8$, the constant term is -8, and its factors are ± 1, ± 2, ± 4, and ± 8. The leading coefficient is 5, and its factors are ± 1 and ± 5. Now you can divide. The possible rational roots are $\frac{\pm 1}{\pm 1}=\pm 1$, $\frac{\pm 2}{\pm 1}=\pm 2$, $\frac{\pm 4}{\pm 1}=\pm 4$, $\frac{\pm 8}{\pm 1}=\pm 8$, $\frac{\pm 1}{\pm 5}=\pm\frac{1}{5}$, $\frac{\pm 2}{\pm 5}=\pm\frac{2}{5}$, $\frac{\pm 4}{\pm 5}=\pm\frac{4}{5}$, and $\frac{\pm 8}{\pm 5}=\pm\frac{8}{5}$.

291. $x+5-\frac{10}{x+5}$

When dividing one polynomial by another, write the divisor outside the division symbol and the dividend inside. The terms should be in descending order of the exponents. If there are any missing terms in the dividend, put zeros in as placeholders. You find the terms in the quotient one at a time by multiplying the divisor by an appropriate term.

$$\begin{array}{r} x+5 \\ x+5\overline{)x^2+10x+15} \\ \underline{x^2+5x} \\ 5x+15 \\ \underline{5x+25} \\ -10 \end{array}$$

To write the remainder as a fraction, place it over the divisor. Written this way, the quotient is

$$x+5-\frac{10}{x+5}$$

Answers 201–300

292. $x^2-2x-4+\frac{1}{3x-1}$

When dividing one polynomial by another, write the divisor outside the division symbol and the dividend inside. The terms should be in descending order of the exponents. If there are any missing terms in the dividend, put zeros in as placeholders. You find the terms in the quotient one at a time by multiplying the divisor by an appropriate term.

$$\begin{array}{r} x^2-2x-4 \\ 3x-1\overline{)3x^3-7x^2-10x+5} \\ \underline{3x^3-x^2} \\ -6x^2-10x \\ \underline{-6x^2+2x} \\ -12x+5 \\ \underline{-12x+4} \\ 1 \end{array}$$

To write the remainder as a fraction, place it over the divisor. Written this way, the quotient is

$$x^2-2x-4+\frac{1}{3x-1}$$

293. $5x-5+\frac{11}{7x+7}$

When dividing one polynomial by another, write the divisor outside the division symbol and the dividend inside. The terms should be in descending order of the exponents. The linear term, x, is missing from the dividend, so insert $0x$ in order to leave space for an x term column. You find the terms in the quotient one at a time by multiplying the divisor by an appropriate term.

$$\begin{array}{r} 5x-5 \\ 7x+7\overline{)35x^2+0x-24} \\ \underline{35x^2+35x} \\ -35x-24 \\ \underline{-35x-35} \\ +11 \end{array}$$

To write the remainder as a fraction, place it over the divisor. Written this way, the quotient is

$$5x-5+\frac{11}{7x+7}$$

294. $4x^2-x-7+\frac{11x+15}{x^2+x+2}$

When dividing one polynomial by another, write the divisor outside the division symbol and the dividend inside. The terms should be in descending order of the exponents. The square term, x^2, is missing from the dividend, so insert $0x^2$ in order to leave space for an x^2 term column. You find the terms in the quotient one at a time by multiplying the divisor by an appropriate term.

$$\begin{array}{r}
4x^2-\ x-7 \\
x^2+x+2\overline{)4x^4+3x^3+0x^2+2x+1} \\
\underline{4x^4+4x^3+8x^2} \\
-x^3-8x^2+2x \\
\underline{-x^3-\ x^2-2x} \\
-7x^2+4x+1 \\
\underline{-7x^2-7x-14} \\
11x+15
\end{array}$$

To write the remainder as a fraction, place it over the divisor. Written this way, the quotient is

$$4x^2-x-7+\frac{11x+15}{x^2+x+2}$$

295. $x+1-\frac{4}{2x^2-1}$

When dividing one polynomial by another, write the divisor outside the division symbol and the dividend inside. The terms should be in descending order of the exponents. If there are any missing terms in the dividend, put zeros in as placeholders. You find the terms in the quotient one at a time by multiplying the divisor by an appropriate term.

$$\begin{array}{r}
x\ +\ 1 \\
2x^2+0x-1\overline{)2x^3+2x^2-x-5} \\
\underline{2x^3+0x^2-x} \\
2x^2+0x-5 \\
\underline{2x^2+0x-1} \\
-4
\end{array}$$

To write the remainder as a fraction, place it over the divisor. Written this way, the quotient is

$$x+1-\frac{4}{2x^2-1}$$

296. quotient: $x+1$; remainder: –20

Synthetic division works only when a polynomial is divided by a linear factor and when the coefficient of the linear factor is 1. If $x-n$ is a factor of the polynomial, then $x=n$ is a zero of the polynomial, and if $x=n$ is a zero, then $x-n$ is a factor.

In this case, $x+5$ may be a factor of $x^2+6x-15$, so $x=-5$ may be a zero. Therefore, write –5 on the outside of the synthetic division sign. Write the coefficients of the polynomial on the inside. Here's the process, step by step:

1. Drop the 1 from the coefficient line to below the division line.
2. Multiply 1 by –5 to get –5 and write that under the 6.
3. Add 6+(–5) to get 1. Write that under the line in the same column.
4. Multiply 1 by –5 to get –5 and write that under the –15.
5. Add –15+(–5) to get –20. Write that under the line in the same column. This amount is the remainder.

$$\begin{array}{r|rrr} -5 & 1 & 6 & -15 \\ & & -5 & -5 \\ \hline & 1 & 1 & -20 \end{array}$$

The first two numbers under the line are the coefficients of the quotient. The answer is always one degree lower than the original polynomial. In this case, the quotient is $x+1$, and the remainder is –20.

297. quotient: $2x^2-5x+7$; remainder: 48

Synthetic division works only when a polynomial is divided by a linear factor and when the coefficient of the linear factor is 1. If $x-n$ is a factor of the polynomial, then $x=n$ is a zero of the polynomial, and if $x=n$ is a zero, then $x-n$ is a factor.

In this case, $x-4$ may be a factor of $2x^3-13x^2+27x+20$, so $x=4$ may be a zero. Therefore, write 4 on the outside of the synthetic division sign. Write the coefficients of the polynomial on the inside. Here's the process, step by step:

1. Drop the 2 from the coefficient line to below the division line.
2. Multiply 4 by 2 to get 8 and write that under the –13.
3. Add –13+8 to get –5. Write that under the line in the same column.
4. Multiply 4 by –5 to get –20 and write that under the 27.
5. Add 27+(–20) to get 7. Write that under the line in the same column.
6. Multiply 4 by 7 to get 28 and write that under the 20.
7. Add 20+28 to get 48. This amount is the remainder.

$$\begin{array}{r|rrrr} 4 & 2 & -13 & 27 & 20 \\ & & 8 & -20 & 28 \\ \hline & 2 & -5 & 7 & 48 \end{array}$$

The first three numbers under the line are the coefficients of the quotient. The answer is always one degree lower than the original polynomial. In this case, the quotient is $2x^2-5x+7$, and the remainder is 48.

298. quotient: $x^2 - 3x + 9$; remainder: –47

Synthetic division works only when a polynomial is divided by a linear factor and when the coefficient of the linear factor is 1. If $x - n$ is a factor of the polynomial, then $x = n$ is a zero of the polynomial, and if $x = n$ is a zero, then $x - n$ is a factor.

In this case, $x + 3$ may be a factor of $x^3 - 20$, so $x = -3$ may be a zero. Therefore, write –3 on the outside of the synthetic division sign. Write the coefficients of the polynomial on the inside, using 0 as the coefficients of the missing x^2 and x terms. Here's the process, step by step:

1. Drop the 1 from the coefficient line to below the division line.
2. Multiply –3 by 1 to get –3 and write that under the first 0.
3. Add 0+(–3) to get –3. Write that under the line in the same column.
4. Multiply –3 by –3 to get 9 and write that under the next 0.
5. Add 0+9 to get 9. Write that under the line in the same column.
6. Multiply –3 by 9 to get –27 and write that under the –20.
7. Add –20+(–27) to get –47. This amount is the remainder.

$$\begin{array}{r|rrrr} -3 & 1 & 0 & 0 & -20 \\ & & -3 & 9 & -27 \\ \hline & 1 & -3 & 9 & -47 \end{array}$$

The first three numbers under the line are the coefficients of the quotient. The answer is always one degree lower than the original polynomial. In this case, the quotient is $x^2 - 3x + 9$, and the remainder is –47.

299. 3

The remainder is 0 when $f(x)$ is divided by $x - a$, where a is a zero of the function. So use synthetic division to test each value.

$$\begin{array}{r|rrr} -3 & 3 & -11 & 6 \\ & & -9 & 60 \\ \hline & 3 & 20 & 66 \end{array}$$

The remainder equals 66, so $x = -3$ isn't a zero.

$$\begin{array}{r|rrr} 3 & 3 & -11 & 6 \\ & & 9 & -6 \\ \hline & 3 & -2 & 0 \end{array}$$

The remainder equals 0, so $x = 3$ is a zero. From this point, you can use a "reduced polynomial" and not the original. The following steps show you the work with the original coefficients.

$$\begin{array}{r|rrr} -6 & 3 & -11 & 6 \\ & & -18 & 162 \\ \hline & 3 & -27 & 168 \end{array}$$

Answers 201–300

The remainder equals 168, so $x = -6$ isn't a zero.

$$\begin{array}{r|rrr} 6 & 3 & -11 & 6 \\ & & 18 & 42 \\ \hline & 3 & 7 & 48 \end{array}$$

The remainder equals 48, so $x = 6$ isn't a zero.

$$\begin{array}{r|rrr} 1 & 3 & -11 & 6 \\ & & 3 & -8 \\ \hline & 3 & -8 & -2 \end{array}$$

The remainder equals –2, so $x = 1$ isn't a zero.

300. –1

The remainder is 0 when $f(x)$ is divided by $x - a$, where a is a zero of the function. So use synthetic division to test each value.

$$\begin{array}{r|rrrrr} -8 & 1 & 2 & -9 & -2 & 8 \\ & & -8 & 48 & -312 & 2{,}512 \\ \hline & 1 & -6 & 39 & -314 & 2{,}520 \end{array}$$

The remainder equals 2,520, so $x = -8$ isn't a zero.

$$\begin{array}{r|rrrrr} -2 & 1 & 2 & -9 & -2 & 8 \\ & & -2 & 0 & 18 & -32 \\ \hline & 1 & 0 & -9 & -16 & -24 \end{array}$$

The remainder equals –24, so $x = -8$ isn't a zero.

$$\begin{array}{r|rrrrr} -1 & 1 & 2 & -9 & -2 & 8 \\ & & -1 & -1 & 10 & -8 \\ \hline & 1 & 1 & -10 & 8 & 0 \end{array}$$

The remainder equals 0, so $x = -1$ is a zero. From this point, you can use a "reduced polynomial" and not the original. The following steps show you the work with the original coefficients.

$$\begin{array}{r|rrrrr} 4 & 1 & 2 & -9 & -2 & 8 \\ & & 4 & 24 & 60 & 232 \\ \hline & 1 & 6 & 15 & 58 & 240 \end{array}$$

The remainder equals 240, so $x = 4$ isn't a zero.

$$\begin{array}{r|rrrrr} 8 & 1 & 2 & -9 & -2 & 8 \\ & & 8 & 80 & 568 & 4{,}528 \\ \hline & 1 & 10 & 71 & 566 & 4{,}536 \end{array}$$

The remainder equals 4,536, so $x = 8$ isn't a zero.

301.

$x^3-2x^2-9x+18$

The zeros of the polynomial are –3, 2, and 3, so the factors of the polynomial are $x-(-3)$, $x-2$, and $x-3$. Multiply the factors to find the polynomial:

$$(x+3)(x-2)(x+3)=x^3-2x^2-9x+18$$

302.

$x^5+4x^4-7x^3-22x^2+24x$

The zeros of the polynomial are –4, –3, 0, 1, and 2, so the factors of the polynomial are $x-(-4)$, $x-(-3)$, $x-0$, $x-1$, and $x-2$. Multiply the factors to find the polynomial:

$$(x+4)(x+3)(x)(x-1)(x-2)=x^5+4x^4-7x^3-22x^2+24x$$

303.

$x^4-3x^3+2x^2+2x-4$

The zeros of the polynomial are $1+i$, $1-i$, –1, and 2, so the factors of the polynomial are $x-(1+i)$, $x-(1-i)$, $x-(-1)$, and $x-2$.

Remember that $\sqrt{-1}=i$ and that imaginary roots occur when a polynomial has a factor of the form x^2+k. When inserting those imaginary roots back into the polynomial form, you'll end up with i^2 terms that you can replace with –1.

Multiply the factors to find the polynomial:

$$\begin{aligned}
&(x-(1+i))\,(x-(1-i))\,(x-(-1))\,(x-2)\\
&=(x-(1+i))\,(x-(1-i))\,(x+1)(x-2)\\
&=\left[x^2-x(1-i)-x(1+i)+(1+i)(1-i)\right](x+1)(x-2)\\
&=\left[x^2-x-ix-x+ix+\left(1-i^2\right)\right](x+1)(x-2)\\
&=\left[x^2-2x+(1-(-1))\right](x+1)(x-2)\\
&=\left(x^2-2x+2\right)(x+1)(x-2)\\
&=x^4-3x^3+2x^2+2x-4
\end{aligned}$$

304.

$x^4+5x^3+2x^2-10x-8$

The zeros of the polynomial are $-\sqrt{2}$, $\sqrt{2}$, –1, and –4, so the factors of the polynomial are $x-\left(-\sqrt{2}\right)$, $x-\sqrt{2}$, $x-(-1)$, and $x-(-4)$. Multiply the factors to find the polynomial:

$$\begin{aligned}
&\left(x-\left(-\sqrt{2}\right)\right)\left(x-\sqrt{2}\right)(x-(-1))\,(x-(-4))\\
&=\left(x+\sqrt{2}\right)\left(x-\sqrt{2}\right)(x+1)(x+4)\\
&=\left(x^2-2\right)(x+1)(x+4)\\
&=x^4+5x^3+2x^2-10x-8
\end{aligned}$$

305. $x^4 - 7x^3 + 15x^2 - 11x + 2$

The zeros of the polynomial are $2-\sqrt{3}$, $2+\sqrt{3}$, 1, and 2, so the factors of the polynomial are $x-\left(2-\sqrt{3}\right)$, $x-\left(2+\sqrt{3}\right)$, $x-1$, and $x-2$. Multiply the factors to find the polynomial:

$$\begin{aligned}&\left[x-\left(2-\sqrt{3}\right)\right]\left[x-\left(2+\sqrt{3}\right)\right](x-1)(x-2)\\&=\left[x^2-x\left(2+\sqrt{3}\right)-x\left(2-\sqrt{3}\right)+\left(2-\sqrt{3}\right)\left(2+\sqrt{3}\right)\right](x-1)(x-2)\\&=\left[x^2-2x-x\sqrt{3}-2x+x\sqrt{3}+4-3\right](x-1)(x-2)\\&=\left(x^2-4x+1\right)(x-1)(x-2)\\&=x^4-7x^3+15x^2-11x+2\end{aligned}$$

306. $2x^3 - 14x + 12$

The zeros of the polynomial are −3, 1, and 2, so the factors are $x-(-3)$, $x-1$, and $x-2$. The polynomial also has a constant factor, a. Therefore, the polynomial is given by

$$\begin{aligned}a\left(x-(-3)\right)(x-1)(x-2)&=a(x+3)(x-1)(x-2)\\&=a\left(x^3-7x+6\right)\end{aligned}$$

The value of the polynomial at $x = -1$ is 24, so plug that in for x and solve for a:

$$\begin{aligned}24&=a\left((-1)^3-7(-1)+6\right)\\24&=a(-1+7+6)\\24&=12a\\a&=\frac{24}{12}\\a&=2\end{aligned}$$

Therefore, the polynomial is

$$2\left(x^3-7x+6\right)=2x^3-14x+12$$

307. $-2x^3 - 6x^2 + 20x + 48$

The zeros of the polynomial are −4, −2, and 3, so the factors are $x-(-4)$, $x-(-2)$, and $x-3$. The polynomial also has a constant factor, a. Therefore, the polynomial is given by

$$\begin{aligned}a\left(x-(-4)\right)\left(x-(-2)\right)(x-3)&=a(x+4)(x+2)(x-3)\\&=a\left(x^3+3x^2-10x-24\right)\end{aligned}$$

The value of the polynomial at $x = 2$ is 48, so plug that in for x and solve for a:

$$48 = a\left((2)^3 + 3(2)^2 - 10(2) - 24\right)$$
$$48 = a(8 + 12 - 20 - 24)$$
$$48 = -24a$$
$$a = \frac{48}{-24}$$
$$a = -2$$

Therefore, the polynomial is

$$-2\left(x^3 + 3x^2 - 10x - 24\right) = -2x^3 - 6x^2 + 20x + 48$$

308. $3x^4 + 6x^3 - 21x^2 - 24x + 36$

The zeros of the polynomial are −3, −2, 1, and 2, so the factors are $x - (-3)$, $x - (-2)$, $x - 1$, and $x - 2$. The polynomial also has a constant factor, a. Therefore, the polynomial is given by

$$a\left(x - (-3)\right)\left(x - (-2)\right)(x - 1)(x - 2)$$
$$= a(x + 3)(x + 2)(x - 1)(x - 2)$$
$$= a\left(x^4 + 2x^3 - 7x^2 - 8x + 12\right)$$

The value of the polynomial at $x = 0$ is 36, so plug that in for x and solve for a:

$$36 = a\left(0^4 + 2(0)^3 - 7(0)^2 - 8(0) + 12\right)$$
$$36 = a(12)$$
$$a = \frac{36}{12}$$
$$a = 3$$

Therefore, the polynomial is

$$3\left(x^4 + 2x^3 - 7x^2 - 8x + 12\right) = 3x^4 + 6x^3 - 21x^2 - 24x + 36$$

309. $4x^4 + 4x^3 - 20x^2 - 12x + 24$

The zeros of the polynomial are $-\sqrt{3}$, $\sqrt{3}$, −2, and 1, so the factors are $x - \left(-\sqrt{3}\right)$, $x - \sqrt{3}$, $x - (-2)$, and $x - 1$. The polynomial also has a constant factor, a. Therefore, the polynomial is given by

$$a\left(x - \left(-\sqrt{3}\right)\right)\left(x - (-2)\right)\left(x - \sqrt{3}\right)(x - 1)$$
$$= a\left(x + \sqrt{3}\right)\left(x - \sqrt{3}\right)(x + 2)(x - 1)$$
$$= a\left(x^2 - 3\right)(x + 2)(x - 1)$$
$$= a\left(x^4 + x^3 - 5x^2 - 3x + 6\right)$$

Answers 301–400

The value of the polynomial at $x = 2$ is 16, so plug that in for x and solve for a:

$$16 = a\left(2^4 + 2^3 - 5(2)^2 - 3(2) + 6\right)$$
$$16 = a(16 + 8 - 20 - 6 + 6)$$
$$16 = a(4)$$
$$a = \frac{16}{4}$$
$$a = 4$$

Therefore, the polynomial is

$$4\left(x^4 + x^3 - 5x^2 - 3x + 6\right) = 4x^4 + 4x^3 - 20x^2 - 12x + 24$$

310. $-2x^4 + 6x^3 + 10x^2 - 58x + 60$

The zeros of the polynomial are $2+i$, $2-i$, -3, and 2, so the factors are $x-(2+i)$, $x-(2-i)$, $x-(-3)$, and $x-2$.

Remember that $\sqrt{-1} = i$ and that imaginary roots occur when a polynomial has a factor of the form $x^2 + k$. When inserting those imaginary roots back into the polynomial form, you'll end up with i^2 terms that you can replace with -1.

The polynomial also has a constant factor, a. Therefore, the polynomial is given by

$$\begin{aligned} & a(x-(2+i))(x-(2-i))(x-(-3))(x-2) \\ &= a(x-(2+i))(x-(2-i))(x+3)(x-2) \\ &= a\left(x^2 - x(2-i) - x(2+i) + (2+i)(2-i)\right)(x+3)(x-2) \\ &= a\left(x^2 - 2x + xi - 2x - xi + 4 - i^2\right)(x+3)(x-2) \\ &= a\left(x^2 - 4x + 4 - (-1)\right)(x+3)(x-2) \\ &= a\left(x^2 - 4x + 5\right)(x+3)(x-2) \\ &= a\left(x^4 - 3x^3 - 5x^2 + 29x - 30\right) \end{aligned}$$

The value of the polynomial at $x = -1$ is 120, so plug that in for x and solve for a:

$$120 = a\left((-1)^4 - 3(-1)^3 + 5(-1)^2 + 29(-1) - 30\right)$$
$$120 = a(1 + 3 - 5 - 29 - 30)$$
$$120 = a(-60)$$
$$a = \frac{120}{-60}$$
$$a = -2$$

Therefore, the polynomial is

$$-2\left(x^4 - 3x^3 - 5x^2 + 29x - 30\right) = -2x^4 + 6x^3 + 10x^2 - 58x + 60$$

311.

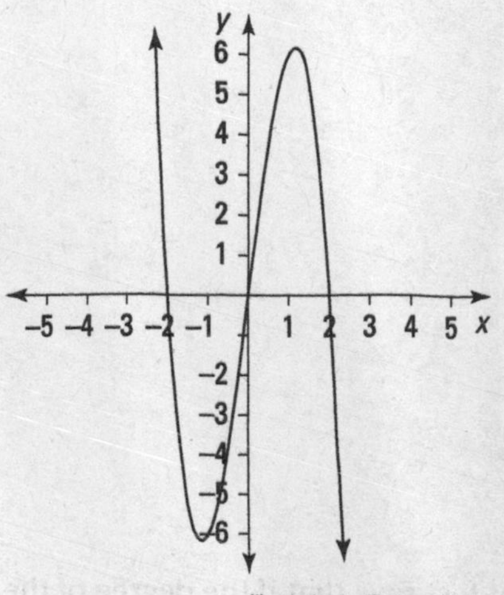

Illustration by Thomson Digital

End behavior: The leading coefficient test says that if the degree of the polynomial is odd and the leading coefficient is negative, then the left side of the graph points up and the right side points down.

Intercepts: Solve $-2x^3+8x=0$ to find the x-intercepts:

$$\begin{aligned}-2x^3+8x&=0\\-2x\left(x^2-4\right)&=0\\-2x(x+2)(x-2)&=0\\x&=0,-2,2\end{aligned}$$

Each zero has multiplicity 1 (meaning the exponent on each factor is 1, which is an odd number), so the graph crosses the x-axis at each zero.

Find the y-intercept by finding $f(0)$:

$$-2(0)^3+8(0)=0$$

Test points: Test points in the intervals $(-\infty,-2)$, $(-2,0)$, $(0,2)$, and $(2,\infty)$.

Interval	Test point, x	$f(x)$ Above or below x-axis.
$(-\infty,-2)$	−3	30; above
$(-2,0)$	−1	−6; below
$(0,2)$	1	6; above
$(2,\infty)$	3	−30; below

Illustration by Thomson Digital

312.

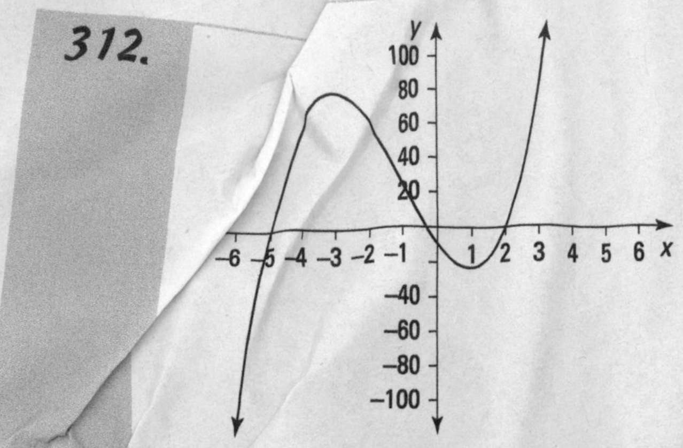

Illustration by Thomson Digital

End behavior: The leading coefficient test says that if the degree of the polynomial is odd and the leading coefficient is positive, then the left side of the graph points down and the right side points up.

Intercepts: Solve $3x^3+10x^2-27x-10=0$ to find the x-intercepts. The rational root theorem says that if you take all the factors of the constant term in a polynomial and divide them by all the factors of the leading coefficient, you produce a list of all the possible rational roots of the polynomial. Therefore, the possible rational zeros are

$$\frac{\pm1}{\pm1}=\pm1,\ \frac{\pm1}{\pm3}=\pm\frac{1}{3},\ \frac{\pm2}{\pm1}=\pm2,\ \frac{\pm2}{\pm3}=\pm\frac{2}{3},\ \frac{\pm5}{\pm1}=\pm5,$$

$$\frac{\pm5}{\pm3}=\pm\frac{5}{3},\ \frac{\pm10}{\pm1}=\pm10,\ \frac{\pm10}{\pm3}=\pm\frac{10}{3}$$

Use synthetic division to find that $x=2$ is one of the zeros:

$$\begin{array}{r|rrrr} 2 & 3 & 10 & -27 & -10 \\ & & 6 & 32 & 10 \\ \hline & 3 & 16 & 5 & 0 \end{array}$$

Now solve the resulting equation, $3x^2+16x+5=0$, to find the remaining zeros:

$$3x^2+16x+5=0$$
$$(3x+1)(x+5)=0$$
$$x=-\frac{1}{3},-5$$

Each zero has multiplicity 1 (the exponent on each factor is 1, which is an odd number), so the graph crosses the x-axis at each zero.

Find the y-intercept by finding $f(0)$:

$$3(0)^3+10(0)^2-27(0)-10=-10$$

311.

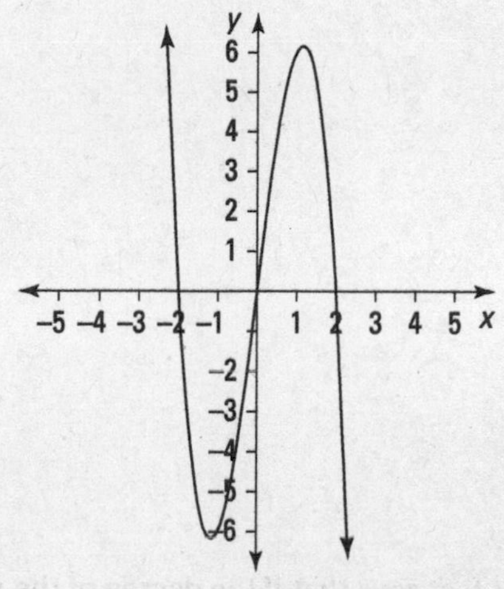

Illustration by Thomson Digital

End behavior: The leading coefficient test says that if the degree of the polynomial is odd and the leading coefficient is negative, then the left side of the graph points up and the right side points down.

Intercepts: Solve $-2x^3+8x=0$ to find the x-intercepts:

$$-2x^3+8x=0$$
$$-2x\left(x^2-4\right)=0$$
$$-2x(x+2)(x-2)=0$$
$$x=0,\,-2,\,2$$

Each zero has multiplicity 1 (meaning the exponent on each factor is 1, which is an odd number), so the graph crosses the x-axis at each zero.

Find the y-intercept by finding $f(0)$:

$$-2(0)^3+8(0)=0$$

Test points: Test points in the intervals $(-\infty,\ -2)$, $(-2,\ 0)$, $(0,\ 2)$, and $(2,\ \infty)$.

Interval	Test point, x	$f(x)$ Above or below x-axis.
$(-\infty, -2)$	-3	30; above
$(-2, 0)$	-1	-6; below
$(0, 2)$	1	6; above
$(2, \infty)$	3	-30; below

Illustration by Thomson Digital

312.

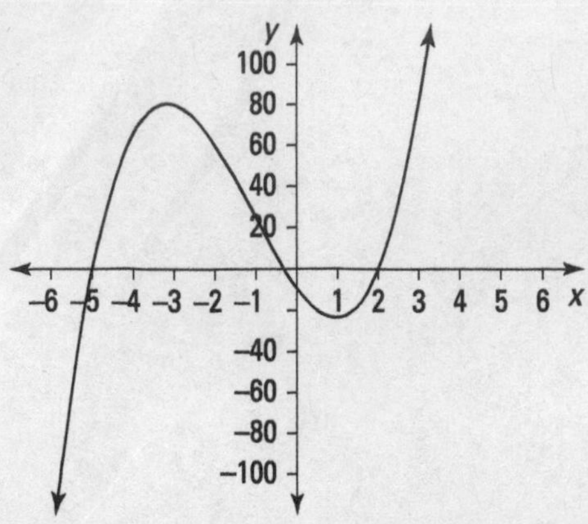

Illustration by Thomson Digital

End behavior: The leading coefficient test says that if the degree of the polynomial is odd and the leading coefficient is positive, then the left side of the graph points down and the right side points up.

Intercepts: Solve $3x^3+10x^2-27x-10=0$ to find the *x*-intercepts. The rational root theorem says that if you take all the factors of the constant term in a polynomial and divide them by all the factors of the leading coefficient, you produce a list of all the possible rational roots of the polynomial. Therefore, the possible rational zeros are

$$\frac{\pm1}{\pm1}=\pm1,\ \frac{\pm1}{\pm3}=\pm\frac{1}{3},\ \frac{\pm2}{\pm1}=\pm2,\ \frac{\pm2}{\pm3}=\pm\frac{2}{3},\ \frac{\pm5}{\pm1}=\pm5,$$
$$\frac{\pm5}{\pm3}=\pm\frac{5}{3},\ \frac{\pm10}{\pm1}=\pm10,\ \frac{\pm10}{\pm3}=\pm\frac{10}{3}$$

Use synthetic division to find that $x = 2$ is one of the zeros:

$$\begin{array}{r|rrrr} 2 & 3 & 10 & -27 & -10 \\ & & 6 & 32 & 10 \\ \hline & 3 & 16 & 5 & 0 \end{array}$$

Now solve the resulting equation, $3x^2+16x+5=0$, to find the remaining zeros:

$$\begin{aligned} 3x^2+16x+5&=0 \\ (3x+1)(x+5)&=0 \\ x&=-\frac{1}{3},\ -5 \end{aligned}$$

Each zero has multiplicity 1 (the exponent on each factor is 1, which is an odd number), so the graph crosses the *x*-axis at each zero.

Find the *y*-intercept by finding $f(0)$:

$$3(0)^3+10(0)^2-27(0)-10=-10$$

Test points: Test points in the intervals $(-\infty,-5)$, $\left(-5,-\frac{1}{3}\right)$, $\left(-\frac{1}{3},2\right)$, and $(2,\infty)$.

Interval	Test point, x	$f(x)$ Above or below x-axis.
$(-\infty,-5)$	–6	–136; below
$\left(-5,-\frac{1}{3}\right)$	–3	80; above
$\left(-\frac{1}{3},2\right)$	1	–24; below
$(2,\infty)$	3	80; above

Illustration by Thomson Digital

313.

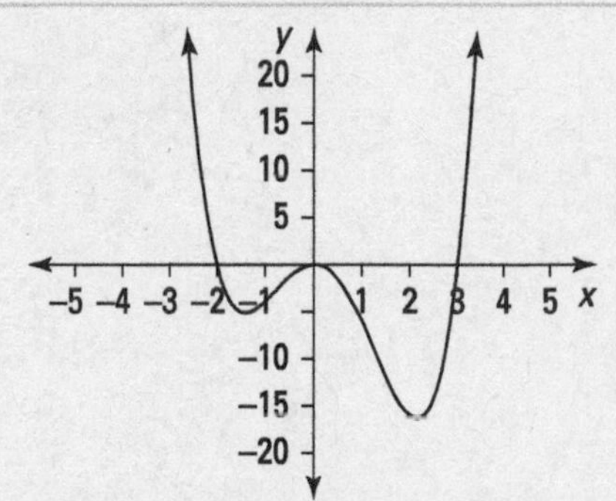

Illustration by Thomson Digital

End behavior: The leading coefficient test says that if the degree of the polynomial is even and the leading coefficient is positive, then both sides of the graph point up.

Intercepts: Solve $x^4-x^3-6x^2=0$ to find the x-intercepts:

$$\begin{aligned}x^4-x^3-6x^2&=0\\ x^2\left(x^2-x-6\right)&=0\\ x^2(x+2)(x-3)&=0\\ x&=0 \text{ (multiplicity 2)}, -2, 3\end{aligned}$$

The zeros at $x = -2$ and $x = 3$ have multiplicity 1 (the exponent on each factor is 1, which is an odd number), so the graph crosses the x-axis at –2 and 3. The zero at $x = 0$ has multiplicity 2 (the exponent of x is an even number, meaning there's no change of sign on either side), so the graph touches the x-axis at $x = 0$ but doesn't cross it.

Find the y-intercept by finding $f(0)$:

$$0^4-0^3-6(0)^2=0$$

Answers 301–400

Test points: Test points in the intervals $(-\infty, -2)$, $(-2, 0)$, $(0, 3)$, and $(3, \infty)$.

Interval	Test point, x	$f(x)$ Above or below x-axis.
$(-\infty, -2)$	-3	54; above
$(-2, 0)$	-1	-4; below
$(0, 3)$	1	-6; below
$(3, \infty)$	4	96; above

Illustration by Thomson Digital

314.

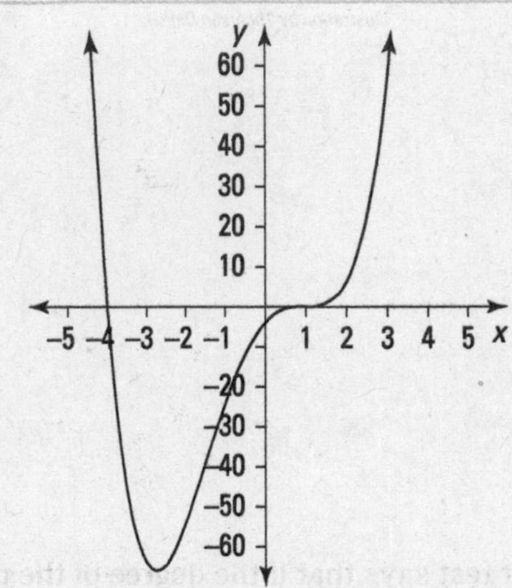

Illustration by Thomson Digital

End behavior: The leading coefficient test says that if the degree of the polynomial is even and the leading coefficient is positive, then both sides of the graph point up.

Intercepts: Solve $x^4 + x^3 - 9x^2 + 11x - 4 = 0$ to find the x-intercepts. The rational root theorem says that if you take all the factors of the constant term in a polynomial and divide them by all the factors of the leading coefficient, you produce a list of all the possible rational roots of the polynomial. Therefore, the possible rational zeros are

$$\frac{\pm 1}{\pm 1} = \pm 1,\ \frac{\pm 2}{\pm 1} = \pm 2,\ \frac{\pm 4}{\pm 1} = \pm 4$$

Use synthetic division to find that $x = 1$ is one of the zeros:

$$\begin{array}{r|rrrrr} 1 & 1 & 1 & -9 & 11 & -4 \\ & & 1 & 2 & -7 & 4 \\ \hline & 1 & 2 & -7 & 4 & 0 \end{array}$$

Now solve the resulting equation, $x^3 + 2x^2 - 7x + 4 = 0$, to find another zero. The possible rational zeros are

$$\frac{\pm 1}{\pm 1} = \pm 1,\ \frac{\pm 2}{\pm 1} = \pm 2,\ \frac{\pm 4}{\pm 1} = \pm 4$$

Use synthetic division to find that $x = 1$ is one of the zeros:

$$\begin{array}{r|rrrr} 1 & 1 & 2 & -7 & 4 \\ & & 1 & 3 & -4 \\ \hline & 1 & 3 & -4 & 0 \end{array}$$

Now solve this resulting equation, $x^2+3x-4=0$, to find the remaining zeros:

$$x^2+3x-4=0$$
$$(x+4)(x-1)=0$$
$$x=-4,1$$

The zero at $x=-4$ has multiplicity 1 (the exponent on the factor is 1, which is an odd number), so the graph crosses the x-axis at -4. The zero at $x = 1$ has multiplicity 3 (the exponent is 3, which is an odd number), so the graph crosses the x-axis at $x = 1$.

Find the y-intercept by finding $f(0)$:

$$0^4+0^3-9(0)^2+11(0)-4=-4$$

Test points: Test points in the intervals $(-\infty, -4)$, $(-4, 1)$, and $(1, \infty)$.

Interval	Test point, x	$f(x)$ Above or below x-axis.
$(-\infty, -4)$	-5	216; above
$(-4, 1)$	-1	-24; below
$(1, \infty)$	2	6; above

Illustration by Thomson Digital

315.

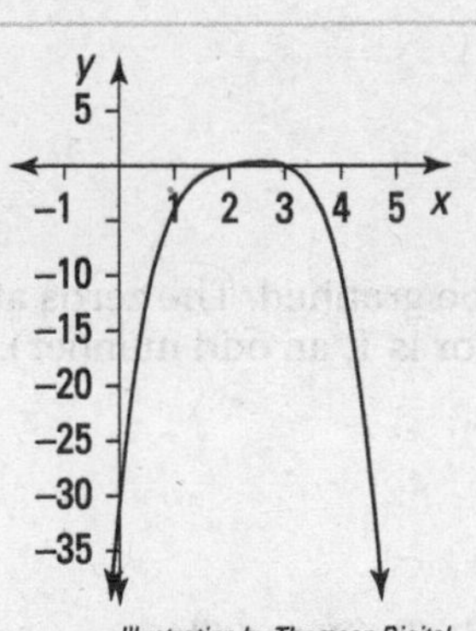

Illustration by Thomson Digital

End behavior: The leading coefficient test says that if the degree of the polynomial is even and the leading coefficient is negative, then both sides of the graph point down.

Intercepts: Solve $-x^4+9x^3-31x^2+49x-30=0$ to find the x-intercepts. The rational root theorem says that if you take all the factors of the constant term in a polynomial and divide them by all the factors of the leading coefficient, you produce a list of all the possible rational roots of the polynomial. Therefore, the possible rational zeros are

$$\frac{\pm 1}{\pm 1}=\pm 1,\ \frac{\pm 2}{\pm 1}=\pm 2,\ \frac{\pm 3}{\pm 1}=\pm 3,\ \frac{\pm 5}{\pm 1}=\pm 5,\ \frac{\pm 6}{\pm 1}=\pm 6,$$
$$\frac{\pm 10}{\pm 1}=\pm 10,\ \frac{\pm 15}{\pm 1}=\pm 15,\ \frac{\pm 30}{\pm 1}=\pm 30$$

Use synthetic division to find that $x = 2$ is one of the zeros:

$$\begin{array}{r|rrrrr} 2 & -1 & 9 & -31 & 49 & -30 \\ & & -2 & 14 & -34 & 30 \\ \hline & -1 & 7 & -17 & 15 & 0 \end{array}$$

Solve the resulting equation, $-x^3+7x^2-17x+15=0$, to find another zero. The possible rational zeros are

$$\frac{\pm 1}{\pm 1}=\pm 1,\ \frac{\pm 3}{\pm 1}=\pm 3,\ \frac{\pm 5}{\pm 1}=\pm 5,\ \frac{\pm 15}{\pm 1}=\pm 15$$

Use synthetic division to find that $x = 3$ is one of the zeros:

$$\begin{array}{r|rrrr} 3 & -1 & 7 & -17 & 15 \\ & & -3 & 12 & -15 \\ \hline & -1 & 4 & -5 & 0 \end{array}$$

Now solve the resulting equation, $-x^2+4x-5=0$, to find the remaining zeros:

$$\begin{aligned} x=\frac{-b\pm\sqrt{b^2-4ac}}{2a} &= \frac{-4\pm\sqrt{4^2-4(-1)(-5)}}{2(-1)} \\ &= \frac{-4\pm\sqrt{16-20}}{-2} \\ &= \frac{-4\pm\sqrt{-4}}{-2} \\ &= \frac{-4\pm 2i}{-2} \\ &= 2\pm i \end{aligned}$$

Therefore, two zeros are imaginary and can't be graphed. The zeros at $x = 2$ and $x = 3$ have multiplicity 1 (the exponent on each factor is 1, an odd number), so the graph crosses the x-axis at 2 and 3.

Find the y-intercept by finding $f(0)$:

$$-0^4+9(0)^3-31(0)^2+49(0)-30=-30$$

Test points: Test points in the intervals $(-\infty, 2)$, $(2, 3)$, and $(3, \infty)$.

Interval	Test point, x	$f(x)$ Above or below x-axis.
$(-\infty, 2)$	1	–4; below
$(2, 3)$	$\frac{5}{2}$	$\frac{5}{16}$; above
$(3, \infty)$	4	–10; below

Illustration by Thomson Digital

316.

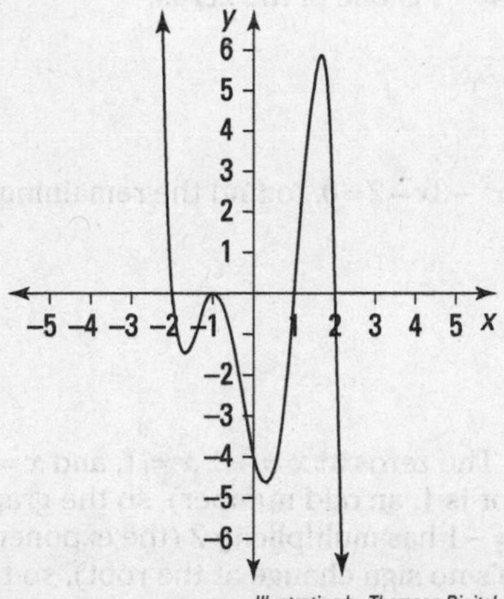

Illustration by Thomson Digital

End behavior: The leading coefficient test says that if the degree of the polynomial is odd and the leading coefficient is negative, then the left side of the graph points up and the right side points down.

Intercepts: Solve $-x^5-x^4+5x^3+5x^2-4x-4=0$ to find the x-intercepts. The rational root theorem says that if you take all the factors of the constant term in a polynomial and divide them by all the factors of the leading coefficient, you produce a list of all the possible rational roots of the polynomial. Therefore, the possible rational zeros are

$$\frac{\pm 1}{\pm 1}=\pm 1,\ \frac{\pm 2}{\pm 1}=\pm 2,\ \frac{\pm 4}{\pm 1}=\pm 4$$

Use synthetic division to find that $x = 2$ is one of the zeros:

$$\begin{array}{r|rrrrrr} 2 & -1 & -1 & 5 & 5 & -4 & -4 \\ & & -2 & -6 & -2 & 6 & 4 \\ \hline & -1 & -3 & -1 & 3 & 2 & 0 \end{array}$$

Solve the resulting equation, $-x^4-3x^3-x^2+3x+2=0$, to find another zero. The possible rational zeros are

$$\frac{\pm 1}{\pm 1}=\pm 1,\ \frac{\pm 2}{\pm 1}=\pm 2$$

Use synthetic division to find that $x = 1$ is one of the zeros:

$$\begin{array}{r|rrrrr} 1 & -1 & -3 & -1 & 3 & 2 \\ & & -1 & -4 & -5 & -2 \\ \hline & -1 & -4 & -5 & -2 & 0 \end{array}$$

Solve the resulting equation, $-x^3-4x^2-5x-2=0$, to find another zero. The possible rational zeros are

$$\frac{\pm 1}{\pm 1}=\pm 1,\ \frac{\pm 2}{\pm 1}=\pm 2$$

Use synthetic division to find that $x = -1$ is one of the zeros:

$$\begin{array}{r|rrrr} -1 & -1 & -4 & -5 & -2 \\ & & 1 & 3 & 5 \\ \hline & -1 & -3 & -2 & 0 \end{array}$$

Now solve the resulting equation, $-x^2-3x-2=0$, to find the remaining zeros:

$$\begin{aligned} -x^2-3x-2&=0 \\ -\left(x^2+3x+2\right)&=0 \\ -(x+2)(x+1)&=0 \\ x&=-2,\,-1 \end{aligned}$$

Thus, the zeros are −2, −1, 1, and 2. The zeros at $x = -2$, $x = 1$, and $x = 2$ have multiplicity 1 (the exponent on each factor is 1, an odd number), so the graph crosses the x-axis at these points. The zero at $x = -1$ has multiplicity 2 (the exponent on the factor is 2, an even number, meaning there's no sign change at the root), so the graph touches but doesn't cross the x-axis at this point.

Find the y-intercept by finding $f(0)$:

$$-(0)^5-0^4+5(0)^3+5(0)^2-4(0)-4=-4$$

Test points: Test points in the intervals $(-\infty, -2)$, $(-2, -1)$, $(-1, 1)$, $(1, 2)$, and $(2, \infty)$.

Interval	Test point, x	$f(x)$ Above or below x-axis.
$(-\infty, -2)$	−3	80; above
$(-2, -1)$	$-\frac{3}{2}$	$-\frac{35}{32}=-1.09375$; below
$(-1, 1)$	0	−4; below
$(1, 2)$	$\frac{3}{2}$	$\frac{175}{32}=5.46875$; above
$(2, \infty)$	3	−160; below

Illustration by Thomson Digital

317. $f(x)=x^3-5x^2+2x+8$

The graph crosses the x-axis at $x = -1$, $x = 2$, and $x = 4$, so the function is given by

$$f(x)=a(x+1)(x-2)(x-4)$$

where a is a constant. The y-intercept is (0, 8), so find a by plugging in these values and solving:

$$\begin{aligned} 8&=a(0+1)(0-2)(0-4) \\ 8&=8a \\ a&=1 \end{aligned}$$

Therefore, the function is

$$f(x)=(x+1)(x-2)(x-4)=x^3-5x^2+2x+8$$

318. $f(x)=-2x^4+26x^2-72$

The graph crosses the x-axis at $x=-3$, $x=-2$, $x=2$, and $x=3$, so the function is given by

$$f(x)=a(x+3)(x+2)(x-2)(x-3)$$

where a is a constant. The y-intercept is (0, –72), so find a by plugging in these values and solving:

$$-72=a(0+3)(0+2)(0-2)(0-3)$$
$$-72=36a$$
$$a=-2$$

Therefore, the function is

$$f(x)=-2(x+3)(x+2)(x-2)(x-3)=-2x^4+26x^2-72$$

319. $f(x)=-x^4-13x^3-55x^2-75x$

The graph crosses the x-axis at $x=0$ and $x=-3$. This means that two factors of $f(x)$ are x and $(x+3)$. The graph touches the x-axis but doesn't cross it at $x=-5$, so $f(x)$ has an even number of factors of $(x+5)$. Therefore, try the following as an equation of the graph:

$$f(x)=ax(x+3)(x+5)^2$$

To find the constant a, substitute the coordinates of the given point, (–1, 32), for x and y and solve:

$$32=a(-1)((-1)+3)((-1)+5)^2$$
$$32=-a(2)(4)^2$$
$$32=-32a$$
$$a=-1$$

Therefore, the function is

$$f(x)=-x(x+3)(x+5)^2=-x^4-13x^3-55x^2-75x$$

320. $f(x)=-x^5-2x^4+6x^3+4x^2-13x+6$

The graph crosses the x-axis at $x=-3$ and $x=-2$, so two factors of $f(x)$ are $(x+3)$ and $(x+2)$. The graph also crosses the x-axis at $x=1$, so $(x-1)$ is also a factor. If $f(x)=a(x+3)(x+2)(x-1)$, then $f(x)$ is a cubic polynomial. But $(x-1)$ must be a factor of $f(x)$ an odd number of times, so try the following as an equation of the graph:

$$f(x)=a(x+3)(x+2)(x-1)^3$$

The y-intercept is (0, 6), so find the constant a by plugging in these values and solving:

$$6=a(0+3)(0+2)(0-1)^3$$
$$6=-6a$$
$$a=-1$$

Therefore, the function is

$$f(x)=-1(x+3)(x+2)(x-1)^3=-x^5-2x^4+6x^3+4x^2-13x+6$$

321. $f(-2)=9, f(1)=\frac{1}{3}$

To evaluate the function, plug in the given values for x and solve. At $f(-2)$, the function equals

$$f(-2)=\left(\frac{1}{3}\right)^{-2}=3^2=9$$

And at $f(1)$, the function equals

$$f(1)=\left(\frac{1}{3}\right)^{1}=\frac{1}{3}$$

322. $f(-3)=-\frac{1}{8}, f(2)=-4$

To evaluate the function, plug in the given values for x and solve. At $f(-3)$, the function equals

$$f(-3)=-2^{-3}=-\frac{1}{2^3}=-\frac{1}{8}$$

And at $f(2)$, the function equals

$$f(2)=-2^2=-4$$

323. $f(0)=-1, f(3)=123$

To evaluate the function, plug in the given values for x and solve. At $f(0)$, the function equals

$$\begin{aligned} f(0)&=5^x-2=5^0-2\\ &=1-2=-1 \end{aligned}$$

And at $f(3)$, the function equals

$$\begin{aligned} f(3)&=5^x-2=5^3-2\\ &=125-2=123 \end{aligned}$$

324. $f(-3)=-22, f(4)=\frac{29}{16}$

To evaluate the function, plug in the given values for x and solve. At $f(-3)$, the function equals

$$\begin{aligned} f(-3)&=-3\left(\frac{1}{2}\right)^{-3}+2\\ &=-3(2)^3+2\\ &=-3(8)+2\\ &=-24+2\\ &=-22 \end{aligned}$$

And at $f(4)$, the function equals

$$\begin{aligned} f(4) &= -3\left(\frac{1}{2}\right)^4 + 2 \\ &= -3\left(\frac{1}{16}\right) + 2 \\ &= -\frac{3}{16} + 2 \\ &= \frac{29}{16} \end{aligned}$$

325. $f(1) = 0, f(5) = -15$

To evaluate the function, plug in the given values for x and solve. At $f(1)$, the function equals

$$\begin{aligned} f(1) &= \frac{1}{2}(-2)^1 + 1 \\ &= \frac{1}{2}(-2) + 1 \\ &= -1 + 1 \\ &= 0 \end{aligned}$$

And at $f(5)$, the function equals

$$\begin{aligned} f(5) &= \frac{1}{2}(-2)^5 + 1 \\ &= \frac{1}{2}(-32) + 1 \\ &= -16 + 1 \\ &= -15 \end{aligned}$$

326.

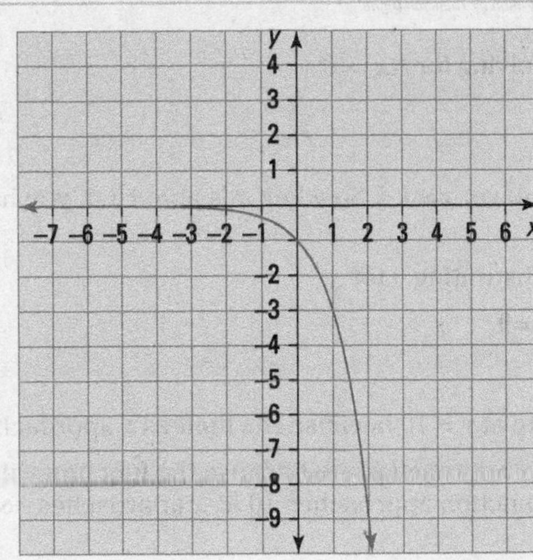

Illustration by Thomson Digital

You find the x-intercepts by solving for $f(x) = 0$. No values of x make the equation true, so there are no x-intercepts.

You find the y-intercept by substituting 0 for x:

$$f(0) = -3^0 = -1$$

So the y-intercept is (0, –1).

There's a horizontal asymptote at $y = 0$ because the limit as x approaches $-\infty$ is 0. The function is decreasing as x approaches $+\infty$ because the values of the function are getting smaller and smaller, and the function approaches 0 as x approaches $-\infty$ because of the horizontal asymptote.

327.

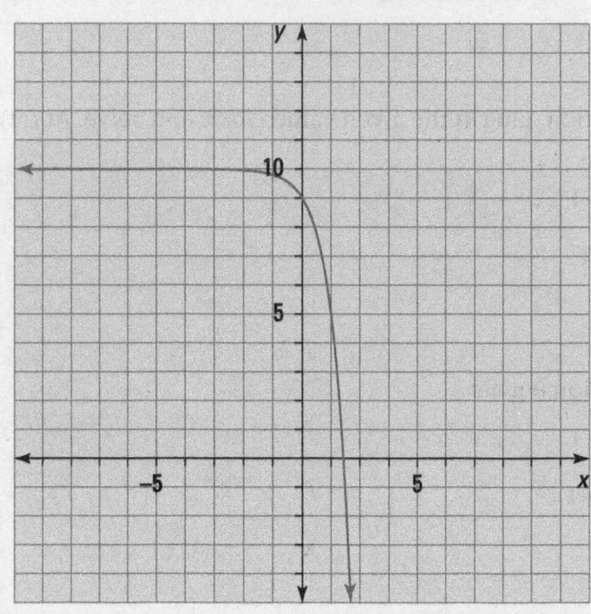

Illustration by Thomson Digital

You find the x-intercepts by solving for $f(x) = 0$:

$$0 = -5^x + 10$$
$$5^x = 10$$

Taking the log of each side, you get $x \approx 1.4$. So when x is about 1.4, you have an x-intercept.

You find the y-intercept by substituting 0 for x:

$$f(0) = -5^0 + 10 = -1 + 10 = 9$$

So the y-intercept is (0, 9).

There's a horizontal asymptote at $y = 10$ because the limit as x approaches $-\infty$ is 10.

The function is decreasing as x approaches $+\infty$ because the function values are getting smaller and smaller, and the function approaches 10 as x approaches $-\infty$ because of the horizontal asymptote.

328.

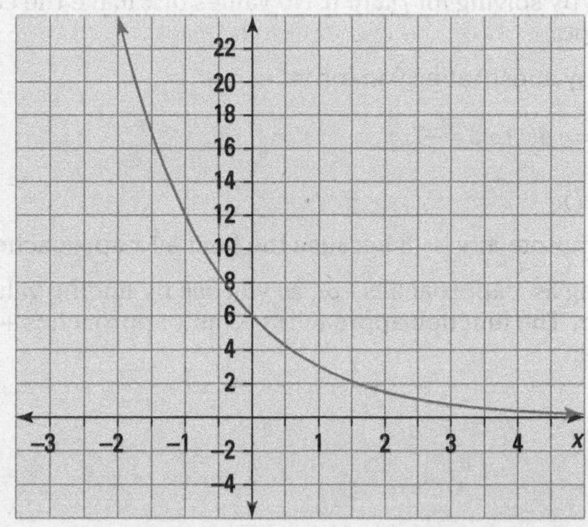

Illustration by Thomson Digital

You find the x-intercepts by solving for $f(x)=0$. No values of x make the equation true, so there are no x-intercepts.

You find the y-intercept by substituting 0 in for x:

$$f(0)=6\left(\frac{1}{2}\right)^0=6(1)=6$$

So the y-intercept is (0, 6).

There's a horizontal asymptote at $y=0$ because the limit as x approaches $+\infty$ is 0.

The function is increasing as x approaches $-\infty$ because the values of the function keep getting larger and larger, and the function approaches 0 as x approaches $+\infty$ because of the horizontal asymptote.

329.

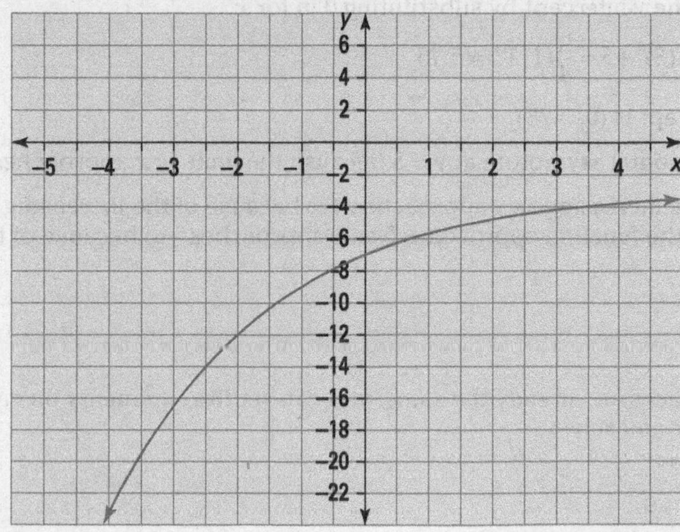

Illustration by Thomson Digital

You find the x-intercepts by solving for $f(x) = 0$. No values of x make the equation true, so there are no x-intercepts.

You find the y-intercept by substituting 0 in for x:

$$f(0) = -4\left(\frac{2}{3}\right)^0 - 3 = -4(1) - 3 = -7$$

So the y-intercept is (0, –7).

There's a horizontal asymptote at $y = -3$ because the limit as x approaches $+\infty$ is –3.

The function is decreasing as x approaches $-\infty$, as you see by finding values of the function as x gets smaller. The function approaches –3 as x approaches $+\infty$ because of the horizontal asymptote.

330.

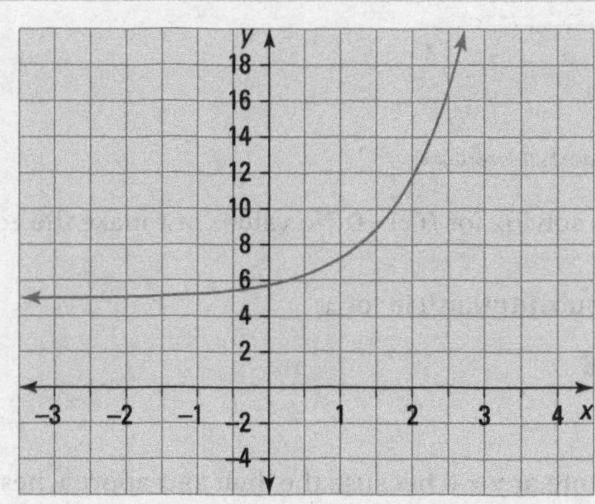

Illustration by Thomson Digital

You find the x-intercepts by solving for $f(x) = 0$. No values of x make the equation true, so there are no x-intercepts.

You can find the y-intercept by substituting 0 in for x:

$$f(x) = \frac{3}{4}(3)^0 + 5 = \frac{3}{4}(1) + 5 = 5.75$$

So the y-intercept is (0, 5.75).

There's a horizontal asymptote at $y = 5$ because the limit as x approaches $-\infty$ is 5.

The function is increasing as x approaches $+\infty$ because of the increasing values of the function, and the function approaches 5 as x approaches $-\infty$ because of the horizontal asymptote.

331. $x = 6$

Because the bases are already the same, you can set the exponents on each side equal to one another and solve:

$$e^{4x} = e^{3x+6}$$
$$4x = 3x + 6$$
$$x = 6$$

332. $x = -5$

First, rewrite the right side of the equation so that it has the same base as the left:

$$2^x = \frac{1}{32}$$
$$2^x = \frac{1}{2^5}$$
$$2^x = 2^{-5}$$

The bases are now the same, so set the exponents on each side equal to each other:

$$x = -5$$

333. $x = -\frac{3}{2}$

First, rewrite the left side to have the same base as the right side:

$$\left(\frac{1}{e}\right)^{2x} = e^3$$
$$\left(e^{-1}\right)^{2x} = e^3$$
$$e^{-2x} = e^3$$

When the bases are the same, you can set the exponents on each side equal to each other and solve for x:

$$-2x = 3$$
$$x = -\frac{3}{2}$$

334. $x = -\frac{2}{3}$

First, rewrite the right side of the equation so that it has the same base as the left side:

$$3^{3x+4} = 9$$
$$3^{3x+4} = 3^2$$

When the bases are the same, set the exponents on each side equal to each other and solve for x:

$$3x + 4 = 2$$
$$3x = -2$$
$$x = -\frac{2}{3}$$

335. $x = \frac{5}{2}$

First, rewrite the left and right sides of the equation so they have the same base:

$$\left(\frac{1}{5}\right)^{4x-6} = \frac{1}{625}$$
$$(5^{-1})^{4x-6} = 5^{-4}$$
$$5^{-4x+6} = 5^{-4}$$

When the bases are the same, set the exponents on each side equal to each other and solve for x:

$$-4x + 6 = -4$$
$$-4x = -10$$
$$x = \frac{5}{2}$$

336. $x = -\frac{3}{4}$

The bases are the same, and so are the coefficients:

$$2e^{3x+8} = 2e^{-9x-1}$$
$$e^{3x+8} = e^{-9x-1}$$

Set the exponents on each side of the equation equal to each other and solve for x:

$$3x + 8 = -9x - 1$$
$$12x + 8 = -1$$
$$12x = -9$$
$$x = -\frac{3}{4}$$

337. $x = 0$

First, add 44 to each side and rewrite the left and right sides of the equation so that they have the same base:

$$12^{-3x+2} - 44 = 100$$
$$12^{-3x+2} = 144$$
$$12^{-3x+2} = 12^2$$

When the bases are the same, set the exponents on each side equal to each other and solve for x:

$$-3x + 2 = 2$$
$$-3x = 0$$
$$x = 0$$

338. $x = -12$

First, subtract 80 from each side and rewrite the left and right sides of the equation so that they have the same base:

$$3^{-x/2} + 80 = 809$$
$$3^{-x/2} = 729$$
$$3^{-x/2} = 3^6$$

When the bases are the same, set the exponents on each side equal to each other and solve for x:

$$-\frac{x}{2} = 6$$
$$x = -12$$

339. $x = \frac{2}{7}$

First, add $\frac{1}{2}$ to each side and rewrite the left and right sides of the equation so that they have the same base:

$$\left(\frac{3}{4}\right)^{7x} - \frac{1}{2} = \frac{1}{16}$$
$$\left(\frac{3}{4}\right)^{7x} = \frac{9}{16}$$
$$\left(\frac{3}{4}\right)^{7x} = \left(\frac{3}{4}\right)^2$$

When the bases are the same, set the exponents on each side equal to each other and solve for x:

$$7x = 2$$
$$x = \frac{2}{7}$$

340. $x = -3$

First, divide each side by 6 and rewrite the left and right sides of the equation so that they have the same base:

$$6\left(6^{-4x-9}\right) = 1{,}296$$
$$6^{-4x-9} = 216$$
$$6^{-4x-9} = 6^3$$

When the bases are the same, set the exponents on each side equal to each other and solve for x:

$$-4x - 9 = 3$$
$$-4x = 12$$
$$x = -3$$

341. $x = 1$

...tract 24 from each side and divide each side of the resulting equation by 2. Rewrite the ...ght and left sides so that the bases are the same:

$$2\left(8^{6x-3}\right)\ 24 = 1{,}048$$

$$2\left(\ ^{x-3}\right) = 1{,}024$$

$$8^{6x-3} = 512$$

$$8^{6x-3} = 8^3$$

Next ...et the exponents on each side equal to each other and solve for x:

$$6x - 3 = 3$$

$$6x = 6$$

$$x = 1$$

342. $x = \frac{13}{6}$

First, subtract 25 from each side of the equation and rewrite the right and left sides so that the bases are the same:

$$\left(\frac{1}{49}\right)^{3x-8} + 25 = 368$$

$$\left(\frac{1}{49}\right)^{3x-8} = 343$$

$$\left(7^{-2}\right)^{3x-8} = 7^3$$

$$7^{-6x+16} = 7^3$$

Then set the exponents on each side equal to each other and solve for x:

$$-6x + 16 = 3$$

$$-6x = -13$$

$$x = \frac{13}{6}$$

343. $x = -2$

First, subtract $\frac{1}{64}$ from each side and rewrite the right and left sides so that the bases are the same:

$$(64)^{-2x-5} + \frac{1}{64} = \frac{1}{32}$$

$$(64)^{-2x-5} = \frac{1}{64}$$

$$(64)^{-2x-5} = 64^{-1}$$

Then set the exponents on each side equal to each other and solve for x:

$$-2x - 5 = -1$$

$$-2x = 4$$

$$x = -2$$

338. $x=-12$

First, subtract 80 from each side and rewrite the left and right sides of the equation so that they have the same base:

$$3^{-x/2}+80=809$$
$$3^{-x/2}=729$$
$$3^{-x/2}=3^6$$

When the bases are the same, set the exponents on each side equal to each other and solve for x:

$$-\frac{x}{2}=6$$
$$x=-12$$

339. $x=\frac{2}{7}$

First, add $\frac{1}{2}$ to each side and rewrite the left and right sides of the equation so that they have the same base:

$$\left(\frac{3}{4}\right)^{7x}-\frac{1}{2}=\frac{1}{16}$$
$$\left(\frac{3}{4}\right)^{7x}=\frac{9}{16}$$
$$\left(\frac{3}{4}\right)^{7x}=\left(\frac{3}{4}\right)^{2}$$

When the bases are the same, set the exponents on each side equal to each other and solve for x:

$$7x=2$$
$$x=\frac{2}{7}$$

340. $x=-3$

First, divide each side by 6 and rewrite the left and right sides of the equation so that they have the same base:

$$6\left(6^{-4x-9}\right)=1{,}296$$
$$6^{-4x-9}=216$$
$$6^{-4x-9}=6^3$$

When the bases are the same, set the exponents on each side equal to each other and solve for x:

$$-4x-9=3$$
$$-4x=12$$
$$x=-3$$

341. $x=1$

First, subtract 24 from each side and divide each side of the resulting equation by 2. Then rewrite the right and left sides so that the bases are the same:

$$\begin{aligned}2\left(8^{6x-3}\right)+24&=1{,}048\\2\left(8^{6x-3}\right)&=1{,}024\\8^{6x-3}&=512\\8^{6x-3}&=8^3\end{aligned}$$

Next, set the exponents on each side equal to each other and solve for x:

$$\begin{aligned}6x-3&=3\\6x&=6\\x&=1\end{aligned}$$

342. $x=\frac{13}{6}$

First, subtract 25 from each side of the equation and rewrite the right and left sides so that the bases are the same:

$$\begin{aligned}\left(\frac{1}{49}\right)^{3x-8}+25&=368\\\left(\frac{1}{49}\right)^{3x-8}&=343\\\left(7^{-2}\right)^{3x-8}&=7^3\\7^{-6x+16}&=7^3\end{aligned}$$

Then set the exponents on each side equal to each other and solve for x:

$$\begin{aligned}-6x+16&=3\\-6x&=-13\\x&=\frac{13}{6}\end{aligned}$$

343. $x=-2$

First, subtract $\frac{1}{64}$ from each side and rewrite the right and left sides so that the bases are the same:

$$\begin{aligned}(64)^{-2x-5}+\frac{1}{64}&=\frac{1}{32}\\(64)^{-2x-5}&=\frac{1}{64}\\(64)^{-2x-5}&=64^{-1}\end{aligned}$$

Then set the exponents on each side equal to each other and solve for x:

$$\begin{aligned}-2x-5&=-1\\-2x&=4\\x&=-2\end{aligned}$$

344. $x = -8, 5$

First, rewrite the right side so that the bases are the same:

$$4^{x^2-40} = \left(\frac{1}{4}\right)^{3x}$$
$$4^{x^2-40} = \left(4^{-1}\right)^{3x}$$
$$4^{x^2-40} = 4^{-3x}$$

Then you can set the exponents on each side equal to each other and solve for x:

$$x^2 - 40 = -3x$$
$$x^2 + 3x - 40 = 0$$
$$(x+8)(x-5) = 0$$
$$x = -8, 5$$

345. $x = -5, 5$

First, rewrite the right and left sides of the equation so that the bases are the same:

$$2^{x^2+8x-15} = 4^{4x+5}$$
$$2^{x^2+8x-15} = \left(2^2\right)^{4x+5}$$
$$2^{x^2+8x-15} = 2^{8x+10}$$

Then set the exponents on each side equal to each other and solve for x:

$$x^2 + 8x - 15 = 8x + 10$$
$$x^2 = 25$$
$$\sqrt{x^2} = \pm\sqrt{25}$$
$$x = -5, 5$$

346. $\log_{10} 100 = x$

To convert an exponential expression to a logarithm, use

$$a^b = c \rightarrow \log_a c = b$$

Therefore, $10^x = 100$ becomes $\log_{10} 100 = x$.

347. $\log_y \frac{1}{27} = 3$

To convert an exponential expression to a logarithm, use

$$a^b = c \rightarrow \log_a c = b$$

Therefore, $y^3 = \frac{1}{27}$ becomes $\log_y \frac{1}{27} = 3$.

Answers 301–400

348. $\log_4 z = 3$

To convert an exponential expression to a logarithm, use

$$a^b = c \to \log_a c = b$$

Therefore, $4^3 = z$ becomes $\log_4 z = 3$.

349. $8^x = 128$

To convert a logarithmic expression to an exponential expression, use

$$\log_a c = b \to a^b = c$$

Therefore, $\log_8 128 = x$ becomes $8^x = 128$.

350. $64^{1/2} = y$

To convert a logarithmic expression to an exponential expression, use

$$\log_a c = b \to a^b = c$$

Therefore, $\log_{64} y = \frac{1}{2}$ becomes $64^{1/2} = y$.

351. $\log 4 + \log x$

Using the logarithm product rule,

$$\log_b MN = \log_b M + \log_b N$$

rewrite the logarithm as a sum:

$$\log(4x) = \log 4 + \log x$$

352. $\log_6 24 - \log_6 y$

Using the logarithm quotient rule,

$$\log_b\left(\frac{M}{N}\right) = \log_b M - \log_b N$$

rewrite the logarithm as a difference:

$$\log_6\left(\frac{24}{y}\right) = \log_6 24 - \log_6 y$$

353. $2\log_8 x + 1$

Using the logarithm power rule and the logarithm of a base rule,

$$\log_b M^a = a\log_b M$$

$$\log_b b = 1$$

rewrite each term in the sum:

$$\log_8 x^2 + \log_8 8 = 2\log_8 x + 1$$

354. 2

Using the logarithm of a base rule and the logarithm of 1 rule,

$$\log_b b^k = k$$

$$\log_b 1 = 0$$

rewrite each term in the sum:

$$\log_5 5^2 + \log_5 1 = 2\log_5 5 + 0 = 2(1)$$

355. $\log_6(8x^3)$

Using the power and product rules of logarithms,

$$\log_b M^a = a\log_b M$$

$$\log_b MN = \log_b M + \log_b N$$

condense the expression:

$$\begin{aligned} 3\log_6 x + \log_6 8 &= \log_6 x^3 + \log_6 8 \\ &= \log_6 (8x^3) \end{aligned}$$

356. $\ln(10y^2)$

Using the power and product rules of logarithms,

$$\log_b M^a = a\log_b M$$

$$\log_b MN = \log_b M + \log_b N$$

condense the expression:

$$\begin{aligned} \ln 10 + 2\ln y &= \ln 10 + \ln y^2 \\ &= \ln(10y^2) \end{aligned}$$

357. –1

Using the power rule and logarithm of a base rule,

$$\log_b M^a = a\log_b M$$

$$\log_b b = 1$$

condense the expression:

$$\begin{aligned} &\log 100 - \log 1{,}000 \\ &= \log 10^2 - \log 10^3 \\ &= 2\log 10 - 3\log 10 \\ &= 2 - 3 = -1 \end{aligned}$$

358. $\log_4\left(\frac{x^5}{y^2}\right)$

Using the power and quotient rules of logarithms,

$$\log_b M^a = a\log_b M$$

$$\log_b\left(\frac{M}{N}\right) = \log_b M - \log_b N$$

condense the expression:

$$\begin{aligned}5\log_4 x - 2\log_4 y &= \log_4 x^5 - \log_4 y^2\\ &= \log_4\left(\frac{x^5}{y^2}\right)\end{aligned}$$

359. $\log_8 w + 2\log_8 x - 2$

Using the product, quotient, and power rules of logarithms, as well as the logarithm of a base rule,

$$\log_b MN = \log_b M + \log_b N$$

$$\log_b\left(\frac{M}{N}\right) = \log_b M - \log_b N$$

$$\log_b M^a = a\log_b M$$

$$\log_b b = 1$$

expand the expression:

$$\begin{aligned}&\log_8\left(\frac{wx^2}{64}\right)\\ &= \log_8 w + \log_8 x^2 - \log_8 64\\ &= \log_8 w + 2\log_8 x - \log_8 8^2\\ &= \log_8 w + 2\log_8 x - 2\log_8 8\\ &= \log_8 w + 2\log_8 x - 2\end{aligned}$$

360. $\ln 10 - 4 - 2\ln y$

Using the product, quotient, and power rules of logarithms, as well as the logarithm of a base rule,

$$\log_b MN = \log_b M + \log_b N$$

$$\log_b\left(\frac{M}{N}\right) = \log_b M - \log_b N$$

$$\log_b M^a = a\log_b M$$

$$\log_b b = 1$$

expand the expression:

$$\ln\left(\frac{10e^{-4}}{y^2}\right)$$
$$= \ln 10 + \ln e^{-4} - \ln y^2$$
$$= \ln 10 - 4\ln e - 2\ln y$$
$$= \ln 10 - 4 - 2\ln y$$

361. $3 + 2\log x + 3\log y$

Using the product and power rules of logarithms, as well as the logarithm of a base rule,

$$\log_b MN = \log_b M + \log_b N$$
$$\log_b M^a = a\log_b M$$
$$\log_b b = 1$$

expand the expression:

$$\log 1000x^2y^3$$
$$= \log 1000 + \log x^2 + \log y^3$$
$$= \log 10^3 + 2\log x + 3\log y$$
$$= 3\log 10 + 2\log x + 3\log y$$
$$= 3 + 2\log x + 3\log y$$

362. $-2 - \log_6 x - 5\log_6 y$

Using the quotient and power rules of logarithms, as well as the logarithm of a base rule and the logarithm of 1 rule,

$$\log_b\left(\frac{M}{N}\right) = \log_b M - \log_b N$$
$$\log_b M^a = a\log_b M$$
$$\log_b b = 1$$
$$\log_b 1 = 0$$

expand the expression:

$$\log_6\left(\frac{1}{36xy^5}\right)$$
$$= \log_6 1 - \log_6 36 - \log_6 x - 5\log_6 y^5$$
$$= 0 - \log_6 6^2 - \log_6 x - 5\log_6 y$$
$$= -2\log_6 6 - \log_6 x - 5\log_6 y$$
$$= -2 - \log_6 x - 5\log_6 y$$

363. $-9+\ln 6+7\ln y-\ln 7-2\ln z$

Using the product, quotient, and power rules of logarithms,

$$\log_b MN=\log_b M+\log_b N$$

$$\log_b\left(\frac{M}{N}\right)=\log_b M-\log_b N$$

$$\log_b M^a=a\log_b M$$

expand the expression:

$$\begin{aligned}&\ln\left(\frac{6e^{-4}y^7}{7e^5z^2}\right)\\&=\ln 6+\ln e^{-4}+\ln y^7-\ln 7-\ln e^5-\ln z^2\\&=\ln 6-4\ln e+7\ln y-\ln 7-5\ln e-2\ln z\\&=\ln 6-4+7\ln y-\ln 7-5-2\ln z\\&=-9+\ln 6+7\ln y-\ln 7-2\ln z\end{aligned}$$

364. $1+4\log_3 x-5\log_3 y-2\log_3 z$

Using the product, quotient, and power rules of logarithms, as well as the logarithm of a base rule,

$$\log_b MN=\log_b M+\log_b N$$

$$\log_b\left(\frac{M}{N}\right)=\log_b M-\log_b N$$

$$\log_b M^a=a\log_b M$$

$$\log_b b=1$$

expand the expression:

$$\begin{aligned}&\log_3\left(\frac{9x^4y^{-5}}{3z^2}\right)\\&=\log_3 9+\log_3 x^4+\log_3 y^{-5}-\log_3 3-\log_3 z^2\\&=\log_3 3^2+4\log_3 x-5\log_3 y-1-2\log_3 z\\&=2+4\log_3 x-5\log_3 y-1-2\log_3 z\\&=1+4\log_3 x-5\log_3 y-2\log_3 z\end{aligned}$$

365. $3-2\log y+\log z-\log 8+4\log x$

Using the product, quotient, and power rules of logarithms, as well as the logarithm of a base rule,

$$\log_b MN=\log_b M+\log_b N$$

$$\log_b\left(\frac{M}{N}\right)=\log_b M-\log_b N$$

$$\log_b M^a = a \log_b M$$
$$\log_b b = 1$$

expand the expression:

$$\begin{aligned}&\log\left(\frac{1{,}000y^{-2}z}{8x^{-4}}\right)\\&=\log 1{,}000+\log y^{-2}+\log z-\log 8-\log x^{-4}\\&=\log 10^3-2\log y+\log z-\log 8+4\log x\\&=3\log 10-2\log y+\log z-\log 8+4\log x\\&=3-2\log y+\log z-\log 8+4\log x\end{aligned}$$

366. 11

First, convert the equation to an exponential format. Then solve for x:

$$\begin{aligned}\log(4x-8)&=2\\6^2&=4x-8\\36&=4x-8\\44&=4x\\11&=x\end{aligned}$$

367. $\frac{e^4+1}{8}$

First, convert the equation to an exponential format (recall that e is the base of the natural log). Then solve for x:

$$\begin{aligned}\ln(8x-1)&=4\\e^4&=8x-1\\e^4+1&=8x\\\frac{e^4+1}{8}&=x\end{aligned}$$

368. $\frac{17}{10}$

First, convert the equation to an exponential format (remember that if the base is unlisted, it's equal to 10). Then solve for x:

$$\begin{aligned}\log(3x-5)&=-1\\10^{-1}&=3x-5\\\frac{1}{10}&=3x-5\\\frac{51}{10}&=3x\\\frac{17}{10}&=x\end{aligned}$$

369. 5

First, rewrite the equation as one logarithm:

$$\log_3(5x+2)+\log_3(x-4)=3$$
$$\log_3(5x+2)(x-4)=3$$
$$\log_3\left(5x^2-18x-8\right)=3$$

Next, change the logarithmic equation to an exponential format and solve for x:

$$3^3=5x^2-18x-8$$
$$27=5x^2-18x-8$$
$$0=5x^2-18x-35$$
$$0=(5x+7)(x-5)$$
$$x=-\frac{7}{5},5$$

The solution $-\frac{7}{5}$ is extraneous for the equation because it creates a negative argument for the second log term.

370. $-\frac{5}{7}$

First, rewrite the equation as one logarithm:

$$\log_5(4x+3)-\log_5(2x+5)=-2$$
$$\log_5\left(\frac{4x+3}{2x+5}\right)=-2$$

Then convert the equation to an exponential format and solve for x:

$$5^{-2}=\frac{4x+3}{2x+5}$$
$$\frac{1}{25}=\frac{4x+3}{2x+5}$$
$$2x+5=25(4x+3)$$
$$2x+5=100x+75$$
$$-98x=70$$
$$x=-\frac{70}{98}=-\frac{5}{7}$$

371. $\frac{5e^3+4}{3}$

First, rewrite the equation as one logarithm.

$$\ln(6x-8)-\ln 10=3$$
$$\ln\left(\frac{6x-8}{10}\right)=3$$
$$\ln\left(\frac{3x-4}{5}\right)=3$$

Then convert the equation to an exponential format and solve for x (recall that the base of the natural log is e):

$$e^3 = \frac{3x-4}{5}$$
$$5e^3 = 3x-4$$
$$5e^3+4 = 3x$$
$$\frac{5e^3+4}{3} = x$$

372. 3

First, rewrite the equation as one logarithm:

$$\log_2(x+5)+\log_2(x-1)=4$$
$$\log_2(x+5)(x-1)=4$$
$$\log_2\left(x^2+4x-5\right)=4$$

Then convert the equation to an exponential format and solve for x:

$$2^4=x^2+4x-5$$
$$16=x^2+4x-5$$
$$0=x^2+4x-21$$
$$0=(x+7)(x-3)$$
$$x=-7,3$$

The solution −7 is extraneous for the equation because it makes both of the arguments negative in the original log terms.

373. −3, 15

Because the two logarithms have the same base (base 10), their arguments are equal. Set those two arguments equal to one another. Then solve the equation by using inverse operations.

$$\log\left(x^2-8x\right)=\log(4x+45)$$
$$x^2-8x=4x+45$$
$$x^2-12x-45=0$$
$$(x-15)(x+3)=0$$
$$x=15,-3$$

374. no solution

First, write each side of the equation as a power of e to eliminate the natural log. Then solve the equation by using inverse operations.

$$\ln(x^2+16x)=\ln(4x-32)$$
$$e^{\ln(x^2+16x)}=e^{\ln(4x-32)}$$
$$x^2+16x=4x-32$$
$$x^2+12x+32=0$$
$$(x+4)(x+8)=0$$
$$x=-4,-8$$

The solutions –4 and –8 are extraneous for the equation because the arguments of logs have to be positive. These numbers create negative and zero arguments, so this equation has no solution.

375. 10

First, rewrite the left side of the equation as one logarithm:

$$\log_3(3x)-\log_3(x+5)=\log_3 2$$
$$\log_3\left(\frac{3x}{x+5}\right)=\log_3 2$$

Next, write each side of the equation as a power of 3 to eliminate the log. Then use inverse operations to solve the equation.

$$3^{\log_3\left(\frac{3x}{x+5}\right)}=3^{\log_3 2}$$
$$\frac{3x}{x+5}=2$$
$$3x=2(x+5)$$
$$3x=2x+10$$
$$x=10$$

376. $f(x)=\log(x)+4$

The graph appears to be raised 4 units above where the graph of the original function would be. The function transformation $f(x)+k$ raises the original graph by k units. So in this case, using $f(x)=\log(x)+4$, the point (1, 0) in the original graph moves to (1, 4). The x-intercept of $f(x)=\log(x)+4$ is found by setting the function equal to 0 and solving for x:

$$0=\log(x)+4$$
$$-4=\log(x)$$

Then change the log expression to its equivalent exponential expression:

$$10^{-4}=x$$

So the x-intercept is (0.0001, 0).

The vertical asymptote is still $x=0$, because this is a vertical slide.

377. $f(x)=\log(x+1)$

The graph appears to have slid 1 unit to the left of where the graph of the original function would be.

The function transformation $f(x+k)$ slides the original graph k units to the left. Using $f(x)=\log(x+1)$, the point (1, 0) on the original graph moves to (0, 0). The vertical asymptote also slides to the left and is now $x=-1$.

378. $f(x)=3\log(-x)$

The graph appears to be a reflection of the original graph over a vertical line, and it seems slightly steeper.

The function transformation $f(-x)$ reflects the original graph over a vertical line. Using $f(x)=\log(-x)$, the point (1, 0) on the original graph flips over the vertical axis to (–1, 0), and the vertical asymptote stays the same because it's the line of reflection.

The multiplier of 3 makes the graph slightly steeper than the original. For example, when $x=-5$ on the graph of $f(x)=\log(-x)$, then $y=\log(5)\approx 0.69897$. Multiplying that by 3, the point on the new graph becomes about (–5, 2.09691).

379. $f(x)=\log\left(\frac{1}{x}\right)$

The graph appears to be a reflection of the original graph over a horizontal line.

The function transformation $-f(x)$ reflects the original graph over a horizontal line, the x-axis. Using $f(x)=-\log(x)$, the point (1, 0) on the original graph stays in place because it's on the line of reflection; also, the vertical asymptote stays the same. The signs of the function values are all reversed — positive to negative and vice versa. However, this function doesn't appear to be one of the choices.

Investigating the choice $f(x)=\log\left(\frac{1}{x}\right)$, first apply the law of logarithms involving a quotient and rewrite the function rule as $f(x)=\log(1)-\log(x)$, which simplifies to $f(x)=-\log(x)$. This is the desired function.

380. $f(x)=\log\left(\frac{x}{10}\right)$

The graph appears to have slid 1 unit down from where the graph of the original function would be.

The function transformation $f(x)-k$ slides the original graph k units down. Using $f(x)=\log(x)-1$, the point (1, 0) on the original graph moves down to (1, –1). The vertical asymptote stays the same. However, this function doesn't appear to be one of the choices.

Investigating the choice $f(x)=\log\left(\frac{x}{10}\right)$, first apply the law of logarithms involving a quotient and rewrite the function rule as $f(x)=\log(x)-\log(10)$, which simplifies to $f(x)=\log(x)-1$. This is the desired function.

381. 3

To find the pH, substitute 0.001 into the formula and simplify:

$$\begin{aligned} pH &= -\log(0.001) \\ &= -\log\left(10^{-3}\right) \\ &= -(-3)\log 10 = 3 \end{aligned}$$

382. $10^{-11.6}$

To find the H^+, plug 11.6 into the formula for pH:

$$\begin{aligned} 11.6 &= -\log\left(H^+\right) \\ -11.6 &= \log\left(H^+\right) \end{aligned}$$

Using the log equivalence, $x = \log(y)$ becomes $10^x = y$, so you have

$$10^{-11.6} = H^+$$

383. $5,225.79

Substitute the following values into the equation: P = $5,000, r = 0.01475, n = 4, and t = 3. Then simplify the equation to find the amount in the account:

$$\begin{aligned} A &= 5{,}000\left(1 + \frac{0.01475}{4}\right)^{4(3)} \\ &= 5{,}000(1.0036875)^{12} \\ &= 5{,}225.79 \end{aligned}$$

384. 21.33 years

Substitute the following values into the equation: P = $1000, r = 0.0325, and A = $2,000. Then simplify the equation:

$$\begin{aligned} 2{,}000 &= 1{,}000e^{0.0325t} \\ 2 &= e^{0.0325t} \end{aligned}$$

To eliminate the e, take the natural log of both sides of the equation; then solve for t:

$$\begin{aligned} \ln 2 &= \ln e^{0.0325t} \\ \ln 2 &= 0.0325t \ln e \\ \ln 2 &= 0.0325t \\ 21.33 &= t \end{aligned}$$

385. $11,077.61

Use the exponential model, $y = Ce^{kt}$. Substitute the following values into the equation: C = $18,995 (the value of the car when you drive off the lot), k = –0.10785, and t = 5. Then simplify the equation to find the value of the car:

$$
\begin{aligned}
y &= Ce^{kt} \\
&= 18{,}995e^{(-0.10785)5} \\
&= 18{,}995e^{-0.53925} \\
&= 11{,}077.61
\end{aligned}
$$

386. 10^{-6}

Substitute $L = 60$ into the decibel equation. Then simplify the equation and expand the right side into two logs.

$$
\begin{aligned}
60 &= 10\log\left(\frac{I}{10^{-12}}\right) \\
6 &= \log\left(\frac{I}{10^{-2}}\right) \\
6 &= \log I - \log 10^{-12} \\
6 &= \log I - (-12)\log 10 \\
6 &= \log I + 12 \\
-6 &= \log I
\end{aligned}
$$

Using the log equivalence, $x = \log(y)$ becomes $10^x = y$, so you have

$$10^{-6} = I$$

387. $I_1 = 10{,}000I_2$

You have two equations to use for the two earthquakes. Earthquake 1 has $M = 8.0$, and Earthquake 2 has $M = 4.0$. Substitute these values into each equation and then expand the log:

$$
\begin{aligned}
8 &= \log\left(\frac{I_1}{I_0}\right) & \qquad 4 &= \log\left(\frac{I_2}{I_0}\right) \\
8 &= \log I_1 - \log I_0 & \qquad 4 &= \log I_2 - \log I_0
\end{aligned}
$$

Solve the resulting system of equations to eliminate the $\log I_0$:

$$
\begin{aligned}
8 &= \log I_1 - \log I_0 \\
-4 &= -\log I_2 + \log I_0 \\
\hline
4 &= \log I_1 - \log I_2 \\
4 &= \log\left(\frac{I_1}{I_2}\right)
\end{aligned}
$$

Using the log equivalence, $x = \log(y)$ becomes $10^x = y$, so you have

$$
\begin{aligned}
10^4 &= \frac{I_1}{I_2} \\
10{,}000 &= \frac{I_1}{I_2} \\
10{,}000I_2 &= I_1
\end{aligned}
$$

The intensity of the first earthquake is 10,000 times that of the second earthquake.

388. 10,174

Substitute the following values into the equation: $C = 6{,}250$, $y = 8{,}125$, and $t = 2010 - 1975 = 35$. Then simplify the equation to find the growth rate:

$$\begin{aligned} y &= Ce^{kt} \\ 8{,}125 &= 6{,}250e^{k35} \\ 1.3 &= e^{35k} \end{aligned}$$

To eliminate the e, take the natural log on both sides of the equation. Then solve the problem for k:

$$\begin{aligned} \ln 1.3 &= \ln e^{35k} \\ \ln 1.3 &= 35k \ln e \\ \ln 1.3 &= 35k \\ 0.007496 &= k \end{aligned}$$

Now, to find the population in the year 2040, substitute the values $C = 6{,}250$, $k = 0.007496$, and $t = 2040 - 1975 = 65$ into the equation:

$$\begin{aligned} y &= Ce^{kt} \\ &= 6{,}250e^{(0.007496)65} \\ &= 6{,}250e^{0.48724} \\ &= 10{,}174 \end{aligned}$$

389. \$119,639

Substitute the following values into the equation: $C = 179{,}900$, $y = 138{,}800$, and $t = 2013 - 2000 = 13$. Then solve for k to find the decay rate:

$$\begin{aligned} y &= Ce^{kt} \\ 138{,}800 &= 179{,}900e^{k13} \\ \left(\frac{1{,}380}{1{,}799}\right) &= e^{13k} \end{aligned}$$

To eliminate the e, take the natural log on both sides of the equation. Then solve the problem for k:

$$\begin{aligned} \left(\frac{1{,}380}{1{,}799}\right) &= e^{13k} \\ \ln\left(\frac{1{,}380}{1{,}799}\right) &= 13k \ln e \\ \ln\left(\frac{1{,}380}{1{,}799}\right) &= 13k \\ -0.020396 &= k \end{aligned}$$

Now, to find the value of the house in the year 2020, plug the values $C = 179{,}900$, $k = -0.020396$, and $t = 2020 - 2000 = 20$ into the exponential growth equation:

$$\begin{aligned} y &= Ce^{kt} \\ &= 179{,}900e^{(-0.020396)20} \\ &= 179{,}900e^{-0.40792} \\ &= 119{,}639 \end{aligned}$$

390. 11

First find y, the required number of sick students and employees, by taking 45% of 8,500:

$$y = 8{,}500(0.45) = 3{,}825$$

Plug 8,500 into the equation for y. Then solve for t:

$$\begin{aligned} 3{,}825 &= \frac{8{,}500}{1+999e^{-0.6t}} \\ 3{,}825\left(1+999e^{-0.6t}\right) &= 8{,}500 \\ 1+999e^{-0.6t} &= \frac{20}{9} \\ 999e^{-0.6t} &= \frac{11}{9} \\ e^{-0.6t} &= \frac{11}{8{,}991} \end{aligned}$$

To eliminate the e, take the natural log on both sides of the equation. Solve for t, the time in days:

$$\begin{aligned} \ln e^{-0.6t} &= \ln\left(\frac{11}{8{,}991}\right) \\ -0.6t \ln e &= \ln\left(\frac{11}{8{,}991}\right) \\ -0.6t &= \ln\left(\frac{11}{8{,}991}\right) \\ t &= \frac{\ln\left(\frac{11}{8{,}991}\right)}{-0.6} = 11.17681 \approx 11 \end{aligned}$$

391. $\frac{3}{5}$

Using the right-triangle definition for the sine of an angle, determine the measure of the side opposite the angle in question and divide that measure by the length of the hypotenuse of the triangle:

$$\sin A = \frac{\text{Opposite}}{\text{Hypotenuse}} = \frac{3}{5}$$

392. $\frac{5}{13}$

Using the right-triangle definition for the cosine of an angle, determine the measure of the side adjacent to the angle in question (not the hypotenuse) and divide that measure by the length of the hypotenuse of the triangle:

$$\cos A = \frac{\text{Adjacent}}{\text{Hypotenuse}} = \frac{5}{13}$$

393. 1

Using the right-triangle definition for the tangent of an angle, determine the measure of the side opposite the angle in question and divide that measure by the length of the side adjacent to the angle (not the hypotenuse):

$$\tan C = \frac{\text{Opposite}}{\text{Adjacent}} = \frac{7}{7} = 1$$

394. $\frac{9}{7}$

Using the right-triangle definition for the cotangent of an angle, determine the measure of the side adjacent to the angle in question and divide that measure by the length of the side opposite the angle:

$$\cot D = \frac{\text{Adjacent}}{\text{Opposite}} = \frac{9}{7}$$

395. $\frac{17}{15}$

Using the right-triangle definition for the secant of an angle, determine the measure of the hypotenuse and divide that measure by the length of the side adjacent to the angle:

$$\sec D = \frac{\text{Hypotenuse}}{\text{Adjacent}} = \frac{17}{15}$$

396. $\frac{\sqrt{61}}{6}$

Using the right-triangle definition for the cosecant of an angle, determine the measure of the hypotenuse and divide that measure by the length of the side opposite the angle.

$$\csc A = \frac{\text{Hypotenuse}}{\text{Opposite}} = \frac{2\sqrt{61}}{12} = \frac{\sqrt{61}}{6}$$

397. 8.1 ft

The diagram represents the situation. You're looking for the length of the ladder, which is the hypotenuse of the right triangle. You're given an angle and the length of the adjacent side, so use $\cos\theta$:

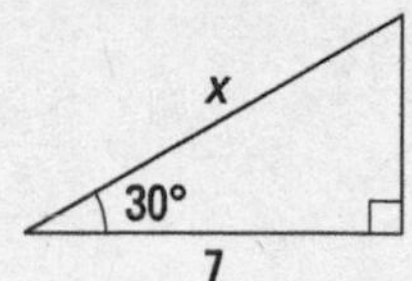

Illustration by Thomson Digital

$$\cos 30° = \frac{7}{x}$$

$$x = \frac{7}{\cos 30°} \approx 8.1$$

398. 172.1 ft

The diagram shows the situation. You're given an angle and the length of the hypotenuse of the right triangle. You're looking for the length of the opposite side, so use $\sin\theta$:

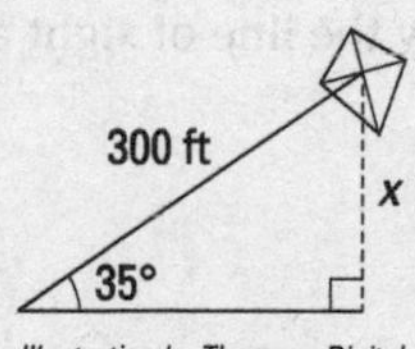

Illustration by Thomson Digital

$$\sin 35° = \frac{x}{300}$$

$$x = 300 \sin 35° \approx 172.1$$

399. 2,525.5 ft

The diagram shows the situation. You're given an angle and the height of the right triangle. You're looking for the length of the adjacent side, so use $\tan\theta$:

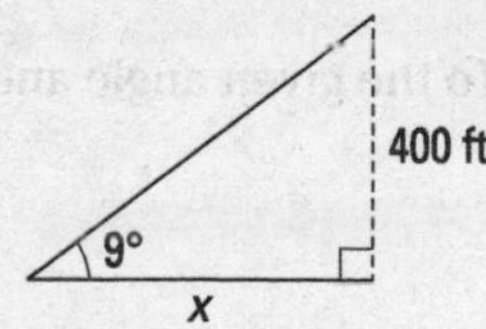

not drawn to scale

Illustration by Thomson Digital

$$\tan 9° = \frac{400}{x}$$

$$x = \frac{400}{\tan 9°} \approx 2{,}525.5$$

400. 368.3 ft

The diagram shows the situation. You're given an angle and the length of the hypotenuse. You're looking for the length of the opposite side, so use sin θ:

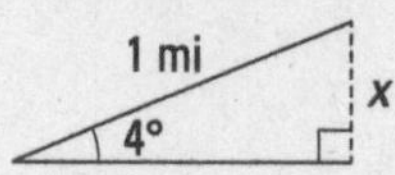

not drawn to scale

Illustration by Thomson Digital

$$\sin 4° = \frac{x}{1 \text{ mi}}$$

$$\sin 4° = \frac{x}{5{,}280 \text{ ft}}$$

$$x = 5{,}280 \sin 4° \approx 368.3$$

The vertical change in feet is about 368.3 feet.

401. 1,286.7 ft

The diagram shows the situation. The angle of depression is outside the triangle. However, the angle of depression is defined as the angle between a horizontal line and the observer's line of sight. The horizontal line is parallel to the ground, so the angle of depression is equal to the angle formed by the line of sight and the ground.

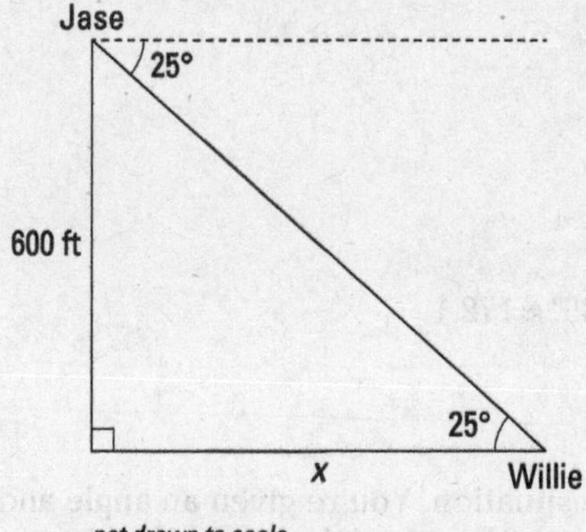

not drawn to scale

Illustration by Thomson Digital

The side you're looking for is adjacent to the given angle and the given side is opposite the angle, so use tan θ:

$$\tan 25° = \frac{600}{x}$$

$$x = \frac{600}{\tan 25°} \approx 1{,}286.7$$

Willie is about 1,286.7 feet from the point on the ground directly below the hot air balloon.

402. 59.6 ft

The diagram shows the situation.

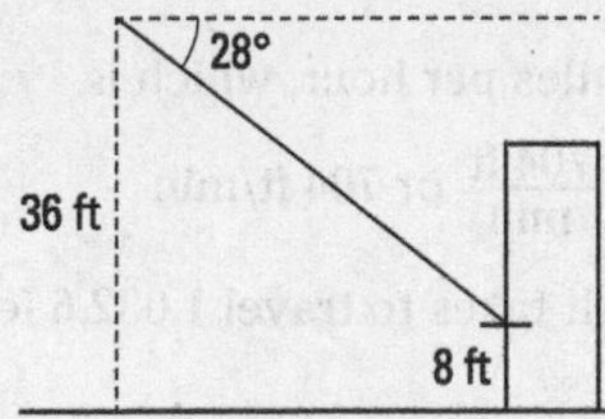

Illustration by Thomson Digital

Because the bird is landing 8 feet above the ground, its height is 28 feet above the windowsill.

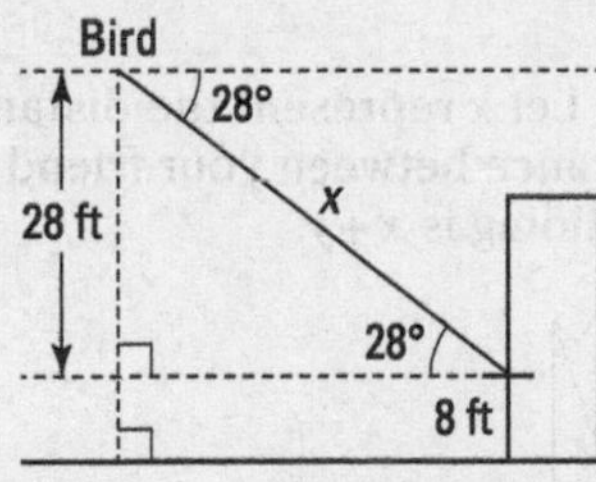

Illustration by Thomson Digital

The angle of depression is outside the triangle. However, the angle of depression is defined as the angle between a horizontal line and the observer's line of sight. The horizontal line is parallel to the ground, so the angle of depression is equal to the angle formed by the line of sight and the ground.

To find how far the bird must fly, use $\sin\theta$:

$$\sin 28° = \frac{28}{x}$$

$$x = \frac{28}{\sin 28°} \approx 59.6$$

The bird must fly about 59.6 feet before it can land on the windowsill.

403. 1.5 min

The diagram shows the situation. You're given an angle and the height of the right triangle. The height represents the depth of the submarine, and the hypotenuse of the triangle is the distance the sub travels to reach that depth.

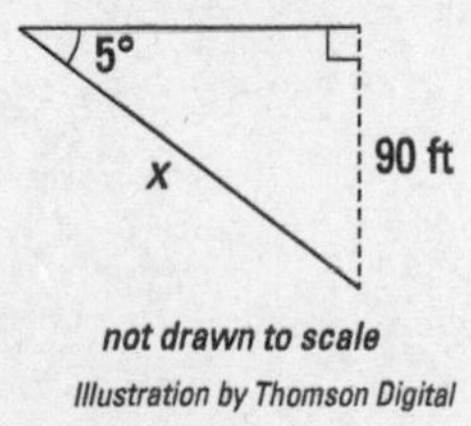

Illustration by Thomson Digital

Answers 401–500

You're looking for the length of the hypotenuse, so use $\sin\theta$:

$$\sin 5^\circ = \frac{90}{x}$$

$$x = \frac{90}{\sin 5^\circ} \approx 1{,}032.6$$

The submarine is traveling at 8 miles per hour, which is

$$\frac{8 \text{ mi}}{\text{hr}} \times \frac{5{,}280 \text{ ft}}{\text{mi}} \times \frac{1 \text{ hr}}{60 \text{ min}} = \frac{704 \text{ ft}}{\text{min}} \text{ or } 704 \text{ ft/min}$$

At that rate, to find out how long it takes to travel 1,032.6 feet, divide:

$$\frac{1{,}032.6 \text{ ft}}{704 \text{ ft/min}} \approx 1.5 \text{ min}$$

The submarine will take about 1.5 minutes to reach a depth of 90 feet.

404. 628 ft

The diagram shows the situation. Let x represent the distance between you and your friend and let y represent the distance between your friend and the building. Then the distance between you and the building is $x+y$.

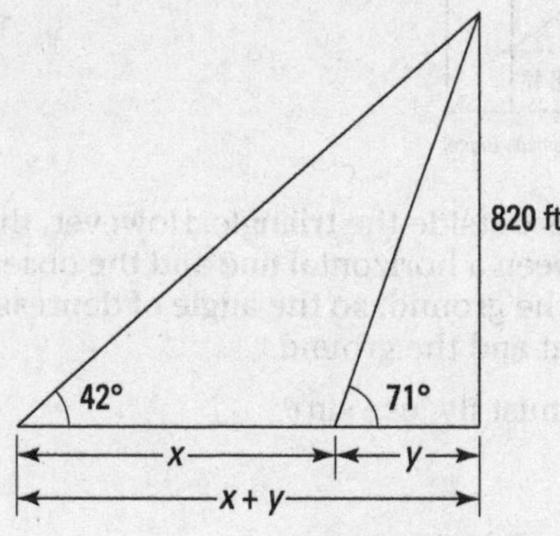

Illustration by Thomson Digital

The side represented by y is adjacent to the 71° angle. You know the length of the side opposite the 71° angle, so use $\tan\theta$:

$$\tan 71^\circ = \frac{820}{y}$$

$$y = \frac{820}{\tan 71^\circ}$$

The side represented by $x+y$ is adjacent to the 42° angle, so use $\tan\theta$ to find $x+y$:

$$\tan 42^\circ = \frac{820}{x+y}$$

$$x+y = \frac{820}{\tan 42^\circ}$$

Now substitute $y = \frac{820}{\tan 71^\circ}$ into the expression for $x+y$ and solve for x:

$$x+y = \frac{820}{\tan 42^\circ}$$

$$x + \frac{820}{\tan 71^\circ} = \frac{820}{\tan 42^\circ}$$

$$x = \frac{820}{\tan 42^\circ} - \frac{820}{\tan 71^\circ} \approx 628$$

You're about 628 feet from your friend.

405. 155.6 ft

The diagram shows the situation.

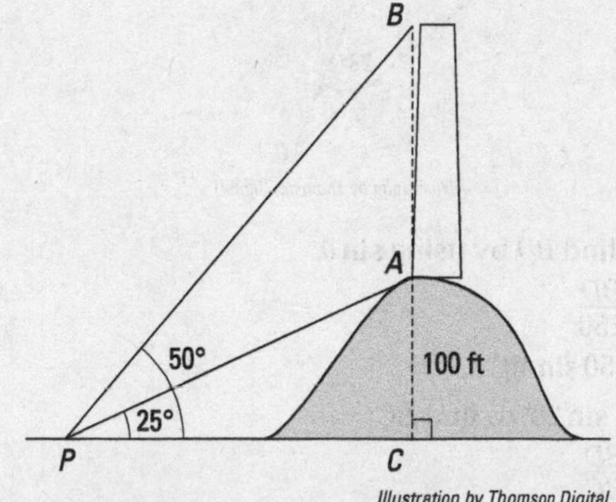

Illustration by Thomson Digital

First, find the length of PC in triangle APC by using $\tan\theta$:

$$\tan 25^\circ = \frac{100}{PC}$$

$$PC = \frac{100}{\tan 25^\circ}$$

Next, find AB by finding the length of CB in triangle PCB. Again, use $\tan\theta$:

$$\tan 50^\circ = \frac{CB}{PC}$$

$$\tan 50^\circ = \frac{CB}{\frac{100}{\tan 25^\circ}}$$

$$CB = \tan 50^\circ \left(\frac{100}{\tan 25^\circ}\right) \approx 255.6$$

Now, subtract 100 from the length of CB to find the height of the lighthouse, AB:

$$AB = CB - 100 = 255.6 - 100 = 155.6$$

The lighthouse is about 155.6 feet tall.

406. 342 ft

Draw BD perpendicular to AC. In triangle ABC,

$$m\angle C = 180° - (m\angle A + m\angle B) = 180° - (40° + 112°)$$
$$= 180° - 152° = 28°$$

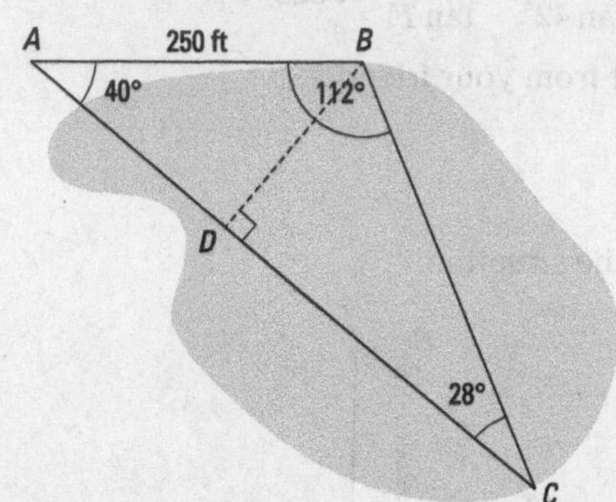

Illustration by Thomson Digital

In triangle ABD, find BD by using $\sin\theta$:

$$\sin 40° = \frac{BD}{250}$$
$$BD = 250 \sin 40°$$

Now use BD and $\sin 28°$ to find BC:

$$\sin 28° = \frac{BD}{BC}$$
$$\sin 28° = \frac{250 \sin 40°}{BC}$$
$$BC = \frac{250 \sin 40°}{\sin 28°} \approx 342$$

The distance across the lake is about 342 feet.

407. $9\sqrt{2}$

In a 45°-45°-90° triangle, the lengths of the two legs are equal, and the length of the hypotenuse equals $\sqrt{2}$ times the length of a leg.

408. $\frac{15\sqrt{2}}{2}$

In a 45°-45°-90° triangle, the lengths of the two legs are equal, and the length of the hypotenuse equals $\sqrt{2}$ times the length of a leg. The length of the hypotenuse is given, so you find the length of a leg as follows:

$$x\sqrt{2} = 15$$
$$x = \frac{15}{\sqrt{2}}$$
$$x = \frac{15\sqrt{2}}{2}$$

409. $15\sqrt{3}$

In a 30°-60°-90° triangle where the length of the shorter leg is a, the length of the longer leg is $a\sqrt{3}$, and the length of the hypotenuse is $2a$. In this case, you're given the length of the shorter leg and are looking for the length of the longer leg. Substitute 15 for a in $a\sqrt{3}$ to get $15\sqrt{3}$.

410. $4\sqrt{3}$

In a 30°-60°-90° triangle where the length of the shorter leg is a, the length of the longer leg is $a\sqrt{3}$, and the length of the hypotenuse is $2a$. In this case, you're given the length of the hypotenuse and are looking for the length of the longer leg. First, find the length of the shorter leg, a:

$$a = \frac{\text{Hypotenuse}}{2} = \frac{8}{2} = 4$$

Now find the length of the longer leg by substituting 4 for a in $a\sqrt{3}$ to get $4\sqrt{3}$.

411. 120°

To convert radians to degrees, multiply the radian measure by $\left(\frac{180}{\pi}\right)^\circ$:

$$\frac{2\pi}{3}\left(\frac{180}{\pi}\right)^\circ = 120^\circ$$

412. 2,160°

To convert radians to degrees, multiply the radian measure by $\left(\frac{180}{\pi}\right)^\circ$:

$$12\pi\left(\frac{180}{\pi}\right)^\circ = 12(180)^\circ = 2{,}160^\circ$$

413. –330°

To convert radians to degrees, multiply the radian measure by $\left(\frac{180}{\pi}\right)^\circ$:

$$-\frac{11\pi}{6}\left(\frac{180}{\pi}\right)^\circ = -330^\circ$$

414. 171.89°

To convert radians to degrees, multiply the radian measure by $\left(\frac{180}{\pi}\right)^\circ$:

$$3\left(\frac{180}{\pi}\right)^\circ = \left(\frac{540}{\pi}\right)^\circ \approx 171.89^\circ$$

415. 53.29°

To convert radians to degrees, multiply the radian measure by $\left(\frac{180}{\pi}\right)^\circ$.

$$0.93\left(\frac{180}{\pi}\right)^\circ = \left(\frac{167.4}{\pi}\right)^\circ \approx 53.29^\circ$$

416. $\frac{5\pi}{4}$

To convert degrees to radians, multiply the degree measure by $\frac{\pi}{180}$:

$$225\left(\frac{\pi}{180}\right)=\frac{5\pi}{4}$$

417. $\frac{\pi}{3}$

To convert degrees to radians, multiply the degree measure by $\frac{\pi}{180}$:

$$60\left(\frac{\pi}{180}\right)=\frac{\pi}{3}$$

418. $-\frac{9\pi}{4}$

To convert degrees to radians, multiply the degree measure by $\frac{\pi}{180}$:

$$-405\left(\frac{\pi}{180}\right)=-\frac{9\pi}{4}$$

419. 0.63

To convert degrees to radians, multiply the degree measure by $\frac{\pi}{180}$:

$$36\left(\frac{\pi}{180}\right)=\frac{\pi}{5}\approx 0.63$$

420. 2.91

To convert degrees to radians, multiply the degree measure by $\frac{\pi}{180}$:

$$167\left(\frac{\pi}{180}\right)=\frac{167\pi}{180}\approx 2.91$$

421. 60°

This is a special value that you should memorize. Refer to the unit circle.

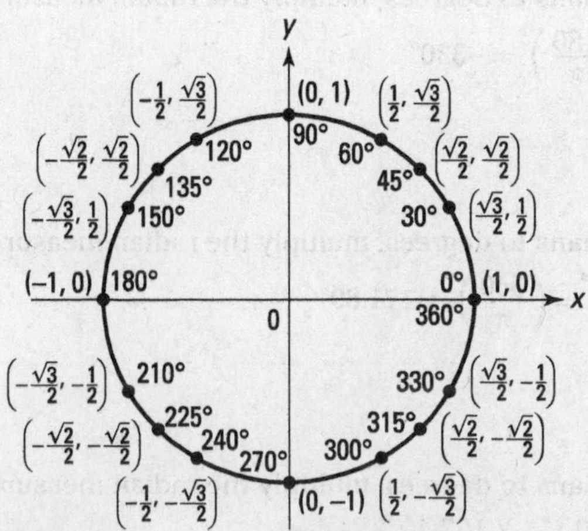

Illustration by Thomson Digital

Answers 401–500

422. 210°

This is a special value that you should memorize. Refer to the unit circle.

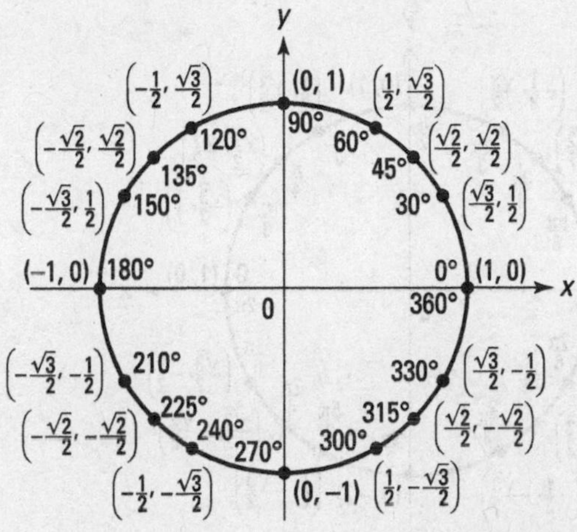

Illustration by Thomson Digital

423. $\frac{2\pi}{3}$

This is a special value that you should memorize. Refer to the unit circle.

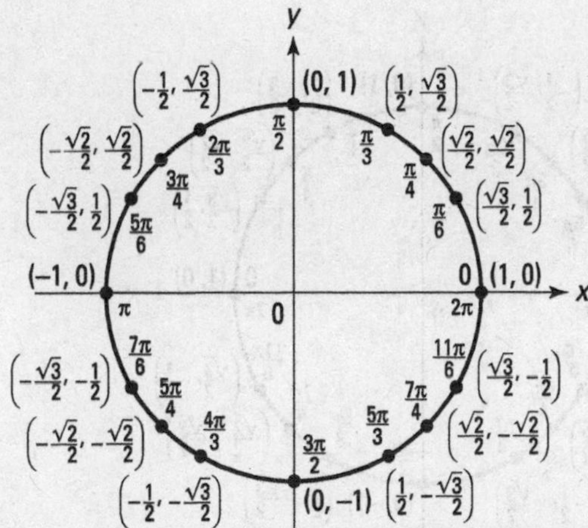

Illustration by Thomson Digital

424. $\frac{5\pi}{4}$

This is a special value that you should memorize. Refer to the unit circle.

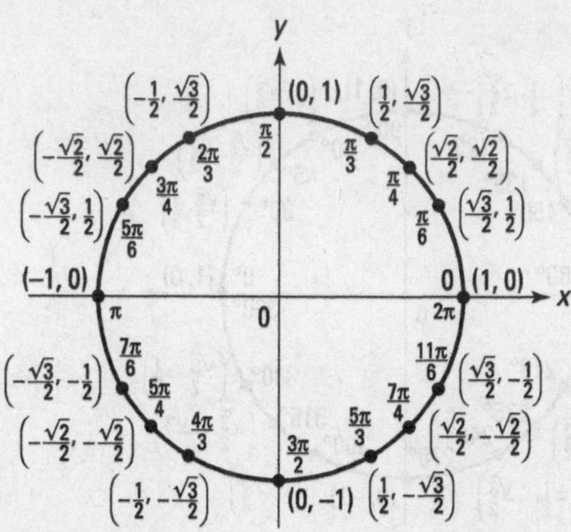

Illustration by Thomson Digital

425. $\frac{5\pi}{6}$

This is a special value that you should memorize. Refer to the unit circle.

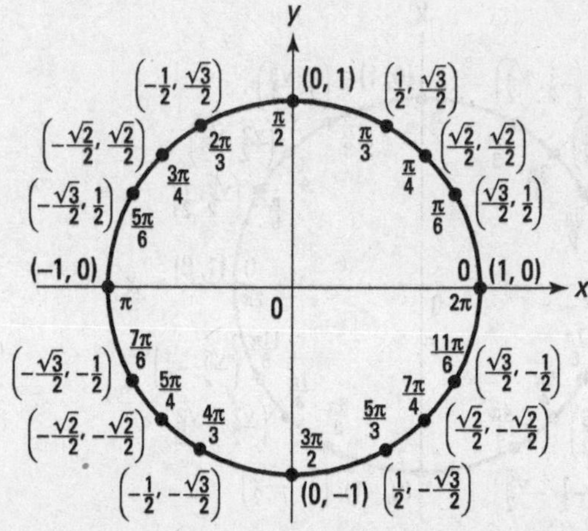

Illustration by Thomson Digital

426. 13°

Because 167° lies in quadrant II, its reference angle is $180° - 167° = 13°$.

427. 18°

Because 342° lies in quadrant IV, its reference angle is $360° - 342° = 18°$.

428. 85°

Because 265° lies in quadrant III, its reference angle is $265° - 180° = 85°$.

429. 72°

Because 792° > 360°, subtract 360° from 792° until the result is less than 360°:

$$792° - 360° = 432°$$
$$432° - 360° = 72°$$

An angle with a measure of 72° lies in quadrant I, so the reference angle is 72°.

430. 28°

Because –748° is less than 360°, add 360° to –748° until the result is greater than 0°.

$$-748° + 360° = -388°$$
$$-388° + 360° = -28°$$
$$-28° + 360° = 332°$$

An angle with a measure of 332° lies in quadrant IV, so the reference angle is $360° - 332° = 28°$.

431. $-\frac{\sqrt{2}}{2}$

The terminal side of a 225° angle in standard position on the unit circle contains the point $\left(-\frac{\sqrt{2}}{2}, -\frac{\sqrt{2}}{2}\right)$:

$$\sin\theta = \frac{y}{r}$$

$$\sin 225° = \frac{-\frac{\sqrt{2}}{2}}{1}$$
$$= -\frac{\sqrt{2}}{2}$$

432. $-\frac{\sqrt{3}}{3}$

The terminal side of a 330° angle in standard position on the unit circle contains the point $\left(\frac{\sqrt{3}}{2}, -\frac{1}{2}\right)$:

$$\tan\theta = \frac{y}{x}$$

$$\tan 330° = \frac{-\frac{1}{2}}{\frac{\sqrt{3}}{2}} = -\frac{1}{\sqrt{3}} = -\frac{\sqrt{3}}{3}$$

433. $\frac{\sqrt{2}}{2}$

First, find the reference angle. Because 405° > 360°, subtract 360° from 405° until the result is less than 360°.

$$405° - 360° = 45°$$

The reference angle is 45°.The terminal side of a 45° angle in standard position on the unit circle contains the point $\left(\frac{\sqrt{2}}{2}, \frac{\sqrt{2}}{2}\right)$:

$$\cos\theta = \frac{x}{r}$$

$$\cos 405° = \cos 45° = \frac{\frac{\sqrt{2}}{2}}{1} = \frac{\sqrt{2}}{2}$$

434. undefined

The terminal side of an angle with a measure of $\frac{3\pi}{2}$ in standard position on the unit circle contains the point (0, –1)

$$\sec\theta = \frac{r}{x}$$

$$\sec\frac{3\pi}{2} = \frac{1}{0}$$

which is undefined.

435. $\frac{12}{5}$

Recall that $\cos\theta = \frac{x}{r}$. Use the Pythagorean theorem to find the length of the missing side, *y*:

$$5^2+y^2=13^2$$
$$y^2=169-25$$
$$y^2=144$$
$$y=\pm 12$$

Because θ is in quadrant I, $y=+12$. Draw a right triangle on the coordinate plane with the vertex of θ located at the origin.

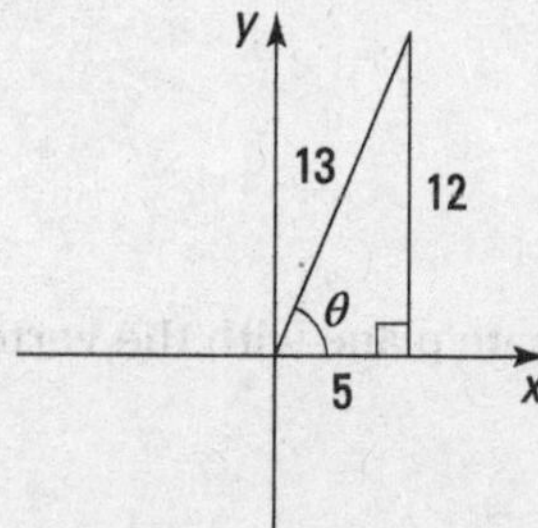

Illustration by Thomson Digital

$$\tan\theta=\frac{y}{x}=\frac{12}{5}$$

436. $-\frac{2\sqrt{2}}{3}$

Recall that $\sin\theta=\frac{y}{r}$. Use the Pythagorean theorem to find the length of the missing side, x:

$$x^2+1^2=3^2$$
$$x^2=9-1$$
$$x^2=8$$
$$x=\pm\sqrt{8}$$
$$x=\pm 2\sqrt{2}$$

Because θ is in quadrant II, $x=-2\sqrt{2}$. Draw a right triangle on the coordinate plane with the vertex of θ located at the origin:

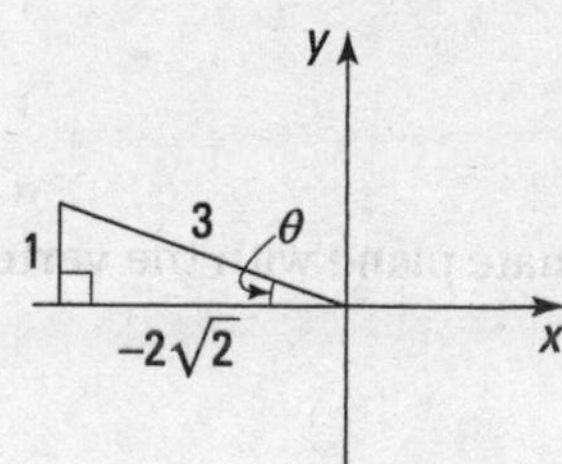

Illustration by Thomson Digital

$$\cos\theta=\frac{x}{r}=-\frac{2\sqrt{2}}{3}$$

437. $-\frac{2\sqrt{13}}{13}$

Recall that $\tan\theta = \frac{y}{x}$. Use the Pythagorean theorem to find the length of the missing side, r. Angle θ is in quadrant IV, so y is negative.

$$3^2 + (-2)^2 = r^2$$
$$9 + 4 = r^2$$
$$13 = r^2$$
$$r = \sqrt{13}$$

By definition, r is always positive.

Draw a right triangle on the coordinate plane with the vertex of θ located at the origin:

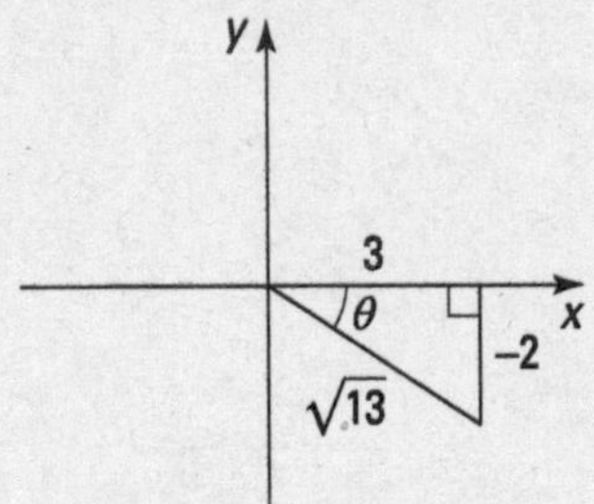

Illustration by Thomson Digital

$$\sin\theta = \frac{x}{r} = -\frac{2}{\sqrt{13}} = -\frac{2\sqrt{13}}{13}$$

438. $-\frac{9\sqrt{17}}{17}$

Recall that $\sec\theta = \frac{r}{x}$. Use the Pythagorean theorem to find the length of the missing side, y. Angle θ is in quadrant III, so x and y are both negative.

$$(-8)^2 + y^2 = 9^2$$
$$y^2 = 81 - 64$$
$$y^2 = 17$$
$$y = -\sqrt{17}$$

Draw a right triangle on the coordinate plane with the vertex of θ located at the origin:

Answers 401–500

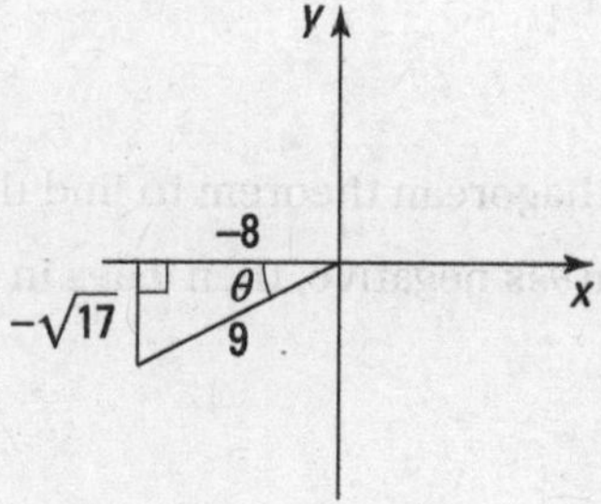

Illustration by Thomson Digital

$$\csc\theta = \frac{r}{x} = -\frac{9}{\sqrt{17}} = -\frac{9\sqrt{17}}{17}$$

439. $-\frac{\sqrt{3}}{3}$

Recall that $\csc\theta = \frac{r}{y}$. Use the Pythagorean theorem to find the length of the missing side, x. If csc θ is positive and sec θ is negative, then θ lies in quadrant II, and x is negative while y is positive.

$$x^2 + 1^2 = 2^2$$
$$x^2 = 4 - 1$$
$$x^2 = 3$$
$$x = -\sqrt{3}$$

Draw a right triangle on the coordinate plane with the vertex of θ located at the origin.

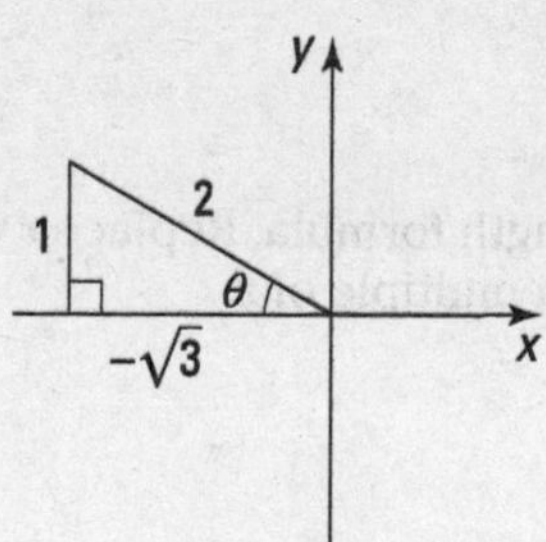

Illustration by Thomson Digital

$$\tan\theta = \frac{y}{x} = -\frac{1}{\sqrt{3}} = -\frac{\sqrt{3}}{3}$$

440. $\frac{8}{17}$

Recall that $\cot\theta = \frac{x}{y}$. Use the Pythagorean theorem to find the length of the missing side, x. If $\cot\theta$ is negative and $\sin\theta$ is negative, then θ lies in quadrant IV, and x is positive while y is negative.

$$8^2 + (-15)^2 = r^2$$
$$64 + 225 = r^2$$
$$289 = r^2$$
$$r = 17$$

By definition, r is always positive.

Draw a right triangle on the coordinate plane with the vertex of θ located at the origin.

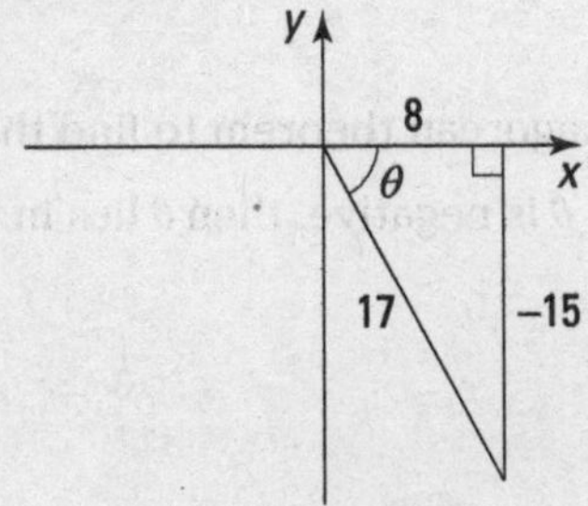

Illustration by Thomson Digital

$$\cos\theta = \frac{x}{r} = \frac{8}{17}$$

441. 25π cm

Plug the given values into the arc length formula. Replace r with 20 and θ with $\frac{5\pi}{4}$. Simplify and express the answer as a multiple of π.

$$s = r\theta$$
$$s = 20\left(\frac{5\pi}{4}\right)$$
$$s = 25\pi$$

442. $\frac{17\pi}{6}$ cm

First change 85° to radians:

$$85\left(\frac{\pi}{180}\right) = \frac{17\pi}{36}$$

Plug the given values into the arc length formula. Replace r with 6 and θ with $\frac{17\pi}{36}$. Simplify and express the answer as a multiple of π.

$$s = r\theta$$
$$s = 6\left(\frac{17\pi}{36}\right)$$
$$s = \frac{17\pi}{6}$$

443. $\frac{3\pi}{4}$

Plug the given values into the arc length formula. Replace s with $\frac{39\pi}{4}$ and r with 13. Then solve for θ.

$$s = r\theta$$
$$\frac{39\pi}{4} = 13\theta$$
$$\theta = \frac{39\pi}{4}\left(\frac{1}{13}\right)$$
$$\theta = \frac{3\pi}{4}$$

444. 15π inches

Plug the given values into the arc length formula:

$$s = r\theta$$
$$s = 18\,(150^\circ)\left(\frac{\pi}{180}\right)$$
$$s = 15\pi$$

445. 5π cm

Ten minutes is $\frac{1}{6}$ of an hour, so the minute hand moves through an angle with radian measure $\frac{1}{6}(2\pi) = \frac{\pi}{3}$. Plug this value and the given length into the arc length formula:

$$s = r\theta$$
$$s = 15\left(\frac{\pi}{3}\right)$$
$$s = 5\pi$$

446. $\frac{\pi}{3}$

You solve the equation $y = \sin^{-1}(x)$ by writing the corresponding equation $\sin(y) = x$ and evaluating by using the domain of the inverse function:

$$y = \sin^{-1}\left(\frac{\sqrt{3}}{2}\right)$$
$$\sin y = \frac{\sqrt{3}}{2}$$
$$y = \frac{\pi}{3}, \frac{2\pi}{3}$$

The domain of the inverse sine function is $[-1, 1]$, and the range is $\left[-\frac{\pi}{2}, \frac{\pi}{2}\right]$. Therefore, $-\frac{\pi}{2} \le y \le \frac{\pi}{2}$. So $y = \frac{\pi}{3}$.

447. $\frac{2\pi}{3}$

You solve the equation $y=\cos^{-1}(x)$ by writing the corresponding equation $\cos(y)=x$ and evaluating by using the domain of the inverse function:

$$y=\cos^{-1}\left(-\frac{1}{2}\right)$$
$$\cos y=-\frac{1}{2}$$
$$y=\frac{2\pi}{3},\frac{4\pi}{3}$$

The domain of the inverse cosine function is $[-1, 1]$, and the range is $[0, \pi]$. Therefore, $0\leq y\leq\pi$. So $y=\frac{2\pi}{3}$.

448. $\frac{5\pi}{3}$

You solve the equation $y=\tan^{-1}(x)$ by writing the corresponding equation $\tan(y)=x$ and evaluating by using the domain of the inverse function:

$$y=\tan^{-1}\left(-\sqrt{3}\right)$$
$$\tan y=-\sqrt{3}$$
$$y=\frac{2\pi}{3},\frac{5\pi}{3}$$

The domain of the inverse tangent function is $(-\infty, \infty)$, and the range is $\left(-\frac{\pi}{2},\frac{\pi}{2}\right)$. Therefore, $-\frac{\pi}{2}<y<\frac{\pi}{2}$. So $y=\frac{5\pi}{3}$, or $-\frac{\pi}{3}$.

449. 0

You solve the equation $y=\sin^{-1}(x)$ by writing the corresponding equation $\sin(y)=x$ and evaluating by using the domain of the inverse function. Because $\sin\pi=0$, you know that

$$y=\sin^{-1}(\sin\pi)$$
$$y=\sin^{-1}0$$
$$\sin y=0$$
$$y=0,\pi,2\pi$$

The domain of the inverse sine function is $[-1, 1]$, and the range is $\left[-\frac{\pi}{2},\frac{\pi}{2}\right]$. Therefore, $-\frac{\pi}{2}\leq y\leq\frac{\pi}{2}$. So $y=0$.

450. $\frac{\sqrt{2}}{2}$

You solve the equation $y = \tan^{-1}(x)$ by writing the corresponding equation $\tan(y) = x$ and evaluating by using the domain of the inverse function:

$$\theta=\tan^{-1}(-1)$$
$$\tan\theta=-1$$
$$\theta=-\frac{\pi}{4},\frac{3\pi}{4}$$

The domain of the inverse tangent function is $(-\infty, \infty)$ and the range is $\left(-\frac{\pi}{2}, \frac{\pi}{2}\right)$. Therefore, $-\frac{\pi}{2} < \theta < \frac{\pi}{2}$. So $\theta = -\frac{\pi}{4}$, and you can plug that into the given equation:

$$y = \cos\left(-\frac{\pi}{4}\right)$$
$$y = \cos\left(2\pi - \frac{\pi}{4}\right)$$
$$y = \cos\left(\frac{7\pi}{4}\right)$$
$$y = \frac{\sqrt{2}}{2}$$

451. {120°, 240°}

Solve the equation for cos *x* by isolating that term on the left:

$$2\cos x + 6 = 5$$
$$2\cos x = -1$$
$$\cos x = -\frac{1}{2}$$

The two angles for which the cosine is $-\frac{1}{2}$ on the interval are in the second and third quadrants: 120° and 240°.

452. {30°, 210°}

Solve the equation for tan *x* by isolating that term on the left and then simplifying on the right:

$$-1 + \tan x = \frac{-3+\sqrt{3}}{3}$$
$$\tan x = \frac{-3+\sqrt{3}}{3} + 1$$
$$\tan x = \frac{-3+\sqrt{3}+3}{3}$$
$$\tan x = \frac{\sqrt{3}}{3}$$

The two angles for which the tangent is $\frac{\sqrt{3}}{3}$ on the interval are in the first and third quadrants: 30° and 210°.

453. {30°, 150°, 210°, 330°}

Solve the equation for sin *x* by isolating the trig term on the left and then using the square-root rule for quadratic functions:

$$4\sin^2 x - 1 = 0$$
$$\sin^2 x = \frac{1}{4}$$
$$\sin x = \pm\sqrt{\frac{1}{4}}$$
$$\sin x = \pm\frac{1}{2}$$

The four angles for which the sine is $\pm\frac{1}{2}$ on the interval are 30°, 150°, 210°, and 330°.

454. {90°, 210°, 330°}

Solve the equation for sin *x*. The first thing you have to do is factor the given equation:

$$2\sin^2 x - \sin x - 1 = 0$$
$$(2\sin x + 1)(\sin x - 1) = 0$$

Then set each factor equal to zero and solve. First factor:

$$2\sin x + 1 = 0$$
$$2\sin x = -1$$
$$\sin x = -\frac{1}{2}$$
$$x = 210°, 330°$$

Second factor:

$$\sin x - 1 = 0$$
$$\sin x = 1$$
$$x = 90°$$

So the complete set of solutions includes 90°, 210°, and 330°.

455. {120°, 240°}

Solve the equation for cos *x*. First, rearrange the given equation as a quadratic equation and then factor it:

$$2\cos^2 x - 2 = 3\cos x$$
$$2\cos^2 x - 3\cos x - 2 = 0$$
$$(2\cos x + 1)(\cos x - 2) = 0$$

Then set each factor equal to zero and solve. First factor:

$$2\cos x + 1 = 0$$
$$2\cos x = -1$$
$$\cos x = -\frac{1}{2}$$
$$x = 120°, 240°$$

Second factor:

$$\cos x - 2 = 0$$
$$\cos x = 2$$

There are no solutions for the second factor because $-1 \le \cos x \le 1$. So the complete set of solutions includes just 120° and 240°.

456. $\left\{\frac{\pi}{6}, \frac{11\pi}{6}\right\}$

First, solve the equation for cos *x*:

$$5\cos x - \sqrt{3} = 3\cos x$$
$$2\cos x = \sqrt{3}$$
$$\cos x = \frac{\sqrt{3}}{2}$$

Because cos *x* is positive in quadrants I and IV, there are two solutions in the interval $[0, 2\pi)$:

$$x = \frac{\pi}{6}, \frac{11\pi}{6}$$

457. $\left\{\frac{\pi}{6}, \frac{5\pi}{6}, \frac{7\pi}{6}, \frac{11\pi}{6}\right\}$

First, solve the equation for tan *x*:

$$3\tan^2 x - 1 = 0$$
$$3\tan^2 x = 1$$
$$\tan^2 x = \frac{1}{3}$$
$$\tan x = \pm\sqrt{\frac{1}{3}}$$
$$\tan x = \pm\frac{\sqrt{3}}{3}$$

Because the period of $\tan x$ is π and the instructions ask for solutions in the interval $[0, 2\pi)$, there are four solutions. If $\tan x = \frac{\sqrt{3}}{3}$, then $x = \frac{\pi}{6}$ and $x = \frac{7\pi}{6}$. If $\tan x = -\frac{\sqrt{3}}{3}$, then $x = \frac{5\pi}{6}$ and $x = \frac{11\pi}{6}$.

458. $\left\{0, \frac{2\pi}{3}, \frac{4\pi}{3}\right\}$

First, solve the equation for cos *x* by rearranging the given equation as a quadratic equation and factoring it:

$$2\cos^2 x - \cos x = 1$$
$$2\cos^2 x - \cos x - 1 = 0$$
$$(2\cos x + 1)(\cos x - 1) = 0$$

Then set each factor equal to zero and solve. First factor:

$$2\cos x+1=0$$
$$2\cos x=-1$$
$$\cos x=-\frac{1}{2}$$

Second factor:

$$\cos x-1=0$$
$$\cos x=1$$

If $\cos x=-\frac{1}{2}$, then there are two solutions in the interval $[0, 2\pi)$, $x=\frac{2\pi}{3}$ and $x=\frac{4\pi}{3}$. If $\cos x=1$, then $x=0$.

459. $\left\{\frac{\pi}{4}, \frac{3\pi}{4}, \frac{5\pi}{4}, \frac{7\pi}{4}, \frac{7\pi}{6}, \frac{11\pi}{6}\right\}$

First solve the equation for $\sin x$ by factoring (factor by grouping):

$$4\sin^3 x+2\sin^2 x-2\sin x-1=0$$
$$(4\sin^3 x+2\sin^2 x)-(2\sin x+1)=0$$
$$2\sin^2 x(2\sin x+1)-1(2\sin x+1)=0$$
$$(2\sin^2 x-1)(2\sin x+1)=0$$

Then set each factor equal to zero and solve. First factor:

$$2\sin^2 x-1=0$$
$$2\sin^2 x=1$$
$$\sin^2 x=\frac{1}{2}$$
$$\sin x=\pm\sqrt{\frac{1}{2}}$$
$$\sin x=\pm\frac{\sqrt{2}}{2}$$

Second factor:

$$2\sin x+1=0$$
$$2\sin x=-1$$
$$\sin x=-\frac{1}{2}$$

If $\sin x=\frac{\sqrt{2}}{2}$, then there are two solutions in the interval $[0, 2\pi)$, $x=\frac{\pi}{4}$ and $x=\frac{3\pi}{4}$. If $\sin x=-\frac{\sqrt{2}}{2}$, then there are two solutions in the interval $[0, 2\pi)$, $x=\frac{5\pi}{4}$ and $x=\frac{7\pi}{4}$. If $\sin x=-\frac{1}{2}$, then there are two solutions in the interval $[0, 2\pi)$, $x=\frac{7\pi}{6}$ and $x=\frac{11\pi}{6}$.

460. $\left\{0, \frac{\pi}{3}, \pi, \frac{5\pi}{3}\right\}$

Factor the left side of the equation:

$$2\sin x\cos x - \sin x = 0$$
$$\sin x(2\cos x - 1) = 0$$

Then use the zero product property to solve for $\sin x$ and $\cos x$. (The zero product property says that if $a \cdot b = 0$, then either a or b is equal to 0.)

$$\sin x = 0 \qquad 2\cos x - 1 = 0$$
$$2\cos x = 1$$
$$\cos x = \frac{1}{2}$$

If $\sin x = 0$, then $x = 0$ and $x = \pi$. If $\cos x = \frac{1}{2}$, then there are two solutions in the interval $[0, 2\pi)$, $x = \frac{\pi}{3}$ and $x = \frac{5\pi}{3}$.

461. $f(x) = 2\sin(4x)$

Use the standard equation $f(x) = A\sin Bx$, where A is the amplitude and $\frac{2\pi}{|B|}$ is the period. The trigonometric function has a maximum value of 2 and a minimum value of −2, for a difference of $2-(-2)=4$. This means that the amplitude of the function is half that difference, or 2, so $A = 2$. The period of the function is about 1.57 units, or $\frac{\pi}{2}$, so $B = 4$.

462. $f(x) = 3\cos(2x)$

Use the standard equation $f(x) = A\cos Bx$, where A is the amplitude and $\frac{2\pi}{|B|}$ is the period. The trigonometric function has a maximum value of 3 and a minimum value of −3, for a difference of $3-(-3)=6$. This means that the amplitude of the function is half that difference, or 3, so $A = 3$. The period of the function is about 3.14 units, or exactly π, so $B = 2$.

463. $f(x) = -\frac{1}{2}\sin(\pi x)$

Use the standard equation $f(x) = A\sin Bx$, where A is the amplitude and $\frac{2\pi}{|B|}$ is the period. The trigonometric function has a maximum value of $\frac{1}{2}$ and a minimum value of $-\frac{1}{2}$, for a difference of $\frac{1}{2} - \left(-\frac{1}{2}\right) = 1$. This means that the amplitude of the function is half that difference, or $\frac{1}{2}$, so $A = \frac{1}{2}$. Because the standard sine graph is flipped over the x-axis, the function is multiplied by −1. The period of the function is exactly 2 units, so $B = \pi$.

464. $f(x) = \frac{4}{3}\cos\left(\frac{\pi}{2}x\right)$

Use the standard equation $f(x) = A\cos Bx$, where A is the amplitude and $\frac{2\pi}{|B|}$ is the period. The trigonometric function has a maximum value of $\frac{4}{3}$ and a minimum value of $-\frac{4}{3}$, for a difference of $\frac{4}{3} - \left(-\frac{4}{3}\right) = \frac{8}{3}$. This means that the amplitude of the function is half the difference, or $\frac{4}{3}$, so $A = \frac{4}{3}$. The period of the function is 4 units, so $B = \frac{\pi}{2}$.

465. $f(x) = -8\sin\left(\frac{\pi}{4}x\right)$

Use the standard equation $f(x) = A\sin Bx$, where A is the amplitude and $\frac{2\pi}{|B|}$ is the period. The trigonometric function has a maximum value of 8 and a minimum value of −8, for a difference of $8 - (-8) = 16$. This means that the amplitude of the function is half that difference, or 8, so $A = 8$. Because the standard sine graph is flipped over the x-axis, the function is multiplied by −1. The period of the function is exactly 8 units, so $B = \frac{\pi}{4}$.

466.

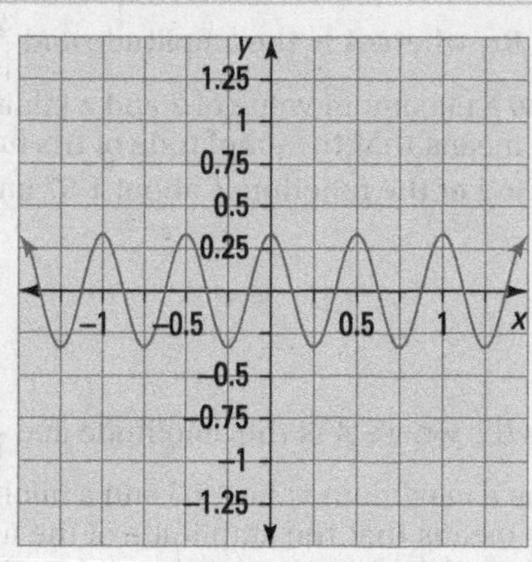

Illustration by Thomson Digital

The given function is

$$f(x) = \frac{1}{3}\cos(4\pi x)$$

Using $f(x) = A\cos Bx$, the amplitude of the function is A. Therefore, the amplitude of the graph is $\frac{1}{3}$. The period of the function is determined by $\frac{2\pi}{|B|}$; $B = 4\pi$, so the period of the function is $\frac{2\pi}{4\pi} = \frac{1}{2}$.

467.

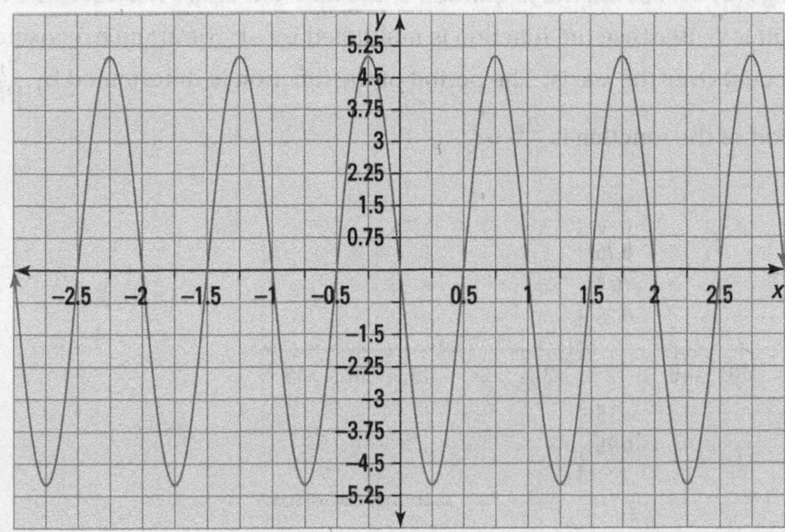

Illustration by Thomson Digital

The given function is

$$f(x) = -5\sin(2\pi x)$$

Using $f(x) = A\sin Bx$, the amplitude of the function is A. Therefore, the amplitude of the graph is 5. The period of the function is determined by $\frac{2\pi}{|B|}$; $B = 2\pi$, so the period of the function is 1. Because the function is multiplied by –1, the standard sine graph is reflected over the x-axis.

468.

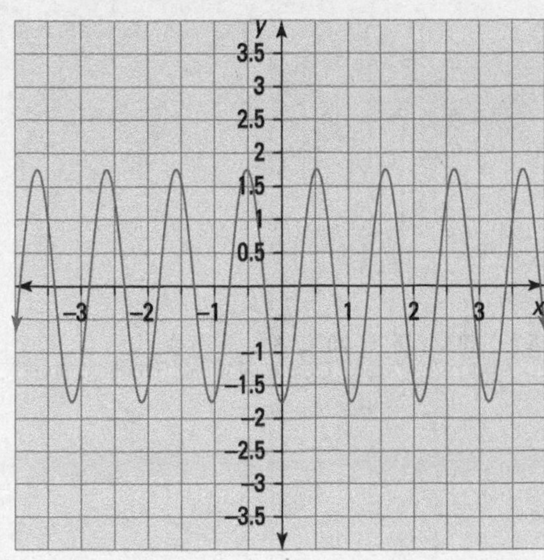

Illustration by Thomson Digital

The given function is

$$f(x) = -\frac{7}{4}\cos(6x)$$

Using $f(x) = A\cos Bx$, the amplitude of the function is A. Therefore, the amplitude of the graph is $\frac{7}{4}$. Because the function is multiplied by –1, the standard cosine graph is reflected over the x-axis. The period of the function is determined by $\frac{2\pi}{|B|}$; $B = 6$, so the period of the function is $\frac{2\pi}{|6|} = \frac{\pi}{3}$.

469.

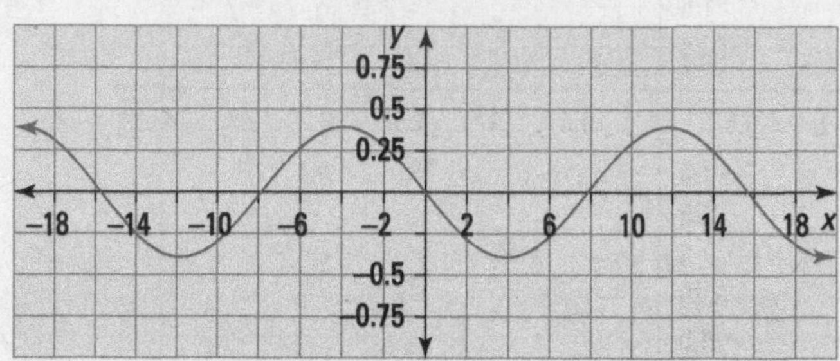

Illustration by Thomson Digital

The given function is

$$f(x) = \frac{\pi}{8}\sin\left(-\frac{2}{5}x\right)$$

Using $f(x) = A\sin Bx$, the amplitude of the function is A. Therefore, the amplitude of the graph is $\frac{\pi}{8}$. The period of the function is determined by $\frac{2\pi}{|B|}$; $B = -\frac{2}{5}$, so the period of the function is $\frac{2\pi}{\left|-2/5\right|} = \frac{2\pi}{1}\cdot\frac{5}{2} = 5\pi$. The negative multiplier of the angle reflects the graph over the y-axis.

470.

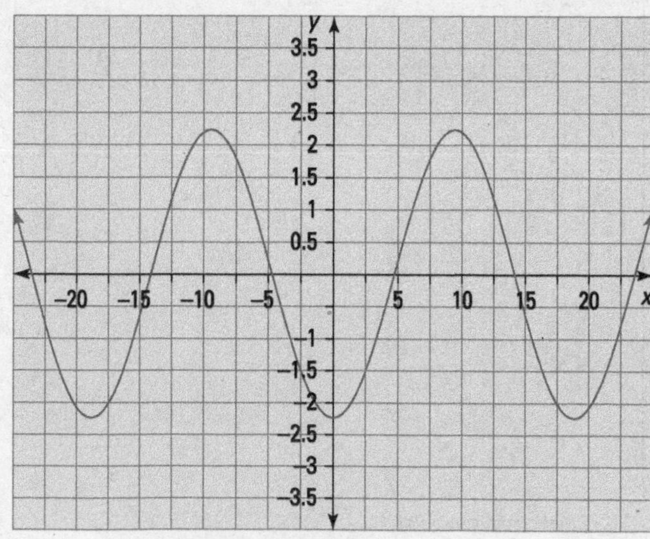

Illustration by Thomson Digital

The given function is

$$f(x) = -\sqrt{5}\cos\left(\frac{1}{3}x\right)$$

Using $f(x) = A\cos Bx$, the amplitude of the function is A. Therefore, the amplitude of the graph is $\sqrt{5}$. Because the function is multiplied by –1, the standard cosine graph is reflected over the x-axis. The period of the function is determined by $\frac{2\pi}{|B|}$; $B = \frac{1}{3}$, so the period of the function is $\frac{2\pi}{|1/3|} = \frac{2\pi}{1} \cdot \frac{3}{1} = 6\pi$.

471. right $\frac{\pi}{3}$, up 3

The transformed function is

$$g(x) = \sin\left(x - \frac{\pi}{3}\right) + 3$$

Using $g(x) = A\sin B(x + C) + D$, C represents a horizontal shift and D represents a vertical shift. In this function, C is $-\frac{\pi}{3}$, which is a shift to the right, and D is 3, which is a shift upward.

472. left $\frac{\pi}{4}$, down 2

The transformed function is

$$g(x) = \cos\left(x + \frac{\pi}{4}\right) - 2$$

Using $g(x) = A\cos B(x + C) + D$, C represents a horizontal shift and D represents a vertical shift. In this function, C is $+\frac{\pi}{4}$, which is a shift to the left, and D is –2, which is a shift downward.

473. right $\frac{\pi}{8}$, down $\frac{\pi}{4}$; period: $\frac{\pi}{2}$

Using $g(x) = A\sin B(x + C) + D$, C represents a horizontal shift and D represents a vertical shift. Rewrite $g(x)$ to match the standard form:

$$\begin{aligned} g(x) &= \sin\left(4x - \frac{\pi}{2}\right) - \frac{\pi}{4} \\ &= \sin 4\left(x - \frac{\pi}{8}\right) - \frac{\pi}{4} \end{aligned}$$

In this function, C is $-\frac{\pi}{8}$, which is a shift to the right, and D is $-\frac{\pi}{4}$, which is a shift downward. In the general form $g(x) = A\sin B(x + C) + D$, the period of sine is $\frac{2\pi}{|B|}$; therefore, the 4 multiplier changes the period to $\frac{\pi}{2}$ because $\frac{2\pi}{4} = \frac{\pi}{2}$.

474. right $\frac{5}{2}$, up π; reflected over y-axis; period: π

The transformed function is

$$g(x) = \cos(-2x + 5) + \pi$$

Using $g(x) = A\cos B(x + C) + D$, C represents a horizontal shift and D represents a vertical shift. Rewrite $g(x)$ to match the standard form:

$$g(x) = \cos 2\left(-\left(x - \frac{5}{2}\right)\right) + \pi$$

In this function, C is $-\frac{5}{2}$, which is a shift to the right, and D is $+\pi$, which is a shift upward. The negative multiplier on the angle reflects the graph over the y-axis. In the general form $g(x)=A\cos B(x+C)+D$, the period of cosine is $\frac{2\pi}{|B|}$; therefore, the 2 multiplier changes the period to π because $\frac{2\pi}{2}=\pi$.

475. left $\frac{7}{4}$, up 8; period: $\frac{\pi}{2}$

The transformed function is

$$g(x)=\sin(4x+7)+8$$

Using $g(x)=A\sin B(x+C)+D$, C represents a horizontal shift and D represents a vertical shift. Rewrite $g(x)$ to match the standard form:

$$g(x)=\sin 4\left(x+\frac{7}{4}\right)+8$$

In this function, C is $+\frac{7}{4}$, which is a shift to the left, and D is 8, which is a shift upward. In the general form $g(x)=A\sin B(x+C)+D$, the period of sine is $\frac{2\pi}{|B|}$; therefore, the 4 multiplier changes the period to $\frac{2\pi}{4}=\frac{\pi}{2}$.

476. $g(x)=\cos(x+\pi)+5$

Use $g(x)=A\cos B(x+C)+D$, where C represents a horizontal shift and D represents a vertical shift. In this case, $C=+\pi$ and $D=+5$. Inserting the values, you have

$$g(x)=\cos(x+\pi)+5$$

477. $g(x)=\sin\left(x-\frac{\pi}{8}\right)-\pi$

Use $g(x)=A\sin B(x+C)+D$, where C represents a horizontal shift and D represents a vertical shift. In this case, $C=-\frac{\pi}{8}$ and $D=-\pi$. Inserting the values, you have

$$g(x)=\sin\left(x-\frac{\pi}{8}\right)-\pi$$

478. $g(x)=\cos\left(x-\frac{2\pi}{3}\right)-2\pi$

Use $g(x)=A\cos B(x+C)+D$, where C represents a horizontal shift and D represents a vertical shift. In this case, $C=-\frac{2\pi}{3}$ and $D=-2\pi$. Inserting the values, you have

$$g(x)=\cos\left(x-\frac{2\pi}{3}\right)-2\pi$$

479. $g(x)=\sin\left(x+\frac{7}{2}\right)-3$

Use $g(x)=A\sin B(x+C)+D$, where C represents a horizontal shift and D represents a vertical shift. In this case, $C=+\frac{7}{2}$ and $D=-3$. Inserting the values, you have

$$g(x)=\sin\left(x+\frac{7}{2}\right)-3$$

480. $g(x)=\cos\left(x-\frac{4}{5}\right)+\frac{\pi}{6}$

Use $g(x)=A\cos B(x+C)+D$, where C represents a horizontal shift and D represents a vertical shift. In this case, $C=-\frac{4}{5}$ and $D=+\frac{\pi}{6}$. Inserting the values, you have

$$g(x)=\cos\left(x-\frac{4}{5}\right)+\frac{\pi}{6}$$

481.

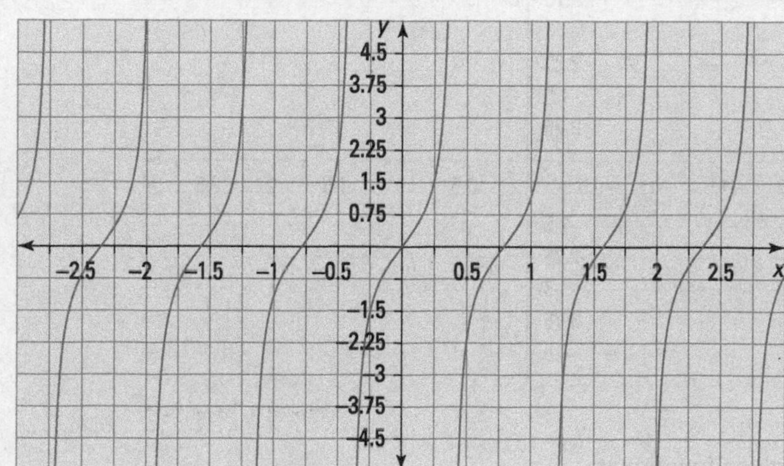

Illustration by Thomson Digital

The given function is

$$f(x)=\tan(4x)$$

Using $f(x)=A\tan Bx$, the period of the function is determined by $\frac{\pi}{|B|}$; therefore, the graph is the standard tangent function, except with a period of $\frac{\pi}{4}$.

482.

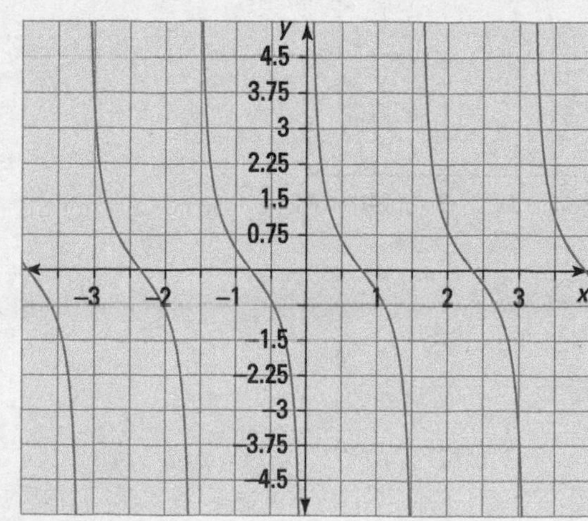

Illustration by Thomson Digital

The given function is

$$f(x) = \cot(2x)$$

Using $f(x) = A\cot Bx$, the period of the function is determined by $\frac{\pi}{|B|}$; therefore, the graph is the standard cotangent function, except with a period of $\frac{\pi}{2}$.

483.

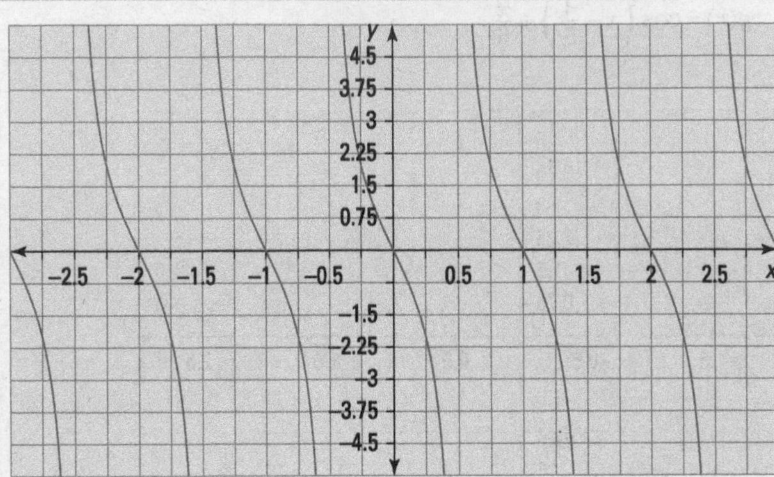

Illustration by Thomson Digital

The given function is

$$f(x) = -2\tan(\pi x)$$

Using $f(x) = A\tan Bx$, the multiplier $A = 2$ makes the graph steeper, and the negative sign reflects the graph over the x-axis. The period of the function is determined by $\frac{\pi}{|B|}$; because $B = \pi$, the graph has a period of 1.

484.

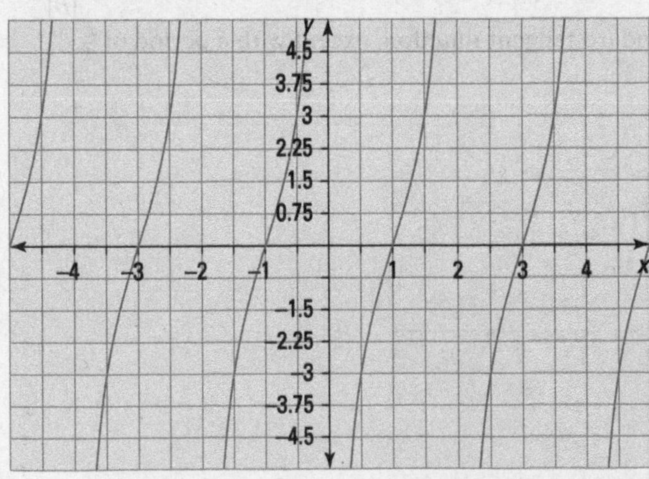

Illustration by Thomson Digital

The given function is

$$f(x) = 3\cot\left(-\frac{\pi}{2}x\right)$$

Using $f(x) = A \cot Bx$, the multiplier $A = 3$ makes the graph steeper. The period of the function is determined by $\frac{\pi}{|B|}$, so the graph has a period of 2. The negative multiplier on the angle reflects the graph over the y-axis.

485. period: $\frac{\pi}{6}$; amplitude: $\frac{1}{2}$; reflected over the x-axis

The transformed function is

$$g(x) = -\frac{1}{2}\tan(6x)$$

Using $g(x) = A \tan B(x + C) + D$, A represents a multiplier that steepens or flattens the curve, B affects the period, C represents a horizontal shift, and D represents a vertical shift. With $A = -\frac{1}{2}$, the curve flattens, because $\frac{1}{2}$ is smaller than 1. The negative multiplier creates a reflection over the x-axis, and $B = 6$ results in a period of $\frac{\pi}{6}$. The period of tangent or cotangent is found by dividing π by the absolute value of B.

486. period: $\frac{1}{3}$; amplitude: 2

The transformed function is

$$g(x) = 2\tan(3\pi x)$$

Using $g(x) = A \tan B(x + C) + D$, A represents a multiplier that steepens or flattens the curve, B affects the period, C represents a horizontal shift, and D represents a vertical shift.

With $A = 2$, the curve steepens. You find the period of tangent or cotangent by dividing π by the absolute value of B, so $B = 3\pi$ results in a period of $\frac{\pi}{3\pi} = \frac{1}{3}$.

487. period: 3; amplitude: $\sqrt{3}$

The transformed function is

$$g(x) = \sqrt{3}\tan\left(\frac{\pi}{3}x\right)$$

Using $g(x) = A \tan B(x + C) + D$, A represents a multiplier that steepens or flattens the curve, B affects the period, C represents a horizontal shift, and D represents a vertical shift.

With $A = \sqrt{3}$, the curve steepens, because $\sqrt{3}$ is greater than 1. $B = \frac{\pi}{3}$ results in a period of $\frac{\pi}{\pi/3} = 3$.

488. period: 6; amplitude: $\frac{\pi}{4}$; reflected over the y-axis

The transformed function is

$$g(x) = \frac{\pi}{4}\tan\left(-\frac{\pi}{6}x\right)$$

Using $g(x) = A \tan B(x + C) + D$, A represents a multiplier that steepens or flattens the curve, B affects the period, C represents a horizontal shift, and D represents a vertical shift.

With $A=\frac{\pi}{4}$, the curve flattens, because $\frac{\pi}{4}$ is smaller than 1. You find the period of tangent or cotangent by dividing π by the absolute value of B, so $B=-\frac{\pi}{6}$ results in a period of $\frac{\pi}{\left|-\pi/6\right|}=6$. The negative sign on B results in a reflection over the y-axis.

489. right $\frac{\pi}{4}$, up 3; period: $\frac{\pi}{4}$

The transformed function is

$$g(x)=\tan(4x-\pi)+3$$

Using $g(x)=A\tan B(x+C)+D$, A represents a multiplier that steepens or flattens the curve, B affects the period, C represents a horizontal shift, and D represents a vertical shift.

First, rewrite the function equation as

$$g(x)=\tan 4\left(x-\frac{\pi}{4}\right)+3$$

You find the period of tangent or cotangent by dividing π by the absolute value of B, so with $B=4$, the period is $\frac{\pi}{4}$. The value $C=-\frac{\pi}{4}$ means a shift to the right, and with $D=3$, the curve moves up 3 units.

490. left $\frac{\pi}{6}$, down 1; period: $\frac{\pi}{3}$

The transformed function is

$$g(x)=\cot\left(3x+\frac{\pi}{2}\right)-1$$

Using $g(x)=A\cot B(x+C)+D$, A represents a multiplier that steepens or flattens the curve, B affects the period, C represents a horizontal shift, and D represents a vertical shift.

First, rewrite the function equation as

$$g(x)=\cot 3\left(x+\frac{\pi}{6}\right)-1$$

You find the period of tangent or cotangent by dividing π by the absolute value of B, so with $B=3$, the period is $\frac{\pi}{3}$. The value $C=+\frac{\pi}{6}$ means a shift to the left, and with $D=-1$, the curve moves down 1 unit.

491. right 1, up π; period: $\frac{\pi}{2}$; reflected over the y-axis

The transformed function is

$$g(x)=\tan(-2x+2)+\pi$$

Using $g(x)=A\tan B(x+C)+D$, A represents a multiplier that steepens or flattens the curve, B affects the period, C represents a horizontal shift, and D represents a vertical shift.

First, rewrite the function equation as

$$g(x)=\tan(-2(x-1))+\pi$$

Answers 401–500

You find the period of tangent or cotangent by dividing π by the absolute value of B, so with $B = -2$, the period is $\frac{\pi}{2}$. The value $C = -1$ means a shift to the right, and with $D = \pi$, the curve moves up π units. The negative sign on $B = -2$ indicates a reflection over the y-axis.

492. left $\frac{4}{3}$, up $\frac{\pi}{2}$; period: $\frac{\pi}{3}$; reflected over the y-axis

The transformed function is

$$g(x) = \cot(-3x - 4) + \frac{\pi}{2}$$

Using $g(x) = A\cot B(x + C) + D$, A represents a multiplier that steepens or flattens the curve, B affects the period, C represents a horizontal shift, and D represents a vertical shift.

First, rewrite the function equation as

$$g(x) = \cot\left(-3\left(x + \frac{4}{3}\right)\right) + \frac{\pi}{2}$$

You find the period of tangent or cotangent by dividing π by the absolute value of B, so with $B = -3$, the period is $\frac{\pi}{3}$. Because B is negative, the function is reflected over the y-axis. The value $C = +\frac{4}{3}$ means a shift to the left, and with $D = +\frac{\pi}{2}$, the curve moves up $\frac{\pi}{2}$ units.

493. left $\frac{5}{2}$, down $\frac{\pi}{4}$; period: $\frac{\pi}{2}$; reflected over the x-axis

The transformed function is

$$g(x) = -3\tan(2x + 5) - \frac{\pi}{4}$$

Using $g(x) = A\tan B(x + C) + D$, A represents a multiplier that steepens or flattens the curve, B affects the period, C represents a horizontal shift, and D represents a vertical shift.

First, rewrite the function equation as

$$g(x) = -3\tan 2\left(x + \frac{5}{2}\right) - \frac{\pi}{4}$$

With $A = -3$, the curve is steeper, because 3 is greater than 1. There's also a reflection over a horizontal axis — in this case, $y = -\frac{\pi}{4}$ — because of the negative sign.

You find the period of tangent or cotangent by dividing π by the absolute value of B, so with $B = 2$, the period is $\frac{\pi}{2}$. The value $C = +\frac{5}{2}$ means a shift to the left, and with $D = -\frac{\pi}{4}$, the curve moves down $\frac{\pi}{4}$ units.

494. right $\frac{2\pi}{5}$, up 4; period: $\frac{\pi}{5}$; reflected over the x-axis

The transformed function is

$$g(x) = -\frac{1}{2}\cot(5x - 2\pi) + 4$$

Using $g(x) = A\cot B(x + C) + D$, A represents a multiplier that steepens or flattens the curve, B affects the period, C represents a horizontal shift, and D represents a vertical shift.

Answers 401–500

First, rewrite the function equation as

$$g(x) = -\frac{1}{2}\cot 5\left(x - \frac{2\pi}{5}\right) + 4$$

With $A = -\frac{1}{2}$, the curve is flatter, because $\frac{1}{2}$ is less than 1. There's also a reflection over a horizontal axis — in this case, $y = 4$ — because of the negative sign.

You find the period of tangent or cotangent by dividing π by the absolute value of B, so with $B = 5$, the period is $\frac{\pi}{5}$. The value $C = -\frac{2\pi}{5}$ means a shift to the right, and with $D = 4$, the curve moves up 4 units.

495. right $\frac{2\pi}{3}$, down π; period: 2π

The transformed function is

$$g(x) = \frac{3}{4}\tan\left(\frac{1}{2}x - \frac{\pi}{3}\right) - \pi$$

Using $g(x) = A\tan B(x + C) + D$, A represents a multiplier that steepens or flattens the curve, B affects the period, C represents a horizontal shift, and D represents a vertical shift.

First, rewrite the function equation as

$$g(x) = \frac{3}{4}\tan\frac{1}{2}\left(x - \frac{2\pi}{3}\right) - \pi$$

With $A = \frac{3}{4}$, the curve is flatter, because $\frac{3}{4}$ is less than 1.

You find the period of tangent or cotangent by dividing π by the absolute value of B, so with $B = \frac{1}{2}$, the period is 2π. The value $C = -\frac{2\pi}{3}$ means a shift to the right, and with $D = -\pi$, the curve moves down π units.

496. $x = \frac{2k+1}{2}$

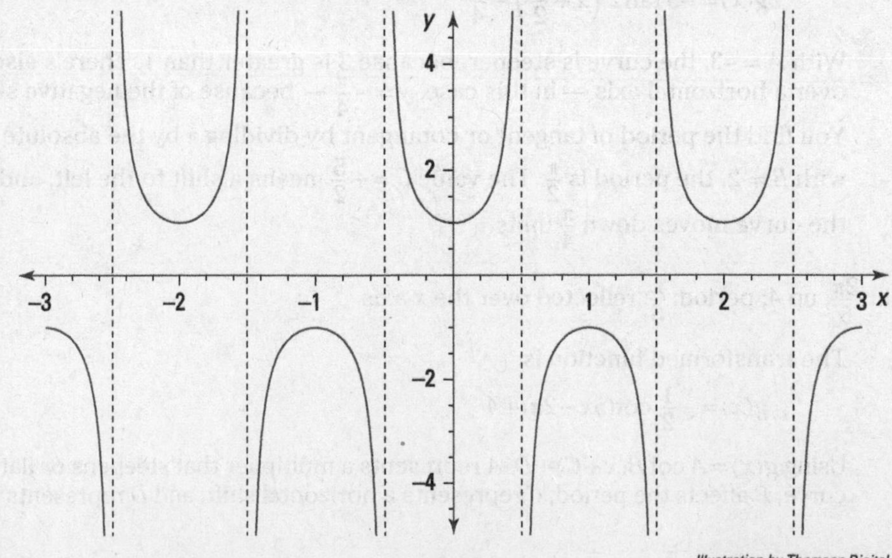

Illustration by Thomson Digital

Use $g(x)=A\sin B(x+C)+D$, where A is the amplitude, $\frac{2\pi}{|B|}$ is the period, C represents a horizontal shift, and D represents a vertical shift. For $f(x)=\sec(\pi x)$, the period is $\frac{2\pi}{|B|}=\frac{2\pi}{\pi}=2$.

The asymptotes are found where the reciprocal of the secant, $f(x)=\cos(\pi x)$, is equal to 0: $\cos(\pi x)=0$ when

$$\pi x=\ldots,-\frac{3\pi}{2},-\frac{\pi}{2},\frac{\pi}{2},\frac{3\pi}{2},\ldots$$

Solving for x, you divide each term by π to get

$$x=\ldots,-\frac{3}{2},-\frac{1}{2},\frac{1}{2},\frac{3}{2},\ldots$$

Letting k be an integer, the general rule for the equations of the asymptotes is

$$x=\frac{2k+1}{2}$$

497. $x=\frac{k}{2}$

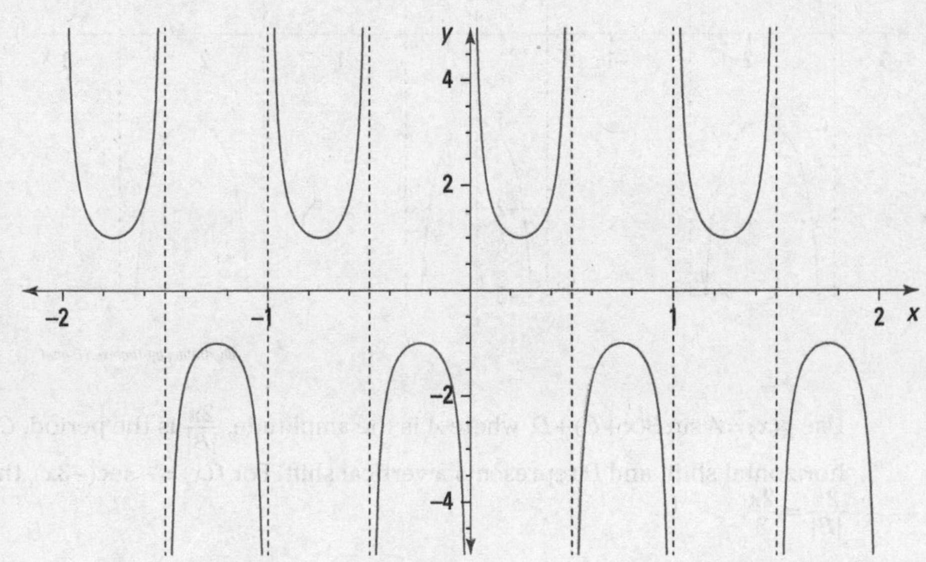

Illustration by Thomson Digital

Use $g(x)=A\sin B(x+C)+D$, where A is the amplitude, $\frac{2\pi}{|B|}$ is the period, C represents a horizontal shift, and D represents a vertical shift. For $f(x)=\csc(2\pi x)$, the period is $\frac{2\pi}{|B|}=\frac{2\pi}{2\pi}=1$.

The asymptotes are found where the reciprocal of the cosecant, $f(x)=\sin(2\pi x)$, is equal to 0: $\sin(2\pi x)=0$ when

$$2\pi x=\ldots,-2\pi,-\pi,0,\pi,2\pi,\ldots$$

Solving for x, you divide each term by 2π to get

$$2\pi x = \ldots, -1, -\frac{1}{2}, 0, \frac{1}{2}, 1, \ldots$$

Letting k be an integer, the general rule for the equations of the asymptotes is

$$x = \frac{k}{2}$$

498. $x = \frac{(2k+1)\pi}{6}$

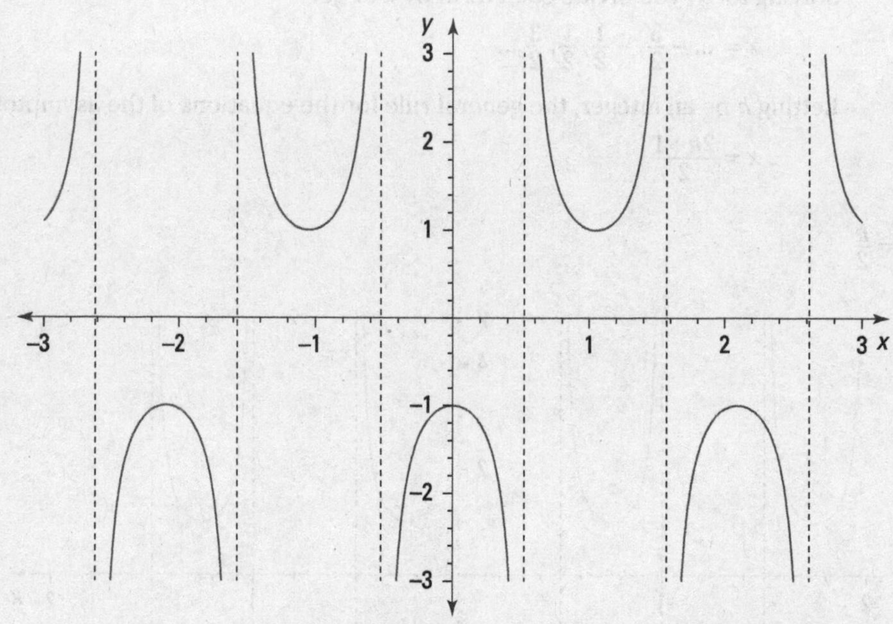

Illustration by Thomson Digital

Use $g(x) = A \sin B(x + C) + D$, where A is the amplitude, $\frac{2\pi}{|B|}$ is the period, C represents a horizontal shift, and D represents a vertical shift. For $f(x) = -\sec(-3x)$, the period is $\frac{2\pi}{|B|} = \frac{2\pi}{3}$.

The asymptotes are found where the reciprocal of the secant, $f(x) = \cos(-3x)$, is equal to 0: $\cos(-3x) = 0$ when

$$-3x = \ldots, -\frac{3\pi}{2}, -\frac{\pi}{2}, \frac{\pi}{2}, \frac{3\pi}{2}, \ldots$$

Solving for x, you divide each term by −3 to get

$$x = \ldots, \frac{\pi}{2}, \frac{\pi}{6}, -\frac{\pi}{6}, -\frac{\pi}{2}, \ldots$$

Letting k be an integer, the general rule for the equations of the asymptotes is

$$x = \frac{(2k+1)\pi}{6}$$

499. $x=\frac{2k+1}{10}$

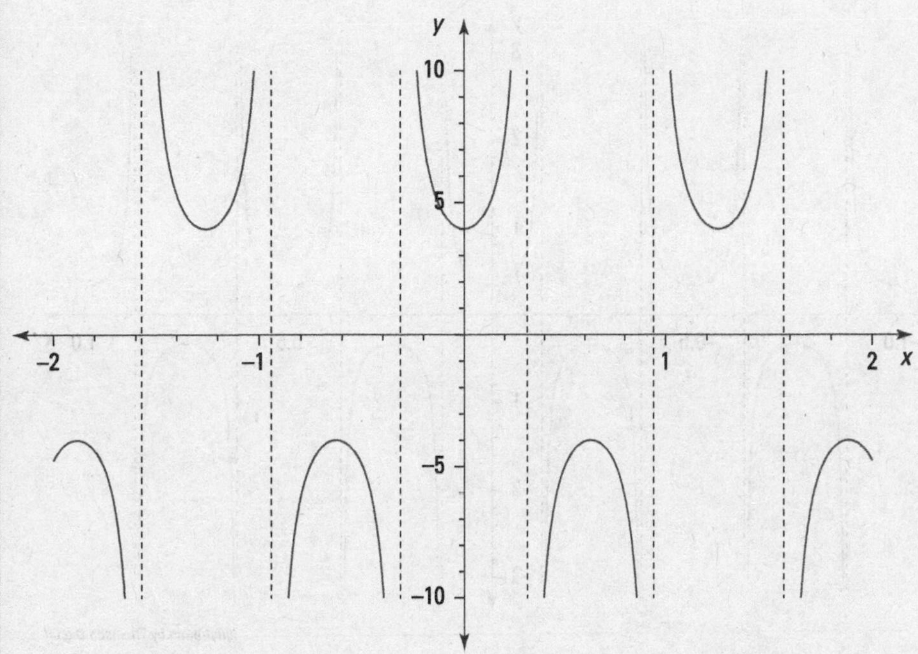

Illustration by Thomson Digital

Use $g(x)=A\sin B(x+C)+D$, where A is the amplitude, $\frac{2\pi}{|B|}$ is the period, C represents a horizontal shift, and D represents a vertical shift. For $f(x)=4\sec(5x)$, the period is $\frac{2\pi}{|B|}=\frac{2\pi}{5}$. The multiplier 4 brings the upper curves down to 4 and the lower curves up to −4.

The asymptotes are found where the reciprocal of the secant, $f(x)=4\cos(5x)$, is equal to 0: $\cos(5x)=0$ when

$$5x=\ldots,-\frac{3\pi}{2},-\frac{\pi}{2},\frac{\pi}{2},\frac{3\pi}{2},\ldots$$

Solving for x, you divide each term by 5 to get

$$x=\ldots,-\frac{3\pi}{10},-\frac{\pi}{10},\frac{\pi}{10},\frac{3\pi}{10},\ldots$$

Letting k be an integer, the general rule for the equations of the asymptotes is

$$x=\frac{2k+1}{10}$$

500. $x = \frac{2k+1}{8}$

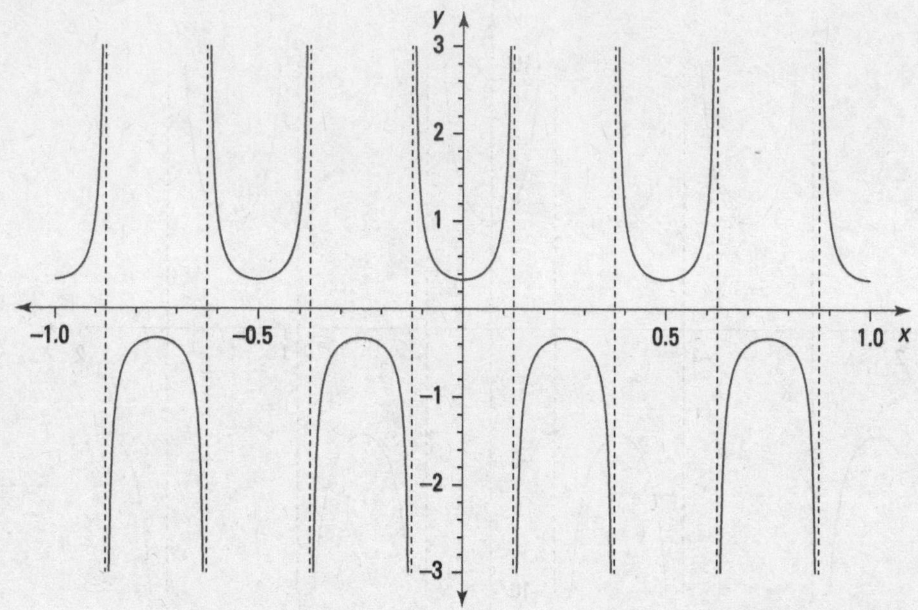

Illustration by Thomson Digital

Use $g(x) = A \sin B(x + C) + D$, where A is the amplitude, $\frac{2\pi}{|B|}$ is the period, C represents a horizontal shift, and D represents a vertical shift. For $f(x) = \frac{1}{3}\sec(4\pi x)$, the period is $\frac{2\pi}{|B|} = \frac{2\pi}{4\pi} = \frac{1}{2}$. The multiplier $\frac{1}{3}$ brings the upper curves down to $\frac{1}{3}$ and the lower curves up to $-\frac{1}{3}$.

The asymptotes are found where the reciprocal of the secant, $f(x) = \cos(4\pi x)$, is equal to 0: $\cos(4\pi x) = 0$ when

$$4\pi x = \ldots, -\frac{3\pi}{2}, -\frac{\pi}{2}, \frac{\pi}{2}, \frac{3\pi}{2}, \ldots$$

Solving for x, you divide each term by 4π to get

$$x = \ldots, -\frac{3}{8}, -\frac{1}{8}, \frac{1}{8}, \frac{3}{8}, \ldots$$

Letting k be an integer, the general rule for the equations of the asymptotes is

$$x = \frac{2k+1}{8}$$

501. $x = \frac{k}{8}\pi$

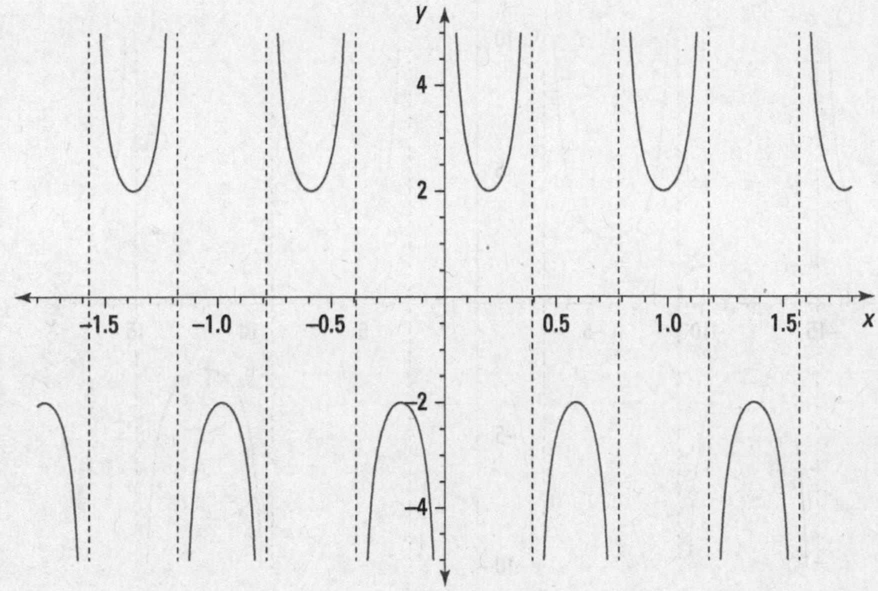

Illustration by Thomson Digital

Use $g(x) = A \sin B(x + C) + D$, where A is the amplitude, $\frac{2\pi}{|B|}$ is the period, C represents a horizontal shift, and D represents a vertical shift. For $f(x) = -2\csc(-8x)$, the period is $\frac{2\pi}{|B|} = \frac{2\pi}{8} = \frac{\pi}{4}$. The multiplier –2 brings the upper curves down to 2 and the lower curves up to –2. The negative multiplier of the angle reflects the standard curve over the *y*-axis.

The asymptotes are found where the reciprocal of the cosecant, $f(x) = -2\sin(-8x)$, is equal to 0: $\sin(-8x) = 0$ when

$$-8x = \ldots, -2\pi, -\pi, 0, \pi, 2\pi, \ldots$$

Solving for x, you divide each term by –8 to get

$$x = \ldots, \frac{\pi}{4}, \frac{\pi}{8}, 0, -\frac{\pi}{8}, -\frac{\pi}{4}, \ldots$$

Letting k be an integer, the general rule for the equations of the asymptotes is

$$x = \frac{k}{8}\pi$$

502. $x = 6k + 3$

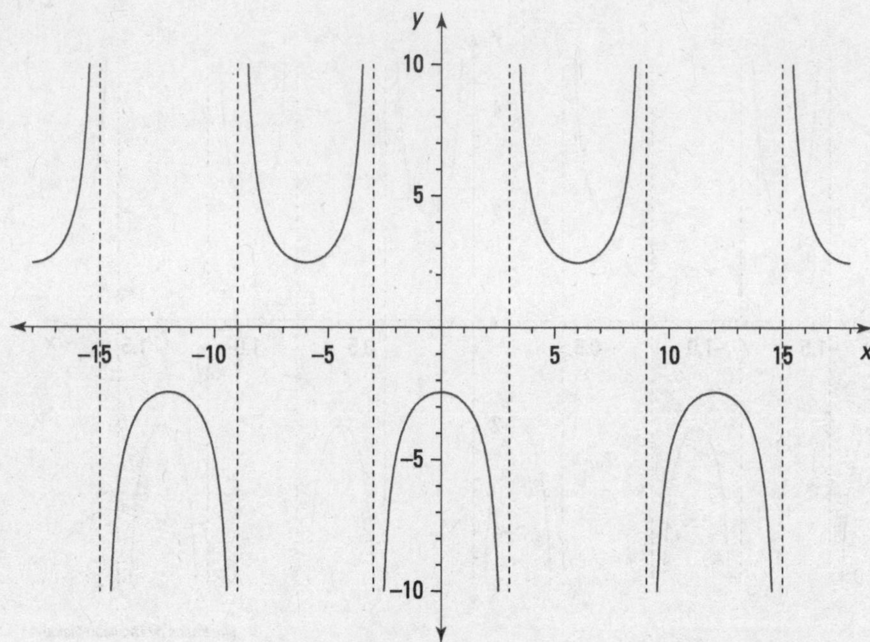

Illustration by Thomson Digital

Use $g(x) = A\sin B(x + C) + D$, where A is the amplitude, $\frac{2\pi}{|B|}$ is the period, C represents a horizontal shift, and D represents a vertical shift. For $f(x) = -\sqrt{6}\sec\left(\frac{\pi}{6}x\right)$, the period is $\frac{2\pi}{|B|} = \frac{2\pi}{\frac{\pi}{6}} = 12$. The multiplier $-\sqrt{6}$ brings the upper curves down to $\sqrt{6}$ and the lower curves up to $-\sqrt{6}$. The negative multiplier reflects the standard curve over the x-axis.

The asymptotes are found where the reciprocal of the secant, $f(x) = \cos\left(\frac{\pi}{6}x\right)$, is equal to 0: $\cos\left(\frac{\pi}{6}x\right) = 0$ when

$$\frac{\pi}{6}x = \ldots, -\frac{3\pi}{2}, -\frac{\pi}{2}, \frac{\pi}{2}, \frac{3\pi}{2}, \ldots$$

Solving for x, you divide each term by $\frac{\pi}{6}$ to get

$$x = \ldots, -9, -3, 3, 9, \ldots$$

Letting k be an integer, the general rule for the equations of the asymptotes is

$$x = 6k + 3$$

503. $x = 4k$

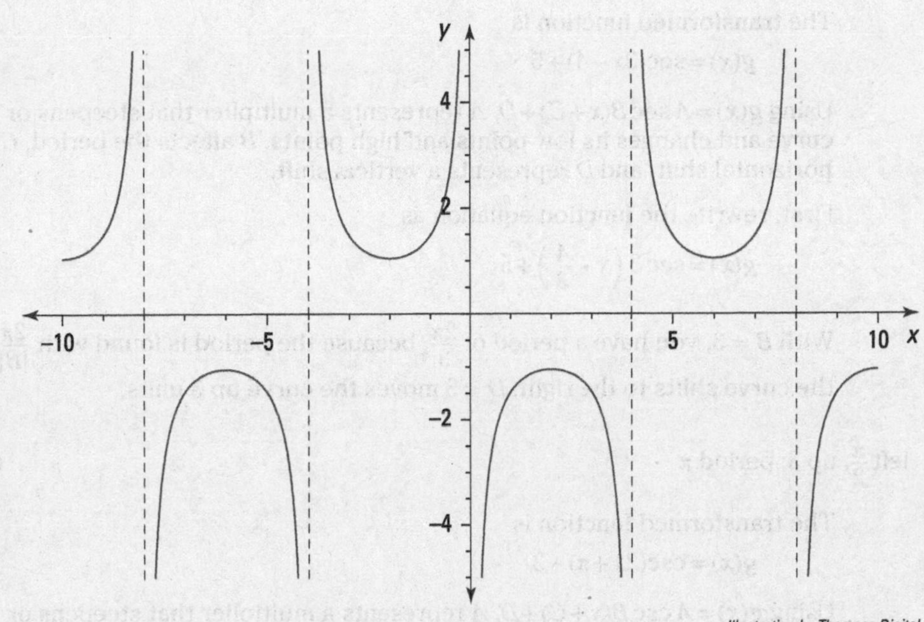

Illustration by Thomson Digital

Use $g(x) = A \sin B(x + C) + D$, where A is the amplitude, $\frac{2\pi}{|B|}$ is the period, C represents a horizontal shift, and D represents a vertical shift. For $f(x) = \frac{\pi}{3} \csc\left(-\frac{\pi}{4}x\right)$, the period is $\frac{2\pi}{|B|} = \frac{2\pi}{\pi/4} = 8$. The multiplier $\frac{\pi}{3}$ brings the upper curves down to $\frac{\pi}{3}$ and the lower curves up to $-\frac{\pi}{3}$. The negative multiplier of the angle reflects the standard curve over the *y*-axis.

The asymptotes are found where the reciprocal of the cosecant, $f(x) = \frac{\pi}{3} \sin\left(-\frac{\pi}{4}x\right)$, is equal to 0: $\sin\left(-\frac{\pi}{4}x\right) = 0$ when

$$-\frac{\pi}{4}x = \ldots, -2\pi, -\pi, 0, \pi, 2\pi, \ldots$$

Solving for x, you divide each term by $-\frac{\pi}{4}$ to get

$$x = \ldots, 8, 4, 0, -4, -8, \ldots$$

Letting k be an integer, the general rule for the equations of the asymptotes is

$$x = 4k$$

504. right $\frac{4}{3}$, up 5; period: $\frac{2\pi}{3}$

The transformed function is

$$g(x)=\sec(3x-4)+5$$

Using $g(x)=A\sec B(x+C)+D$, A represents a multiplier that steepens or flattens the curve and changes its low points and high points, B affects the period, C represents a horizontal shift, and D represents a vertical shift.

First, rewrite the function equation as

$$g(x)=\sec 3\left(x-\frac{4}{3}\right)+5$$

With $B=3$, you have a period of $\frac{2\pi}{3}$, because the period is found with $\frac{2\pi}{|B|}$. $C=-\frac{4}{3}$, so the curve shifts to the right. $D=5$ moves the curve up 5 units.

505. left $\frac{\pi}{2}$, up 3; period π

The transformed function is

$$g(x)=\csc(2x+\pi)+3$$

Using $g(x)=A\csc B(x+C)+D$, A represents a multiplier that steepens or flattens the curve and changes its low points and high points, B affects the period, C represents a horizontal shift, and D represents a vertical shift.

First, rewrite the function equation as

$$g(x)=\csc 2\left(x+\frac{\pi}{2}\right)+3$$

With $B=2$, you have a period of $\frac{2\pi}{2}=\pi$, because the period is found with $\frac{2\pi}{|B|}$. $C=+\frac{\pi}{2}$, so the curve shifts to the left. $D=3$ moves the curve up 3 units.

506. right 8, down π

The transformed function is

$$g(x)=\sec(x-8)-\pi$$

Using $g(x)=A\sec B(x+C)+D$, A represents a multiplier that steepens or flattens the curve and changes its low points and high points, B affects the period, C represents a horizontal shift, and D represents a vertical shift.

$C=-8$, so the curve shifts to the right. $D=-\pi$ moves the curve down π units.

507. right 2π, up $\frac{\pi}{4}$; period: 2π; reflected over the *y*-axis

The transformed function is

$$g(x)=\csc(-x+2\pi)+\frac{\pi}{4}$$

Using $g(x)=A\csc B(x+C)+D$, A represents a multiplier that steepens or flattens the curve and changes its low points and high points, B affects the period, C represents a horizontal shift, and D represents a vertical shift.

First, rewrite the function equation as

$$g(x)=\csc(-(x-2\pi))+\frac{\pi}{4}$$

With $B=-1$, you have a period of 2π, because the period is found with $\frac{2\pi}{|B|}$. The negative sign reflects the curve over the *y*-axis. $C=-2\pi$, so the curve shifts to the right. $D=+\frac{\pi}{4}$ moves the curve up $\frac{\pi}{4}$ units.

508. right $\frac{\pi}{4}$, down $\frac{3}{2}$; period: π; reflected over the *x*-axis

The transformed function is

$$g(x)=-2\sec\left(-2x+\frac{\pi}{2}\right)-\frac{3}{2}$$

Using $g(x)=A\sec B(x+C)+D$, A represents a multiplier that steepens or flattens the curve and changes its low points and high points, B affects the period, C represents a horizontal shift, and D represents a vertical shift.

First, rewrite the function equation as

$$g(x)=-2\sec\left(-2\left(x-\frac{\pi}{4}\right)\right)-\frac{3}{2}$$

With $A=-2$, the curve is steeper, and the distance between the upper and lower curves is greater because of the 2 multiplier. The negative sign reflects the curve over the *x*-axis. With $B=-2$, you have a period of $\frac{2\pi}{2}=\pi$, because the period is found with $\frac{2\pi}{|B|}$. The negative sign on B also reflects the graph over the *y*-axis. $C=-\frac{\pi}{4}$, so the curve shifts to the right. $D=-\frac{3}{2}$ moves the curve down $\frac{3}{2}$ units.

509. left $\frac{3}{\pi}$, down 5; period: 2

The transformed function is

$$g(x)=\frac{1}{2}\csc(\pi x+3)-5$$

Using $g(x) = A\csc B(x+C)+D$, A represents a multiplier that steepens or flattens the curve and changes its low points and high points, B affects the period, C represents a horizontal shift, and D represents a vertical shift.

First, rewrite the function equation as

$$g(x)=\frac{1}{2}\csc\pi\left(x+\frac{3}{\pi}\right)-5$$

With $A=\frac{1}{2}$, the curve is flatter, and the distance between the upper and lower curves is smaller because of the $\frac{1}{2}$ multiplier. With $B=\pi$, you have a period of 2, because the period is found with $\frac{2\pi}{|B|}$. $C=+\frac{3}{\pi}$, so the curve shifts to the left. $D=-5$ moves the curve down 5 units.

510. left 3, down $\frac{\pi}{2}$; period: 4π

The transformed function is

$$g(x)=\frac{2}{3}\sec\left(\frac{1}{2}x+\frac{3}{2}\right)-\frac{\pi}{2}$$

Using $g(x) = A\sec B(x+C)+D$, A represents a multiplier that steepens or flattens the curve and changes its low points and high points, B affects the period, C represents a horizontal shift, and D represents a vertical shift.

First, rewrite the function equation as

$$g(x)=\frac{2}{3}\sec\frac{1}{2}(x+3)-\frac{\pi}{2}$$

With $A=\frac{2}{3}$, the curve is slightly flatter, and the distance between the upper and lower curves is smaller because of the $\frac{2}{3}$ multiplier. With $B=\frac{1}{2}$, you have a period of $\frac{2\pi}{1/2}=4\pi$. $C=+3$, so the curve shifts to the left. $D=-\frac{\pi}{2}$ moves the curve down $\frac{\pi}{2}$ units.

511. use the reciprocal identity

Because each term contains a function and its reciprocal, using reciprocal identities will simplify the terms quickly.

Replace $\csc\theta$ with its reciprocal identity, $\frac{1}{\sin\theta}$, and $\cot\theta$ with its reciprocal identity, $\frac{1}{\tan\theta}$:

$$\sin\theta\csc\theta+\cot\theta\tan\theta=2$$

$$\sin\theta\cdot\frac{1}{\sin\theta}+\frac{1}{\tan\theta}\cdot\tan\theta=2$$

Multiply and simplify:

$$\cancel{\sin\theta}\cdot\frac{1}{\cancel{\sin\theta}}+\frac{1}{\cancel{\tan\theta}}\cdot\cancel{\tan\theta}=2$$

$$1+1=2$$

Answers 501–600

512. use the reciprocal identity

Because each term contains a function and its reciprocal, using reciprocal identities will simplify the terms quickly.

First, replace $\csc\theta$ with its reciprocal identity, $\frac{1}{\sin\theta}$, and $\sec\theta$ with its reciprocal identity, $\frac{1}{\cos\theta}$. Then simplify the complex fractions.

$$\frac{\sin\theta}{\csc\theta}+\frac{\cos\theta}{\sec\theta}=1$$

$$\frac{\sin\theta}{\frac{1}{\sin\theta}}+\frac{\cos\theta}{\frac{1}{\cos\theta}}=1$$

$$\frac{\sin\theta}{1}\cdot\frac{\sin\theta}{1}+\frac{\cos\theta}{1}\cdot\frac{\cos\theta}{1}=1$$

$$\sin^2\theta+\cos^2\theta=1$$

Finally, replace $\sin^2\theta+\cos^2\theta$ with 1, using the Pythagorean identity:

$$1=1$$

513. use the ratio identity

The term on the left can actually be rewritten using the ratio identity directly, but the following shows how to change to positive angles first.

Replace $\sin(-\theta)$ with $-\sin(\theta)$ and $\cos(-\theta)$ with $\cos(\theta)$ using the two even-odd identities:

$$\frac{\sin(-\theta)}{\cos(-\theta)}=\tan(-\theta)$$

$$\frac{-\sin(\theta)}{\cos(\theta)}=\tan(-\theta)$$

Use the ratio identity to replace $\frac{\sin\theta}{\cos\theta}$ with $\tan\theta$:

$$-\frac{\sin(\theta)}{\cos(\theta)}=\tan(-\theta)$$

$$-\tan(\theta)=\tan(-\theta)$$

Replace $-\tan(\theta)$ with $\tan(-\theta)$ using the even-odd identity:

$$-\tan(\theta)=-\tan(\theta)$$

514. use the co function identity

You want both trig terms to use the same angle measure, so first apply the co-function identity.

Rewrite the first term on the left as the power of the sine:

$$\sin^2\left(\frac{\pi}{2}-x\right)+\sin^2 x=1$$

$$\left[\sin\left(\frac{\pi}{2}-x\right)\right]^2+\sin^2 x=1$$

Replace $\sin\left(\frac{\pi}{2}-x\right)$ with $\cos x$ using the co-function identity:

$$[\cos x]^2+\sin^2 x=1$$

Rewrite the first term on the left:

$$\cos^2 x+\sin^2 x=1$$

Replace $\cos^2 x+\sin^2 x$ with 1, using the Pythagorean identity:

$$1=1$$

515. use the periodicity identity

You want the term on the left to have the same angle as that on the right (simplifying the more complex expression), so apply the periodicity identity.

Rewrite $\sin(2\pi-x)$ as the sine of a negative angle:

$$\sin(2\pi-x)=-\sin x$$
$$\sin(2\pi-x)=\sin(-(x-2\pi))$$
$$\sin(-(x-2\pi))=-\sin x$$

Use the even-odd identity to replace $\sin(-(x-2\pi))$ with $-\sin(x-2\pi)$:

$$-\sin(x-2\pi)=-\sin x$$

Replace $-\sin(x-2\pi)$ with $-\sin x$ using the periodicity identity:

$$-\sin x=-\sin x$$

Another method would be to use the difference formula, although it might not be as intuitive that it will work with two different angle measures.

Replace the left side using the difference formula and substitute in the function values.

$$\sin(2\pi-x)=-\sin x$$
$$\sin 2\pi\cos x-\cos 2\pi\sin x=-\sin x$$
$$0(\cos x)-1(\sin x)=-\sin x$$
$$-\sin x=-\sin x$$

516. use the reciprocal identity

The more complex term in this identity is the fraction on the left. Because it contains a function and its reciprocal, apply a reciprocal identity.

Replace $\cos\theta$ with its reciprocal identity, $\frac{1}{\sec\theta}$. Then simplify the complex fraction.

$$\frac{\sec\theta}{\cos\theta}=\tan^2\theta+1$$
$$\frac{\sec\theta}{\frac{1}{\sec\theta}}=\tan^2\theta+1$$
$$\frac{\sec\theta}{1}\cdot\frac{\sec\theta}{1}=\tan^2\theta+1$$
$$\sec^2\theta=\tan^2\theta+1$$

Answers 501–600

Use the Pythagorean identity to replace $\sec^2\theta$ with $\tan^2\theta+1$:

$$\tan^2\theta+1=\tan^2\theta+1$$

517. use the Pythagorean identity

Even though you could split the term on the left into two fractions and simplify, a quicker move is to apply the Pythagorean identity to the numerator to create one term.

Replace $\sin^2 x+\cos^2 x$ with 1 using the Pythagorean identity:

$$\frac{\sin^2 x+\cos^2 x}{\cos^2 x}=\sec^2 x$$

$$\frac{1}{\cos^2 x}=\sec^2 x$$

Replace $\frac{1}{\cos^2 x}$ with $\sec^2 x$ using the reciprocal identity:

$$\sec^2 x=\sec^2 x$$

518. use the even-odd identity

You want the angle measures in the trig terms to be the same, so apply the even-odd identity to the first term.

Rewrite the first term on the left as the power of the sine:

$$\sin^2(-\theta)+\cos^2(\theta)=1$$

$$[\sin(-\theta)]^2+\cos^2(\theta)=1$$

Use the even-odd identity to replace $\sin(-\theta)$ with $-\sin(\theta)$:

$$[-\sin(\theta)]^2+\cos^2(\theta)=1$$

Square the first term:

$$[-\sin(\theta)]^2+\cos^2(\theta)=1$$

$$(-1)^2\sin^2(\theta)+\cos^2(\theta)=1$$

$$\sin^2(\theta)+\cos^2(\theta)=1$$

Replace $\sin^2(\theta)+\cos^2(\theta)$ with 1 using the Pythagorean identity:

$$1=1$$

519. use the co-function identity

You want the angle measures in the trig terms to be the same, so apply the co-function identity to the first term on the left.

Rewrite the first term on the left as the power of the cosine:

$$\cos^2\left(\frac{\pi}{2}-x\right)+\sin x=\sin x(\sin x+1)$$

$$\left[\cos\left(\frac{\pi}{2}-x\right)\right]^2+\sin x=\sin x(\sin x+1)$$

Replace $\cos\left(\frac{\pi}{2}-x\right)$ with $\sin x$ using the co-function identity:

$$[\sin x]^2+\sin x=\sin x(\sin x+1)$$

Factor the terms on the left:

$$\sin x(\sin x+1)=\sin x(\sin x+1)$$

520. use the ratio identity

Before distributing through the terms on the left, rewrite the factor outside the parentheses by using the ratio identity.

Use the ratio identity to replace $\frac{\sin x}{\cos x}$ with $\tan x$:

$$\frac{\sin x}{\cos x}(\tan x+\cot x)=\sec^2 x$$

$$\tan x(\tan x+\cot x)=\sec^2 x$$

Replace $\cot x$ with $\frac{1}{\tan x}$ using the reciprocal identity. Then distribute $\tan x$.

$$\tan x\left(\tan x+\frac{1}{\tan x}\right)=\sec^2 x$$

$$\tan x\cdot\tan x+\frac{\cancel{\tan x}}{1}\cdot\frac{1}{\cancel{\tan x}}=\sec^2 x$$

$$\tan^2 x+1=\sec^2 x$$

Replace $\tan^2 x+1$ with $\sec^2 x$ using the Pythagorean identity:

$$\sec^2 x=\sec^2 x$$

521. use the ratio identity

This equation contains three different trig functions, so you want to rewrite it with trig functions that are more closely related. Taking advantage of the fact that cotangent is cosine over sine — and planning on using that ratio in a future multiplication — use the ratio identity first.

Replace $\cot\theta$ with $\frac{\cos\theta}{\sin\theta}$ using the ratio identity, and replace $\frac{1}{\sec\theta}$ with $\cos\theta$ using the reciprocal identity. Multiply on the left.

$$\sin\theta\cot\theta=\frac{1}{\sec\theta}$$

$$\cancel{\sin\theta}\cdot\frac{\cos\theta}{\cancel{\sin\theta}}=\cos\theta$$

$$\cos\theta=\cos\theta$$

522. use the Pythagorean identity

Looking ahead to the product you'll get on the right after multiplying the binomials, use the Pythagorean identity to rewrite the single term on the left.

Take the Pythagorean identity, $\sin^2 x + \cos^2 x = 1$, and subtract $\sin^2 x$ from each side:

$$\begin{aligned} \sin^2 x + \cos^2 x &= 1 \\ \cos^2 x &= 1 - \sin^2 x \end{aligned}$$

Replace $\cos^2 x$ with $1 - \sin^2 x$ in the identity you're proving:

$$\begin{aligned} \cos^2 x &= (1 - \sin x)(1 + \sin x) \\ 1 - \sin^2 x &= (1 - \sin x)(1 + \sin x) \end{aligned}$$

Factor the binomial as the difference of two squares:

$$(1 - \sin x)(1 + \sin x) = (1 - \sin x)(1 + \sin x)$$

Another option is to first multiply the two terms on the right, add $\sin^2 x$ to each side, and then apply the Pythagorean identity, as follows. It all depends on how you see the problem.

$$\begin{aligned} \cos^2 x &= (1 - \sin x)(1 + \sin x) \\ \cos^2 x &= 1 - \sin^2 x \\ \sin^2 x + \cos^2 x &= 1 \\ 1 &= 1 \end{aligned}$$

523. use the reciprocal identity

To change the ratio on the left to a single term, first use the reciprocals of each factor and simplify.

Replace $\csc(-\theta)$ and $\sec(-\theta)$ using reciprocal identities:

$$\frac{\csc(-\theta)}{\sec(-\theta)} = -\cot(\theta)$$

$$\frac{\frac{1}{\sin(-\theta)}}{\frac{1}{\cos(-\theta)}} = -\cot(\theta)$$

Simplify the complex fraction:

$$\frac{\frac{1}{\sin(-\theta)}}{\frac{1}{\cos(-\theta)}} = -\cot(\theta)$$

$$\frac{1}{\sin(-\theta)} \cdot \frac{\cos(-\theta)}{1} = -\cot(\theta)$$

$$\frac{\cos(-\theta)}{\sin(-\theta)} = -\cot(\theta)$$

Use the even-odd identities to replace $\cos(-\theta)$ with $\cos(\theta)$ and $\sin(-\theta)$ with $-\sin(\theta)$:

$$\frac{\cos(\theta)}{-\sin(\theta)} = -\cot(\theta)$$

Now use the ratio identity to replace $\frac{\cos(\theta)}{\sin(\theta)}$ with $\cot(\theta)$:

$$\frac{\cos(\theta)}{-\sin(\theta)} = -\cot(\theta)$$
$$-\cot(\theta) = -\cot(\theta)$$

524. use the even-odd identity

You want the angles of the two terms to be the same so you can apply the even-odd identity.

Factor –1 from the two terms of the tangent's angle:

$$\tan\left(\frac{\pi}{2} - x\right) = -\tan\left(x - \frac{\pi}{2}\right)$$
$$\tan\left(-1\left(x - \frac{\pi}{2}\right)\right) = -\tan\left(x - \frac{\pi}{2}\right)$$

Use the even-odd identity to replace the expression on the left:

$$\tan\left(-1\left(x - \frac{\pi}{2}\right)\right) = -\tan\left(x - \frac{\pi}{2}\right)$$
$$-\tan\left(x - \frac{\pi}{2}\right) = -\tan\left(x - \frac{\pi}{2}\right)$$

525. use the reciprocal identity

The two terms in the parentheses on the left are reciprocals of the multiplier outside the parentheses, so first rewrite the tangent by using the reciprocal identity.

Replace $\tan x$ on the left with $\frac{1}{\cot x}$ using the reciprocal identity:

$$\frac{1}{\cot x}(\cot x + \cot^3 x) = \csc^2 x$$

Distribute $\frac{1}{\cot x}$ through the parentheses. Then multiply and simplify terms.

$$\frac{1}{\cot x} \cdot \cot x + \frac{1}{\cot x} \cdot \cot^3 x = \csc^2 x$$
$$\frac{1}{\cancel{\cot x}} \cdot \frac{\cancel{\cot x}}{1} + \frac{1}{\cancel{\cot x}} \cdot \frac{\cot^{\cancel{3}2} x}{1} = \csc^2 x$$
$$1 + \cot^2 x = \csc^2 x$$

Use the Pythagorean identity to replace $1 + \cot^2 x$ with $\csc^2 x$:

$$\csc^2 x = \csc^2 x$$

Answers 501–600

526. use the ratio identity

The equation has four different trig identities with no apparent quick connections, so first distribute the tangent over the two terms and then use the ratio identity on the tangent to simplify the terms.

Distribute $\tan\theta$:

$$\tan\theta\cos\theta - \tan\theta\cot\theta = \sin\theta - 1$$

Use the ratio identity to replace $\tan\theta$ with $\frac{\sin\theta}{\cos\theta}$; then use the reciprocal identity to replace $\cot\theta$ with $\frac{1}{\tan\theta}$. Multiply and simplify.

$$\frac{\sin\theta}{\cos\theta}\cdot\cos\theta - \tan\theta\cdot\frac{1}{\tan\theta} = \sin\theta - 1$$

$$\frac{\sin\theta}{\cancel{\cos\theta}}\cdot\frac{\cancel{\cos\theta}}{1} - \frac{\cancel{\tan\theta}}{1}\cdot\frac{1}{\cancel{\tan\theta}} = \sin\theta - 1$$

$$\sin\theta - 1 = \sin\theta - 1$$

527. use the Pythagorean identity

When one side of an identity is more complex because of products or quotients, it's a good idea to multiply or divide to see the results. In this case, the product produces terms that belong in a Pythagorean identity.

Multiply the two binomials:

$$(\cot x - \csc x)(\cot x + \csc x) + 1 = 0$$

$$\cot^2 x + \cot x\csc x - \cot x\csc x - \csc^2 x + 1 = 0$$

$$\cot^2 x - \csc^2 x + 1 = 0$$

Use the Pythagorean identity to replace $\cot^2 x + 1$ with $\csc^2 x$:

$$\cot^2 x - \csc^2 x + 1 = 0$$

$$\cot^2 x + 1 - \csc^2 x = 0$$

$$\csc^2 x - \csc^2 x = 0$$

$$0 = 0$$

528. use the even-odd identity

You want the angles in the four trig functions to be the same, so first deal with the $\cos(-\theta)$ by using an even-odd identity.

Replace $\cos(-\theta)$ with $\cos\theta$ using the even-odd identity:

$$\tan\theta[\cos(-\theta) + \cos\theta] = 2\sin\theta$$

$$\tan\theta[\cos\theta + \cos\theta] = 2\sin\theta$$

$$\tan\theta[2\cos\theta] = 2\sin\theta$$

Use the ratio identity to replace $\tan\theta$ with $\frac{\sin\theta}{\cos\theta}$:

$$\frac{\sin\theta}{\cos\theta}[2\cos\theta]=2\sin\theta$$

Multiply and simplify:

$$\frac{\sin\theta}{\cancel{\cos\theta}}\cdot\frac{2\cancel{\cos\theta}}{1}=2\sin\theta$$

$$2\sin\theta=2\sin\theta$$

529. use the co-function identity

You want the angles in the three trig functions to be the same, so first deal with the tangent by using a co-function identity.

Rewrite the second term on the left as a power of the tangent:

$$\cot^2 x+\tan^2\left(\frac{\pi}{2}-x\right)=2\left(\csc^2 x-1\right)$$

$$\cot^2 x+\left[\tan\left(\frac{\pi}{2}-x\right)\right]^2=2\left(\csc^2 x-1\right)$$

Use the co-function identity to replace $\tan\left(\frac{\pi}{2}-x\right)$ with $\cot x$:

$$\cot^2 x+[\cot x]^2=2\left(\csc^2 x-1\right)$$

Rewrite the second term and combine the terms on the left:

$$\cot^2 x+\cot^2 x=2\left(\csc^2 x-1\right)$$

$$2\cot^2 x=2\left(\csc^2 x-1\right)$$

Rewrite the Pythagorean identity, $\cot^2 x+1=\csc^2 x$, by subtracting 1 from each side to get $\cot^2 x=\csc^2 x-1$. Replace $\cot^2 x$ in the identity being proven with $\csc^2 x-1$.

$$2\cot^2 x=2\left(\csc^2 x-1\right)$$

$$2\left(\csc^2 x-1\right)=2\left(\csc^2 x-1\right)$$

530. use the ratio and reciprocal identities

You have four different trig functions and need to distribute over the two terms on the left. Before doing that, change the two terms in the parentheses to functions involving just sine and cosine.

Replace $\cot x$ with $\frac{\cos x}{\sin x}$ using the ratio identity, and replace $\csc x$ with $\frac{1}{\sin x}$ using the reciprocal identity:

$$\sin x(\cot x-\csc x)=\cos x-1$$

$$\sin x\left(\frac{\cos x}{\sin x}-\frac{1}{\sin x}\right)=\cos x-1$$

Distribute $\sin x$. Then multiply and simplify.

$$\sin x\cdot\frac{\cos x}{\sin x}-\sin x\cdot\frac{1}{\sin x}=\cos x-1$$

$$\frac{\cancel{\sin x}}{1}\cdot\frac{\cos x}{\cancel{\sin x}}-\frac{\cancel{\sin x}}{1}\cdot\frac{1}{\cancel{\sin x}}=\cos x-1$$

$$\cos x-1=\cos x-1$$

531. use the reciprocal identity

You have four different trig functions in the identity, two of which are reciprocals of sine and cosine.

Use the reciprocal identities to replace $\csc\theta$ with $\dfrac{1}{\sin\theta}$ and $\sec\theta$ with $\dfrac{1}{\cos\theta}$. Then simplify the complex fractions and multiply.

$$\frac{\tan\theta}{\csc\theta}\cdot\frac{1}{\csc^2\theta\sec\theta}=\sin^4\theta$$

$$\frac{\tan\theta}{\frac{1}{\sin\theta}}\cdot\frac{1}{\frac{1}{\sin^2\theta}\cdot\frac{1}{\cos\theta}}=\sin^4\theta$$

$$\frac{\tan\theta}{1}\cdot\frac{\sin\theta}{1}\cdot\frac{\sin^2\theta\cos\theta}{1}=\sin^4\theta$$

$$\tan\theta\sin^3\theta\cos\theta=\sin^4\theta$$

Replace $\tan\theta$ with $\dfrac{\sin\theta}{\cos\theta}$ using the ratio identity and then simplify:

$$\frac{\sin\theta}{\cancel{\cos\theta}}\cdot\frac{\sin^3\theta\cancel{\cos\theta}}{1}=\sin^4\theta$$

$$\sin^4\theta=\sin^4\theta$$

532. use the ratio identity

The left side of the identity is more complicated than the right. To simplify on the left, distribute through the two terms and then apply a ratio identity to rewrite the first term in the result.

Distribute $\sin^2 x$. Then multiply and simplify.

$$\sin^2 x\left(\frac{1}{\cos^2 x}+\frac{1}{\sin^2 x}\right)=\sec^2 x$$

$$\sin^2 x\cdot\frac{1}{\cos^2 x}+\sin^2 x\cdot\frac{1}{\sin^2 x}=\sec^2 x$$

$$\frac{\sin^2 x}{\cos^2 x}\frac{\sin^2 x}{\sin^2 x}=\sec^2 x$$

$$\frac{\sin^2 x}{\cos^2 x}+1=\sec^2 x$$

Write the first term as the power of a fraction. Then use the ratio identity to replace the fraction with $\tan x$.

$$\left(\frac{\sin x}{\cos x}\right)^2+1=\sec^2 x$$

$$(\tan x)^2+1=\sec^2 x$$

$$\tan^2 x+1=\sec^2 x$$

Finally, use the Pythagorean identity to replace $\tan^2 x+1$ with $\sec^2 x$:

$$\sec^2 x=\sec^2 x$$

533. use the co-function identity

You want the angles of the trig functions to be the same, so use a co-function identity to rewrite the angle of the sine function on the left.

Use the co-function identity to replace the sine of the difference:

$$\cos x \tan x = \sin x$$

Now use the ratio identity to replace $\tan x$ with $\frac{\sin x}{\cos x}$. Multiply and simplify.

$$\cos x \cdot \frac{\sin x}{\cos x} = \sin x$$

$$\frac{\cancel{\cos x}}{1} \cdot \frac{\sin x}{\cancel{\cos x}} = \sin x$$

$$\sin x = \sin x$$

534. use the even-odd identity

You want the angles of the four trig functions to be the same, so rewrite the functions with negative angles by using even-odd identities.

Rewrite each function with a negative angle:

$$\cot(-\theta)\tan(-\theta) + \sec(-\theta)\cos(\theta) = 2$$

$$(-\cot\theta)(-\tan\theta) + (\sec\theta)(\cos\theta) = 2$$

Use the reciprocal identities to replace $\cot\theta$ with $\frac{1}{\tan\theta}$ and $\sec\theta$ with $\frac{1}{\cos\theta}$. Multiply and simplify.

$$\left(-\frac{1}{\tan\theta}\right)(-\tan\theta) + \left(\frac{1}{\cos\theta}\right)(\cos\theta) = 2$$

$$\left(-\frac{1}{\cancel{\tan\theta}}\right)\left(-\frac{\cancel{\tan\theta}}{1}\right) + \left(\frac{1}{\cancel{\cos\theta}}\right)\left(\frac{\cancel{\cos\theta}}{1}\right) = 2$$

$$1 + 1 = 2$$

535. use the Pythagorean identity

Usually, you want to change two terms to one term. But in this case, because of the shared functions on both sides of the equation, a good move is to rewrite the cosecant term on the left by using a Pythagorean identity and then to distribute to create two terms.

Use the Pythagorean identity to replace $\csc^2 x$ with $\cot^2 x + 1$. Then distribute $\sec^2 x$.

$$\sec^2 x \csc^2 x = \csc^2 x + \sec^2 x$$

$$\sec^2 x\left(\cot^2 x + 1\right) = \csc^2 x + \sec^2 x$$

$$\sec^2 x \cdot \cot^2 x + \sec^2 x = \csc^2 x + \sec^2 x$$

Replace the $\sec^2 x$ in the first term using the reciprocal identity, and replace the $\cot^2 x$ using the ratio identity. Multiply and simplify.

$$\frac{1}{\cancel{\cos^2 x}} \cdot \frac{\cancel{\cos^2 x}}{\sin^2 x} + \sec^2 x \cdot 1 = \csc^2 x + \sec^2 x$$

$$\frac{1}{\sin^2 x} + \sec^2 x \cdot 1 = \csc^2 x + \sec^2 x$$

Finally, use the reciprocal identity to replace $\frac{1}{\sin^2 x}$ with $\csc^2 x$:

$$\csc^2 x + \sec^2 x \cdot 1 = \csc^2 x + \sec^2 x$$

536. $\frac{1}{\sin^2 x}$

Use the reciprocal identities to replace $\csc^2 x$ and $\sec^2 x$. Use the ratio identities to replace $\tan^2 x$ and $\cot^2 x$.

$$\csc^2 x \tan^2 x + \sec^2 x \cot^2 x = \frac{1}{\cos^2 x} + \boxed{}$$

$$\frac{1}{\sin^2 x} \cdot \frac{\sin^2 x}{\cos^2 x} + \frac{1}{\cos^2 x} \cdot \frac{\cos^2 x}{\sin^2 x} = \frac{1}{\cos^2 x} + \boxed{}$$

Reduce the fractions and simplify:

$$\frac{1}{\cancel{\sin^2 x}} \cdot \frac{\cancel{\sin^2 x}}{\cos^2 x} + \frac{1}{\cancel{\cos^2 x}} \cdot \frac{\cancel{\cos^2 x}}{\sin^2 x} = \frac{1}{\cos^2 x} + \boxed{\frac{1}{\sin^2 x}}$$

The missing term is $\frac{1}{\sin^2 x}$.

537. $\cos x$

Use the reciprocal identities to replace $\sec x$ and $\csc x$. Use the ratio identities to replace $\tan x$ and $\cot x$.

$$\frac{\tan x}{\sec x} + \frac{\cot x}{\csc x} = \sin x + \boxed{}$$

$$\frac{\frac{\sin x}{\cos x}}{\frac{1}{\cos x}} + \frac{\frac{\cos x}{\sin x}}{\frac{1}{\sin x}} = \sin x + \boxed{}$$

Simplify the complex fractions by multiplying the numerator by the reciprocal of the denominator:

$$\frac{\sin x}{\cancel{\cos x}} \cdot \frac{\cancel{\cos x}}{1} + \frac{\cos x}{\cancel{\sin x}} \cdot \frac{\cancel{\sin x}}{1} = \sin x + \boxed{\cos x}$$

The missing term is $\cos x$.

538. 1

Use the reciprocal identity to replace $\csc^2 x$, and use the ratio identity to replace $\tan^2 x$:

$$\tan^2 x\left(\csc^2 x - 1\right) = \boxed{}$$

$$\frac{\sin^2 x}{\cos^2 x}\left(\frac{1}{\sin^2 x} - 1\right) = \boxed{}$$

Distribute $\frac{\sin^2 x}{\cos^2 x}$, simplify, and then combine the two terms:

$$\frac{\cancel{\sin^2 x}}{\cos^2 x}\cdot\frac{1}{\cancel{\sin^2 x}}-\frac{\sin^2 x}{\cos^2 x}\cdot 1=\boxed{}$$

$$\frac{1}{\cos^2 x}-\frac{\sin^2 x}{\cos^2 x}=\frac{1-\sin^2 x}{\cos^2 x}=\boxed{}$$

Rewrite the Pythagorean identity, $\sin^2 x+\cos^2 x=1$, by subtracting $\sin^2 x$ from each side to get $\cos^2 x=1-\sin^2 x$. Replace the numerator of the fraction in the identity with $\cos^2 x$:

$$\frac{\cos^2 x}{\cos^2 x}=\boxed{1}$$

The missing term is 1.

539. $(1-\sin x)$

Use the reciprocal identity to replace sec x. Use the ratio identities to replace $\tan x$ and $\cot^2 x$.

$$\cot^2 x\,(\sec x-\tan x)=\frac{\cos x\boxed{}}{\sin^2 x}$$

$$\frac{\cos^2 x}{\sin^2 x}\left(\frac{1}{\cos x}-\frac{\sin x}{\cos x}\right)=\frac{\cos x\boxed{}}{\sin^2 x}$$

Combine the two fractions in the parentheses; multiply by the first factor:

$$\frac{\cos^{\cancel{2}} x}{\sin^2 x}\left(\frac{1-\sin x}{\cancel{\cos x}}\right)=\frac{\cos x\boxed{(1-\sin x)}}{\sin^2 x}$$

The missing factor is $(1-\sin x)$.

540. $(1+\cot x)$

Use the reciprocal identities to replace $\sec x$ and $\csc x$. Use the ratio identity to replace $\tan x$.

$$\cot x(\sec x+\csc x)=\csc x\boxed{}$$

$$\frac{\cos x}{\sin x}\left(\frac{1}{\cos x}+\frac{1}{\sin x}\right)=\csc x\boxed{}$$

Add the two fractions in the parentheses after finding a common denominator:

$$\frac{\cos x}{\sin x}\left(\frac{\sin x}{\sin x\cos x}+\frac{\cos x}{\sin x\cos x}\right)=\csc x\boxed{}$$

$$\frac{\cos x}{\sin x}\left(\frac{\sin x+\cos x}{\sin x\cos x}\right)=\csc x\boxed{}$$

Divide out the $\cos x$ factor; then multiply:

$$\frac{\cancel{\cos x}}{\sin x}\left(\frac{\sin x+\cos x}{\sin x\cancel{\cos x}}\right)=\csc x\boxed{}$$

$$\left(\frac{\sin x+\cos x}{\sin^2 x}\right)=\csc x\boxed{}$$

Write the fraction as two separate fractions. Reduce the fraction on the left, and write the fraction on the right as a product.

$$\frac{\cancel{\sin x}}{\sin^{\cancel{2}} x}+\frac{\cos x}{\sin^2 x}=\csc x\boxed{}$$

$$\frac{1}{\sin x}+\frac{1}{\sin x}\cdot\frac{\cos x}{\sin x}=\csc x\boxed{}$$

Use the reciprocal identity to replace $\frac{1}{\sin x}$. Use the ratio identity to replace $\frac{\cos x}{\sin x}$. Finally, factor out $\csc x$.

$$\csc x+\csc x\cdot\cot x=\csc x\boxed{(1+\cot x)}$$

The missing factor is $(1+\cot x)$.

541. $1-\cos x$

Multiply both the numerator and the denominator by $1-\sec x$:

$$\frac{\cos x}{1+\sec x}=\frac{\boxed{}}{\tan^2 x}$$

$$\frac{\cos x}{1+\sec x}\cdot\frac{1-\sec x}{1-\sec x}=\frac{\boxed{}}{\tan^2 x}$$

$$\frac{\cos x(1-\sec x)}{(1+\sec x)(1-\sec x)}=\frac{\boxed{}}{\tan^2 x}$$

Distribute $\cos x$ in the numerator and multiply in the denominator:

$$\frac{\cos x-\cos x\sec x}{1-\sec^2 x}=\frac{\boxed{}}{\tan^2 x}$$

Replace $\sec x$ in the numerator by using the reciprocal identity. Replace $\sec^2 x$ in the denominator by using the Pythagorean identity. Simplify.

$$\frac{\cos x-\cancel{\cos x}\cdot\frac{1}{\cancel{\cos x}}}{1-(\tan^2 x+1)}=\frac{\boxed{}}{\tan^2 x}$$

$$\frac{\cos x-1}{1-\tan^2 x-1}=\frac{\cos x-1}{-\tan^2 x}=\frac{\boxed{}}{\tan^2 x}$$

Multiply numerator and denominator by –1:

$$\frac{\cos x-1}{-\tan^2 x}\cdot\frac{-1}{-1}=\frac{\boxed{1-\cos x}}{\tan^2 x}$$

The missing factor is $1-\cos x$.

Answers 501–600

542. $1+\tan x$

Multiply both the numerator and the denominator by $\cot x+1$:

$$\frac{\tan x}{\cot x-1}=\frac{\boxed{}}{\csc^2 x-2}$$

$$\frac{\tan x}{\cot x-1}\cdot\frac{\cot x+1}{\cot x+1}=\frac{\boxed{}}{\csc^2 x-2}$$

$$\frac{\tan x(\cot x+1)}{(\cot x-1)(\cot x+1)}=\frac{\boxed{}}{\csc^2 x-2}$$

Distribute $\cos x$ in the numerator and multiply in the denominator:

$$\frac{\tan x\cot x+\tan x}{\cot^2 x-1}=\frac{\boxed{}}{\csc^2 x-2}$$

Replace $\cot x$ in the numerator using the reciprocal identity. Replace $\cot^2 x$ in the denominator with $\csc^2 x-1$ after subtracting 1 from each side of the Pythagorean identity, $\cot^2 x+1=\csc^2 x$. Simplify.

$$\frac{\tan x\cdot\frac{1}{\tan x}+\tan x}{(\csc^2 x-1)-1}=\frac{1+\tan x}{\csc^2 x-1-1}=\frac{\boxed{1+\tan x}}{\csc^2 x-2}$$

The missing factor is $1+\tan x$.

543. $\sin x-\cos x$

Multiply both the numerator and the denominator by $\sin x-\cos x$:

$$\frac{1}{\sin x+\cos x}=\frac{\boxed{}}{1-2\cos^2 x}$$

$$\frac{1}{\sin x+\cos x}\cdot\frac{\sin x-\cos x}{\sin x-\cos x}=\frac{\boxed{}}{1-2\cos^2 x}$$

$$\frac{\sin x-\cos x}{(\sin x+\cos x)(\sin x-\cos x)}=\frac{\boxed{}}{1-2\cos^2 x}$$

Multiply in the denominator:

$$\frac{\sin x-\cos x}{\sin^2 x-\cos^2 x}=\frac{\boxed{}}{1-2\cos^2 x}$$

Rewrite the Pythagorean identity, $\sin^2 x+\cos^2 x=1$, by subtracting $\cos^2 x$ from each side. Then substitute $1-\cos^2 x$ for $\sin^2 x$ in the denominator. Simplify.

$$\frac{\sin x-\cos x}{(1-\cos^2 x)-\cos^2 x}=\frac{\boxed{\sin x-\cos x}}{1-2\cos^2 x}$$

The missing factor is $\sin x-\cos x$.

544. $\csc^2 x$

Multiply both the numerator and the denominator of the first fraction by $1+\cos x$, and multiply the numerator and denominator of the second fraction by $1-\cos x$:

$$\frac{1}{1-\cos x}+\frac{1}{1+\cos x}=2\boxed{}$$

$$\frac{1}{1-\cos x}\cdot\frac{1+\cos x}{1+\cos x}+\frac{1}{1+\cos x}\cdot\frac{1-\cos x}{1-\cos x}=2\boxed{}$$

$$\frac{1+\cos x}{(1-\cos x)(1+\cos x)}+\frac{1-\cos x}{(1+\cos x)(1-\cos x)}=2\boxed{}$$

$$\frac{1+\cos x}{1-\cos^2 x}+\frac{1-\cos x}{1-\cos^2 x}=2\boxed{}$$

Add the two fractions together. Then replace the 1 in the denominator with $\sin^2 x+\cos^2 x$ from the Pythagorean identity.

$$\frac{1+\cos x+1-\cos x}{1-\cos^2 x}=2\boxed{}$$

$$\frac{2}{1-\cos^2 x}=2\boxed{}$$

$$\frac{2}{\sin^2 x+\cos^2 x-\cos^2 x}=2\boxed{}$$

$$\frac{2}{\sin^2 x}=2\boxed{}$$

Replace $\sin^2 x$ with its reciprocal and then simplify:

$$\frac{2}{\frac{1}{\csc^2 x}}=\frac{2}{1}\cdot\frac{\csc^2 x}{1}=2\boxed{\csc^2 x}$$

The missing factor is $\csc^2 x$.

545. $\dfrac{1+\sin x}{\cos x}$

Multiply both the numerator and the denominator of the first fraction by $1+\cos x$, and multiply the numerator and denominator of the second fraction by $1+\sin x$:

$$\frac{\sin x}{1-\cos x}+\frac{\cos x}{1-\sin x}=\frac{1+\cos x}{\sin x}+\boxed{}$$

$$\frac{\sin x}{1-\cos x}\cdot\frac{1+\cos x}{1+\cos x}+\frac{\cos x}{1-\sin x}\cdot\frac{1+\sin x}{1+\sin x}=\frac{1+\cos x}{\sin x}+\boxed{}$$

$$\frac{\sin x(1+\cos x)}{(1-\cos x)(1+\cos x)}+\frac{\cos x(1+\sin x)}{(1-\sin x)(1+\sin x)}=\frac{1+\cos x}{\sin x}+\boxed{}$$

Leave the numerators factored. Multiply in the denominators.

$$\frac{\sin x(1+\cos x)}{1-\cos^2 x}+\frac{\cos x(1+\sin x)}{1-\sin^2 x}=\frac{1+\cos x}{\sin x}+\boxed{}$$

Use the Pythagorean identity to replace each 1 in the denominators with $\sin^2 x+\cos^2 x$. Simplify.

$$\frac{\sin x(1+\cos x)}{\sin^2 x+\cos^2 x-\cos^2 x}+\frac{\cos x(1+\sin x)}{\sin^2 x+\cos^2 x-\sin^2 x}=\frac{1+\cos x}{\sin x}+\boxed{}$$

$$\frac{\sin x(1+\cos x)}{\sin^2 x}+\frac{\cos x(1+\sin x)}{\cos^2 x}=\frac{1+\cos x}{\sin x}+\boxed{}$$

Reduce each term:

$$\frac{\cancel{\sin x}(1+\cos x)}{\sin^{\cancel{2}} x}+\frac{\cancel{\cos x}(1+\sin x)}{\cos^{\cancel{2}} x}=\frac{1+\cos x}{\sin x}+\boxed{\frac{1+\sin x}{\cos x}}$$

The missing term is $\frac{1+\sin x}{\cos x}$.

546. $\cos^2 x$

Square the binomial:

$$(1+\sin x)^2=2+2\sin x-\boxed{}$$

$$1+2\sin x+\sin^2 x=2+2\sin x-\boxed{}$$

Take the Pythagorean identity, $\sin^2 x+\cos^2 x=1$, and subtract $\cos^2 x$ from each side. Then replace $\sin^2 x$ in the problem with $1-\cos^2 x$. Simplify.

$$1+2\sin x+(1-\cos^2 x)=2+2\sin x-\boxed{}$$

$$2+2\sin x-\cos^2 x=2+2\sin x-\boxed{\cos^2 x}$$

The missing term is $\cos^2 x$.

547. $\sec^2 x$

Square the binomial:

$$(\cot x+\tan x)^2=\csc^2 x+\boxed{}$$

$$\cot^2 x+2\cot x\tan x+\tan^2 x=\csc^2 x+\boxed{}$$

Use the reciprocal identity to replace $\cot x$ with $\frac{1}{\tan x}$. Simplify.

$$\cot^2 x+2\frac{1}{\cancel{\tan x}}\cdot\cancel{\tan x}+\tan^2 x=\csc^2 x+\boxed{}$$

$$\cot^2 x+2+\tan^2 x=\csc^2 x+\boxed{}$$

Take the Pythagorean identity, $\cot^2 x+1=\csc^2 x$, and subtract 1 from each side of the equation. Then replace $\cot^2 x$ in the problem with $\csc^2 x-1$. Simplify.

$$(\csc^2 x-1)+2+\tan^2 x=\csc^2 x+\boxed{}$$

$$\csc^2 x+1+\tan^2 x=\csc^2 x+\boxed{}$$

Using the Pythagorean identity, replace $1+\tan^2 x$ with $\sec^2 x$:

$$\csc^2 x+\sec^2 x=\csc^2 x+\boxed{\sec^2 x}$$

The missing term is $\sec^2 x$.

548. 1

Square the binomial:

$$(\sin x+\cos x)^2=2\sin x\cos x+\boxed{\quad}$$

$$\sin^2 x+2\sin x\cos x+\cos^2 x=2\sin x\cos x+\boxed{\quad}$$

Using the Pythagorean identity, replace $\sin^2 x+\cos^2 x$ with 1:

$$2\sin x\cos x+1=2\sin x\cos x+\boxed{1}$$

The missing term is 1.

549. $3\csc x$

Cube the binomial:

$$(\sin x+\csc x)^3=\sin^3 x+3\sin x+\boxed{\quad}+\csc^3 x$$

$$\sin^3 x+3\sin^2 x\csc x+3\sin x\csc^2 x+\csc^3 x=\sin^3 x+3\sin x+\boxed{\quad}+\csc^3 x$$

Replace $\csc x$ with its reciprocal, $\frac{1}{\sin x}$, and $\sin x$ with its reciprocal, $\frac{1}{\csc x}$, in the second and third terms. Multiply and simplify.

$$\sin^3 x+3\sin^{\cancel{2}\,1} x\cdot\frac{1}{\cancel{\sin x}}+3\frac{1}{\cancel{\csc x}}\cdot\csc^{\cancel{2}\,1} x+\csc^3 x=\sin^3 x+3\sin x+\boxed{3\csc x}+\csc^3 x$$

The missing term is $3\csc x$.

550. $4\sin x\cos x$

Raise the binomial to the fourth power. Using Pascal's triangle for the coefficients of the terms, you'll have 1, 4, 6, 4, and 1 for the coefficients, decreasing powers of sine, and increasing powers of cosine:

$$(\sin x-\cos x)^4=\sin^4 x-4\sin^3 x\cos x+6\sin^2 x\cos^2 x-4\sin x\cos^3 x+\cos^4 x$$

The problem becomes

$$\sin^4 x-4\sin^3 x\cos x+6\sin^2 x\cos^2 x-4\sin x\cos^3 x+\cos^4 x$$
$$=1-4\cos^4 x+4\cos^2 x-\boxed{\quad}$$

Replace $\sin^2 x$ factor with $1-\cos^2 x$, obtained from subtracting $\cos^2 x$ from each side of the Pythagorean identity:

$$(1-\cos^2 x)(1-\cos^2 x)-4\sin x(1-\cos^2 x)\cos x+6(1-\cos^2 x)\cos^2 x$$
$$-4\sin x\cos^3 x+\cos^4 x=1-4\cos^4 x+4\cos^2 x-\boxed{\quad}$$

Multiply and simplify:

$$1-2\cos^2 x+\cos^4 x-4\sin x\cos x+4\sin x\cos^3 x+6\cos^2 x-6\cos^4 x$$
$$-4\sin x\cos^3 x+\cos^4 x=1-4\cos^4 x+4\cos^2 x-\boxed{\quad}$$

Now combine like terms:

$$1-4\cos^4 x+4\cos^2 x-4\sin x\cos x=1-4\cos^4 x+4\cos^2 x-\boxed{4\sin x\cos x}$$

The missing term is $4\sin x\cos x$.

551. $\tan x$

Factor $\sin x$ from the two terms:

$$\sin x\tan^2 x+\sin x=\sec x\boxed{}$$

$$\sin x\left(\tan^2 x+1\right)=\sec x\boxed{}$$

Replace $\tan^2 x+1$ with $\sec^2 x$ using the Pythagorean identity:

$$\sin x\left(\sec^2 x\right)=\sec x\boxed{}$$

Replace $\sec^2 x$ with its reciprocal, $\frac{1}{\cos^2 x}$; then multiply and rewrite:

$$\sin x\left(\frac{1}{\cos^2 x}\right)=\frac{\sin x}{\cos^2 x}=\frac{\sin x}{\cos x}\cdot\frac{1}{\cos x}=\sec x\boxed{}$$

Using the ratio identity, replace $\frac{\sin x}{\cos x}$ with $\tan x$, and change $\frac{1}{\cos x}$ back to its reciprocal:

$$\frac{\sin x}{\cos x}\cdot\frac{1}{\cos x}=\sec x\boxed{}$$

$$\tan x\cdot\sec x=\sec x\boxed{\tan x}$$

The missing factor is $\tan x$.

552. $\tan^2 x$

Factor $\tan^2 x$ from each term:

$$\tan^2 x\sin^2 x+\tan^2 x\cos^2 x=\boxed{}$$

$$\tan^2 x\left(\sin^2 x+\cos^2 x\right)=\boxed{}$$

Use the Pythagorean identity to replace $\sin^2 x+\cos^2 x$ with 1:

$$\tan^2 x(1)=\boxed{\tan^2 x}$$

The missing term is $\tan^2 x$.

553. $\tan x$

Factor the numerator as the square of a binomial:

$$\frac{\sin^2 x+2\sin x\tan x+\tan^2 x}{\sin x+\tan x}=\sin x+\boxed{}$$

$$\frac{(\sin x+\tan x)^2}{\sin x+\tan x}=\sin x+\boxed{}$$

Answers 501–600

Reduce the fraction:

$$\frac{(\sin x+\tan x)^{2}}{\sin x+\tan x}=\sin x+\boxed{\tan x}$$

The missing term is $\tan x$.

554. $\cos x$

Factor the numerator of the fraction as the difference of squares. Then reduce the fraction.

$$\frac{\sin^2 x-\cos^2 x}{\sin x+\cos x}=\sin x-\boxed{}$$

$$\frac{(\sin x+\cos x)(\sin x-\cos x)}{\sin x+\cos x}=\sin x-\boxed{}$$

$$\frac{\cancel{(\sin x+\cos x)}(\sin x-\cos x)}{\cancel{\sin x+\cos x}}=\sin x-\boxed{}$$

$$\sin x-\cos x=\sin x-\boxed{\cos x}$$

The missing term is $\cos x$.

555. 1

Factor the terms by grouping:

$$\sin^2 x\tan x+\sin^2 x+\cos^2 x\tan x+\cos^2 x=\tan x+\boxed{}$$

$$\sin^2 x(\tan x+1)+\cos^2 x(\tan x+1)=\tan x+\boxed{}$$

Factor $(\tan x+1)$ from each term:

$$(\tan x+1)\left(\sin^2 x+\cos^2 x\right)=\tan x+\boxed{}$$

Using the Pythagorean identity, replace $\sin^2 x+\cos^2 x$ with 1:

$$(\tan x+1)(1)=\tan x+\boxed{1}$$

The missing term is 1.

556. $\csc x$

The common denominator is $\sin x\cos x$. Write both fractions with that denominator:

$$\frac{\tan x}{\sin x}+\frac{\cot x}{\cos x}=\sec x+\boxed{}$$

$$\frac{\tan x}{\sin x}\cdot\frac{\cos x}{\cos x}+\frac{\cot x}{\cos x}\cdot\frac{\sin x}{\sin x}=\sec x+\boxed{}$$

$$\frac{\tan x\cos x}{\sin x\cos x}+\frac{\cot x\sin x}{\sin x\cos x}=\sec x+\boxed{}$$

Rewrite $\tan x$ and $\cot x$ using their ratio identities. Multiply and simplify.

$$\frac{\frac{\sin x}{\cancel{\cos x}}\cdot\cancel{\cos x}}{\sin x\cos x}+\frac{\frac{\cos x}{\cancel{\sin x}}\cdot\cancel{\sin x}}{\sin x\cos x}=\sec x+\boxed{}$$

$$\frac{\sin x}{\sin x\cos x}+\frac{\cos x}{\sin x\cos x}=\sec x+\boxed{}$$

Reduce each fraction. Then replace $\frac{1}{\cos x}$ and $\frac{1}{\sin x}$ with their respective reciprocal identities.

$$\frac{\cancel{\sin x}}{\cancel{\sin x}\cos x}+\frac{\cancel{\cos x}}{\sin x\cancel{\cos x}}=\sec x+\boxed{}$$

$$\frac{1}{\cos x}+\frac{1}{\sin x}=\sec x+\boxed{}$$

$$\sec x+\csc x=\sec x+\boxed{\csc x}$$

The missing term is $\csc x$. As it turns out, this problem would be much shorter if you just worked on the two fractions separately and didn't add them together. There are many ways to solve most identities.

557. $\sec x$

The common denominator is $\sin x\cos x$. Write both fractions with that denominator:

$$\frac{\sin x}{\cos x}+\frac{\cos x}{\sin x}=\csc x\boxed{}$$

$$\frac{\sin x}{\cos x}\cdot\frac{\sin x}{\sin x}+\frac{\cos x}{\sin x}\cdot\frac{\cos x}{\cos x}=\csc x\boxed{}$$

$$\frac{\sin^2 x}{\sin x\cos x}+\frac{\cos^2 x}{\sin x\cos x}=\csc x\boxed{}$$

Add the fractions. Then replace $\sin^2 x+\cos^2 x$ with 1 from the Pythagorean identity.

$$\frac{\sin^2 x+\cos^2 x}{\sin x\cos x}=\frac{1}{\sin x\cos x}=\csc x\boxed{}$$

Write the fraction as the product of two fractions. Then replace $\frac{1}{\sin x}$ with its reciprocal identity and $\frac{1}{\cos x}$ with its reciprocal identity.

$$\frac{1}{\sin x}\cdot\frac{1}{\cos x}=\csc x\boxed{}$$

$$\csc x\cdot\sec x=\csc x\boxed{\sec x}$$

The missing factor is $\sec x$.

558. $\cos x$

The common denominator is $\sec x\csc x$. Write both fractions with that denominator:

$$\frac{\sin x}{\sec x}+\frac{\cos x}{\csc x}=2\sin x\boxed{}$$

$$\frac{\sin x}{\sec x}\cdot\frac{\csc x}{\csc x}+\frac{\cos x}{\csc x}\cdot\frac{\sec x}{\sec x}=2\sin x\boxed{}$$

$$\frac{\sin x\csc x}{\sec x\csc x}+\frac{\cos x\sec x}{\sec x\csc x}=2\sin x\boxed{}$$

Write the fraction as the product of two fractions. Then replace $\csc x$ with its reciprocal identity and $\sec x$ with its reciprocal identity. Simplify the numerators.

$$\frac{\cancel{\sin x}\cdot\frac{1}{\cancel{\sin x}}}{\sec x\csc x}+\frac{\cancel{\cos x}\cdot\frac{1}{\cancel{\cos x}}}{\sec x\csc x}=2\sin x\boxed{}$$

$$\frac{1}{\sec x\csc x}+\frac{1}{\sec x\csc x}=2\sin x\boxed{}$$

Add the two fractions. Then rewrite the fraction as a product.

$$\frac{2}{\sec x\csc x}=2\sin x\boxed{}$$

$$\frac{2}{1}\cdot\frac{1}{\sec x}\cdot\frac{1}{\csc x}=2\sin x\boxed{}$$

Replace $\frac{1}{\sec x}$ with its reciprocal identity and $\frac{1}{\csc x}$ with its reciprocal identity:

$$\frac{2}{1}\cdot\cos x\cdot\sin x=2\sin x\boxed{\cos x}$$

The missing factor is $\cos x$.

559. 3

The common denominator is $\sin x\cos x$. Write both fractions with that denominator:

$$\frac{\sin x+\csc x}{\cos x}+\frac{\cos x+\sec x}{\sin x}=\frac{\boxed{}}{\sin x\cos x}$$

$$\frac{\sin x+\csc x}{\cos x}\cdot\frac{\sin x}{\sin x}+\frac{\cos x+\sec x}{\sin x}\cdot\frac{\cos x}{\cos x}=\frac{\boxed{}}{\sin x\cos x}$$

$$\frac{\sin^2 x+\csc x\sin x}{\sin x\cos x}+\frac{\cos^2 x+\sec x\cos x}{\sin x\cos x}=\frac{\boxed{}}{\sin x\cos x}$$

Replace $\csc x$ with its reciprocal and $\sec x$ with its reciprocal. Multiply. Then add the fractions together.

$$\frac{\sin^2 x+\frac{1}{\cancel{\sin x}}\cdot\cancel{\sin x}}{\sin x\cos x}+\frac{\cos^2 x+\frac{1}{\cancel{\cos x}}\cdot\cancel{\cos x}}{\sin x\cos x}=\frac{\boxed{}}{\sin x\cos x}$$

$$\frac{\sin^2 x+1}{\sin x\cos x}+\frac{\cos^2 x+1}{\sin x\cos x}=\frac{\boxed{}}{\sin x\cos x}$$

$$\frac{\sin^2 x+1+\cos^2 x+1}{\sin x\cos x}=\frac{\boxed{}}{\sin x\cos x}$$

Replace $\sin^2 x+\cos^2 x$ with 1 from the Pythagorean identity:

$$\frac{1+1+1}{\sin x\cos x}=\frac{\boxed{3}}{\sin x\cos x}$$

The missing factor is 3.

560. $\tan x$

The common denominator is $\tan^2 x \cot^2 x$. Write both fractions with that denominator:

$$\frac{\sin x \sec x}{\tan^2 x}+\frac{\cos x \csc x}{\cot^2 x}=\cot x+\boxed{}$$

$$\frac{\sin x \sec x}{\tan^2 x}\cdot\frac{\cot^2 x}{\cot^2 x}+\frac{\cos x \csc x}{\cot^2 x}\cdot\frac{\tan^2 x}{\tan^2 x}=\cot x+\boxed{}$$

$$\frac{\sin x \sec x \cot^2 x}{\tan^2 x \cot^2 x}+\frac{\cos x \csc x \tan^2 x}{\tan^2 x \cot^2 x}=\cot x+\boxed{}$$

Because $\tan x$ and $\cot x$ are reciprocal functions, their product is 1. Both denominators have products of 1:

$$\frac{\sin x \sec x \cot^2 x}{1}+\frac{\cos x \csc x \tan^2 x}{1}=\cot x+\boxed{}$$

$$\sin x \sec x \cot^2 x+\cos x \csc x \tan^2 x=\cot x+\boxed{}$$

Rewrite all the functions in terms of $\sin x$ and $\cos x$:

$$\sin x\cdot\frac{1}{\cos x}\cdot\frac{\cos^2 x}{\sin^2 x}+\cos x\cdot\frac{1}{\sin x}\cdot\frac{\sin^2 x}{\cos^2 x}=\cot x+\boxed{}$$

$$\frac{\cancel{\sin x}}{1}\cdot\frac{1}{\cancel{\cos x}}\cdot\frac{\cos^{\cancel{2}1} x}{\sin^{\cancel{2}1} x}+\frac{\cancel{\cos x}}{1}\cdot\frac{1}{\cancel{\sin x}}\cdot\frac{\sin^{\cancel{2}1} x}{\cos^{\cancel{2}1} x}=\cot x+\boxed{}$$

$$\frac{\cos x}{\sin x}+\frac{\sin x}{\cos x}=\cot x+\boxed{}$$

Now use the ratio identities to rewrite $\frac{\cos x}{\sin x}$ and $\frac{\sin x}{\cos x}$:

$$\cot x+\tan x=\cot x+\boxed{\tan x}$$

The missing term is $\tan x$.

Another way to solve the problem is to apply reciprocal identities to $\sec x$ and $\csc x$ in the original fractions:

$$\frac{\sin x \sec x}{\tan^2 x}+\frac{\cos x \csc x}{\cot^2 x}=\cot x+\boxed{}$$

$$\frac{\sin x\cdot\frac{1}{\cos x}}{\tan^2 x}+\frac{\cos x\cdot\frac{1}{\sin x}}{\cot^2 x}=\cot x+\boxed{}$$

$$\frac{\frac{\sin x}{\cos x}}{\tan^2 x}+\frac{\frac{\cos x}{\sin x}}{\cot^2 x}=\cot x+\boxed{}$$

$$\frac{\tan x}{\tan^2 x}+\frac{\cot x}{\cot^2 x}=\cot x+\boxed{}$$

$$\frac{1}{\tan x}+\frac{1}{\cot x}=\cot x+\boxed{}$$

Then use the reciprocal identities on the two terms on the left to get $\cot x+\tan x$.

Answers 501–600

561. $\sec x + \sec x \tan^2 x - \sec^3 x = 0$

The directions say to distribute on the right and move all terms to the left. On the right side, distribute the $\sec x$:

$$\sec x \tan x + \sec x \tan^2 x - \sec^3 x = \sec x(\tan x - 1)$$
$$\sec x \tan x + \sec x \tan^2 x - \sec^3 x = \sec x \tan x - \sec x$$

Subtract $\sec x \tan x$ from both sides and add $\sec x$ to both sides:

$$\sec x + \sec x \tan^2 x - \sec^3 x = 0$$

To finish solving the identity, factor $\sec x$ from each term:

$$\sec x\left(1 + \tan^2 x - \sec^2 x\right) = 0$$

Replace $1 + \tan^2 x$ with $\sec^2 x$ using the Pythagorean identity:

$$\sec x\left(\sec^2 x - \sec^2 x\right) = 0$$
$$\sec x(0) = 0$$
$$0 = 0$$

562. $\dfrac{1}{\tan x} + \dfrac{\sec x}{\tan x} - \dfrac{\tan x}{\sec x} = \cot x + \cot x \cos x$

Split up the first fraction on the left and distribute on the right:

$$\frac{1 + \sec x}{\tan x} - \frac{\tan x}{\sec x} = \cot x(1 + \cos x)$$
$$\frac{1}{\tan x} + \frac{\sec x}{\tan x} - \frac{\tan x}{\sec x} = \cot x + \cot x \cos x$$

To finish solving the identity, replace $\dfrac{1}{\tan x}$ with its reciprocal identity, $\cot x$. Then subtract $\cot x$ from both sides of the equation:

$$\cot x + \frac{\sec x}{\tan x} - \frac{\tan x}{\sec x} = \cot x + \cot x \cos x$$
$$\frac{\sec x}{\tan x} - \frac{\tan x}{\sec x} = \cot x \cos x$$

Find a common denominator on the left and subtract; replace $\cot x$ by using the reciprocal identity, replace $\cos x$ by using the reciprocal identity, and then multiply:

$$\frac{\sec x}{\tan x} \cdot \frac{\sec x}{\sec x} - \frac{\tan x}{\sec x} \cdot \frac{\tan x}{\tan x} = \frac{1}{\tan x} \cdot \frac{1}{\sec x}$$
$$\frac{\sec^2 x}{\tan x \sec x} - \frac{\tan^2 x}{\tan x \sec x} = \frac{1}{\tan x} \cdot \frac{1}{\sec x}$$
$$\frac{\sec^2 x - \tan^2 x}{\tan x \sec x} = \frac{1}{\tan x \sec x}$$

Replace $\sec^2 x$ with $\tan^2 x + 1$ using the Pythagorean identity. Simplify the numerator.

$$\frac{\tan^2 x + 1 - \tan^2 x}{\tan x \sec x} = \frac{1}{\tan x \sec x}$$
$$\frac{1}{\tan x \sec x} = \frac{1}{\tan x \sec x}$$

563. $\sin x \cos x - \cos^2 x + \cos x = \sin^2 x + \sin x \cos x + \cos x - 1$

Cross-multiply:

$$\frac{\sin x - \cos x + 1}{\sin x + \cos x - 1} = \frac{\sin x + 1}{\cos x}$$

$$\cos x(\sin x - \cos x + 1) = (\sin x + \cos x - 1)(\sin x + 1)$$

$$\sin x \cos x - \cos^2 x + \cos x = \sin^2 x + \sin x \cos x + \cos x - 1$$

To finish solving the identity, move all the terms to the right by subtracting $\sin x \cos x$, adding $\cos^2 x$, and subtracting $\cos x$:

$$0 = \sin^2 x + \cos^2 x - 1$$

Replace $\sin^2 x + \cos^2 x$ with 1 using the Pythagorean identity. Simplify.

$$0 = 1 - 1$$

$$0 = 0$$

564. $\sin^2 x - \cos^2 x = \cos^2 x \tan^2 x - \cos^2 x$

Cross-multiply:

$$\frac{\sin x - \cos x}{\cos^2 x} = \frac{\tan^2 x - 1}{\sin x + \cos x}$$

$$(\sin x - \cos x)(\sin x + \cos x) = \cos^2 x\,(\tan^2 x - 1)$$

$$\sin^2 x - \cos^2 x = \cos^2 x \tan^2 x - \cos^2 x$$

To finish solving the identity, add $\cos^2 x$ to each side. Then replace $\tan^2 x$ with the ratio identity.

$$\sin^2 x = \cancel{\cos^2 x} \cdot \frac{\sin^2 x}{\cancel{\cos^2 x}}$$

$$\sin^2 x = \sin^2 x$$

565. $\sin x + 1 = \dfrac{\cos^2 x}{1 - \sin x}$

Factor the numerator on the left and reduce the fraction:

$$\frac{\sin^2 x + 4\sin x + 3}{\sin x + 3} = \frac{\cos^2 x}{1 - \sin x}$$

$$\frac{(\sin x + 1)\cancel{(\sin x + 3)}}{\cancel{\sin x + 3}} = \frac{\cos^2 x}{1 - \sin x}$$

$$\sin x + 1 = \frac{\cos^2 x}{1 - \sin x}$$

To finish solving the identity, cross-multiply:

$$\frac{\sin x + 1}{1} = \frac{\cos^2 x}{1 - \sin x}$$

$$(\sin x + 1)(1 - \sin x) = \cos^2 x$$

$$1 - \sin^2 x = \cos^2 x$$

Answers 501–600

Replace the 1 with $\sin^2 x+\cos^2 x$ from the Pythagorean identity:

$$\sin^2 x+\cos^2 x-\sin^2 x=\cos^2 x$$
$$\cos^2 x=\cos^2 x$$

566. $\tan x+\tan x\cot x-1-\cot x=\tan x-\tan x\cot x+1-\cot x$

Cross-multiply:

$$\frac{\tan x-1}{\tan x+1}=\frac{1-\cot x}{1+\cot x}$$
$$(\tan x-1)(1+\cot x)=(\tan x+1)(1-\cot x)$$
$$\tan x+\tan x\cot x-1-\cot x=\tan x-\tan x\cot x+1-\cot x$$

To finish solving the identity, note that the product of $\tan x\cot x$ is equal to 1 because $\cot x$ is the reciprocal function of $\tan x$. Replace $\tan x\cot x$ with 1 and then simplify:

$$\tan x+1-1-\cot x=\tan x-1+1-\cot x$$
$$\tan x-\cot x=\tan x-\cot x$$

567. $\dfrac{\tan x\cos x}{1+\sin x}=\dfrac{1}{\csc x+1}$

Add the middle fraction to both sides, and then add the two fractions on the right:

$$\frac{\tan x\cos x}{1+\sin x}-\frac{1-\sin x}{\csc x}=\frac{\sin^2 x}{\csc x+1}$$
$$\frac{\tan x\cos x}{1+\sin x}=\frac{\sin^2 x}{\csc x+1}+\frac{1-\sin x}{\csc x}$$
$$\frac{\tan x\cos x}{1+\sin x}=\frac{\sin^2 x}{\csc x+1}\cdot\frac{\csc x}{\csc x}+\frac{1-\sin x}{\csc x}\cdot\frac{\csc x+1}{\csc x+1}$$
$$\frac{\tan x\cos x}{1+\sin x}=\frac{\sin^2 x\csc x}{(\csc x+1)\csc x}+\frac{(1-\sin x)(\csc x+1)}{(\csc x+1)\csc x}$$
$$\frac{\tan x\cos x}{1+\sin x}=\frac{\sin^2 x\csc x+(1-\sin x)(\csc x+1)}{(\csc x+1)\csc x}$$
$$\frac{\tan x\cos x}{1+\sin x}=\frac{\sin^2 x\csc x+\csc x+1-\sin x\csc x-\sin x}{(\csc x+1)\csc x}$$

Replace the two $\csc x$ terms in the numerator with their reciprocal, multiply, and reduce the fractions:

$$\frac{\tan x\cos x}{1+\sin x}=\frac{\sin^{\cancel{2}\,1} x\cdot\frac{1}{\cancel{\sin x}}+\csc x+1-\cancel{\sin x}\cdot\frac{1}{\cancel{\sin x}}-\sin x}{(\csc x+1)\csc x}$$
$$\frac{\tan x\cos x}{1+\sin x}=\frac{\sin x+\csc x+1-1-\sin x}{(\csc x+1)\csc x}$$
$$\frac{\tan x\cos x}{1+\sin x}=\frac{\cancel{\csc x}}{(\csc x+1)\cancel{\csc x}}$$
$$\frac{\tan x\cos x}{1+\sin x}=\frac{1}{(\csc x+1)}$$

To finish solving the identity, cross-multiply:

$$\frac{\tan x\cos x}{1+\sin x}=\frac{1}{\csc x+1}$$
$$(\tan x\cos x)(\csc x+1)=(1+\sin x)(1)$$
$$\tan x\cos x\csc x+\tan x\cos x=1+\sin x$$

Replace $\tan x$ with its ratio identity and $\csc x$ with its reciprocal identity. Multiply and simplify.

$$\left(\frac{\sin x}{\cancel{\cos x}}\cdot\cancel{\cos x}\cdot\frac{1}{\cancel{\sin x}}\right)+\left(\frac{\sin x}{\cancel{\cos x}}\cdot\cancel{\cos x}\right)=1+\sin x$$
$$\sin x+1=1+\sin x$$

568. $\dfrac{\sin x}{1+\cos x}=\dfrac{1-\cos x}{\sin x}$

The directions say to combine the two terms on the right. First, replace $\csc x$ with its reciprocal identity and $\cot x$ with its ratio identity; then subtract:

$$\frac{\sin x}{1+\cos x}=\csc x-\cot x$$
$$\frac{\sin x}{1+\cos x}=\frac{1}{\sin x}-\frac{\cos x}{\sin x}$$
$$\frac{\sin x}{1+\cos x}=\frac{1-\cos x}{\sin x}$$

To finish solving the identity, cross-multiply:

$$\frac{\sin x}{1+\cos x}=\frac{1-\cos x}{\sin x}$$
$$\sin^2 x=(1+\cos x)(1-\cos x)$$
$$\sin^2 x=1-\cos^2 x$$

Replace the 1 with $\sin^2 x+\cos^2 x$ from the Pythagorean identity:

$$\sin^2 x=\sin^2 x+\cos^2 x-\cos^2 x$$
$$\sin^2 x=\sin^2 x$$

569. $\dfrac{2\tan x}{\sec^2 x-1}=2\cot x$

The directions say to subtract the two terms on the left. Find a common denominator and then subtract:

$$\frac{\tan x}{\sec x-1}-\frac{\tan x}{\sec x+1}=2\cot x$$
$$\frac{\tan x}{\sec x-1}\cdot\frac{\sec x+1}{\sec x+1}-\frac{\tan x}{\sec x+1}\cdot\frac{\sec x-1}{\sec x-1}=2\cot x$$
$$\frac{\tan x(\sec x+1)}{(\sec x-1)(\sec x+1)}-\frac{\tan x(\sec x-1)}{(\sec x-1)(\sec x+1)}=2\cot x$$
$$\frac{\tan x\sec x+\tan x-\tan x\sec x+\tan x}{\sec^2 x-1}=2\cot x$$
$$\frac{2\tan x}{\sec^2 x-1}=2\cot x$$

To finish solving the identity, replace $\sec^2 x$ with $\tan^2 x+1$ from the Pythagorean identity. On the right, replace $\cot x$ with its reciprocal identity.

Answers 501–600

$$\frac{2\tan x}{\tan^2 x+1-1}=2\cdot\frac{1}{\tan x}$$
$$\frac{2\tan x}{\tan^2 x}=\frac{2}{\tan x}$$

Reduce the fraction on the left:

$$\frac{2\cancel{\tan x}}{\tan^{\cancel{2}\,1} x}=\frac{2}{\tan x}$$
$$\frac{2}{\tan x}=\frac{2}{\tan x}$$

570. $\frac{1-2\cot x+\cot^2 x}{\csc^2 x}=1-2\sin x\cos x$

Square both sides of the equation:

$$\left(\frac{1-\cot x}{\csc x}\right)^2=\left(\sqrt{1-2\sin x\cos x}\right)^2$$
$$\frac{(1-\cot x)^2}{\csc^2 x}=1-2\sin x\cos x$$
$$\frac{1-2\cot x+\cot^2 x}{\csc^2 x}=1-2\sin x\cos x$$

To finish solving the identity, replace $\csc^2 x$ with its reciprocal identity. Then simplify the complex fraction.

$$\frac{1-2\cot x+\cot^2 x}{\frac{1}{\sin^2 x}}=1-2\sin x\cos x$$
$$\frac{\sin^2 x}{1}\cdot\frac{1-2\cot x+\cot^2 x}{1}=1-2\sin x\cos x$$
$$\frac{\sin^2 x-2\cot x\sin^2 x+\cot^2 x\sin^2 x}{1}=1-2\sin x\cos x$$

Replace $\cot x$ with its ratio identity:

$$\sin^2 x-2\cdot\frac{\cos x}{\cancel{\sin x}}\cdot\sin^{\cancel{2}\,1} x+\frac{\cos^2 x}{\cancel{\sin^2 x}}\cdot\cancel{\sin^2 x}=1-2\sin x\cos x$$
$$\sin^2 x-2\cos x\sin x+\cos^2 x=1-2\sin x\cos x$$

Replace $\sin^2 x+\cos^2 x$ with 1 from the Pythagorean identity:

$$1-2\cos x\sin x=1-2\sin x\cos x$$

571. $\cos\theta$

Use the sine-of-a-sum identity:

$$\sin(45°+\theta)=\frac{\sqrt{2}}{2}\left(\sin\theta+\boxed{}\right)$$
$$\sin 45°\cos\theta+\cos 45°\sin\theta=\frac{\sqrt{2}}{2}\left(\sin\theta+\boxed{}\right)$$
$$\frac{\sqrt{2}}{2}\cos\theta+\frac{\sqrt{2}}{2}\sin\theta=\frac{\sqrt{2}}{2}\left(\sin\theta+\boxed{\cos\theta}\right)$$

Answers 501–600

572. $\tan\theta$

Use the tangent-of-a-sum identity:

$$\tan(45°+\theta)=\frac{1+\boxed{}}{1-\tan\theta}$$

$$\frac{\tan 45°+\tan\theta}{1-\tan 45°\tan\theta}=\frac{1+\boxed{}}{1-\tan\theta}$$

$$\frac{1+\tan\theta}{1-1\cdot\tan\theta}=\frac{1+\boxed{\tan\theta}}{1-\tan\theta}$$

573. $\sqrt{3}\sin\theta$

Use the cosine-of-a-difference identity:

$$\cos(60°-\theta)=\frac{1}{2}\left(\cos\theta+\boxed{}\right)$$

$$\cos 60°\cos\theta+\sin 60°\sin\theta=\frac{1}{2}\left(\cos\theta+\boxed{}\right)$$

$$\frac{1}{2}\cos\theta+\frac{\sqrt{3}}{2}\sin\theta=\frac{1}{2}\left(\cos\theta+\boxed{\sqrt{3}\sin\theta}\right)$$

574. $\sin\theta\cos\theta$

Change the double angle to the sum of two angles:

$$\sin 2\theta=2\boxed{}$$

$$\sin(\theta+\theta)=2\boxed{}$$

Now apply the sine-of-a-sum identity:

$$\sin\theta\cos\theta+\cos\theta\sin\theta=2\boxed{}$$

$$2\sin\theta\cos\theta=2\boxed{\sin\theta\cos\theta}$$

575. $\tan^2\theta$

Change the double angle to the sum of two angles:

$$\tan 2\theta=\frac{2\tan\theta}{1-\boxed{}}$$

$$\tan(\theta+\theta)=\frac{2\tan\theta}{1-\boxed{}}$$

Now apply the tangent-of-a-sum identity:

$$\frac{\tan\theta+\tan\theta}{1-\tan\theta\tan\theta}=\frac{2\tan\theta}{1-\boxed{}}$$

$$\frac{2\tan\theta}{1-\tan^2\theta}=\frac{2\tan\theta}{1-\boxed{\tan^2\theta}}$$

576. $\cos\theta$

Replace sin 2θ with the double-angle identity:

$$\sin 2\theta + 2\sin^2\theta = 2\sin\theta\left(\sin\theta + \boxed{}\right)$$

$$2\sin\theta\cos\theta + 2\sin^2\theta = 2\sin\theta\left(\sin\theta + \boxed{}\right)$$

Factor:

$$2\sin\theta(\cos\theta + \sin\theta) = 2\sin\theta\left(\sin\theta + \boxed{\cos\theta}\right)$$

577. 0

Replace cos 2θ with the double-angle identity involving the cosine:

$$\cos^2\theta - \cos 2\theta = \sin^2\theta + \boxed{}$$

$$\cos^2\theta - (2\cos^2\theta - 1) = \sin^2\theta + \boxed{}$$

$$\cos^2\theta - 2\cos^2\theta + 1 = \sin^2\theta + \boxed{}$$

$$1 - \cos^2\theta = \sin^2\theta + \boxed{}$$

Replace the 1 with $\sin^2\theta + \cos^2\theta$ from the Pythagorean identity:

$$\sin^2\theta + \cos^2\theta - \cos^2\theta = \sin^2\theta + \boxed{}$$

$$\sin^2\theta = \sin^2\theta + \boxed{0}$$

578. 1

Rewrite the Pythagorean identity $\sin^2 2\theta + \cos^2 2\theta = 1$ as $\cos^2 2\theta = 1 - \sin^2 2\theta$ by subtracting $\sin^2 2\theta$ from each side of the equation. Then replace $\cos^2 2\theta$ with $1 - \sin^2 2\theta$:

$$\cos^2 2\theta + 4\sin^2\theta\cos^2\theta = \boxed{}$$

$$1 - \sin^2 2\theta + 4\sin^2\theta\cos^2\theta = \boxed{}$$

Rewrite $\sin^2 2\theta$ as the square of sin 2θ. Then replace sin 2θ with its double-angle identity. Square and simplify:

$$1 - (\sin 2\theta)^2 + 4\sin^2\theta\cos^2\theta = \boxed{}$$

$$1 - (2\sin\theta\cos\theta)^2 + 4\sin^2\theta\cos^2\theta = \boxed{}$$

$$1 - 4\sin^2\theta\cos^2\theta + 4\sin^2\theta\cos^2\theta = \boxed{}$$

$$1 = \boxed{1}$$

579. 1

Replace $\cos(4\theta)$ with the cosine of a sum:

$$\cos(2\theta+2\theta)=8\cos^4\theta-8\cos^2\theta+\boxed{}$$

Replace the cosine of the sum with its identity:

$$\cos 2\theta\cos 2\theta-\sin 2\theta\sin 2\theta=8\cos^4\theta-8\cos^2\theta+\boxed{}$$

Now replace the cosine double angles with the identity involving cosine, and replace the sine double angles with their identity:

$$(2\cos^2\theta-1)(2\cos^2\theta-1)-(2\sin\theta\cos\theta)^2=8\cos^4\theta-8\cos^2\theta+\boxed{}$$

Multiply and simplify:

$$4\cos^4\theta-4\cos^2\theta+1-4\sin^2\theta\cos^2\theta=8\cos^4\theta-8\cos^2\theta+\boxed{}$$

Replace $\sin^2\theta$ with $1-\cos^2\theta$ using the Pythagorean identity:

$$4\cos^4\theta-4\cos^2\theta+1-4(1-\cos^2\theta)\cos^2\theta=8\cos^4\theta-8\cos^2\theta+\boxed{}$$

$$4\cos^4\theta-4\cos^2\theta+1-4\cos^2\theta+4\cos^4\theta=8\cos^4\theta-8\cos^2\theta+\boxed{}$$

$$8\cos^4\theta-8\cos^2\theta+1=8\cos^4\theta-8\cos^2\theta+\boxed{1}$$

580. $(1-2\sin^2\theta)$

Replace $\sin(4\theta)$ with the sine of a sum:

$$\sin(4\theta)=4\sin\theta\cos\theta\boxed{}$$

$$\sin(2\theta+2\theta)=4\sin\theta\cos\theta\boxed{}$$

Replace the sine of the sum with its identity:

$$\sin(2\theta)\cos(2\theta)+\cos(2\theta)\sin(2\theta)=4\sin\theta\cos\theta\boxed{}$$

$$2\sin(2\theta)\cos(2\theta)=4\sin\theta\cos\theta\boxed{}$$

Now replace $\sin(2\theta)$ and $\cos(2\theta)$ with their respective double-angle identities; for the cosine, use the identity involving sine:

$$2[2\sin\theta\cos\theta][1-2\sin^2\theta]=4\sin\theta\cos\theta\boxed{}$$

$$4\sin\theta\cos\theta[1-2\sin^2\theta]=4\sin\theta\cos\theta\boxed{(1-2\sin^2\theta)}$$

581. $\cos\theta$

Begin with the identity $\cos 2x=1-2\sin^2 x$. Replace the $2x$ with θ and replace the x with $\frac{\theta}{2}$:

$$\cos\theta=1-2\sin^2\left(\frac{\theta}{2}\right)$$

Answers 501–600

Add $2\sin^2\left(\frac{\theta}{2}\right)$ to each side and subtract $\cos\theta$ from each side of the equation:

$$2\sin^2\left(\frac{\theta}{2}\right)=1-\cos\theta$$

Divide each side by 2:

$$\sin^2\left(\frac{\theta}{2}\right)=\frac{1-\cos\theta}{2}$$

Take the square root of each side:

$$\sqrt{\sin^2\left(\frac{\theta}{2}\right)}=\pm\sqrt{\frac{1-\cos\theta}{2}}$$

$$\sin\left(\frac{\theta}{2}\right)=\pm\sqrt{\frac{1-\cos\theta}{2}}$$

$$=\pm\sqrt{\frac{1-\boxed{\cos\theta}}{2}}$$

582. $\cos\theta$

Begin with the identity $\cos 2x=2\cos^2 x-1$. Replace the $2x$ with θ and replace the x with $\frac{\theta}{2}$:

$$\cos\theta=2\cos^2\left(\frac{\theta}{2}\right)-1$$

Add 1 to each side of the equation:

$$\cos\theta+1=2\cos^2\left(\frac{\theta}{2}\right)$$

Divide each side by 2:

$$\frac{\cos\theta+1}{2}=\cos^2\left(\frac{\theta}{2}\right)$$

Take the square root of each side:

$$\pm\sqrt{\frac{\cos\theta+1}{2}}=\sqrt{\cos^2\left(\frac{\theta}{2}\right)}$$

$$\cos\left(\frac{\theta}{2}\right)=\pm\sqrt{\frac{1+\cos\theta}{2}}$$

$$=\pm\sqrt{\frac{1+\boxed{\cos\theta}}{2}}$$

583. $\sin\theta, \sin\theta$

Use the ratio identity to replace $\tan\left(\frac{\theta}{2}\right)$ with identities in sine and cosine:

$$\tan\left(\frac{\theta}{2}\right)=\frac{\sin\left(\frac{\theta}{2}\right)}{\cos\left(\frac{\theta}{2}\right)}$$

Answers 501–600

Replace $\sin\left(\frac{\theta}{2}\right)$ and $\cos\left(\frac{\theta}{2}\right)$ with their respective identities; then simplify:

$$\tan\left(\frac{\theta}{2}\right)=\frac{\sin\left(\frac{\theta}{2}\right)}{\cos\left(\frac{\theta}{2}\right)}=\frac{\pm\sqrt{\frac{1-\cos\theta}{2}}}{\pm\sqrt{\frac{1+\cos\theta}{2}}}$$

$$=\pm\sqrt{\frac{\frac{1-\cos\theta}{2}}{\frac{1+\cos\theta}{2}}}=\pm\sqrt{\frac{1-\cos\theta}{2}\cdot\frac{2}{1+\cos\theta}}$$

$$=\pm\sqrt{\frac{1-\cos\theta}{1+\cos\theta}}$$

Inside the radical, multiply the numerator and denominator by $1-\cos\theta$:

$$\tan\left(\frac{\theta}{2}\right)=\pm\sqrt{\frac{1-\cos\theta}{1+\cos\theta}\cdot\frac{1-\cos\theta}{1-\cos\theta}}=\pm\sqrt{\frac{(1-\cos\theta)^2}{1-\cos^2\theta}}$$

Replace the 1 in the denominator with $\sin^2\theta+\cos^2\theta$ from the Pythagorean identity; then simplify:

$$\tan\left(\frac{\theta}{2}\right)=\pm\sqrt{\frac{(1-\cos\theta)^2}{\sin^2\theta+\cos^2\theta-\cos^2\theta}}$$

$$=\pm\sqrt{\frac{(1-\cos\theta)^2}{\sin^2\theta}}=\pm\frac{1-\cos\theta}{\sin\theta}$$

You can drop the $\pm$ sign because the numerator will always be a positive number or 0 and the denominator will be the sign of $\sin\theta$. ***Remember:*** When θ is an angle between 0° and 180°, the sine is positive — corresponding to the tangent positive between 0° and 90° (half 0° to 180°). When θ is an angle between 180° and 360°, the sine is negative — corresponding to the tangent negative between 90° and 180° (half 180° to 360°). The sign will be determined by the angle being evaluated.

$$\tan\left(\frac{\theta}{2}\right)=\frac{1-\cos\theta}{\sin\theta}=\boxed{\frac{1-\cos\theta}{\sin\theta}}$$

For the other form of the identity, begin with $\pm\sqrt{\frac{1-\cos\theta}{1+\cos\theta}}$ and, inside the radical, multiply the numerator and the denominator by $1+\cos\theta$:

$$\tan\left(\frac{\theta}{2}\right)=\pm\sqrt{\frac{1-\cos\theta}{1+\cos\theta}\cdot\frac{1+\cos\theta}{1+\cos\theta}}=\pm\sqrt{\frac{1-\cos^2\theta}{(1+\cos\theta)^2}}$$

Replace the 1 in the numerator with $\sin^2\theta+\cos^2\theta$ from the Pythagorean identity; then simplify:

$$\tan\left(\frac{\theta}{2}\right)=\pm\sqrt{\frac{\sin^2\theta+\cos^2\theta-\theta\cos^2\theta}{(1+\cos\theta)^2}}$$

$$=\pm\sqrt{\frac{\sin^2\theta}{(1+\cos\theta)^2}}=\pm\frac{\sin\theta}{1+\cos\theta}$$

As before, you can drop the ± sign:

$$\tan\left(\frac{\theta}{2}\right)=\frac{\sin\theta}{1+\cos\theta}=\frac{\boxed{\sin\theta}}{1+\cos\theta}$$

584. $\cos\theta$

Rewrite the denominator as $\sec^2\left(\frac{\theta}{2}\right)$ using the Pythagorean identity. Then replace $\sec^2\left(\frac{\theta}{2}\right)$ with its reciprocal identity:

$$\frac{1-\tan^2\left(\frac{\theta}{2}\right)}{1+\tan^2\left(\frac{\theta}{2}\right)}=\frac{1-\tan^2\left(\frac{\theta}{2}\right)}{\sec^2\left(\frac{\theta}{2}\right)}=\boxed{}$$

$$\frac{1-\tan^2\left(\frac{\theta}{2}\right)}{\frac{1}{\cos^2\left(\frac{\theta}{2}\right)}}=\boxed{}$$

Multiply the numerator by the reciprocal of the denominator:

$$\left[1-\tan^2\left(\frac{\theta}{2}\right)\right]\cdot\left[\frac{\cos^2\left(\frac{\theta}{2}\right)}{1}\right]=\boxed{}$$

$$\cos^2\left(\frac{\theta}{2}\right)-\tan^2\left(\frac{\theta}{2}\right)\cos^2\left(\frac{\theta}{2}\right)=\boxed{}$$

Replace the tangent term using its ratio identity; then multiply:

$$\cos^2\left(\frac{\theta}{2}\right)-\frac{\sin^2\left(\frac{\theta}{2}\right)}{\cancel{\cos^2\left(\frac{\theta}{2}\right)}}\cdot\cancel{\cos^2\left(\frac{\theta}{2}\right)}=\boxed{}$$

$$\cos^2\left(\frac{\theta}{2}\right)-\sin^2\left(\frac{\theta}{2}\right)=\boxed{}$$

Use the Pythagorean identity to replace $\sin^2\left(\frac{\theta}{2}\right)$ with $1-\cos^2\left(\frac{\theta}{2}\right)$:

$$\cos^2\left(\frac{\theta}{2}\right)-\left[1-\cos^2\left(\frac{\theta}{2}\right)\right]=\boxed{}$$

$$2\cos^2\left(\frac{\theta}{2}\right)-1=\boxed{}$$

The expression on the right is one of the double-angle forms for $\cos 2\theta$. So if $2\cos^2 x-1=\cos 2x$, then twice $\frac{\theta}{2}$ is equal to θ and the left side of the identity equals $\cos\theta$:

$$2\cos^2\left(\frac{\theta}{2}\right)-1=\boxed{\cos\theta}$$

Answers 501–600

585. $\tan\theta$

Replace $\cos\left(\frac{\theta}{2}\right)$ and $\sin\left(\frac{\theta}{2}\right)$ with their respective half-angle identities, using the positive version of each identity:

$$\frac{\cos\left(\frac{\theta}{2}\right)+\sin\left(\frac{\theta}{2}\right)}{\cos\left(\frac{\theta}{2}\right)-\sin\left(\frac{\theta}{2}\right)}=\sec\theta+\boxed{}$$

$$\frac{\sqrt{\frac{1+\cos\theta}{2}}+\sqrt{\frac{1-\cos\theta}{2}}}{\sqrt{\frac{1+\cos\theta}{2}}-\sqrt{\frac{1-\cos\theta}{2}}}=\sec\theta+\boxed{}$$

Multiply numerator and denominator by $\sqrt{\frac{1+\cos\theta}{2}}+\sqrt{\frac{1-\cos\theta}{2}}$:

$$\frac{\sqrt{\frac{1+\cos\theta}{2}}+\sqrt{\frac{1-\cos\theta}{2}}}{\sqrt{\frac{1+\cos\theta}{2}}-\sqrt{\frac{1-\cos\theta}{2}}}\cdot\frac{\sqrt{\frac{1+\cos\theta}{2}}+\sqrt{\frac{1-\cos\theta}{2}}}{\sqrt{\frac{1+\cos\theta}{2}}+\sqrt{\frac{1-\cos\theta}{2}}}=\sec\theta+\boxed{}$$

$$\frac{\left(\frac{1+\cos\theta}{2}\right)+2\sqrt{\frac{1+\cos\theta}{2}\cdot\frac{1-\cos\theta}{2}}+\left(\frac{1-\cos\theta}{2}\right)}{\left(\frac{1+\cos\theta}{2}\right)-\left(\frac{1-\cos\theta}{2}\right)}=\sec\theta+\boxed{}$$

Multiply the terms under the radical. Replace the result in the numerator with $\sin^2\theta$ using the Pythagorean identity. Then simplify:

$$\frac{\left(\frac{1+\cos\theta}{2}\right)+2\sqrt{\frac{1-\cos^2\theta}{4}}+\left(\frac{1-\cos\theta}{2}\right)}{\left(\frac{1+\cos\theta}{2}\right)-\left(\frac{1-\cos\theta}{2}\right)}=\sec\theta+\boxed{}$$

$$\frac{\left(\frac{1+\cos\theta}{2}\right)+2\sqrt{\frac{\sin^2\theta}{4}}+\left(\frac{1-\cos\theta}{2}\right)}{\left(\frac{1+\cos\theta}{2}\right)-\left(\frac{1-\cos\theta}{2}\right)}=\sec\theta+\boxed{}$$

$$\frac{\left(\frac{1+\cos\theta}{2}\right)+2\left(\frac{\sin\theta}{2}\right)+\left(\frac{1-\cos\theta}{2}\right)}{\left(\frac{1+\cos\theta}{2}\right)-\left(\frac{1-\cos\theta}{2}\right)}=\sec\theta+\boxed{}$$

Multiply the numerator and denominator by 2 and simplify:

$$\frac{\left(\frac{1+\cos\theta}{2}\right)+2\left(\frac{\sin\theta}{2}\right)+\left(\frac{1-\cos\theta}{2}\right)}{\left(\frac{1+\cos\theta}{2}\right)-\left(\frac{1-\cos\theta}{2}\right)}\cdot\frac{2}{2}=\sec\theta+\boxed{}$$

$$\frac{(1+\cos\theta)+2(\sin\theta)+(1-\cos\theta)}{(1+\cos\theta)-(1-\cos\theta)}=\sec\theta+\boxed{}$$

$$\frac{2+2\sin\theta}{2\cos\theta}=\sec\theta+\boxed{}$$

$$\frac{1+\sin\theta}{\cos\theta}=\sec\theta+\boxed{}$$

Rewrite the fraction as the sum of two fractions. Then replace the first term with its reciprocal identity and the second term with its ratio identity:

$$\frac{1}{\cos\theta}+\frac{\sin\theta}{\cos\theta}=\sec\theta+\boxed{}$$

$$\sec\theta+\tan\theta=\sec\theta+\boxed{\tan\theta}$$

586. $\cos b$

Working on the right side of the identity, replace the sine of the sum and sine of the difference with their respective identities. Then simplify:

$$\sin a\boxed{}=\frac{1}{2}\left[\sin(a+b)+\sin(a-b)\right]$$

$$\sin a\boxed{}=\frac{1}{2}\left[\sin a\cos b+\cos a\sin b+\sin a\cos b-\cos a\sin b\right]$$

$$\sin a\boxed{\cos b}=\frac{1}{2}\left[2\sin a\cos b\right]=\sin a\cos b$$

587. $\cos b$

Working on the right side of the identity, replace the cosine of the sum and cosine of the difference with their respective identities. Then simplify:

$$\cos a\boxed{}=\frac{1}{2}\left[\cos(a+b)+\cos(a-b)\right]$$

$$\cos a\boxed{}=\frac{1}{2}\left[\cos a\cos b-\sin a\sin b+\cos a\cos b+\sin a\sin b\right]$$

$$\cos a\boxed{\cos b}=\frac{1}{2}\left[2\cos a\cos b\right]=\cos a\cos b$$

588. $\sin b$

Working on the right side of the identity, replace the cosine of the difference and cosine of the sum with their respective identities. Then simplify:

$$\sin a\boxed{}=\frac{1}{2}\left[\cos(a-b)-\cos(a+b)\right]$$

$$\sin a\boxed{}=\frac{1}{2}\left[\cos a\cos b+\sin a\sin b-(\cos a\cos b-\sin a\sin b)\right]$$

$$\sin a\boxed{}=\frac{1}{2}\left[\cos a\cos b+\sin a\sin b-\cos a\cos b+\sin a\sin b\right]$$

$$\sin a\boxed{\sin b}=\frac{1}{2}\left[2\sin a\sin b\right]=\sin a\sin b$$

589. $\frac{2+\sqrt{3}}{4}$

Use the identity $\sin a \cos b = \frac{1}{2}[\sin(a+b)+\sin(a-b)]$, letting $a = 75$ and $b = 15$:

$$\begin{aligned}\sin(75°)\cos(15°) &= \frac{1}{2}[\sin(75+15)+\sin(75-15)]\\ &= \frac{1}{2}[\sin(90)+\sin(60)]\\ &= \frac{1}{2}\left[1+\frac{\sqrt{3}}{2}\right] = \frac{1}{2}\left[\frac{2}{2}+\frac{\sqrt{3}}{2}\right] = \frac{2+\sqrt{3}}{4}\end{aligned}$$

590. $-\frac{1}{4}$

Use the identity $\cos a \cos b = \frac{1}{2}[\cos(a+b)+\cos(a-b)]$, letting $a = 120$ and $b = 60$:

$$\begin{aligned}\cos a \cos b &= \frac{1}{2}[\cos(120+60)+\cos(120-60)]\\ &= \frac{1}{2}[\cos(180)+\cos(60)]\\ &= \frac{1}{2}\left[-1+\frac{1}{2}\right] = \frac{1}{2}\left[-\frac{1}{2}\right] = -\frac{1}{4}\end{aligned}$$

591. $\frac{\sqrt{6}}{2}$

Use the sum-to-product identity $\sin x + \sin y = 2\sin\left(\frac{x+y}{2}\right)\cos\left(\frac{x-y}{2}\right)$:

$$\begin{aligned}\sin 75 + \sin 15 &= 2\sin\left(\frac{75+15}{2}\right)\cos\left(\frac{75-15}{2}\right)\\ &= 2\sin\left(\frac{90}{2}\right)\cos\left(\frac{60}{2}\right) = 2\sin(45)\cos(30)\\ &= 2\left(\frac{\sqrt{2}}{2}\right)\left(\frac{\sqrt{3}}{2}\right) = \frac{\sqrt{6}}{2}\end{aligned}$$

592. $-\frac{1}{2}$

Use the sum-to-product identity $\sin x - \sin y = 2\cos\left(\frac{x+y}{2}\right)\sin\left(\frac{x-y}{2}\right)$:

$$\begin{aligned}\sin 150 - \sin 90 &= 2\cos\left(\frac{150+90}{2}\right)\sin\left(\frac{150-90}{2}\right)\\ &= 2\cos\left(\frac{240}{2}\right)\sin\left(\frac{60}{2}\right)\\ &= 2\cos(120)\sin(30) = 2\left(-\frac{1}{2}\right)\left(\frac{1}{2}\right) = -\frac{1}{2}\end{aligned}$$

593. $-\frac{\sqrt{2}}{2}$

Use the sum-to-product identity $\cos x+\cos y=2\cos\left(\frac{x+y}{2}\right)\cos\left(\frac{x-y}{2}\right)$:

$$\begin{aligned}\cos 195+\cos 75&=2\cos\left(\frac{195+75}{2}\right)\cos\left(\frac{195-75}{2}\right)\\&=2\cos\left(\frac{270}{2}\right)\cos\left(\frac{120}{2}\right)\\&=2\cos(135)\cos(60)\\&=2\left(-\frac{\sqrt{2}}{2}\right)\left(\frac{1}{2}\right)=-\frac{\sqrt{2}}{2}\end{aligned}$$

594. $-\frac{\sqrt{2}}{2}$

Use the sum-to-product identity $\cos x-\cos y=-2\sin\left(\frac{x+y}{2}\right)\sin\left(\frac{x-y}{2}\right)$:

$$\begin{aligned}\cos 75-\cos 15&=-2\sin\left(\frac{75+15}{2}\right)\sin\left(\frac{75-15}{2}\right)\\&=-2\sin\left(\frac{90}{2}\right)\sin\left(\frac{60}{2}\right)\\&=-2\sin(45)\sin(30)\\&=-2\left(\frac{\sqrt{2}}{2}\right)\left(\frac{1}{2}\right)=-\frac{\sqrt{2}}{2}\end{aligned}$$

595. $-\frac{1}{2}$

Use the sum-to-product identity $\cos x+\cos y=2\cos\left(\frac{x+y}{2}\right)\cos\left(\frac{x-y}{2}\right)$:

$$\begin{aligned}\cos 120+\cos 90&=2\cos\left(\frac{120+90}{2}\right)\cos\left(\frac{120-90}{2}\right)\\&=2\cos\left(\frac{210}{2}\right)\cos\left(\frac{30}{2}\right)\\&=2\cos(105)\cos(15)\end{aligned}$$

Now use the cosine-of-a-sum identity to find cos(105):

$$\begin{aligned}\cos(105)=\cos(60+45)&=\cos 60\cos 45-\sin 60\sin 45\\&=\frac{1}{2}\left(\frac{\sqrt{2}}{2}\right)-\frac{\sqrt{3}}{2}\left(\frac{\sqrt{2}}{2}\right)=\frac{\sqrt{2}-\sqrt{6}}{4}\end{aligned}$$

And use the cosine-of-a-difference identity to find cos(15):

$$\begin{aligned}\cos(15)=\cos(45-30)&=\cos 45\cos 30+\sin 45\sin 30\\&=\frac{\sqrt{2}}{2}\left(\frac{\sqrt{3}}{2}\right)+\left(\frac{\sqrt{2}}{2}\right)\left(\frac{1}{2}\right)\\&=\frac{\sqrt{6}+\sqrt{2}}{4}\end{aligned}$$

Now use these values to finish the problem:

$$2\cos(105)\cos(15)=2\left(\frac{\sqrt{2}-\sqrt{6}}{4}\right)\left(\frac{\sqrt{2}+\sqrt{6}}{4}\right)$$
$$=\frac{\sqrt{4}-\sqrt{36}}{8}=\frac{2-6}{8}=\frac{-4}{8}=-\frac{1}{2}$$

596. $\cos\theta$

Use the power-reducing identity $\sin^2 x=\frac{1-\cos 2x}{2}$:

$$\sin^2\left(\frac{\theta}{2}\right)=\frac{1-\cos 2\left(\frac{\theta}{2}\right)}{2}=\frac{1-\boxed{\cos\theta}}{2}$$

597. $\cos 8\theta$

Use the power-reducing identities $\cos^2 x=\frac{1+\cos 2x}{2}$ and $\sin^2 x=\frac{1-\cos 2x}{2}$, replacing x with 4θ and $2x$ with 8θ. Then simplify:

$$\cos^2 4\theta-\sin^2 4\theta=\boxed{}$$
$$\frac{1+\cos 8\theta}{2}-\frac{1-\cos 8\theta}{2}=\boxed{}$$
$$\frac{1+\cos 8\theta-(1-\cos 8\theta)}{2}=\boxed{}$$
$$\frac{1+\cos 8\theta-1+\cos 8\theta}{2}=\boxed{}$$
$$\frac{2\cos 8\theta}{2}=\boxed{\cos 8\theta}$$

598. 1

Use the power-reducing identity $\tan^2 x=\frac{1-\cos 2x}{1+\cos 2x}$:

$$\tan^2(2\theta)=\sec^2 2\theta-\boxed{}$$
$$\frac{1-\cos 2(2\theta)}{1+\cos 2(2\theta)}=\sec^2 2\theta-\boxed{}$$

Now use the double-angle identity involving the cosine to replace $\cos 2(2\theta)$:

$$\frac{1-\left(2\cos^2(2\theta)-1\right)}{1+\left(2\cos^2(2\theta)-1\right)}=\sec^2 2\theta-\boxed{}$$
$$\frac{2-2\cos^2(2\theta)}{2\cos^2(2\theta)}=\sec^2 2\theta-\boxed{}$$
$$\frac{1-\cos^2(2\theta)}{\cos^2(2\theta)}=\sec^2 2\theta-\boxed{}$$

Rewrite the fraction as the difference of two fractions. Then use the reciprocal identity for $\frac{1}{\cos^2(2\theta)}$:

$$\frac{1}{\cos^2(2\theta)} - \frac{\cos^2(2\theta)}{\cos^2(2\theta)} = \sec^2 2\theta - \boxed{}$$

$$\sec^2(2\theta) - 1 = \sec^2 2\theta - \boxed{1}$$

599. 1

Use the reciprocal of the power-reducing identity $\cos^2 x = \frac{1+\cos 2x}{2}$ and also use $\tan^2 x = \frac{1-\cos 2x}{1+\cos 2x}$:

$$\sec^2\theta - \tan^2\theta = \boxed{}$$

$$\frac{2}{1+\cos 2\theta} - \frac{1-\cos 2\theta}{1+\cos 2\theta} = \boxed{}$$

$$\frac{1+\cos 2\theta}{1+\cos 2\theta} = \boxed{1}$$

As an alternative, if you prefer not to use power-reducing identities, you can use the Pythagorean identity $\tan^2\theta + 1 = \sec^2\theta$ in the original statement and get to the result even more quickly.

600. $\cos(2\theta)$

Use the reciprocal of the power-reducing identity $\tan^2 x = \frac{1-\cos 2x}{1+\cos 2x}$ and also use $\cos^2 x = \frac{1+\cos 2x}{2}$:

$$\cot^2\theta - \cos^2\theta = \frac{\left(1+\boxed{}\right)^2}{2-2\cos(2\theta)}$$

$$\frac{1+\cos(2\theta)}{1-\cos(2\theta)} - \frac{1+\cos(2\theta)}{2} = \frac{\left(1+\boxed{}\right)^2}{2-2\cos(2\theta)}$$

Find a common denominator, change the fractions, and subtract:

$$\frac{1+\cos(2\theta)}{1-\cos(2\theta)}\cdot\frac{2}{2} - \frac{1+\cos(2\theta)}{2}\cdot\frac{1-\cos(2\theta)}{1-\cos(2\theta)} = \frac{\left(1+\boxed{}\right)^2}{2-2\cos(2\theta)}$$

$$\frac{2+2\cos(2\theta)}{2(1-\cos(2\theta))} - \frac{1-\cos^2(2\theta)}{2(1-\cos(2\theta))} = \frac{\left(1+\boxed{}\right)^2}{2-2\cos(2\theta)}$$

$$\frac{1+2\cos(2\theta)+\cos^2(2\theta)}{2-2\cos(2\theta)} = \frac{\left(1+\boxed{}\right)^2}{2-2\cos(2\theta)}$$

$$\frac{(1+\cos(2\theta))^2}{2-2\cos(2\theta)} = \frac{\left(1+\boxed{\cos(2\theta)}\right)^2}{2-2\cos(2\theta)}$$

601. $\dfrac{\sqrt{2+\sqrt{3}}}{2}$

Use the half-angle identity:

$$\sin 75° = \sin\left(\frac{150}{2}\right) = \pm\sqrt{\frac{1-\cos 150}{2}} = \pm\sqrt{\frac{1-\left(-\frac{\sqrt{3}}{2}\right)}{2}}$$

$$= \pm\sqrt{\frac{1+\frac{\sqrt{3}}{2}}{2}} = \pm\sqrt{\frac{\frac{2}{2}+\frac{\sqrt{3}}{2}}{2}} = \pm\sqrt{\frac{\frac{2+\sqrt{3}}{2}}{2}}$$

$$= \pm\sqrt{\frac{2+\sqrt{3}}{4}} = \pm\frac{\sqrt{2+\sqrt{3}}}{\sqrt{4}} = \pm\frac{\sqrt{2+\sqrt{3}}}{2}$$

Because an angle of 75° is in the first quadrant, the sine is positive.

You can also solve this problem using the angle-sum identity with angles of 45° and 30°.

$$\sin 75° = \sin(45° + 30°)$$
$$= \sin 45° \cos 30° + \cos 45° \sin 30°$$
$$= \frac{\sqrt{2}}{2}\cdot\frac{\sqrt{3}}{2} + \frac{\sqrt{2}}{2}\cdot\frac{1}{2} = \frac{\sqrt{6}+\sqrt{2}}{4}$$

The two answers are equivalent, even though they don't look the same. Here's the math:

$$\frac{\sqrt{2+\sqrt{3}}}{\cancel{2}^{1}} = \frac{\sqrt{6}+\sqrt{2}}{\cancel{4}^{2}}$$
$$2\sqrt{2+\sqrt{3}} = \sqrt{6}+\sqrt{2}$$
$$\left(2\sqrt{2+\sqrt{3}}\right)^2 = \left(\sqrt{6}+\sqrt{2}\right)^2$$
$$4\left(2+\sqrt{3}\right) = 6+2\sqrt{12}+2$$
$$8+4\sqrt{3} = 8+4\sqrt{3}$$

602. $\dfrac{\sqrt{2+\sqrt{3}}}{2}$

Use the half-angle identity:

$$\cos 15° = \cos\left(\frac{30}{2}\right) = \pm\sqrt{\frac{1+\cos 30}{2}} = \pm\sqrt{\frac{1+\frac{\sqrt{3}}{2}}{2}}$$

$$= \pm\sqrt{\frac{1+\frac{\sqrt{3}}{2}}{2}} = \pm\sqrt{\frac{\frac{2}{2}+\frac{\sqrt{3}}{2}}{2}} = \pm\sqrt{\frac{\frac{2+\sqrt{3}}{2}}{2}}$$

$$= \pm\sqrt{\frac{2+\sqrt{3}}{4}} = \pm\frac{\sqrt{2+\sqrt{3}}}{\sqrt{4}} = \pm\frac{\sqrt{2+\sqrt{3}}}{2}$$

Because an angle of 15° is in the first quadrant, the cosine is positive.

You can also solve this problem using the angle-difference identity with angles of 45° and 30°.

603. $\sqrt{3}-2$

Use the angle-sum identity with angles of 120° and 45°:

$$\tan 165° = \frac{\tan 120 + \tan 45}{1 - \tan 120 \tan 45} = \frac{-\sqrt{3}+1}{1-(-\sqrt{3})(1)} = \frac{1-\sqrt{3}}{1+\sqrt{3}}$$

Then rationalize the denominator:

$$\frac{1-\sqrt{3}}{1+\sqrt{3}} \cdot \frac{1-\sqrt{3}}{1-\sqrt{3}} = \frac{1-2\sqrt{3}+\sqrt{9}}{1-\sqrt{9}} = \frac{1-2\sqrt{3}+3}{1-3}$$

$$= \frac{4-2\sqrt{3}}{-2} = -2+\sqrt{3}$$

You can also solve this problem using the half-angle identity with an angle of 330°.

604. $\dfrac{\sqrt{2-\sqrt{3}}}{2}$

Use the half-angle identity:

$$\sin\frac{\pi}{12} = \sin\left(\frac{\pi/6}{2}\right) = \pm\sqrt{\frac{1-\cos \pi/6}{2}} = \pm\sqrt{\frac{1-\sqrt{3}/2}{2}} = \pm\sqrt{\frac{\frac{2}{2}-\frac{\sqrt{3}}{2}}{2}}$$

$$= \pm\sqrt{\frac{\frac{2-\sqrt{3}}{2}}{2}} = \pm\sqrt{\frac{2-\sqrt{3}}{4}} = \pm\frac{\sqrt{2-\sqrt{3}}}{\sqrt{4}} = \pm\frac{\sqrt{2-\sqrt{3}}}{2}$$

Answers 601–700

Because an angle of $\frac{\pi}{12}$ radians is in the first quadrant, the sine is positive.

You can also solve this problem using the angle-difference identity with angles of $\frac{\pi}{4}$ radians and $\frac{\pi}{6}$ radians:

$$\begin{aligned}\sin\frac{\pi}{12}&=\sin\left(\frac{\pi}{4}-\frac{\pi}{6}\right)=\sin\frac{\pi}{4}\cos\frac{\pi}{6}-\cos\frac{\pi}{4}\sin\frac{\pi}{6}\\&=\frac{\sqrt{2}}{2}\cdot\frac{\sqrt{3}}{2}-\frac{\sqrt{2}}{2}\cdot\frac{1}{2}=\frac{\sqrt{6}-\sqrt{2}}{4}\end{aligned}$$

Showing that the two answers are equivalent:

$$\begin{aligned}\frac{\sqrt{2-\sqrt{3}}}{\cancel{2}^{1}}&=\frac{\sqrt{6}-\sqrt{2}}{\cancel{4}^{2}}\\2\sqrt{2-\sqrt{3}}&=\sqrt{6}-\sqrt{2}\\\left(2\sqrt{2-\sqrt{3}}\right)^2&=\left(\sqrt{6}-\sqrt{2}\right)^2\\4\left(2-\sqrt{3}\right)&=6-2\sqrt{12}+2\\8-4\sqrt{3}&=8-4\sqrt{3}\end{aligned}$$

605. $\frac{\sqrt{2}-\sqrt{6}}{4}$

Use the angle-sum identity:

$$\begin{aligned}\cos\frac{7\pi}{12}&=\cos\left(\frac{\pi}{4}+\frac{\pi}{3}\right)=\cos\frac{\pi}{4}\cos\frac{\pi}{3}-\sin\frac{\pi}{4}\sin\frac{\pi}{3}\\&=\left(\frac{\sqrt{2}}{2}\right)\left(\frac{1}{2}\right)-\left(\frac{\sqrt{2}}{2}\right)\left(\frac{\sqrt{3}}{2}\right)=\frac{\sqrt{2}-\sqrt{6}}{4}\end{aligned}$$

606. $2-\sqrt{3}$

Use the angle-difference identity to find $\tan\frac{\pi}{12}$:

$$\tan\frac{\pi}{12}=\tan\left(\frac{\pi}{3}-\frac{\pi}{4}\right)=\frac{\tan\frac{\pi}{3}-\tan\frac{\pi}{4}}{1+\left(\tan\frac{\pi}{3}\right)\left(\tan\frac{\pi}{4}\right)}=\frac{\sqrt{3}-1}{1+\left(\sqrt{3}\right)(1)}=\frac{\sqrt{3}-1}{\sqrt{3}+1}$$

Then rationalize the denominator:

$$\frac{\sqrt{3}-1}{\sqrt{3}+1}\cdot\frac{\sqrt{3}-1}{\sqrt{3}-1}=\frac{\sqrt{9}-2\sqrt{3}+1}{\sqrt{9}-1}=\frac{4-2\sqrt{3}}{2}=2-\sqrt{3}$$

607. $2-\sqrt{3}$

First find tan 75° using the angle-sum identity:

$$\tan 75° = \tan(45+30) = \frac{\tan 45 + \tan 30}{1-(\tan 45)(\tan 30)}$$

$$= \frac{1+\frac{\sqrt{3}}{3}}{1-(1)\left(\frac{\sqrt{3}}{3}\right)} = \frac{1+\frac{\sqrt{3}}{3}}{1-\frac{\sqrt{3}}{3}} = \frac{\frac{3}{3}+\frac{\sqrt{3}}{3}}{\frac{3}{3}-\frac{\sqrt{3}}{3}} = \frac{\frac{3+\sqrt{3}}{3}}{\frac{3-\sqrt{3}}{3}} = \frac{3+\sqrt{3}}{3-\sqrt{3}}$$

Before rationalizing the denominator, use the reciprocal identity to determine cot 75°:

$$\cot 75° = \frac{1}{\tan 75} = \frac{1}{\frac{3+\sqrt{3}}{3-\sqrt{3}}} = \frac{3-\sqrt{3}}{3+\sqrt{3}}$$

Now rationalize the denominator:

$$\frac{3-\sqrt{3}}{3+\sqrt{3}} \cdot \frac{3-\sqrt{3}}{3-\sqrt{3}} = \frac{9-6\sqrt{3}+\sqrt{9}}{9-\sqrt{9}} = \frac{9-6\sqrt{3}+3}{9-3} = \frac{12-6\sqrt{3}}{6} = 2-\sqrt{3}$$

608. $\sqrt{2}-\sqrt{6}$

First find cos 165° using the angle-sum identity with angles of 120° and 45°:

$$\cos 165° = \cos(120+45) = \cos 120 \cos 45 - \sin 120 \sin 45$$

$$= \left(-\frac{1}{2}\right)\left(\frac{\sqrt{2}}{2}\right) - \left(\frac{\sqrt{3}}{2}\right)\left(\frac{\sqrt{2}}{2}\right) = \frac{-\sqrt{2}-\sqrt{6}}{4}$$

Now use the reciprocal identity with the value you just determined:

$$\sec 165° = \frac{1}{\cos 165} = \frac{1}{\frac{-\sqrt{2}-\sqrt{6}}{4}} = \frac{4}{-\sqrt{2}-\sqrt{6}}$$

Rationalize the denominator:

$$\frac{4}{-\sqrt{2}-\sqrt{6}} \cdot \frac{-\sqrt{2}+\sqrt{6}}{-\sqrt{2}+\sqrt{6}} = \frac{4\left(-\sqrt{2}+\sqrt{6}\right)}{\sqrt{4}-\sqrt{36}} = \frac{4\left(-\sqrt{2}+\sqrt{6}\right)}{2-6}$$

$$= \frac{4\left(-\sqrt{2}+\sqrt{6}\right)}{-4} = \sqrt{2}-\sqrt{6}$$

609. $\dfrac{2\sqrt{2}}{\sqrt{4-\sqrt{6}+\sqrt{2}}}$

First, find $\cos\frac{5\pi}{12}$ using the angle-sum identity and angles $\frac{\pi}{6}$ and $\frac{\pi}{4}$:

$$\cos\left(\frac{5\pi}{12}\right)=\cos\left(\frac{\pi}{6}+\frac{\pi}{4}\right)=\cos\frac{\pi}{6}\cos\frac{\pi}{4}-\sin\frac{\pi}{6}\sin\frac{\pi}{4}$$
$$=\left(\frac{\sqrt{3}}{2}\right)\left(\frac{\sqrt{2}}{2}\right)-\left(\frac{1}{2}\right)\left(\frac{\sqrt{2}}{2}\right)=\frac{\sqrt{6}-\sqrt{2}}{4}$$

Now use the half-angle identity to find $\sin\left(\frac{5\pi}{24}\right)$:

$$\sin\left(\frac{5\pi}{24}\right)=\sin\left(\frac{\frac{5\pi}{12}}{2}\right)=\pm\sqrt{\frac{1-\cos\frac{5\pi}{12}}{2}}$$

Next, use the value you just determined and the positive radical (because the angle is in the first quadrant):

$$\sqrt{\frac{1-\cos\frac{5\pi}{12}}{2}}=\sqrt{\frac{1-\frac{\sqrt{6}-\sqrt{2}}{4}}{2}}=\sqrt{\frac{\frac{4}{4}-\frac{\sqrt{6}-\sqrt{2}}{4}}{2}}=\sqrt{\frac{\frac{4-\sqrt{6}+\sqrt{2}}{4}}{2}}$$
$$=\sqrt{\frac{4-\sqrt{6}+\sqrt{2}}{8}}=\frac{\sqrt{4-\sqrt{6}+\sqrt{2}}}{\sqrt{8}}=\frac{\sqrt{4-\sqrt{6}+\sqrt{2}}}{2\sqrt{2}}$$

Finally, use the reciprocal identity:

$$\csc\frac{5\pi}{24}=\frac{1}{\sin\frac{5\pi}{24}}=\frac{1}{\frac{\sqrt{4-\sqrt{6}+\sqrt{2}}}{2\sqrt{2}}}=\frac{2\sqrt{2}}{\sqrt{4-\sqrt{6}+\sqrt{2}}}$$

Rationalizing this answer requires multiplying by three different conjugates — and it gets pretty messy.

610. $\dfrac{2-\sqrt{2+\sqrt{3}}}{\sqrt{2-\sqrt{3}}}$

First, find $\sin\frac{\pi}{12}$ using the half-angle identity:

$$\sin\frac{\pi}{12}=\sin\left(\frac{\pi/6}{2}\right)=\pm\sqrt{\frac{1-\cos\pi/6}{2}}=\pm\sqrt{\frac{1-\sqrt{3}/2}{2}}=\pm\sqrt{\frac{\frac{2}{2}-\frac{\sqrt{3}}{2}}{2}}$$
$$=\pm\sqrt{\frac{\frac{2-\sqrt{3}}{2}}{2}}=\pm\sqrt{\frac{2-\sqrt{3}}{4}}=\pm\frac{\sqrt{2-\sqrt{3}}}{\sqrt{4}}=\pm\frac{\sqrt{2-\sqrt{3}}}{2}$$

Because $\frac{\pi}{12}$ is in the first quadrant, the sine is positive.

Now solve for $\cos\frac{\pi}{12}$ using $\sin\frac{\pi}{12}$ and the Pythagorean identity:

$$\left(\sin\frac{\pi}{12}\right)^2+\left(\cos\frac{\pi}{12}\right)^2=1$$

$$\left(\frac{\sqrt{2-\sqrt{3}}}{2}\right)^2+\left(\cos\frac{\pi}{12}\right)^2=1$$

$$\left(\cos\frac{\pi}{12}\right)^2=1-\frac{2-\sqrt{3}}{4}=\frac{4}{4}-\frac{2-\sqrt{3}}{4}=\frac{2+\sqrt{3}}{4}$$

$$\cos\frac{\pi}{12}=\pm\sqrt{\frac{2+\sqrt{3}}{4}}=\pm\frac{\sqrt{2+\sqrt{3}}}{2}$$

Use the positive radical because the angle is in the first quadrant.

Now, armed with the sine and cosine of $\frac{\pi}{12}$, use the half-angle identity for the tangent:

$$\tan\frac{\pi}{24}=\tan\left(\frac{\pi/12}{2}\right)=\frac{1-\cos\pi/12}{\sin\pi/12}=\frac{1-\frac{\sqrt{2+\sqrt{3}}}{2}}{\frac{\sqrt{2-\sqrt{3}}}{2}}$$

$$=\frac{\frac{2}{2}-\frac{\sqrt{2+\sqrt{3}}}{2}}{\frac{\sqrt{2-\sqrt{3}}}{2}}=\frac{\frac{2-\sqrt{2+\sqrt{3}}}{2}}{\frac{\sqrt{2-\sqrt{3}}}{2}}=\frac{2-\sqrt{2+\sqrt{3}}}{\sqrt{2-\sqrt{3}}}$$

Yes, these radicals within radicals are messy. You can get a simpler answer using angle sum formulas — after using half-angles on sine and cosine. It's your choice.

611. $\cos 2\theta$

Replace $1+\tan^2\theta$ with $\sec^2\theta$ using the Pythagorean identity. Then break the fraction into two terms:

$$\frac{1-\tan^2\theta}{1+\tan^2\theta}=\square$$

$$\frac{1\quad\tan^2\theta}{\sec^2\theta}=\square$$

$$\frac{1}{\sec^2\theta}-\frac{\tan^2\theta}{\sec^2\theta}=\square$$

Replace $\sec^2\theta$ using the reciprocal identity. Then simplify each complex fraction:

$$\frac{1}{\frac{1}{\cos^2\theta}} - \frac{\tan^2\theta}{\frac{1}{\cos^2\theta}} = \boxed{}$$

$$\cos^2\theta - \tan^2\theta\cos^2\theta = \boxed{}$$

Replace $\tan^2\theta$ using the ratio identity, multiply, and simplify:

$$\cos^2\theta - \frac{\sin^2\theta}{\cancel{\cos^2\theta}} \cdot \cancel{\cos^2\theta} = \boxed{}$$

$$\cos^2\theta - \sin^2\theta = \boxed{}$$

Replace $\sin^2\theta$ with $1-\cos^2\theta$ using the Pythagorean identity:

$$\cos^2\theta - (1-\cos^2\theta) = \boxed{}$$

$$2\cos^2\theta - 1 = \boxed{\cos 2\theta}$$

The resulting expression is a version of the double-angle identity for cosine.

612. $2\csc\theta$

Rewrite the fractions with a common denominator, add, and simplify:

$$\frac{\sin\theta}{1-\cos\theta} + \frac{1-\cos\theta}{\sin\theta} = \boxed{}$$

$$\frac{\sin\theta}{1-\cos\theta}\cdot\frac{\sin\theta}{\sin\theta} + \frac{1-\cos\theta}{\sin\theta}\cdot\frac{1-\cos\theta}{1-\cos\theta} = \boxed{}$$

$$\frac{\sin^2\theta}{\sin\theta(1-\cos\theta)} + \frac{(1-\cos\theta)^2}{\sin\theta(1-\cos\theta)} = \boxed{}$$

$$\frac{\sin^2\theta}{\sin\theta(1-\cos\theta)} + \frac{1-2\cos\theta+\cos^2\theta}{\sin\theta(1-\cos\theta)} = \boxed{}$$

$$\frac{\sin^2\theta+1-2\cos\theta+\cos^2\theta}{\sin\theta(1-\cos\theta)} = \boxed{}$$

Replace $\sin^2\theta+\cos^2\theta$ with 1 from the Pythagorean identity. Then simplify:

$$\frac{1+1-2\cos\theta}{\sin\theta(1-\cos\theta)} = \boxed{}$$

$$\frac{2-2\cos\theta}{\sin\theta(1-\cos\theta)} = \boxed{}$$

$$\frac{2\cancel{(1-\cos\theta)}}{\sin\theta\cancel{(1-\cos\theta)}} = \boxed{}$$

Replace $\frac{1}{\sin\theta}$ with the reciprocal identity:

$$2\csc\theta = \boxed{2\csc\theta}$$

613. $\sin\theta$

Multiply the two binomials together:

$$(1+\cos\theta)(\csc\theta-\cot\theta)=\boxed{}$$
$$\csc\theta-\cot\theta+\cos\theta\csc\theta-\cos\theta\cot\theta=\boxed{}$$

Replace $\csc\theta$ with its reciprocal identity and $\cot\theta$ with its ratio identity. Multiply, combine terms, and simplify:

$$\frac{1}{\sin\theta}-\frac{\cos\theta}{\sin\theta}+\cos\theta\cdot\frac{1}{\sin\theta}-\cos\theta\cdot\frac{\cos\theta}{\sin\theta}=\boxed{}$$
$$\frac{1}{\sin\theta}-\frac{\cos\theta}{\sin\theta}+\frac{\cos\theta}{\sin\theta}-\frac{\cos^2\theta}{\sin\theta}=\boxed{}$$
$$\frac{1}{\sin\theta}-\frac{\cos^2\theta}{\sin\theta}=\boxed{}$$
$$\frac{1-\cos^2\theta}{\sin\theta}=\boxed{}$$

Replace the 1 with $\sin^2\theta+\cos^2\theta$ from the Pythagorean identity:

$$\frac{\sin^2\theta+\cos^2\theta-\cos^2\theta}{\sin\theta}=\boxed{}$$
$$\frac{\sin^{\not{2}}\theta}{\not{\sin\theta}}=\boxed{\sin\theta}$$

614. $\csc^4\theta$

Factor the trinomial as the square of a binomial:

$$\cot^4\theta+2\cot^2\theta+1=\boxed{}$$
$$(\cot^2\theta+1)^2=\boxed{}$$

Replace $\cot^2\theta+1$ with $\csc^2\theta$ from the Pythagorean identity:

$$(\csc^2\theta)^2=\boxed{}$$
$$\csc^4\theta=\boxed{\csc^4\theta}$$

Answers 601–700

615. 0

Use the double-angle identity, $\sin 2x=2\sin x\cos x$, letting $x=\frac{\theta}{4}$:

$$\sin 2\left(\frac{\theta}{4}\right)=2\sin\left(\frac{\theta}{4}\right)\cos\left(\frac{\theta}{4}\right)+\boxed{}$$
$$\sin\not{2}\left(\frac{\theta}{\not{4}^{2}}\right)=2\sin\left(\frac{\theta}{4}\right)\cos\left(\frac{\theta}{4}\right)+\boxed{}$$
$$\sin\left(\frac{\theta}{2}\right)=2\sin\left(\frac{\theta}{4}\right)\cos\left(\frac{\theta}{4}\right)+\boxed{0}$$

616. $\cot\theta$

Replace $\tan\theta$ with its reciprocal identity. Combine the terms in the denominator; then simplify the complex fraction:

$$\frac{1-\cot\theta}{\tan\theta-1}=\frac{1-\cot\theta}{\frac{1}{\cot\theta}-1}=\frac{1-\cot\theta}{\frac{1}{\cot\theta}-\frac{\cot\theta}{\cot\theta}}=\frac{1-\cot\theta}{\frac{1-\cot\theta}{\cot\theta}}=\boxed{}$$

$$\frac{\cancel{1-\cot\theta}}{1}\cdot\frac{\cot\theta}{\cancel{1-\cot\theta}}=\boxed{\cot\theta}$$

617. $\sec\theta$

Factor the terms in the denominator on the left:

$$\frac{1}{\cos\theta-\cos^2\theta}=\frac{1}{\cos\theta(1-\cos\theta)}=\frac{\boxed{}}{1-\cos\theta}$$

Replace the first $\cos\theta$ in the denominator with its reciprocal identity:

$$\frac{1}{\frac{1}{\sec\theta}(1-\cos\theta)}=\frac{\boxed{}}{1-\cos\theta}$$

Simplify the complex fraction:

$$\frac{\sec\theta}{1}\cdot\frac{1}{(1-\cos\theta)}=\frac{\boxed{\sec\theta}}{1-\cos\theta}$$

618. 1

Factor the terms by grouping:

$$\cot^2\theta-\cos^2\theta\cot^2\theta+1-\cos^2\theta=\boxed{}$$
$$\cot^2\theta\left(1-\cos^2\theta\right)+1\left(1-\cos^2\theta\right)=\boxed{}$$
$$\left(1-\cos^2\theta\right)\left(\cot^2\theta+1\right)=\boxed{}$$

Use the Pythagorean identity to replace $1-\cos^2\theta$ with $\sin^2\theta$. Also, replace $\cot^2\theta+1$ with $\csc^2\theta$, using its Pythagorean identity:

$$\left(\sin^2\theta\right)\left(\csc^2\theta\right)=\boxed{}$$

Replace $\csc^2\theta$ with its reciprocal identity:

$$\left(\sin^2\theta\right)\left(\frac{1}{\sin^2\theta}\right)=\boxed{1}$$

619. $\csc\theta$

Replace $\cot\theta$ in the denominator with its ratio identity and $\csc\theta$ in the denominator with its reciprocal identity. Multiply and simplify:

$$\frac{\cot\theta+4\sec\theta}{\cot\theta\sin\theta}=4\sec^2\theta+\boxed{}$$

$$\frac{\cot\theta+4\sec\theta}{\frac{\cos\theta}{\cancel{\sin\theta}}\cdot\frac{\cancel{\sin\theta}}{1}}=4\sec^2\theta+\boxed{}$$

$$\frac{\cot\theta+4\sec\theta}{\cos\theta}=4\sec^2\theta+\boxed{}$$

Split the term on the left into two fractions:

$$\frac{\cot\theta}{\cos\theta}+\frac{4\sec\theta}{\cos\theta}=4\sec^2\theta+\boxed{}$$

Replace $\cot\theta$ with its ratio identity and $\cos\theta$ with its reciprocal identity. Simplify both complex fractions:

$$\frac{\frac{\cos\theta}{\sin\theta}}{\cos\theta}+\frac{4\sec\theta}{\frac{1}{\sec\theta}}=4\sec^2\theta+\boxed{}$$

$$\frac{\cancel{\cos\theta}}{\sin\theta}\cdot\frac{1}{\cancel{\cos\theta}}+\frac{4\sec\theta}{1}\cdot\frac{\sec\theta}{1}=4\sec^2\theta+\boxed{}$$

$$\frac{1}{\sin\theta}+4\sec^2\theta=4\sec^2\theta+\boxed{}$$

Replace $\frac{1}{\sin\theta}$ with its reciprocal identity:

$$\csc\theta+4\sec^2\theta=4\sec^2\theta+\boxed{\csc\theta}$$

620. $\cos\theta$

Replace $\csc\theta$ with its reciprocal identity. Then find a common denominator and combine the terms in the denominator:

$$\frac{\cot\theta}{\frac{1}{\sin\theta}-\sin\theta}=\frac{\cot\theta}{\frac{1}{\sin\theta}-\frac{\sin^2\theta}{\sin\theta}}=\frac{\cot\theta}{\frac{1-\sin^2\theta}{\sin\theta}}=\frac{1}{\boxed{}}$$

Replace $\cot\theta$ with its ratio identity. Also replace $1-\sin^2\theta$ with $\cos^2\theta$ using the Pythagorean identity:

$$\frac{\frac{\cos\theta}{\sin\theta}}{\frac{\cos^2\theta}{\sin\theta}}=\frac{1}{\boxed{}}$$

Simplify the complex fraction:

$$\frac{\cancel{\cos\theta}}{\cancel{\sin\theta}}\cdot\frac{\cancel{\sin\theta}}{\cos^{\cancel{2}}\theta}=\frac{1}{\boxed{\cos\theta}}$$

621. $3\cos\theta$

Use the cosine angle-sum identity with angles of 2θ and θ:

$$\cos 3\theta=4\cos^3\theta-\boxed{}$$

$$\cos(2\theta+\theta)=4\cos^3\theta-\boxed{}$$

$$\cos 2\theta\cos\theta-\sin 2\theta\sin\theta=4\cos^3\theta-\boxed{}$$

Now use the cosine double-angle identity (cosine version) and the sine double-angle identity to replace $\cos 2\theta$ and $\sin 2\theta$:

$$(2\cos^2\theta - 1)\cos\theta - (2\sin\theta\cos\theta)\sin\theta = 4\cos^3\theta - \boxed{}$$
$$2\cos^3\theta - \cos\theta - 2\sin^2\theta\cos\theta = 4\cos^3\theta - \boxed{}$$

Replace $\sin^2\theta$ with $1 - \cos^2\theta$ using the Pythagorean identity:

$$2\cos^3\theta - \cos\theta - 2(1-\cos^2\theta)\cos\theta = 4\cos^3\theta - \boxed{}$$
$$2\cos^3\theta - \cos\theta - 2\cos\theta + 2\cos^3\theta = 4\cos^3\theta - \boxed{}$$
$$4\cos^3\theta - 3\cos\theta = 4\cos^3\theta - \boxed{3\cos\theta}$$

622. 1

Use the cosine's double-angle identity involving sine to replace $\cos 20\theta$:

$$\cos 20\theta = 8\sin^4 5\theta - 8\sin^2 5\theta + \boxed{}$$
$$1 - 2\sin^2 10\theta = 8\sin^4 5\theta - 8\sin^2 5\theta + \boxed{}$$

Now replace $\sin^2 10\theta$ using the double-angle identity for sine after rewriting the factor as $(\sin 10\theta)^2$. Then simplify:

$$1 - 2(\sin 10\theta)^2 = 8\sin^4 5\theta - 8\sin^2 5\theta + \boxed{}$$
$$1 - 2(2\sin 5\theta\cos 5\theta)^2 = 8\sin^4 5\theta - 8\sin^2 5\theta + \boxed{}$$
$$1 - 2(4\sin^2 5\theta\cos^2 5\theta) = 8\sin^4 5\theta - 8\sin^2 5\theta + \boxed{}$$
$$1 - 8\sin^2 5\theta\cos^2 5\theta = 8\sin^4 5\theta - 8\sin^2 5\theta + \boxed{}$$

Use the Pythagorean identity to replace $\cos^2 5\theta$ with $1 - \sin^2 5\theta$:

$$1 - 8\sin^2 5\theta(1 - \sin^2 5\theta) = 8\sin^4 5\theta - 8\sin^2 5\theta + \boxed{}$$
$$1 - 8\sin^2 5\theta + 8\sin^4 5\theta = 8\sin^4 5\theta - 8\sin^2 5\theta + \boxed{1}$$

623. 1

Factor the numerator. Then reduce the fraction:

$$\frac{(\sin\theta + \cos\theta)(\sin^2\theta - \sin\theta\cos\theta + \cos^2\theta)}{\sin\theta + \cos\theta} = \boxed{} - \frac{1}{2}\sin 2\theta$$
$$\frac{\cancel{(\sin\theta + \cos\theta)}(\sin^2\theta - \sin\theta\cos\theta + \cos^2\theta)}{\cancel{\sin\theta + \cos\theta}} = \boxed{} - \frac{1}{2}\sin 2\theta$$
$$\sin^2\theta - \sin\theta\cos\theta + \cos^2\theta = \boxed{} - \frac{1}{2}\sin 2\theta$$

Replace $\sin^2\theta + \cos^2\theta$ with 1:

$$1 - \sin\theta\cos\theta = \boxed{} - \frac{1}{2}\sin 2\theta$$

Replace $\sin\theta\cos\theta$ using the double-angle identity for sine. Recall that $\sin 2\theta = 2\sin\theta\cos\theta$. Dividing both sides of the equation by 2 gives you $\frac{1}{2}\sin 2\theta = \sin\theta\cos\theta$, so

$$1 - \frac{1}{2}\sin 2\theta = \boxed{1} - \frac{1}{2}\sin 2\theta$$

624. 0

Subtract the two fractions after finding a common denominator. Then simplify:

$$\frac{\cos\theta}{\cos\theta-\sin\theta}\cdot\frac{\cos\theta+\sin\theta}{\cos\theta+\sin\theta}-\frac{\sin\theta}{\cos\theta+\sin\theta}\cdot\frac{\cos\theta-\sin\theta}{\cos\theta-\sin\theta}=\sec 2\theta+\boxed{}$$

$$\frac{\cos\theta(\cos\theta+\sin\theta)}{(\cos\theta-\sin\theta)(\cos\theta+\sin\theta)}-\frac{\sin\theta(\cos\theta-\sin\theta)}{(\cos\theta-\sin\theta)(\cos\theta+\sin\theta)}=\sec 2\theta+\boxed{}$$

$$\frac{\cos\theta(\cos\theta+\sin\theta)-\sin\theta(\cos\theta-\sin\theta)}{(\cos\theta-\sin\theta)(\cos\theta+\sin\theta)}=\sec 2\theta+\boxed{}$$

$$\frac{\cos^2\theta+\sin\theta\cos\theta-\sin\theta\cos\theta+\sin^2\theta}{(\cos\theta-\sin\theta)(\cos\theta+\sin\theta)}=\sec 2\theta+\boxed{}$$

$$\frac{\cos^2\theta+\sin^2\theta}{(\cos\theta-\sin\theta)(\cos\theta+\sin\theta)}=\sec 2\theta+\boxed{}$$

$$\frac{1}{(\cos\theta-\sin\theta)(\cos\theta+\sin\theta)}=\sec 2\theta+\boxed{}$$

Multiply the binomials in the denominator:

$$\frac{1}{\cos^2\theta-\sin^2\theta}=\sec 2\theta+\boxed{}$$

Use the Pythagorean identity to replace $\cos^2\theta$ with $1-\sin^2\theta$:

$$\frac{1}{1-\sin^2\theta-\sin^2\theta}=\sec 2\theta+\boxed{}$$

$$\frac{1}{1-2\sin^2\theta}=\sec 2\theta+\boxed{}$$

The denominator is one of the double-angle identities for cosine, so replace the denominator with $\cos 2\theta$:

$$\frac{1}{\cos 2\theta}=\sec 2\theta+\boxed{}$$

Finally, use the reciprocal identity to replace $\cos 2\theta$:

$$\sec 2\theta=\sec 2\theta+\boxed{0}$$

If you recognized that $\cos^2\theta-\sin^2\theta$ is the double-angle identity for cosine, you could've used that directly to replace the denominator instead of going through the Pythagorean identity.

625. 8

Square the binomials and combine like terms:

$$(\sec\theta+2\cos\theta)^2-(\sec\theta-2\cos\theta)^2=\boxed{}$$

$$\sec^2\theta+4\sec\theta\cos\theta+4\cos^2\theta-(\sec^2\theta-4\sec\theta\cos\theta+4\cos^2\theta)=\boxed{}$$

$$\sec^2\theta+4\sec\theta\cos\theta+4\cos^2\theta-\sec^2\theta+4\sec\theta\cos\theta-4\cos^2\theta=\boxed{}$$

$$8\sec\theta\cos\theta=\boxed{}$$

Replace $\sec\theta$ with its reciprocal identity:

$$8\cdot\frac{1}{\cancel{\cos\theta}}\cdot\cancel{\cos\theta}=\boxed{8}$$

626. 2

Replace $\sin 2\theta$ and $\cos 2\theta$ with their double-angle identities. It doesn't really matter which double-angle identity you use for cosine.

$$\frac{\sin 2\theta+\cos 2\theta}{\sin\theta\cos\theta}=\cot\theta-\tan\theta+\boxed{}$$

$$\frac{2\sin\theta\cos\theta+2\cos^2\theta-1}{\sin\theta\cos\theta}=\cot\theta-\tan\theta+\boxed{}$$

Replace the 1 with $\sin^2\theta+\cos^2\theta$ using the Pythagorean identity:

$$\frac{2\sin\theta\cos\theta+2\cos^2\theta-(\sin^2\theta+\cos^2\theta)}{\sin\theta\cos\theta}=\cot\theta-\tan\theta+\boxed{}$$

$$\frac{2\sin\theta\cos\theta+2\cos^2\theta-\sin^2\theta-\cos^2\theta}{\sin\theta\cos\theta}=\cot\theta-\tan\theta+\boxed{}$$

$$\frac{2\sin\theta\cos\theta+\cos^2\theta-\sin^2\theta}{\sin\theta\cos\theta}=\cot\theta-\tan\theta+\boxed{}$$

Break up the left side into three fractions. Then reduce:

$$\frac{2\cancel{\sin\theta\cos\theta}}{\cancel{\sin\theta\cos\theta}}+\frac{\cos^{\cancel{2}\,1}\theta}{\sin\theta\cancel{\cos\theta}}-\frac{\sin^{\cancel{2}\,1}\theta}{\cancel{\sin\theta}\cos\theta}=\cot\theta-\tan\theta+\boxed{}$$

$$2+\frac{\cos\theta}{\sin\theta}-\frac{\sin\theta}{\cos\theta}=\cot\theta-\tan\theta+\boxed{}$$

Use ratio identities to replace the two fractions:

$$2+\cot\theta-\tan\theta=\cot\theta-\tan\theta+\boxed{2}$$

627. 1

Replace $\sin 2\theta$ and $\cos 2\theta$ with their double-angle identities. Then simplify:

$$\frac{1}{2}\sin^2 2\theta+\cos 2\theta+2\sin^4\theta=\boxed{}$$

$$\frac{1}{2}(\sin 2\theta)^2+\cos 2\theta+2\sin^4\theta=\boxed{}$$

$$\frac{1}{2}(2\sin\theta\cos\theta)^2+\left(1-2\sin^2\theta\right)+2\sin^4\theta=\boxed{}$$

$$\frac{1}{2}\left(4\sin^2\theta\cos^2\theta\right)+\left(1-2\sin^2\theta\right)+2\sin^4\theta=\boxed{}$$

$$2\sin^2\theta\cos^2\theta+1-2\sin^2\theta+2\sin^4\theta=\boxed{}$$

$$2\sin^2\theta\cos^2\theta+1-2\sin^2\theta+2\sin^4\theta=\boxed{}$$

Replace $\cos^2\theta$ with $1-\sin^2\theta$ using the Pythagorean identity. Then simplify:

$$2\sin^2\theta\left(1-\sin^2\theta\right)+1-2\sin^2\theta+2\sin^4\theta=\boxed{}$$

$$2\sin^2\theta-2\sin^4\theta+1-2\sin^2\theta+2\sin^4\theta=\boxed{}$$

$$1=\boxed{1}$$

628. 0

Replace sin 2θ and cos 2θ with their double-angle identities. Then simplify:

$$\frac{2\sin 2\theta}{(\cos 2\theta - 1)^2} = \csc^2\theta \cot\theta + \boxed{}$$

$$\frac{2(2\sin\theta\cos\theta)}{(1-2\sin^2\theta - 1)^2} = \csc^2\theta \cot\theta + \boxed{}$$

$$\frac{4\sin\theta\cos\theta}{(-2\sin^2\theta)^2} = \csc^2\theta \cot\theta + \boxed{}$$

$$\frac{4\sin\theta\cos\theta}{4\sin^4\theta} = \csc^2\theta \cot\theta + \boxed{}$$

$$\frac{\sin\theta\cos\theta}{\sin^4\theta} = \csc^2\theta \cot\theta + \boxed{}$$

Rewrite the fraction as the product of two fractions and then reduce:

$$\frac{\sin\theta}{\sin^3\theta}\cdot\frac{\cos\theta}{\sin\theta} = \csc^2\theta \cot\theta + \boxed{}$$

$$\frac{\cancel{\sin\theta}}{\sin^{\cancel{3}2}\theta}\cdot\frac{\cos\theta}{\sin\theta} = \csc^2\theta \cot\theta + \boxed{}$$

$$\frac{1}{\sin^2\theta}\cdot\frac{\cos\theta}{\sin\theta} = \csc^2\theta \cot\theta + \boxed{}$$

Replace $\frac{1}{\sin^2\theta}$ with its reciprocal identity and $\frac{\cos\theta}{\sin\theta}$ with its ratio identity:

$$\csc^2\theta\cdot\cot\theta = \csc^2\theta \cot\theta + \boxed{0}$$

629. $\cot\theta$

Multiply both numerator and denominator by $\sqrt{1+\cos\theta}$ and then simplify:

$$\sqrt{\frac{1+\cos\theta}{1-\cos\theta}} = \csc\theta + \boxed{}$$

$$\frac{\sqrt{1+\cos\theta}}{\sqrt{1-\cos\theta}}\cdot\frac{\sqrt{1+\cos\theta}}{\sqrt{1+\cos\theta}} = \csc\theta + \boxed{}$$

$$\frac{\sqrt{(1+\cos\theta)^2}}{\sqrt{1-\cos^2\theta}} = \csc\theta + \boxed{}$$

$$\frac{1+\cos\theta}{\sqrt{1-\cos^2\theta}} = \csc\theta + \boxed{}$$

Replace $1-\cos^2\theta$ with $\sin^2\theta$ using the Pythagorean identity. Then simplify:

$$\frac{1+\cos\theta}{\sqrt{\sin^2\theta}} = \csc\theta + \boxed{}$$

$$\frac{1+\cos\theta}{\sin\theta} = \csc\theta + \boxed{}$$

Answers 601–700

Write the fraction as the sum of two fractions:

$$\frac{1}{\sin\theta}+\frac{\cos\theta}{\sin\theta}=\csc\theta+\boxed{}$$

Replace $\frac{1}{\sin\theta}$ with its reciprocal identity and $\frac{\cos\theta}{\sin\theta}$ with its ratio identity:

$$\csc\theta+\cot\theta=\csc\theta+\boxed{\cot\theta}$$

630. 1

Square the binomial and simplify:

$$(\sin\theta-\sec\theta)^2-\sin^2\theta=\left(\tan\theta-\boxed{}\right)^2$$

$$\sin^2\theta-2\sin\theta\sec\theta+\sec^2\theta-\sin^2\theta=\left(\tan\theta-\boxed{}\right)^2$$

$$-2\sin\theta\sec\theta+\sec^2\theta=\left(\tan\theta-\boxed{}\right)^2$$

In the first term, replace $\sec\theta$ with its reciprocal identity and then multiply. In the second term, replace $\sec^2\theta$ with $\tan^2\theta+1$ using the Pythagorean identity:

$$-2\sin\theta\cdot\frac{1}{\cos\theta}+(\tan^2\theta+1)=\left(\tan\theta-\boxed{}\right)^2$$

$$-2\frac{\sin\theta}{\cos\theta}+\tan^2\theta+1=\left(\tan\theta-\boxed{}\right)^2$$

Replace $\frac{\sin\theta}{\cos\theta}$ with its ratio identity:

$$-2\tan\theta+\tan^2\theta+1=\left(\tan\theta-\boxed{}\right)^2$$

$$\tan^2\theta-2\tan\theta+1=\left(\tan\theta-\boxed{}\right)^2$$

Factor the binomial:

$$(\tan\theta-1)^2=\left(\tan\theta-\boxed{1}\right)^2$$

631. 16.4

You know two angles and a consecutive side, so this is an angle-angle-side (AAS) problem. Find the missing side by using the Law of Sines, $\frac{a}{\sin A}=\frac{c}{\sin C}$:

$$\frac{9}{\sin 26°}=\frac{c}{\sin 53°}$$

$$c=\frac{9\sin 53°}{\sin 26°}\approx 16.4$$

632. 14.0

You know two angles and a consecutive side, so this is an angle-angle-side (AAS) problem. Find the missing side by using the Law of Sines, $\frac{a}{\sin A} = \frac{b}{\sin B}$.

You're looking for side *b*, so you need to know the measure of angle *B*. You know the measures of angles *A* and *C*, so find the measure of angle *B* by subtracting the sum of *A* and *C* from 180°:

$$\begin{aligned} m\angle B &= 180° - (m\angle A + m\angle C) \\ &= 180° - (40° + 118°) \\ &= 180° - 158° \\ &= 22° \end{aligned}$$

Now use the Law of Sines to solve for *b*:

$$\begin{aligned} \frac{24}{\sin 40°} &= \frac{b}{\sin 22°} \\ b &= \frac{24 \sin 22°}{\sin 40°} \approx 14.0 \end{aligned}$$

633. 24.1

You know two angles and the side between them, so this is an angle-side-angle (ASA) problem. Find the missing side by using the Law of Sines, $\frac{a}{\sin A} = \frac{c}{\sin C}$.

You know the length of side *a* but not the measure of angle *A*. You know the measures of angles *B* and *C*, so find the measure of angle *A* by subtracting the sum of *B* and *C* from 180°:

$$\begin{aligned} m\angle B &= 180° - (m\angle A + m\angle C) \\ &= 180° - (24° + 73°) \\ &= 180° - 97° \\ &= 83° \end{aligned}$$

Now use the Law of Sines to solve for *c*:

$$\begin{aligned} \frac{25}{\sin 83°} &= \frac{c}{\sin 73°} \\ c &= \frac{25 \sin 73°}{\sin 83°} \approx 24.1 \end{aligned}$$

634. 43.1°

You know two sides and a consecutive angle, so this is a side-side-angle (SSA) problem.

Find the measure of angle *A* by using the Law of Sines, $\frac{\sin A}{a} = \frac{\sin C}{c}$:

$$\begin{aligned} \frac{\sin A}{20} &= \frac{\sin 82°}{29} \\ \sin A &= \frac{20 \sin 82°}{29} \\ A &= \sin^{-1}\left(\frac{20 \sin 82°}{29}\right) \approx 43.1° \end{aligned}$$

635. 52.4°

You know two sides and a consecutive angle, so this is a side-side-angle (SSA) problem. Find the measure of angle C by using the Law of Sines, $\frac{\sin B}{b} = \frac{\sin C}{c}$:

$$\frac{\sin 48°}{19.7} = \frac{\sin C}{21}$$

$$\sin C = \frac{21 \sin 48°}{19.7}$$

$$C = \sin^{-1}\left(\frac{21 \sin 48°}{19.7}\right) \approx 52.4°$$

636. 15.0°

You know two sides and a consecutive angle, so this is a side-side-angle (SSA) problem. Find the measure of angle C by using the Law of Sines, $\frac{\sin A}{a} = \frac{\sin C}{c}$:

$$\frac{\sin 36°}{25} = \frac{\sin C}{11}$$

$$\sin c = \frac{11 \sin 36°}{25}$$

$$C = \sin^{-1}\left(\frac{11 \sin 36°}{25}\right) \approx 15.0°$$

637. 113 feet

Always start by drawing and labeling a diagram to represent the problem:

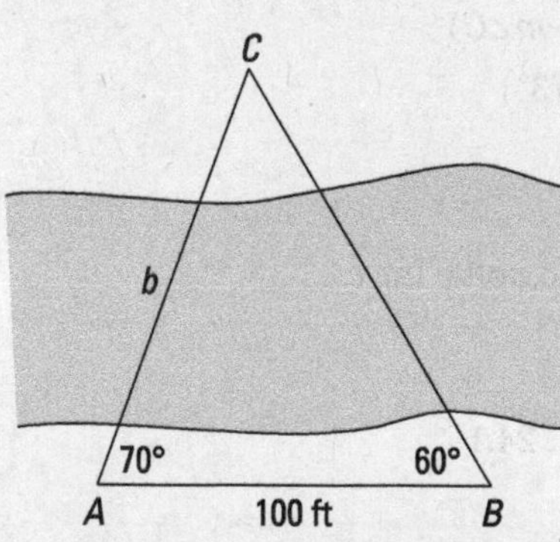

Illustration by Thomson Digital

You can see from the diagram that you know two angles and the side between them, so this is an angle-side-angle (ASA) problem.

You know the length of side c, so you need to find the measure of angle C. You know the measures of angles A and B, so find the measure of angle C by subtracting the sum of A and B from 180°:

Answers 601–700

$$
\begin{aligned}
m\angle C &= 180° - (m\angle A + m\angle B) \\
&= 180° - (70° + 60°) \\
&= 180° - 130° \\
&= 50°
\end{aligned}
$$

The side from point *A* to point *C* is *b*, so use the Law of Sines to solve for *b:*

$$
\begin{aligned}
\frac{b}{\sin B} &= \frac{c}{\sin C} \\
\frac{b}{\sin 60°} &= \frac{100}{\sin 50°} \\
b &= \frac{100 \sin 60°}{\sin 50°} \approx 113
\end{aligned}
$$

The distance from point *A* to point *C* is approximately 113 feet.

638. 85.3°

Always start by drawing and labeling a diagram to represent the problem:

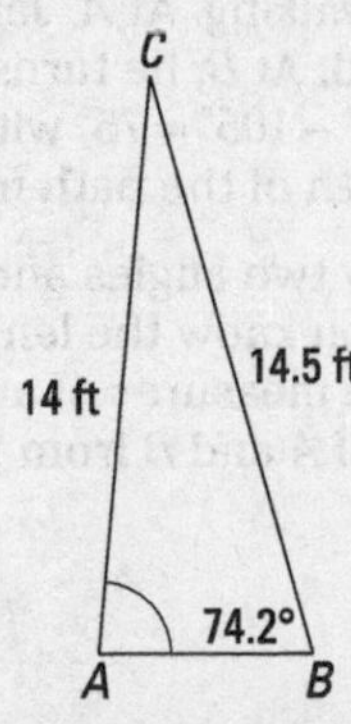

Illustration by Thomson Digital

You can see from the diagram that you know two sides and a consecutive angle, so this is a side-side-angle (SSA) problem. You know the lengths of sides *a* and *b*, and you need to find the measure of angle *A*. Use the Law of Sines:

$$
\begin{aligned}
\frac{\sin A}{a} &= \frac{\sin B}{b} \\
\frac{\sin A}{14.5} &= \frac{\sin 74.2°}{14} \\
\sin A &= \frac{14.5 \sin 74.2°}{14} \\
A &= \sin^{-1}\left(\frac{14.5 \sin 74.2°}{14}\right) \approx 85.3°
\end{aligned}
$$

The pole makes an angle of about 85.3° with the ground.

639. 1,356 feet

Always start by drawing and labeling a diagram to represent the problem:

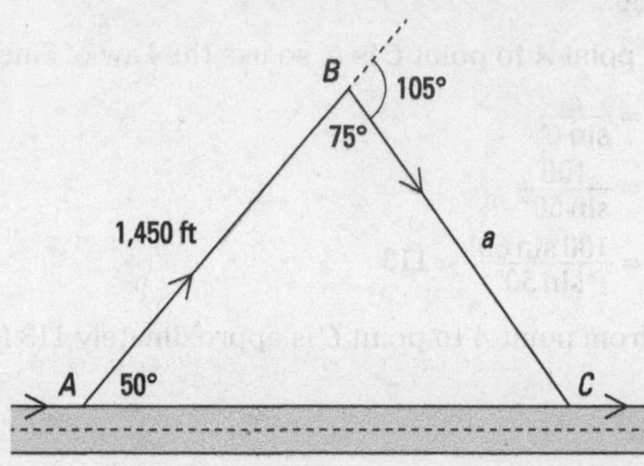

Illustration by Thomson Digital

The arrows show the direction that Jake is walking. At A, Jake turns and walks along a path that makes an angle of 50° with the road. At B, he turns 105° to walk back to the road. This redirection makes an angle of $180° - 105° = 75°$ with the path, so the measure of angle $ABC = 75°$. You want to find the length of the path from B to C, which is a.

You can see from the diagram that you know two angles and a side between them, so this is an angle-side-angle (ASA) problem. You know the length of side c, so you need to find the measure of angle C. You know the measures of angles A and B, so find the measure of angle C by subtracting the sum of A and B from 180°:

$$\begin{aligned} m\angle C &= 180° - (m\angle A + m\angle B) \\ &= 180° - (50° + 75°) \\ &= 180° - 125° \\ &= 55° \end{aligned}$$

Now use the Law of Sines to solve for a:

$$\frac{a}{\sin A} = \frac{c}{\sin C}$$

$$\frac{a}{\sin 50°} = \frac{1{,}450}{\sin 55°}$$

$$a = \frac{1{,}450 \sin 50°}{\sin 55°} \approx 1{,}356$$

Jake has to walk about 1,356 feet to get back to the road. He takes more than a half-mile detour!

640. 27 feet

Always start by drawing and labeling a diagram to represent the problem:

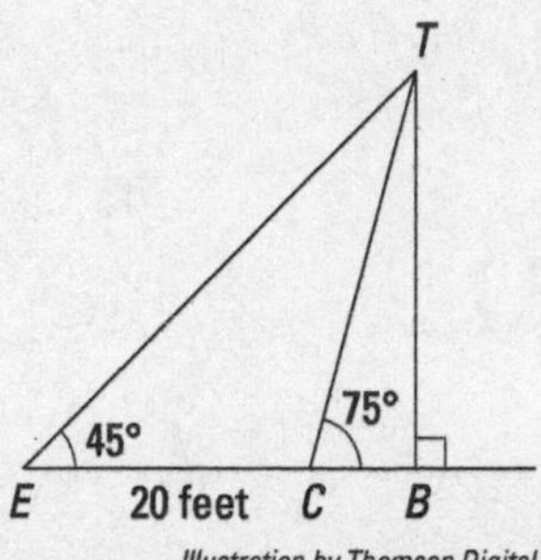

Illustration by Thomson Digital

The tree is labeled *BT*, Elena is standing at *E*, and Carlos is standing at *C*. There are three triangles in the diagram: triangle *BET*, triangle *CET*, and triangle *BCT*. You need to find length *BT*, but neither triangle containing this side has enough information.

Triangles *BCT* and *BET* are both right triangles with one acute angle labeled. If you can find the length of the hypotenuse in either triangle, you'll have enough information to find *BT*.

Use triangle *CET*. In this triangle, the measure of angle *ECT* is $180° - 75° = 105°$. Then the measure of angle *ETC* is $180° - (45° + 105°) = 30°$. Use the Law of Sines to find the length of $\overline{ET}$, which is the hypotenuse of triangle *BET*:

$$\frac{ET}{\sin \angle ECT} = \frac{EC}{\sin \angle ETC}$$

$$\frac{ET}{\sin 105°} = \frac{20}{\sin 30°}$$

$$ET = \frac{20 \sin 105°}{\sin 30°}$$

Because you need to use this result as-is in later calculations, don't approximate it yet.

Now look at triangle *BET*:

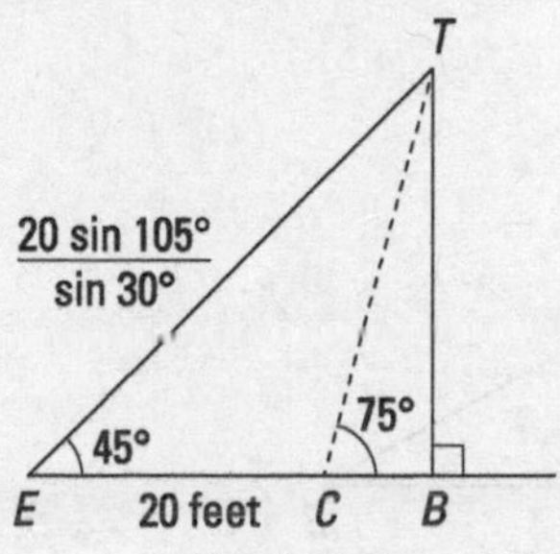

Illustration by Thomson Digital

Because you now have the length of the hypotenuse in triangle *BET*, you have enough information to solve for the length *BT* by using the ratio definition of sine:

$$\sin\angle BET = \frac{\text{Opposite}}{\text{Hypotenuse}} = \frac{BT}{ET}$$

$$\sin 45° = \frac{BT}{\frac{20\sin 105°}{\sin 30°}}$$

Now, replace sin 45° with $\frac{\sqrt{2}}{2}$:

$$\frac{\sqrt{2}}{2} = \frac{BT}{\frac{20\sin 105°}{\sin 30°}}$$

$$BT = \frac{\sqrt{2}}{2}\cdot\frac{20\sin 105°}{\sin 30°} \approx 27$$

The tree is about 27 feet tall.

641. 1

The following chart summarizes the ambiguous case of the Law of Sines. When you're given the lengths of two sides of a triangle and the angle opposite one of them — SSA — various possibilities exist: Two triangles may be possible, one triangle may be possible, or no triangles at all may be possible.

Given Angle	Conditions	Number of Triangles
Acute	Opposite side < altitude	0
Acute	Opposite side = altitude	1
Acute	Altitude < opposite side < adjacent side	2
Acute	Opposite side ≥ adjacent side	1
Obtuse	Opposite side ≤ adjacent side	0
Obtuse	Opposite side > altitude	1

Illustration by Thomson Digital

This figure shows the situation:

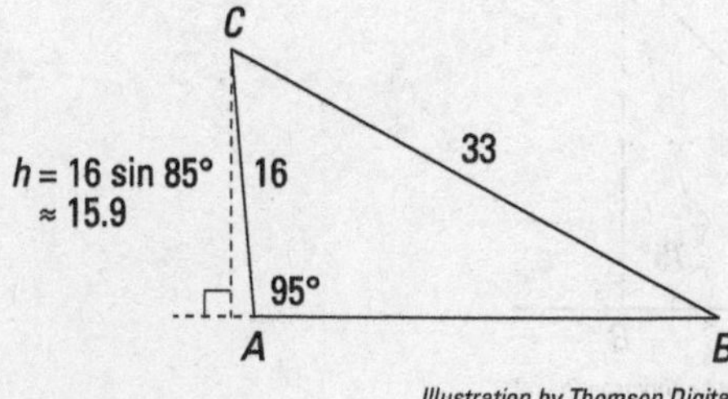

Illustration by Thomson Digital

Drawing an altitude from vertex *C*, you form a right triangle with an acute angle of 85°. You get the 85° by subtracting 180° – 95° because the two angles along the base are supplementary. Then, in that small right triangle formed to the left of *ABC*,

$$\sin 85° = \frac{h}{16}$$
$$h = 16 \sin 85° \approx 15.9$$

Because *A* is obtuse and the length of the altitude is less than that of the opposite side, only one triangle is possible.

642. 0

The following chart summarizes the ambiguous case of the Law of Sines. When you're given the lengths of two sides of a triangle and the angle opposite one of them — SSA — various possibilities exist: Two triangles may be possible, one triangle may be possible, or no triangles at all may be possible.

Given Angle	Conditions	Number of Triangles
Acute	Opposite side < altitude	0
Acute	Opposite side = altitude	1
Acute	Altitude < opposite side < adjacent side	2
Acute	Opposite side ≥ adjacent side	1
Obtuse	Opposite side ≤ adjacent side	0
Obtuse	Opposite side > altitude	1

Illustration by Thomson Digital

This figure shows the situation. Drawing an altitude from angle *B*, you're not sure whether angle *A* is obtuse (making side *c* fold left toward *C*) or acute (making side *c* fold right). Or maybe neither works.

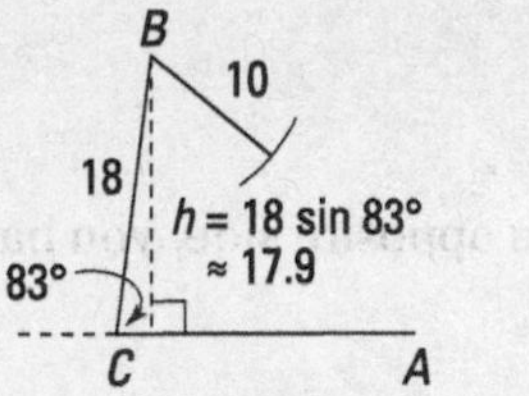

Illustration by Thomson Digital

To compute the altitude, use the following formula.

$$\sin 83° = \frac{h}{18}$$
$$h = 18 \cdot \sin 83° \approx 17.9$$

This situation is impossible because the opposite side can't be shorter than the altitude. Therefore, no triangles are possible.

Answers 601–700

643. 2

The following chart summarizes the ambiguous case of the Law of Sines. When you're given the lengths of two sides of a triangle and the angle opposite one of them — SSA — various possibilities exist: Two triangles may be possible, one triangle may be possible, or no triangles at all may be possible.

Given Angle	Conditions	Number of Triangles
Acute	Opposite side < altitude	0
Acute	Opposite side = altitude	1
Acute	Altitude < opposite side < adjacent side	2
Acute	Opposite side ≥ adjacent side	1
Obtuse	Opposite side ≤ adjacent side	0
Obtuse	Opposite side > altitude	1

Illustration by Thomson Digital

The figure shows the situation:

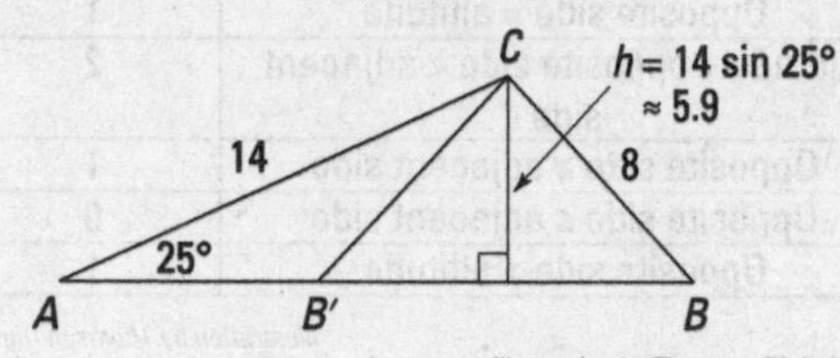

Illustration by Thomson Digital

Drawing an altitude from angle *C*, you're not sure whether angle *B* is obtuse (making side *a* fold left toward *A*) or if angle *B* is acute (making side *a* fold right). Or maybe neither works. To compute the altitude, use the following formula:

$$\sin 25° = \frac{h}{14}$$

$$h = 14 \cdot \sin 25 \approx 5.9$$

Because the altitude is less than the opposite side, you have the possibility of two different triangles.

Answers 601–700

644. Triangle 1: $m\angle B = 32.2°$, $m\angle C = 77.8°$, $b = 14.2$; Triangle 2: $m\angle B = 7.8°$, $m\angle C = 102.2°$, $b = 3.6$

First find the measure of angle *C* by using the Law of Sines:

$$\frac{\sin A}{a} = \frac{\sin C}{c}$$

$$\frac{\sin 70°}{25} = \frac{\sin C}{26}$$

$$\sin C = \frac{26 \sin 70°}{25}$$

$$m\angle C = \sin^{-1}\left(\frac{26 \sin 70°}{25}\right) \approx 77.8°$$

Now find the measure of angle *B*:

$$\begin{aligned} m\angle B &= 180° - (70° + 77.8°) \\ &= 180° - 147.8° \\ &= 32.2° \end{aligned}$$

Finally, find *b* using the Law of Sines. Always use as much of the given information as possible just in case you've made an error elsewhere.

$$\frac{a}{\sin A} = \frac{b}{\sin B}$$

$$\frac{25}{\sin 70°} = \frac{b}{\sin 32.2°}$$

$$b = \frac{25 \sin 32.2°}{\sin 70°} \approx 14.2$$

Because $\sin(180° - \theta) = \sin\theta$, there's another possible angle, C':

$$m\angle C' = 180° - 77.8° = 102.2°$$

This solution creates a second triangle, so find $m\angle B'$:

$$\begin{aligned} m\angle B' &= 180° - (70° + 102.2°) \\ &= 180° - 172.2° \\ &= 7.8° \end{aligned}$$

Now find b' by using the Law of Sines:

$$\frac{a}{\sin A} = \frac{b'}{\sin B'}$$

$$\frac{25}{\sin 70°} = \frac{b'}{\sin 7.8°}$$

$$b' = \frac{25 \sin 7.8°}{\sin 70°} \approx 3.6$$

Here are the two solutions, shown in the following diagram:

Triangle 1: $m\angle B = 32.2°$, $m\angle C = 77.8°$, $b = 14.2$

Triangle 2: $m\angle B = 7.8°$, $m\angle C = 102.2°$, $b = 3.6$

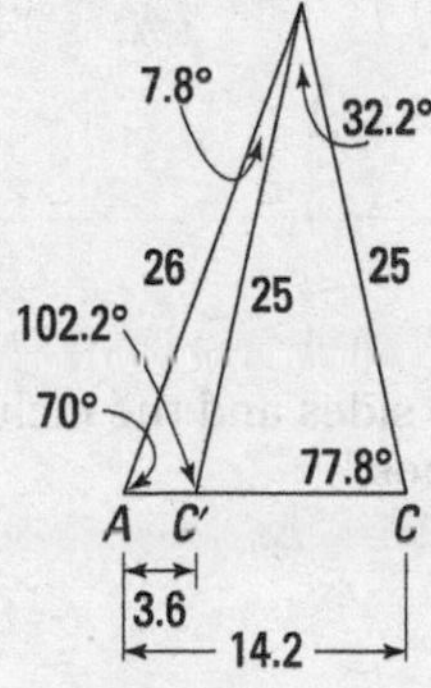

Illustration by Thomson Digital

645. no solution

First, find the measure of angle B by using the Law of Sines:

$$\frac{\sin A}{a}=\frac{\sin B}{b}$$
$$\frac{\sin 46^\circ}{8}=\frac{\sin B}{13}$$
$$\sin B=\frac{13\sin 46^\circ}{8}$$
$$B=\sin^{-1}\left(\frac{13\sin 46^\circ}{8}\right)$$

If you plug $\sin^{-1}\left(\frac{13\sin 46^\circ}{8}\right)$ into your calculator, you get an error message because $\frac{13\sin 46^\circ}{8}\approx 1.2$, and $\sin B$ must be between -1 and 1. Therefore, no triangle exists, and no solution is possible.

646. 17.4

This is an SAS problem (you're given two sides and the included angle), so find the missing side, *a*, by using the Law of Cosines:

$$a^2=b^2+c^2-2bc\cos A$$
$$a^2=19^2+26^2-2(19)(26)\cos 42^\circ$$
$$a=\sqrt{19^2+26^2-2(19)(26)\cos 42^\circ}\approx 17.4$$

647. 38.1

This is an SAS problem (you're given two sides and the included angle), so find the missing side, *c*, by using the Law of Cosines:

$$c^2=a^2+b^2-2ab\cos C$$
$$c^2=18^2+25^2-2(18)(25)\cos 124^\circ$$
$$c=\sqrt{18^2+25^2-2(18)(25)\cos 124^\circ}\approx 38.1$$

648. 19.0

This is an SAS problem (you're given two sides and the included angle), so find the missing side, *b*, by using the Law of Cosines:

$$b^2=a^2+c^2-2ac\cos B$$
$$b^2=25^2+36^2-2(25)(36)\cos 30^\circ$$
$$b=\sqrt{25^2+36^2-2(25)(36)\cos 30^\circ}\approx 19.0$$

649. 41.7°

This is an SSS problem (you're given all three sides), so find the measure of angle C by using the Law of Cosines:

$$c^2 = a^2 + b^2 - 2ab\cos C$$
$$25^2 = 37^2 + 32^2 - 2(37)(32)\cos C$$
$$\cos C = \frac{25^2 - 37^2 - 32^2}{-2(37)(32)}$$
$$C = \cos^{-1}\left(\frac{25^2 - 37^2 - 32^2}{-2(37)(32)}\right) \approx 41.7°$$

650. 137.0°

This is an SSS problem (you're given all three sides), so find the measure of angle A by using the Law of Cosines:

$$a^2 = b^2 + c^2 - 2bc\cos A$$
$$14^2 = 9^2 + 6^2 - 2(9)(6)\cos A$$
$$\cos A = \frac{14^2 - 9^2 - 6^2}{-2(9)(6)}$$
$$A = \cos^{-1}\left(\frac{14^2 - 9^2 - 6^2}{-2(9)(6)}\right) \approx 137.0°$$

651. 72.1°

This is an SSS problem (you're given all three sides), so find the measure of angle B by using the Law of Cosines:

$$b^2 = a^2 + c^2 - 2ac\cos B$$
$$20^2 = 17^2 + 17^2 - 2(17)(17)\cos B$$
$$\cos B = \frac{20^2 - 17^2 - 17^2}{-2(17)(17)}$$
$$B = \cos^{-1}\left(\frac{20^2 - 17^2 - 17^2}{-2(17)(17)}\right) \approx 72.1°$$

652. 347 yards

Always start by drawing and labeling a diagram to represent the problem:

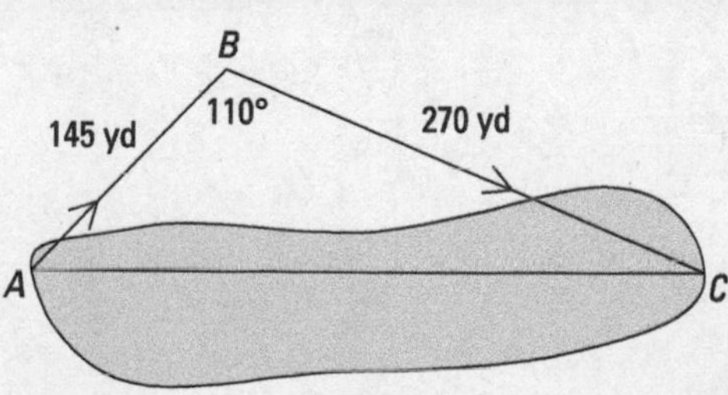

Illustration by Thomson Digital

Answers 601–700

The arrows show the direction that the surveyor is walking. He starts at A, walks 145 yards to B, turns right 110°, and then walks 270 yards to C.

The diagram shows that this is an SAS problem (you're given two sides and the included angle), so use the Law of Cosines to find AC, the length of the lake. Note that AC is opposite angle B, so it's b in the formula for the Law of Cosines.

$$b^2 = a^2 + c^2 - 2ac\cos B$$
$$b^2 = 145^2 + 270^2 - 2(145)(270)\cos 110°$$
$$b = \sqrt{145^2 + 270^2 - 2(145)(270)\cos 110°} \approx 347$$

653. 94°

Always start by drawing and labeling a diagram to represent the problem:

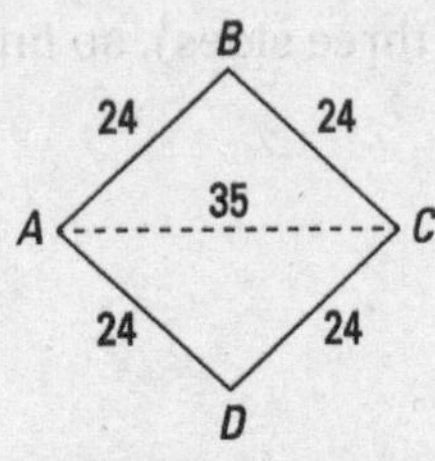

Illustration by Thomson Digital

The diagram shows that this is an SSS problem (you're given all three sides), so use the Law of Cosines to find the measure of angle B, which is the angle opposite the diagonal AC:

$$b^2 = a^2 + c^2 - 2ac\cos B$$
$$35^2 = 24^2 + 24^2 - 2(24)(24)\cos B$$
$$\cos B = \frac{35^2 - 24^2 - 24^2}{-2(24)(24)}$$
$$B = \cos^{-1}\left(\frac{35^2 - 24^2 - 24^2}{-2(24)(24)}\right) \approx 94°$$

Because 94° is an obtuse angle, it's the larger angle of the rhombus.

654. 88 feet

Always start by drawing and labeling a diagram to represent the problem:

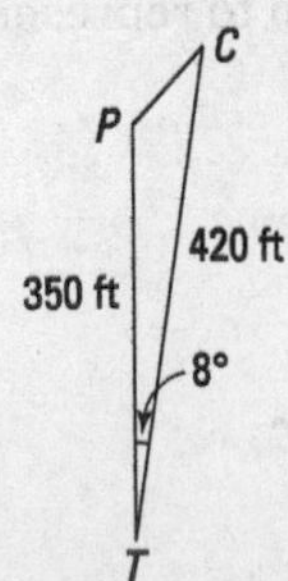

Illustration by Thomson Digital

Answers 601–700

The television camera is located at *T*, the player is originally located at *P*, and he catches the ball at *C*.

The diagram shows that this is an SAS problem (you're given two sides and the included angle), so use the Law of Cosines to find *PC*, the distance the center fielder has to run to catch the ball:

$$PC^2 = PT^2 + CT^2 - 2(PT)(CT)\cos T$$

$$PC^2 = 350^2 + 420^2 - 2(350)(420)\cos 8°$$

$$PC = \sqrt{350^2 + 420^2 - 2(350)(420)\cos 8°} \approx 88$$

The center fielder runs about 88 feet in order to catch the ball.

655. 53°

Always start by drawing and labeling a diagram to represent the problem:

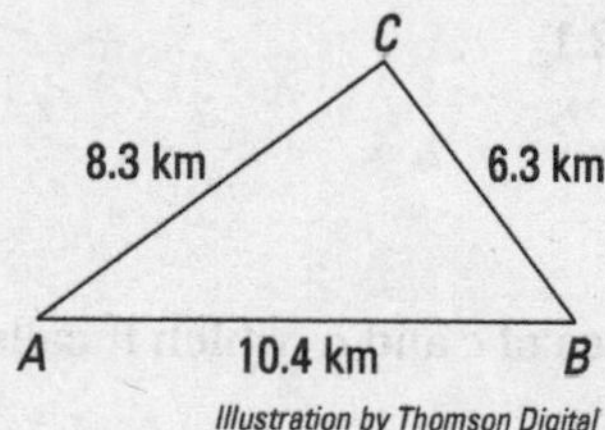

Illustration by Thomson Digital

The race starts at *A*, the next leg starts at *B*, and the final leg starts at *C*.

The diagram shows that this is an SSS problem (you're given all three sides), so use the Law of Cosines to find the measure of angle *B*, which is the angle between the first two legs of the race:

$$b^2 = a^2 + c^2 - 2ac\cos B$$

$$8.3^2 = 6.3^2 + 10.4^2 - 2(6.3)(10.4)\cos B$$

$$\cos B = \frac{8.3^2 - 6.3^2 - 10.4^2}{-2(6.3)(10.4)}$$

$$B = \cos^{-1}\left(\frac{8.3^2 - 6.3^2 - 10.4^2}{-2(6.3)(10.4)}\right) \approx 53°$$

The angle between the first two legs of the race is about 53°.

656. 106.9

You know the measures of two sides and the included angle, so use the formula $\text{Area} = \frac{1}{2}bc\sin A$ to find the area:

$$\begin{aligned}\text{Area} &= \frac{1}{2}bc\sin A \\ &= \frac{1}{2}\cdot 18\cdot 12\sin 98° \approx 106.9\end{aligned}$$

Answers 601–700

657. 120.6

You know the measures of two sides and the included angle, so use the formula $\text{Area} = \frac{1}{2}ac\sin B$ to find the area:

$$\begin{aligned}\text{Area} &= \frac{1}{2}ac\sin B\\ &= \frac{1}{2}\cdot 17\cdot 18\sin 52^\circ \approx 120.6\end{aligned}$$

658. 102.1

You know the measures of two sides and the included angle, so use the formula $\text{Area} = \frac{1}{2}ab\sin C$ to find the area:

$$\begin{aligned}\text{Area} &= \frac{1}{2}ab\sin C\\ &= \frac{1}{2}\cdot 16\cdot 13\sin 79^\circ \approx 102.1\end{aligned}$$

659. 24.0

The problem tells you the measures of c and a, which it calls AB and BC, respectively. So you know that $c = 6$ and $a = 8$.

Because you know the measures of two sides and the included angle, use the formula $\text{Area} = \frac{1}{2}ac\sin B$ to find the area:

$$\begin{aligned}\text{Area} &= \frac{1}{2}ac\sin B\\ &= \frac{1}{2}\cdot 8\cdot 6\sin 87^\circ \approx 24.0\end{aligned}$$

660. 13.7

The problem tells you the measures of b and c, which it calls AC and AB, respectively. So you know that $b = 4$ and $c = 7$.

Because you know the measures of two sides and the included angle, use the formula $\text{Area} = \frac{1}{2}bc\sin A$ to find the area:

$$\begin{aligned}\text{Area} &= \frac{1}{2}bc\sin A\\ &= \frac{1}{2}\cdot 4\cdot 7\sin 102^\circ \approx 13.7\end{aligned}$$

661. 173.9

The problem tells you the measures of b and a; it just calls them AC and BC, respectively. So you know that $b = 20$ and $a = 18$.

Because you know the measures of two sides and the included angle, use the formula $\text{Area}=\frac{1}{2}ab\sin C$ to find the area:

$$\begin{aligned}\text{Area}&=\frac{1}{2}ab\sin C\\&=\frac{1}{2}\cdot 18\cdot 20\sin 75°\approx 173.9\end{aligned}$$

662. 100 square meters

You know the measures of two sides and the included angle, so use the formula $\text{Area}=\frac{1}{2}ab\sin C$ to find the area:

$$\begin{aligned}\text{Area}&=\frac{1}{2}ab\sin C\\&=\frac{1}{2}\cdot 20\cdot 20\sin 30°\\&=200\left(\frac{1}{2}\right)\\&=100\end{aligned}$$

The area of the triangle is 100 square meters.

663. 224 square inches

Start by drawing and labeling a diagram to represent the problem:

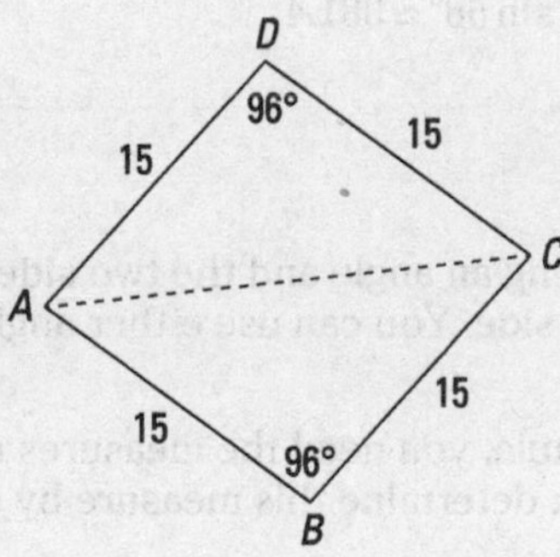

Illustration by Thomson Digital

When you draw the diagonal *AC*, you get two congruent triangles. To find the area of the rhombus, find the area of one of the triangles and then double it.

In triangle *ABC*, you know the measures of two sides and the included angle, so use the formula $\text{Area}=\frac{1}{2}ac\sin B$ to find the triangle's area:

$$\begin{aligned}\text{Area}&=\frac{1}{2}ac\sin B\\&=\frac{1}{2}\cdot 15\cdot 15\sin 96°\\&=224\left(\frac{1}{2}\right)\approx 112\end{aligned}$$

The area of the rhombus is about $112(2)=224$ square inches.

Answers 601–700

664. 381.4

To use the area formula involving an angle and the two sides forming that angle, you need the measure of one more side. You have angle A, so you need the measures of sides $AC = b$ and $AB = c$. You're missing AB, but you can determine this measure by using the Law of Sines.

First, find $m\angle C$, the only angle you don't have yet:

$$m\angle C = 180° - (58° + 43°) = 180° - 101° = 79°$$

Now use the Law of Sines to find AB:

$$\frac{b}{\sin B} = \frac{c}{\sin C}$$
$$\frac{25}{\sin 43°} = \frac{c}{\sin 79°}$$
$$c = \frac{25 \sin 79°}{\sin 43°}$$

Don't bother computing c. Just substitute $\frac{25 \sin 79°}{\sin 43°}$ for c in the area formula. That way, you carry all the decimal places through the calculation so that the answer is as accurate as possible.

Now find the area of the triangle by using $\text{Area} = \frac{1}{2}bc \sin A$:

$$\text{Area} = \frac{1}{2}bc \sin A$$
$$= \frac{1}{2} \cdot 25 \cdot \frac{25 \sin 79°}{\sin 43°} \cdot \sin 58° \approx 381.4$$

665. 72.4

To use the area formula involving an angle and the two sides forming that angle, you need the measure of one more side. You can use either angle B or angle C — your choice.

To use angle B in the area formula, you need the measures of sides $BC = a$ and $AB = c$. You're missing AB, but you can determine this measure by using the Law of Sines.

First, find $m\angle A$:

$$m\angle A = 180° - (101° + 53°) = 180° - 154° = 26°$$

Now use the Law of Sines to find AB:

$$\frac{a}{\sin A} = \frac{c}{\sin C}$$
$$\frac{9}{\sin 26°} = \frac{c}{\sin 53°}$$
$$c = \frac{9 \sin 53°}{\sin 26°}$$

Don't bother computing c. Just substitute $\frac{9 \sin 53°}{\sin 26°}$ for c in the area formula. That way, you carry all the decimal places through the calculation so that the answer is as accurate as possible.

Now find the area of the triangle by using Area$=\frac{1}{2}ac\sin B$:

$$\text{Area}=\frac{1}{2}ac\sin B$$
$$=\frac{1}{2}\cdot 9\cdot\frac{9\sin 53^\circ}{\sin 26^\circ}\cdot\sin 101^\circ\approx 72.4$$

666. 194.4

To use the area formula involving an angle and the two sides forming that angle, you need the measure of one more side. You can use either angle B or angle C — your choice.

To use angle C in the area formula, you need the measures of sides $BC=a$ and $AC=b$. You're missing AC, but you can determine this measure by using the Law of Sines.

First, find $m\angle A$:

$$m\angle A=180^\circ-(115^\circ+32^\circ)=180^\circ-147^\circ=33^\circ$$

Now use the Law of Sines to find AC:

$$\frac{a}{\sin A}=\frac{b}{\sin B}$$
$$\frac{21}{\sin 33^\circ}=\frac{b}{\sin 32^\circ}$$
$$b=\frac{21\sin 32^\circ}{\sin 33^\circ}$$

Don't bother computing c. Just substitute $\frac{21\sin 32^\circ}{\sin 33^\circ}$ for b in the area formula. That way, you carry all the decimal places through the calculation so that the answer is as accurate as possible.

Now find the area of the triangle by using Area$=\frac{1}{2}ab\sin C$:

$$A=\frac{1}{2}ab\sin C$$
$$=\frac{1}{2}\cdot 21\cdot\frac{21\sin 32^\circ}{\sin 33^\circ}\cdot\sin 115^\circ\approx 194.4$$

667. 50.7

To use the area formula involving an angle and the two sides forming that angle, you need the measure of one more side. You can use either angle A or angle C — your choice. Here I use angle C.

To use angle C in the area formula, you need the measures of sides $BC=a$ and $AC=b$. You're missing AC, but you can determine this measure by using the Law of Sines.

First, find $m\angle B$:

$$m\angle B=180^\circ-(69^\circ+47^\circ)=180^\circ-116^\circ=64^\circ$$

Now use the Law of Sines to find AC:

$$\frac{a}{\sin A}=\frac{b}{\sin B}$$
$$\frac{12}{\sin 69^\circ}=\frac{b}{\sin 64^\circ}$$
$$b=\frac{12\sin 64^\circ}{\sin 69^\circ}$$

Don't bother computing b. Just substitute $\frac{12\sin 64°}{\sin 69°}$ for b in the area formula. That way, you carry all the decimal places through the calculation so that the answer is as accurate as possible.

Now find the area of the triangle by using Area $= \frac{1}{2}ab\sin C$:

$$\begin{aligned}\text{Area} &= \frac{1}{2}ab\sin C\\ &= \frac{1}{2}\cdot 12\cdot \frac{12\sin 64°}{\sin 69°}\cdot \sin 47° \approx 50.7\end{aligned}$$

668. 219.2

To use the area formula involving an angle and the two sides forming that angle, you need the measure of angle B, the angle between the two given sides. You can do so by first solving for the measure of angle C — using the Law of Sines — and then subtracting the sum of the measures of angles A and C from 180.

Solve for the measure of angle C:

$$\begin{aligned}\frac{\sin A}{a} &= \frac{\sin C}{c}\\ \frac{\sin 95°}{33} &= \frac{\sin C}{16}\\ \sin C &= \frac{16\sin 95°}{33}\\ C &= \sin^{-1}\left(\frac{16\sin 95°}{33}\right)\end{aligned}$$

Don't bother computing C. Just plug $\sin^{-1}\left(\frac{16\sin 95°}{33}\right)$ into the formula to find the measure of angle B:

$$\begin{aligned}m\angle B &= 180° - \left(95° + \sin^{-1}\left(\frac{16\sin 95°}{33}\right)\right)\\ &= 85° - \sin^{-1}\left(\frac{16\sin 95°}{33}\right) \approx 56.12\end{aligned}$$

Now find the area of the triangle by using Area $= \frac{1}{2}ac\sin B$:

$$\begin{aligned}\text{Area} &= \frac{1}{2}ac\sin B\\ &= \frac{1}{2}\cdot 33\cdot 16\cdot \sin 56.12° \approx 219.2\end{aligned}$$

669. 279.0

To use the area formula involving an angle and the two sides forming that angle, you need the measure of angle C, the angle between the two given sides. You can do so by first solving for the measure of angle A — using the Law of Sines — and then subtracting the sum of the measures of angles A and B from 180.

Find the measure of angle *A*:

$$\frac{\sin A}{a}=\frac{\sin B}{b}$$
$$\frac{\sin A}{31}=\frac{\sin 30^\circ}{18}$$
$$\sin A=\frac{31\sin 30^\circ}{18}$$
$$A=\sin^{-1}\left(\frac{31\sin 30^\circ}{18}\right)$$

Don't bother computing *A*. Just plug $\sin^{-1}\left(\frac{31\sin 30^\circ}{18}\right)$ into the formula to find the measure of angle *C*:

$$\begin{aligned}m\angle C&=180^\circ-\left(30^\circ+\sin^{-1}\left(\frac{31\sin 30^\circ}{18}\right)\right)\\&=150^\circ-\sin^{-1}\left(\frac{31\sin 30^\circ}{18}\right)\approx 90.56^\circ\end{aligned}$$

Now find the area of the triangle by using $\text{Area}=\frac{1}{2}ab\sin C$:

$$\begin{aligned}\text{Area}&=\frac{1}{2}ab\sin C\\&=\frac{1}{2}\cdot 31\cdot 18\cdot \sin 90.56^\circ\approx 279.0\end{aligned}$$

670. 31.3 square inches

In triangle *BCD*, you're given an angle and an adjacent side. You need to find the length of the other adjacent side, *BD*, to find the area of the triangle.

Look at triangle *ABD*. It's a 30°-60°-90° triangle, and you know the length of the hypotenuse. *BD* is the shorter leg, so

$$BD=\frac{1}{2}AD=\frac{1}{2}\cdot 16=8\text{ in.}$$

Now find the area of the triangle by using $\text{Area}=\frac{1}{2}BD\cdot CD\sin D$:

$$\begin{aligned}\text{Area}&=\frac{1}{2}BD\cdot CD\sin D\\&=\frac{1}{2}\cdot 8\cdot 14\sin 34^\circ\approx 31.3\end{aligned}$$

The area of triangle *BCD* is about 31.3 square inches

671. $12\sqrt{5}$

This is an SSS problem (you're given all three sides of the triangle), so use Heron's formula, $\text{Area}=\sqrt{s(s-a)(s-b)(s-c)}$, to find the area.

In Heron's formula, *s* is the semiperimeter (half the perimeter), represented by $s=\frac{1}{2}(a+b+c)$. The semiperimeter of triangle *PQR* is

$$s=\frac{7+8+9}{2}=12$$

Now substitute the value for *s* and the lengths of the sides into the area formula:

$$\begin{aligned}\text{Area}&=\sqrt{s(s-a)(s-b)(s-c)}=\sqrt{12(12-7)(12-8)(12-9)}\\&=\sqrt{12(5)(4)(3)}=\sqrt{720}=\sqrt{144\cdot 5}=12\sqrt{5}\end{aligned}$$

672. $6\sqrt{10}$

This is an SSS problem (you're given all three sides of the triangle), so use Heron's formula, Area $=\sqrt{s(s-a)(s-b)(s-c)}$, to find the area.

In Heron's formula, s is the semiperimeter (half the perimeter), represented by $s=\frac{1}{2}(a+b+c)$. The semiperimeter of triangle ABC is

$$s=\frac{7+6+11}{2}=12$$

Now substitute the value for s and the lengths of the sides into the formula:

$$\begin{aligned}\text{Area}&=\sqrt{s(s-a)(s-b)(s-c)}=\sqrt{12(12-7)(12-6)(12-11)}\\&=\sqrt{12(5)(6)(1)}=\sqrt{360}=\sqrt{36\cdot10}=6\sqrt{10}\end{aligned}$$

673. $30\sqrt{3}$

This is an SSS problem (you're given all three sides of the triangle), so use Heron's formula, Area $=\sqrt{s(s-a)(s-b)(s-c)}$, to find the area.

In Heron's formula, s is the semiperimeter (half the perimeter), represented by $s=\frac{1}{2}(a+b+c)$. The semiperimeter of triangle XYZ is

$$s=\frac{8+13+15}{2}=18$$

Now substitute the value for s and the lengths of the sides into the formula:

$$\begin{aligned}\text{Area}&=\sqrt{s(s-a)(s-b)(s-c)}=\sqrt{18(18-15)(18-8)(18-13)}\\&=\sqrt{18(3)(10)(5)}=\sqrt{2{,}700}=\sqrt{900\cdot3}=30\sqrt{3}\end{aligned}$$

674. 21.3

This is an SSS problem (you're given all three sides of the triangle), so use Heron's formula, Area $=\sqrt{s(s-a)(s-b)(s-c)}$, to find the area.

In Heron's formula, s is the semiperimeter (half the perimeter), represented by $s=\frac{1}{2}(a+b+c)$. The semiperimeter of triangle ABC is

$$s=\frac{6+8+12}{2}=13$$

Now substitute the value for s and the lengths of the sides into the formula:

$$\begin{aligned}\text{Area}&=\sqrt{s(s-a)(s-b)(s-c)}=\sqrt{13(13-6)(13-8)(13-12)}\\&=\sqrt{13(7)(5)(1)}=\sqrt{455}\approx21.3\end{aligned}$$

675. 25.7

This is an SSS problem (you're given all three sides of the triangle), so use Heron's formula, Area $=\sqrt{s(s-a)(s-b)(s-c)}$, to find the area.

In Heron's formula, s is the semiperimeter (half the perimeter), represented by $s=\frac{1}{2}(a+b+c)$. The semiperimeter of triangle ABC is

$$s=\frac{13+4+13}{2}=15$$

Now substitute the value for s and the lengths of the sides into the formula:

$$\begin{aligned}\text{Area}&=\sqrt{s(s-a)(s-b)(s-c)}=\sqrt{15(15-13)(15-4)(15-13)}\\&=\sqrt{15(2)(11)(2)}=\sqrt{660}\approx 25.7\end{aligned}$$

676. 26.5

This is an SSS problem (you're given all three sides of the triangle), so use Heron's formula, $\text{Area}=\sqrt{s(s-a)(s-b)(s-c)}$, to find the area.

In Heron's formula, s is the semiperimeter (half the perimeter), represented by $s=\frac{1}{2}(a+b+c)$. The semiperimeter of triangle ABC is

$$s=\frac{5+15+12}{2}=16$$

Now substitute the value for s and the lengths of the sides into the formula:

$$\begin{aligned}\text{Area}&=\sqrt{s(s-a)(s-b)(s-c)}=\sqrt{16(16-5)(16-15)(16-12)}\\&=\sqrt{16(11)(1)(4)}=\sqrt{704}\approx 26.5\end{aligned}$$

677. 37,382 square feet

This triangle is an SSS problem (you're given all three sides), so use Heron's formula, $\text{Area}=\sqrt{s(s-a)(s-b)(s-c)}$, to find the area.

In Heron's formula, s is the semiperimeter (half the perimeter), represented by $s=\frac{1}{2}(a+b+c)$. The semiperimeter of the given triangle is

$$s=\frac{375+250+300}{2}=462.5$$

Now substitute the value for s and the lengths of the sides into the formula:

$$\begin{aligned}\text{Area}&=\sqrt{s(s-a)(s-b)(s-c)}\\&=\sqrt{462.5(462.5-375)(462.5-250)(462.5-300)}\\&=\sqrt{462.5(87.5)(212.5)(162.5)}\\&=\sqrt{1{,}397{,}436{,}523}\approx 37{,}382\end{aligned}$$

The area of the plot is about 37,382 square feet.

678. 431,600 square miles

This triangle is an SSS problem (you're given all three sides), so use Heron's formula, $\text{Area}=\sqrt{s(s-a)(s-b)(s-c)}$, to find the area.

In Heron's formula, s is the semiperimeter (half the perimeter), represented by $s=\frac{1}{2}(a+b+c)$. The semiperimeter of the given triangle is

$$s=\frac{1{,}040+1{,}000+960}{2}=1{,}500$$

Now substitute the value for s and the lengths of the sides into the formula:

$$\begin{aligned}\text{Area}&==\sqrt{s(s-a)(s-b)(s-c)}\\&=\sqrt{1{,}500(1{,}500-1{,}040)(1{,}500-1{,}000)(1{,}500-960)}\\&=\sqrt{1{,}500(460)(500)(540)}\approx 431{,}600\end{aligned}$$

The area of the Bermuda Triangle is about 431,600 square miles.

Note: When you multiply 1,500(460)(500)(540) on a calculator, you may get an answer like 1.863E11. The product just has more digits than the calculator can handle. One way to deal with this issue is to simplify the radical first:

$$\begin{aligned}\text{Area}&=\sqrt{1{,}500(460)(500)(540)}=\sqrt{(100\cdot 15)(4\cdot 5\cdot 23)(100\cdot 5)(36\cdot 15)}\\&=10^2\cdot 2\cdot 6\cdot 15\cdot 5\sqrt{23}=90{,}000\sqrt{23}\approx 431{,}600\end{aligned}$$

679. 32 square inches

For triangle *BCD*, the diagram shows that $BD=BC$ and $CD=12$ inches. Triangle *ABD* is a 30°-60°-90° triangle with hypotenuse $AD=16$ in. In a 30°-60°-90° triangle, the length of the side opposite the 30° angle is half the length of the hypotenuse, so $BD=8$ inches. That means that $BC=8$ inches as well.

Triangle *BCD* is an SSS problem (you're given all three sides), so use Heron's formula, $\text{Area}=\sqrt{s(s-a)(s-b)(s-c)}$, to find the area.

In Heron's formula, s is the semiperimeter (half the perimeter), represented by $s=\frac{1}{2}(a+b+c)$. The semiperimeter of the given triangle is

$$s=\frac{8+8+12}{2}=14$$

Now substitute the value for s and plug the lengths of the sides into the formula:

$$\begin{aligned}A&=\sqrt{s(s-a)(s-b)(s-c)}\\&=\sqrt{14(14-8)(14-8)(14-12)}\\&=\sqrt{14(6)(6)(2)}\approx 32\end{aligned}$$

The area of triangle *BCD* is about 32 square inches.

680. 12 square units

First, find the length of each side of the triangle using the distance formula $d=\sqrt{(x_1-x_2)^2+(y_1-y_2)^2}$:

$$
\begin{aligned}
d_{AB} &= \sqrt{(-3-1)^2+(1-3)^2} \\
&= \sqrt{(-4)^2+(-2)^2} \\
&= \sqrt{16+4} \\
&= \sqrt{20} \\
&= 2\sqrt{5} \\
d_{BC} &= \sqrt{(1-3)^2+(3-(-2))^2} \\
&= \sqrt{(-2)^2+5^2} \\
&= \sqrt{4+25} \\
&= \sqrt{29} \\
d_{AC} &= \sqrt{(-3-3)^2+(1-(-2))^2} \\
&= \sqrt{(-6)^2+3^2} \\
&= \sqrt{36+9} \\
&= \sqrt{45} \\
&= 3\sqrt{5}
\end{aligned}
$$

This triangle is an SSS problem (you're given all three sides), so use Heron's formula, $\text{Area} = \sqrt{s(s-a)(s-b)(s-c)}$, to find the area.

In Heron's formula, s is the semiperimeter (half the perimeter), represented by $s = \frac{1}{2}(a+b+c)$. The semiperimeter of the given triangle is

$$s = \frac{2\sqrt{5}+\sqrt{29}+3\sqrt{5}}{2} = \frac{5\sqrt{5}+\sqrt{29}}{2}$$

Now substitute the value for s and the lengths of the sides into the formula:

$$
\begin{aligned}
\text{Area} &= \sqrt{s(s-a)(s-b)(s-c)} \\
&= \sqrt{\frac{5\sqrt{5}+\sqrt{29}}{2}\left(\frac{5\sqrt{5}+\sqrt{29}}{2}-2\sqrt{5}\right)\left(\frac{5\sqrt{5}+\sqrt{29}}{2}-\sqrt{29}\right)\left(\frac{5\sqrt{5}+\sqrt{29}}{2}-3\sqrt{5}\right)} \\
&= \sqrt{\frac{5\sqrt{5}+\sqrt{29}}{2}\left(\frac{5\sqrt{5}+\sqrt{29}-4\sqrt{5}}{2}\right)\left(\frac{5\sqrt{5}+\sqrt{29}-2\sqrt{29}}{2}\right)\left(\frac{5\sqrt{5}+\sqrt{29}-6\sqrt{5}}{2}\right)}
\end{aligned}
$$

This result looks really complicated, but it simplifies as follows:

$$
\begin{aligned}
\text{Area} &= \sqrt{\frac{5\sqrt{5}+\sqrt{29}}{2}\left(\frac{\sqrt{5}+\sqrt{29}}{2}\right)\left(\frac{5\sqrt{5}-\sqrt{29}}{2}\right)\left(\frac{-\sqrt{5}+\sqrt{29}}{2}\right)} \\
&= \sqrt{\frac{5\sqrt{5}+\sqrt{29}}{2}\left(\frac{5\sqrt{5}-\sqrt{29}}{2}\right)\left(\frac{\sqrt{29}+\sqrt{5}}{2}\right)\left(\frac{\sqrt{29}-\sqrt{5}}{2}\right)}
\end{aligned}
$$

The numerators have the form $(a+b)(a-b)$. This expression is the special product $(a+b)(a-b)=a^2-b^2$, so you can rewrite the equation as follows:

$$\text{Area}=\sqrt{\frac{5\sqrt{5}+\sqrt{29}}{2}\left(\frac{5\sqrt{5}-\sqrt{29}}{2}\right)\left(\frac{\sqrt{29}+\sqrt{5}}{2}\right)\left(\frac{\sqrt{29}-\sqrt{5}}{2}\right)}$$

$$=\sqrt{\left(\frac{\left(5\sqrt{5}\right)^2-\left(\sqrt{29}\right)^2}{4}\right)\left(\frac{\left(\sqrt{29}\right)^2-\left(\sqrt{5}\right)^2}{4}\right)}$$

$$=\sqrt{\left(\frac{25\cdot 5-29}{4}\right)\left(\frac{29-5}{4}\right)}=\sqrt{\frac{96}{4}\cdot\frac{24}{4}}=\sqrt{24\cdot 6}=\sqrt{144}=12$$

The area of the triangle is 12 square units. (Of course, you can just use a calculator!)

681. 55.6 yards

Always start by drawing and labeling a diagram to represent the problem:

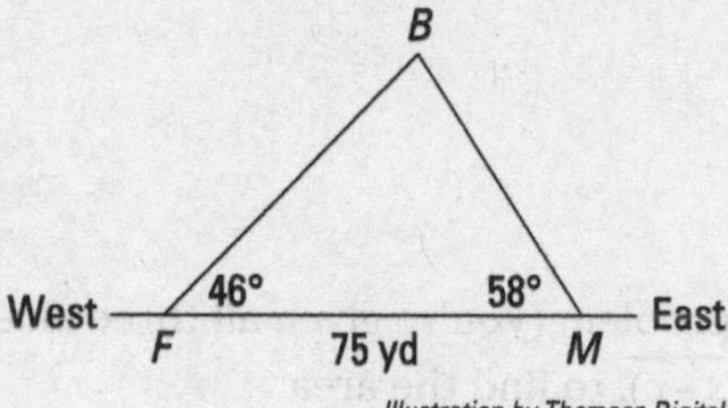

Illustration by Thomson Digital

The woman is located at *F*, the man is located at *M*, and the balloon is located at *B*.

You can see from the diagram that you know two angles and the side between them, so this is an ASA problem. You know the length of side *b*, so you need to find the measure of angle *B*:

$$\begin{aligned} m\angle B &= 180° - (m\angle A + m\angle C) \\ &= 180° - (46° + 58°) \\ &= 180° - 104° \\ &= 76° \end{aligned}$$

The side from point *M* to point *B* is *f*, so use the Law of Sines and solve for *f*:

$$\frac{f}{\sin F}=\frac{b}{\sin B}$$

$$\frac{f}{\sin 46°}=\frac{75}{\sin 76°}$$

$$f=\frac{75\sin 46°}{\sin 76°}\approx 55.6$$

The distance from the man to the balloon is approximately 55.6 yards.

682. 191 meters

The diagram shows that you know two sides and a consecutive angle, so this is an SSA problem. First, find the measure of angle A:

$$\frac{\sin A}{a}=\frac{\sin C}{c}$$
$$\frac{\sin A}{46}=\frac{\sin 68°}{71}$$
$$\sin A=\frac{46\sin 68°}{71}$$
$$A=\sin^{-1}\left(\frac{46\sin 68°}{71}\right)\approx 36.9°$$

Next, find the measure of angle B:

$$\begin{aligned} m\angle B&=180°-(68°+36.9°)\\ &=180°-104.9°\\ &=75.1° \end{aligned}$$

Now use the Law of Sines to find the third side, AC:

$$\frac{b}{\sin B}=\frac{c}{\sin C}$$
$$\frac{b}{\sin 75.1°}=\frac{71}{\sin 68°}$$
$$b=\frac{71\sin 75.1°}{\sin 68°}\approx 74$$

The question asks for the amount of fencing needed to enclose the entire park, so find the perimeter of the park:

$$P=71+46+74=191$$

The park needs 191 meters of fencing.

683. 29.6 miles

Always start by drawing and labeling a diagram to represent the problem:

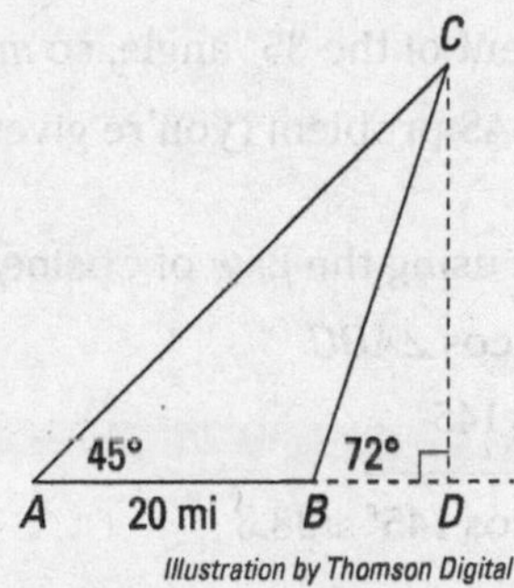

Illustration by Thomson Digital

The tracking stations are located at A and B, and the rocket is located at C. C is located where shown in the diagram because the problem specifically says that the rocket isn't between the two stations. The altitude of the rocket is CD. Triangle ADC is a 45°-45°-90° triangle, so when you know the length of the hypotenuse, AC, you can easily find CD.

Angle *ABC* is supplementary to the 72° angle in triangle *BDC*, so its measure is

$$m\angle ABC = 180° - 72° = 108°$$

Now you know two of the three angles in triangle *ABC*, so you can subtract them from 180° to find the remaining angle:

$$m\angle ACB = 180° - (45° + 108°) = 27°$$

With this information, you can use the Law of Sines to find *AC* in triangle *ABC*:

$$\frac{AC}{\sin \angle ABC} = \frac{AB}{\sin \angle ACB}$$
$$\frac{AC}{\sin 108°} = \frac{20}{\sin 27°}$$
$$AC = \frac{20 \sin 108°}{\sin 27°}$$

Because you'll need to use this result to find the hypotenuse of triangle *ADC*, don't compute the result yet. (Not simplifying yet ensures a more accurate answer later because you don't lose any decimal places.)

Remember that in a 45°-45°-90° triangle, the length of a leg is equal to the length of the hypotenuse divided by $\sqrt{2}$:

$$CD = \frac{\frac{20 \sin 108°}{\sin 27°}}{\sqrt{2}} = \frac{20 \sin 108°}{\sqrt{2} \sin 27°} \approx 29.6$$

The rocket is about 29.6 miles above the earth.

684. 38.3 miles

Always start by drawing and labeling a diagram to represent the problem:

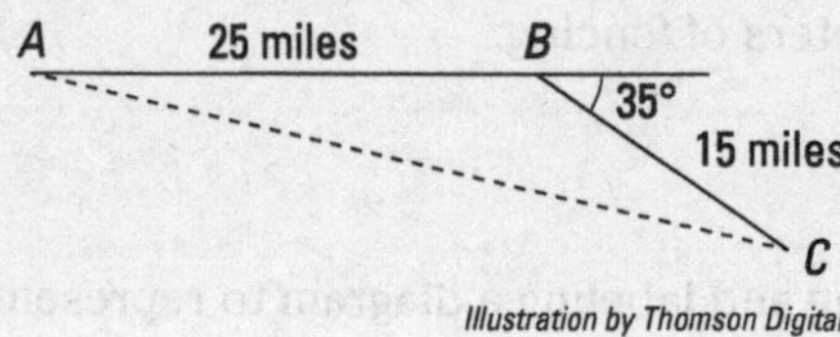

Illustration by Thomson Digital

First, find $m\angle ABC$. It's the supplement of the 35° angle, so $m\angle ABC = 180° - 35° = 145°$.

The diagram shows that this is an SAS problem (you're given two sides and the included angle).

You could use the Law of Sines, but using the Law of Cosines is less work:

$$AC^2 = AB^2 + BC^2 - 2(AB)(BC)\cos \angle ABC$$
$$AC^2 = 25^2 + 15^2 - 2(25)(15)\cos 145°$$
$$AC = \sqrt{25^2 + 15^2 - 2(25)(15)\cos 145°} \approx 38.3$$

The boat is about 38.3 miles from port.

685. 3.6 centimeters

Always start by drawing and labeling a diagram to represent the problem:

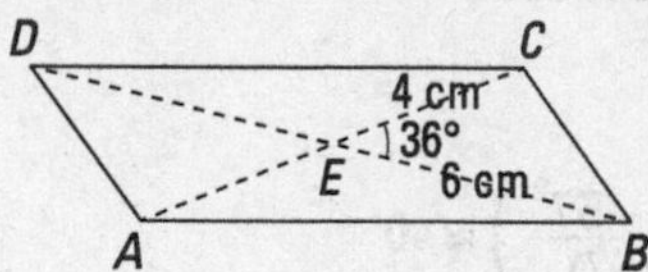

Illustration by Thomson Digital

Remember that the opposite sides of a parallelogram are equal and that the diagonals of a parallelogram bisect each other. So $EC = 4$ centimeters and $EB = 6$ centimeters. You're looking for BC.

The diagram shows that this is an SAS problem (you're given two sides and the included angle). You could use the Law of Sines, but using the Law of Cosines is less work:

$$BC^2 = EB^2 + EC^2 - 2\,(EB)\,(EC)\cos\angle CEB$$

$$BC^2 = 6^2 + 4^2 - 2(6)(4)\cos 36°$$

$$BC = \sqrt{6^2 + 4^2 - 2(6)(4)\cos 36°} \approx 3.6$$

The short sides are about 3.6 centimeters.

686. 40°

Always start by drawing and labeling a diagram to represent the problem:

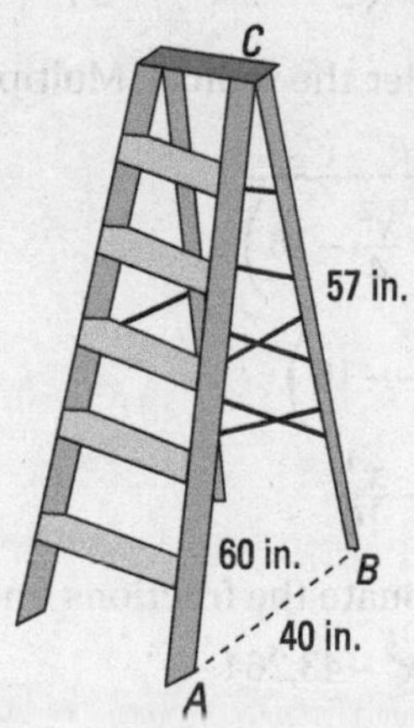

Illustration by Thomson Digital

The diagram shows that this is an SSS problem (you're given all three sides). You're looking for the measure of angle C, so use the Law of Cosines:

$$c^2 = a^2 + b^2 - 2ab\cos\cos C$$
$$40^2 = 60^2 + 57^2 - 2(60)(57)\cos C$$
$$\cos C = \frac{40^2 - 60^2 - 57^2}{-2(60)(57)}$$
$$C = \cos^{-1}\left(\frac{40^2 - 60^2 - 57^2}{-2(60)(57)}\right) \approx 40°$$

The angle between the legs when the ladder is open measures about 40°.

687. 10

To solve this problem, you set the triangle's perimeter equal to its area, using Heron's formula, Area $= \sqrt{s(s-a)(s-b)(s-c)}$, to find the area. In Heron's formula, s is the semiperimeter (half the perimeter). The perimeter of the triangle is $9+17+x$, and half of that is $\frac{26+x}{2}$.

Here's the equation to solve:

$$\text{Perimeter} = \text{Area}$$
$$9+17+x = \sqrt{\frac{26+x}{2}\left(\frac{26+x}{2}-9\right)\left(\frac{26+x}{2}-17\right)\left(\frac{26+x}{2}-x\right)}$$
$$26+x = \sqrt{\left(13+\frac{x}{2}\right)\left(13+\frac{x}{2}-9\right)\left(13+\frac{x}{2}-17\right)\left(13+\frac{x}{2}-x\right)}$$
$$26+x = \sqrt{\left(13+\frac{x}{2}\right)\left(\frac{x}{2}+4\right)\left(\frac{x}{2}-4\right)\left(13-\frac{x}{2}\right)}$$

Notice the nice pairs of binomials under the radical. Multiply them together and then square both sides of the equation:

$$26+x = \sqrt{\left(169-\frac{x^2}{4}\right)\left(\frac{x^2}{4}-16\right)}$$
$$(26+x)^2 = \left(169-\frac{x^2}{4}\right)\left(\frac{x^2}{4}-16\right)$$
$$676+52x+x^2 = 185\cdot\frac{x^2}{4} - 2{,}704 - \frac{x^4}{16}$$

Now multiply each term by 16 to eliminate the fractions and then simplify:

$$10{,}816 + 832x + 16x^2 = 740x^2 - 43{,}264 - x^4$$
$$x^4 - 724x^2 + 832x + 54{,}080 = 0$$

Yes, this is a fourth-degree polynomial that you can solve using the rational root theorem, but you're interested only in finding a solution that could be the side of a triangle. The *triangle inequality theorem* says that any side of a triangle must be shorter than the other two sides added together. So if two of the sides of the triangle are 9 and 17, the third side can be any length between and including 9 and 25. This is assuming, of course, that the measure is an integer.

Answers 601–700

Looking at the polynomial equation, you see from the coefficient of 1 on the x^4 term that any rational roots will be integers. Now look at the constant term. The factors of 54,080 include 1, 2, 4, 5, 8, 10, 13, 16, 20, and so on. These are some of the possible roots/solutions of the equation. The four numbers between 9 and 25 are 10, 13, 16, and 20, so check those four factors. Replacing the *x*'s with 10 in the equation, you have a solution! Neither 13, 16, nor 20 works. So the third side of the triangle measures 10 units.

688. 305,782 square meters

The field consists of two triangles, so find the area of each and add them together to find the area of the field.

In triangle *PQR*, you know the length of one side and the measures of two angles, but the given side isn't between the two angles, and that's the side you need to know to use the area formula. Because you know *PQ* and the measures of $\angle RPQ$ and $\angle QRP$, you need to find *PR* to find the area of the triangle. First, find $m\angle Q$:

$$m\angle Q = 180° - (83° + 56°) = 180° - 139° = 41°$$

Now use the Law of Sines to find *PR*:

$$\frac{PR}{\sin \angle Q} = \frac{PQ}{\sin \angle QRP}$$

$$\frac{PR}{\sin 41°} = \frac{650}{\sin 56°}$$

$$PR = \frac{650 \sin 41°}{\sin 56°}$$

Don't bother computing *PR*. Just substitute $\frac{650 \sin 41°}{\sin 56°}$ for *PR* in the area formula. That way, you carry all the decimal places through the calculation so that the answer is as accurate as possible.

Now find the area of the triangle:

$$\text{Area} = \frac{1}{2}(PR)(PQ)\sin \angle QRP$$

$$= \frac{1}{2}\left(\frac{650 \sin 41°}{\sin 56°}\right)(650)\sin 83° \approx 165{,}927$$

Now you have to find the area of triangle *PSR*. You know the measures of two sides, $RS = 600$ meters and $PR = \frac{650 \sin 41°}{\sin 56°}$, as well as the included angle $PRS = 65°$:

$$\text{Area} = \frac{1}{2}(PR)(RS)\sin \angle PRS$$

$$= \frac{1}{2}\left(\frac{650 \sin 41°}{\sin 56°}\right)(600)\sin 65° \approx 139{,}855$$

The total area of the field is $165{,}927 + 139{,}855 = 305{,}782$ square meters.

689. 5.9 miles

Remember that distance = rate × time. So after three hours, Marie has walked $3(3) = 9$ miles and Don has walked $4(3) = 12$ miles. Now draw a diagram to represent the problem. *S* represents the starting point of both Don and Marie, *M* is Marie's destination, and *D* is Don's stopping point.

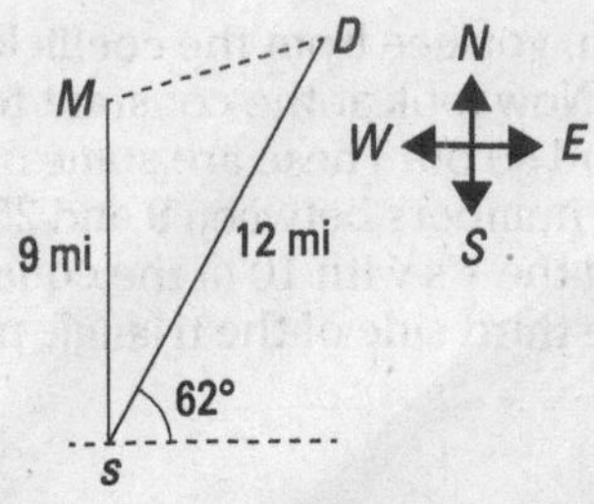

Illustration by Thomson Digital

You need to find length *MD*. In order to do that, first find $m\angle MSD$. Marie is walking due north, which forms a 90° angle with the baseline. Don is walking 62° northeast, so $m\angle MSD = 90° - 62° = 28°$. Now you have an SAS problem (you know two sides and the included angle), so use the Law of Cosines to find *MD*:

$$MD^2 = MS^2 + DS^2 - 2(MS)(DS)\cos\angle MSD$$

$$MD^2 = 9^2 + 12^2 - 2(9)(12)\cos 28°$$

$$MD = \sqrt{9^2 + 12^2 - 2(9)(12)\cos 28°} \approx 5.9$$

They're about 5.9 miles apart after three hours.

690. 309.0 square yards

You can divide the regular octagon into eight congruent triangles. Find the area of one of the triangles and multiply by 8.

Each triangle in the octagon has an angle at the center measuring 45°. Dividing a triangle in half at that center, you have a right triangle with an angle of 22.5° and opposite side measuring 4 yards. Use the tangent of 22.5° to solve for the adjacent side of the triangle, *h*:

$$\tan 22.5° = \frac{4}{h}$$

$$h = \frac{4}{\tan 22.5°} \approx 9.6569$$

Now find the area of one of the eight triangles, which has a base of 8 yards and a height of 9.6569 yards:

$$A = \frac{1}{2}bh = \frac{1}{2} \cdot 8 \cdot 9.6569 = 38.6276$$

So the area of the entire octagon is 8 times that area:

$$8(38.6276) = 309.0208 \text{ square yards}$$

You could do this problem using Heron's formula, which involves finding the length of one of the segments drawn from the center to a vertex of the octagon. That process would involve using the square root. You could also use the trig formula involving the sine of the angle and measures of two sides — the height and the segment to the vertex. That one is too much work. The method that uses right triangles seems to be the easiest. And, finally, if you have the formula in your back pocket, you can find the area of any regular polygon using Area $= \frac{1}{2}$(perimeter)(apothem), where the apothem is the height of the triangle. So many choices!

691. i

Rewrite the exponent as the sum of a multiple of 4 and a number between 0 and 3:

$$i^9 = i^{2(4)+1}$$

Now write the power of i as the product of two powers:

$$i^{2(4)} \cdot i^1$$

The value of i^{4n} is 1, so i^9 is $1 \cdot i^1 = i$.

692. -1

Rewrite the exponent as the sum of a multiple of 4 and a number between 0 and 3:

$$i^{42} = i^{10(4)+2}$$

Now write the power of i as the product of two powers:

$$i^{10(4)} \cdot i^2$$

The value of i^{4n} is 1, and $i^2 = \left(\sqrt{-1}\right)^2 = -1$, so i^{42} is $1 \cdot i^2 = 1(-1) = -1$.

693. 1

Rewrite the exponent as the sum of a multiple of 4 and a number between 0 and 3:

$$i^{100} = i^{25(4)+0}$$

Now write the power of i as the product of two powers:

$$i^{25(4)} \cdot i^0$$

The value of i^{4n} is 1, and $i^0 = 1$, so i^{100} is $1 \cdot 1 = 1$.

694. i

Rewrite the exponent as the sum of a multiple of 4 and a number between 0 and 3:

$$i^{301} = i^{75(4)+1}$$

Now write the power of i as the product of two powers:

$$i^{75(4)} \cdot i^1$$

The value of i^{4n} is 1, so $i^{301} = 1 \cdot i^1 = i$.

695. $-i$

Rewrite the exponent as the sum of a multiple of 4 and a number between 0 and 3:

$$i^{4,003} = i^{1,000(4)+3}$$

Now write the power of i as the product of two powers:

$$i^{1,000(4)} \cdot i^3$$

The value of i^{4n} is 1, and $i^3 = i^2 \cdot i = \left(\sqrt{-1}\right)^2 \cdot i = -1 \cdot i = -i$, so $i^{4,003}$ is $1(-i) = -i$.

696. $1-i$

Add the two real numbers together, and add the two imaginary numbers together:

$$3+(-2)=1$$

$$4i+(-5i)=-1i$$

The answer is $1-i$.

697. $-9+4i$

Subtract the two real numbers, and subtract the two imaginary numbers:

$$-4-5=-9$$

$$-2i-(-6i)=-2i+6i=4i$$

The answer is $-9+4i$.

698. 3

Add the first two real numbers and subtract the third from the sum:

$$4+5-6=9-6=3$$

Also add the first two imaginary numbers together and subtract the third from the sum:

$$3i+(-2i)-i=i-i=0$$

The answer is $3+0=3$.

699. $15-8i$

Add the first two real numbers and add the third to the sum:

$$8+3+4=11+4=15$$

Also add the first two imaginary numbers together:

$$-5i+(-3i)=-8i$$

The answer is $15-8i$.

700. $13-5i$

Subtract the second real number from the first:

$$6-(-7)=6+7=13$$

Subtract the second imaginary from the first and then subtract the third imaginary number from the result:

$$4i-3i-6i=i-6i=-5i$$

The answer is $13-5i$.

701. $11-3i\sqrt{3}$

Add the two real numbers together, and add the two imaginary numbers together:

$$5+6=11$$

$$1i\sqrt{3}+\left(-4i\sqrt{3}\right)=-3i\sqrt{3}$$

The answer is $11-3i\sqrt{3}$.

702. $2+5i\sqrt{2}$

You can't combine the *i* terms unless the radicals match, so first simplify $\sqrt{8}$:

$$\sqrt{8}=\sqrt{4\cdot 2}=2\sqrt{2}$$

The problem now reads

$$\left(6+2i\sqrt{2}\right)-\left(4-3i\sqrt{2}\right)$$

Subtract the two real numbers, and subtract the two imaginary numbers:

$$6-4=2$$

$$2i\sqrt{2}-\left(-3i\sqrt{2}\right)=2i\sqrt{2}+3i\sqrt{2}=5i\sqrt{2}$$

The answer is $2+5i\sqrt{2}$.

703. $-7+i\sqrt{10}$

First simplify $\sqrt{40}$ and $\sqrt{90}$ so you can combine the *i* terms:

$$\sqrt{40}=\sqrt{4\cdot 10}=2\sqrt{10}$$

$$\sqrt{90}=\sqrt{9\cdot 10}=3\sqrt{10}$$

The problem now reads

$$\left(-5-2i\sqrt{10}\right)+\left(-2+3i\sqrt{10}\right)$$

Add the two real numbers together, and add the two imaginary numbers together:

$$-5+(-2)=-7$$

$$-2i\sqrt{10}+3i\sqrt{10}=1i\sqrt{10}$$

The answer is $-7+i\sqrt{10}$.

704. $3+2i$

Subtract the two real numbers, and subtract the two imaginary numbers:

$$-5-(-8)=-5+8=3$$

$$\frac{3}{4}i-\left(-\frac{5}{4}i\right)=\frac{3}{4}i+\frac{5}{4}i=\frac{8}{4}i=2i$$

The answer is $3+2i$.

705. $1-\frac{33}{20}i$

Add the two real numbers together, and add the two imaginary numbers together:

$$-6+7=1$$

$$-\frac{12}{5}i+\frac{3}{4}i=-\frac{48}{20}i+\frac{15}{20}i=-\frac{33}{20}i$$

The answer is $1-\frac{33}{20}i$.

706. $18-12i$

Multiply each term in the parentheses by 6:

$$6(3)=18$$

$$6(-2i)=-12i$$

The answer is $18-12i$.

707. $20-15i$

Multiply each term in the parentheses by –5:

$$-5(-4)=20$$

$$-5(3i)=-15i$$

The answer is $20-15i$.

708. $-16+12i$

Multiply each term in the parentheses by $2i$:

$$2i(6)=12i$$

$$2i(8i)=16i^2=16(-1)=-16$$

The answer is $-16+12i$.

709. $-15-6i$

Multiply each term in the parentheses by $-3i$:

$$-3i(2)=-6i$$

$$-3i(-5i)=15i^2=15(-1)=-15$$

The answer is $-15-6i$.

710. 13

Use FOIL to multiply the binomials:

$$\begin{aligned}&(2-3i)(2+3i)\\&=4+6i-6i-9i^2\\&=4-9(-1)\\&=4+9\\&=13\end{aligned}$$

711. 68

Use FOIL to multiply the binomials:

$$\begin{aligned}&(-8+2i)(-8-2i)\\&=64+16i-16i-4i^2\\&=64-4(-1)\\&=64+4\\&=68\end{aligned}$$

712. $42-9i$

Use FOIL to multiply the binomials:

$$\begin{aligned}&(5-4i)(6+3i)\\&=30+15i-24i-12i^2\\&=30-9i-12(-1)\\&=30-9i+12\\&=42-9i\end{aligned}$$

713. $5-12i$

Rewrite the square of the binomial as the product of two binomials; then multiply:

$$\begin{aligned}&(3-2i)(3-2i)\\&=9-6i-6i+4i^2\\&=9-12i+4(-1)\\&=9-12i-4\\&=5-12i\end{aligned}$$

714. $-46+9i$

To find the cube of the binomial, use Pascal's triangle:

$$\begin{array}{ccccccccccc} &&&&&1&&&&&\\ &&&&1&&1&&&&\\ &&&1&&2&&1&&&\\ &&1&&3&&3&&1&&\\ &1&&4&&6&&4&&1&\\ 1&&5&&10&&10&&5&&1 \end{array}$$

Recall that when using the triangle to determine the coefficients of the power of a binomial, you use the row whose second number is the same as the power.

First write the coefficients from the row whose second number is 3:

$$1\quad 3\quad 3\quad 1$$

The first term in $(2+3i)$ is 2, so write in decreasing powers of 2:

$$1\cdot 2^3\quad 3\cdot 2^2\quad 3\cdot 2^1\quad 1\cdot 2^0$$

Now write in increasing powers of the second term, $3i$:

$$1\cdot 2^3(3i)^0\quad 3\cdot 2^2(3i)^1\quad 3\cdot 2^1(3i)^2\quad 1\cdot 2^0(3i)^3$$

Simplify and write the sum of the terms:

$$\begin{aligned}&1\cdot 8\cdot 1\quad 3\cdot 4\cdot 3i\quad 3\cdot 2\cdot 9i^2\quad 1\cdot 1\cdot 27i^3\\&\to 8+36i+54i^2+27i^3\\&=8+36i+54(-1)+27(-i)\\&=8+36i-54-27i\\&=-46+9i\end{aligned}$$

715. $161-240i$

To find the fourth power of the binomial, use Pascal's triangle:

$$
\begin{array}{ccccccccccc}
 & & & & & 1 & & & & & \\
 & & & & 1 & & 1 & & & & \\
 & & & 1 & & 2 & & 1 & & & \\
 & & 1 & & 3 & & 3 & & 1 & & \\
 & 1 & & 4 & & 6 & & 4 & & 1 & \\
1 & & 5 & & 10 & & 10 & & 5 & & 1
\end{array}
$$

Recall that when using the triangle to determine the coefficients of the power of a binomial, you use the row whose second number is the same as the power. First write the coefficients from the row whose second term is 4:

$$1 \quad 4 \quad 6 \quad 4 \quad 1$$

The first term in $(-1-4i)$ is -1, so write in decreasing powers of -1:

$$1(-1)^4 \quad 4(-1)^3 \quad 6(-1)^2 \quad 4(-1)^1 \quad 1(-1)^0$$

Now write in increasing powers of the second term, $-4i$:

$$1(-1)^4(-4i)^0 \quad 4(-1)^3(-4i)^1 \quad 6(-1)^2(-4i)^2$$
$$4(-1)^1(-4i)^3 \quad 1(-1)^0(-4i)^4$$

Simplify and write the sum of the terms:

$$1\cdot 1\cdot 1 \quad 4(-1)(-4i) \quad 6\cdot 1\cdot 16i^2 \quad 4(-1)\left(-64i^3\right) \quad 1\cdot 1\cdot 256i^4$$
$$\rightarrow 1+16i+96i^2+256i^3+256i^4$$
$$=1+16i+96(-1)+256(-i)+256(1)$$
$$=1+16i-96-256i+256$$
$$=161-240i$$

716. $-\frac{5}{13}-\frac{12}{13}i$

Multiply both the numerator and the denominator by the conjugate of the denominator (the *conjugate* of a binomial is the same two terms with the opposite sign between them):

$$\frac{2-3i}{2+3i}\cdot\frac{2-3i}{2-3i}=\frac{(2-3i)(2-3i)}{(2+3i)(2-3i)}=\frac{4-6i-6i+9i^2}{4-6i+6i-9i^2}$$
$$=\frac{4-12i+9(-1)}{4-9(-1)}=\frac{4-12i-9}{4+9}=\frac{-5-12i}{13}$$

717. $-\frac{13}{20}+\frac{1}{20}i$

Multiply both the numerator and the denominator by the conjugate of the denominator (the *conjugate* of a binomial is the same two terms with the opposite sign between them):

$$\frac{-4-i}{6+2i}\cdot\frac{6-2i}{6-2i}=\frac{-24+8i-6i+2i^2}{36-12i+12i-4i^2}=\frac{-24+2i+2(-1)}{36-4(-1)}$$
$$=\frac{-24+2i-2}{36+4}=\frac{-26+2i}{40}=\frac{-13+i}{20}$$

718. $1+\frac{1}{2}i$

Multiply both the numerator and the denominator by the conjugate of the denominator (the *conjugate* of a binomial is the same two terms with the opposite sign between them):

$$\frac{5}{4-2i}\cdot\frac{4+2i}{4+2i}=\frac{20+10i}{16+8i-8i-4i^2}=\frac{20+10i}{16-4(-1)}$$
$$=\frac{20+10i}{16+4}=\frac{20+10i}{20}=\frac{2+i}{2}$$

719. $-3-3i$

Multiply both the numerator and the denominator by the conjugate of the denominator (the *conjugate* of a binomial is the same two terms with the opposite sign between them):

$$\frac{6i}{-1-i}\cdot\frac{-1+i}{-1+i}=\frac{-6i+6i^2}{1-i+i-i^2}=\frac{-6i+6(-1)}{1-(-1)}$$
$$=\frac{-6i-6}{1+1}=\frac{-6i-6}{2}=-3i-3$$

720. $7-4i$

Multiply both the numerator and the denominator by the conjugate of the denominator (the *conjugate* of a binomial is the same two terms with the opposite sign between them):

$$\frac{4+7i}{i}\cdot\frac{-i}{-i}=\frac{-4i-7i^2}{-i^2}=\frac{-4i-7(-1)}{-(-1)}=\frac{-4i+7}{1}=-4i+7$$

721. $\frac{3}{2}\pm\frac{\sqrt{7}}{2}i$

The quadratic doesn't factor, so use the quadratic formula, $x=\frac{-b\pm\sqrt{b^2-4ac}}{2a}$. In this problem, $a=1$, $b=-3$, and $c=4$:

$$x=\frac{-(-3)\pm\sqrt{(-3)^2-4(1)(4)}}{2(1)}=\frac{3\pm\sqrt{9-16}}{2}$$
$$=\frac{3\pm\sqrt{-7}}{2}=\frac{3\pm\sqrt{(-1)(7)}}{2}=\frac{3\pm i\sqrt{7}}{2}$$

722. $-2\pm 2i$

The quadratic doesn't factor, so use the quadratic formula, $x=\frac{-b\pm\sqrt{b^2-4ac}}{2a}$. In this problem, $a=1$, $b=4$, and $c=8$:

$$x=\frac{-4\pm\sqrt{4^2\ 4(1)(8)}}{2(1)}=\frac{-4\pm\sqrt{16-32}}{2}$$
$$=\frac{-4\pm\sqrt{-16}}{2}=\frac{-4\pm\sqrt{(-1)(16)}}{2}=\frac{-4\pm i\sqrt{16}}{2}$$
$$=\frac{-4\pm 4i}{2}=-2\pm 2i$$

723. $\frac{5}{4} \pm \frac{5\sqrt{7}}{4}i$

The quadratic doesn't factor, so use the quadratic formula, $x = \frac{-b \pm \sqrt{b^2 - 4ac}}{2a}$. In this problem, $a = 2$, $b = -5$, and $c = 25$:

$$x = \frac{-(-5) \pm \sqrt{(-5)^2 - 4(2)(25)}}{2(2)} = \frac{-(-5) \pm \sqrt{25 - 200}}{4}$$
$$= \frac{5 \pm \sqrt{-175}}{4} = \frac{5 \pm \sqrt{(-1)(25)(7)}}{4} = \frac{5 \pm 5i\sqrt{7}}{4}$$

724. $-\frac{1}{8} \pm \frac{3\sqrt{7}}{8}i$

The quadratic doesn't factor, so use the quadratic formula, $x = \frac{-b \pm \sqrt{b^2 - 4ac}}{2a}$. In this problem, $a = 4$, $b = 1$, and $c = 4$:

$$x = \frac{-1 \pm \sqrt{1^2 - 4(4)(4)}}{2(4)} = \frac{-1 \pm \sqrt{1 - 64}}{8}$$
$$= \frac{-1 \pm \sqrt{-63}}{8} = \frac{-1 \pm \sqrt{(-1)(9)(7)}}{8} = \frac{-1 \pm 3i\sqrt{7}}{8}$$

725. $-\frac{1}{5} \pm \frac{\sqrt{14}}{5}i$

The quadratic doesn't factor, so use the quadratic formula, $x = \frac{-b \pm \sqrt{b^2 - 4ac}}{2a}$. In this problem, $a = 5$, $b = 2$, and $c = 3$:

$$x = \frac{-2 \pm \sqrt{2^2 - 4(5)(3)}}{2(5)} = \frac{-2 \pm \sqrt{4 - 60}}{10} = \frac{-2 \pm \sqrt{-56}}{10}$$
$$= \frac{-2 \pm \sqrt{(-1)(4)(14)}}{10} = \frac{-2 \pm 2i\sqrt{14}}{10} = \frac{-1 \pm i\sqrt{14}}{5}$$

726. $3 + 2i$

The point is 3 units to the right on the real axis and 2 units up on the imaginary axis.

727. $-5 + i$

The point is 5 units to the left on the real axis and 1 unit up on the imaginary axis.

728. $-4-4i$

The point is 4 units to the left on the real axis and 4 units down on the imaginary axis.

729. -3

The point is 3 units to the left on the real axis, and it sits on that axis with an imaginary coordinate of 0.

730. $3i$

The point is 3 units up on the imaginary axis, and it sits on that axis with a real coordinate of 0.

731. $\left(2, \frac{2\pi}{3}\right)$

The point is on the 2-unit ring, so the radius is 2. You measure the angle from the positive x-axis in a counterclockwise direction. Each ray represents 15° (or $\frac{\pi}{12}$), so this point is on the ray representing $\frac{2\pi}{3}$. The radius is 2, so the point is 2 units out from the origin.

732. $\left(3, \frac{5\pi}{4}\right)$

The point is on the 3-unit ring, so the radius is 3. You measure the angle from the positive x-axis in a counterclockwise direction. Each ray represents 15° (or $\frac{\pi}{12}$), so this point is on the ray representing $\frac{5\pi}{4}$.

733. $\left(2, \frac{3\pi}{2}\right)$

The point is on the 2-unit ring, so the radius is 2. You measure the angle from the positive x-axis in a counterclockwise direction. Each ray represents 15° (or $\frac{\pi}{12}$), so this point is on the ray representing $\frac{3\pi}{2}$.

734. $\left(1, -\frac{\pi}{6}\right)$

The point is on the 1-unit ring, so the radius is 1. You're measuring the angle from the positive x-axis in a clockwise direction. Each ray represents 15° (or $\frac{\pi}{12}$), so this point is on the ray representing $-\frac{\pi}{6}$. Therefore, the answer is $\left(1, -\frac{\pi}{6}\right)$. If you were to use a positive angle instead, this point would be $\left(1, \frac{11\pi}{6}\right)$.

735. $\left(3, -\frac{5\pi}{3}\right)$

The point is on the 3-unit ring, so the radius is 3. You're measuring the angle from the positive x-axis in a clockwise direction. Each ray represents 15° (or $\frac{\pi}{12}$), so this point is on the ray representing $-\frac{5\pi}{3}$. Therefore, the answer is $\left(3, -\frac{5\pi}{3}\right)$. If you were to use a positive angle instead, this point would be $\left(3, \frac{\pi}{3}\right)$.

736. (0, –2)

You find the x coordinate with $x = r\cos\theta$, so

$$x = 2\cos\left(\frac{3\pi}{2}\right) = 2(0) = 0$$

You find the y coordinate with $y = r\sin\theta$, so

$$y = 2\sin\left(\frac{3\pi}{2}\right) = 2(-1) = -2$$

737. $\left(-\frac{3}{2}, \frac{3\sqrt{3}}{2}\right)$

You find the x coordinate with $x = r\cos\theta$, so

$$x = 3\cos\left(\frac{2\pi}{3}\right) = 3\left(-\frac{1}{2}\right) = -\frac{3}{2}$$

You find the y coordinate with $y = r\sin\theta$, so

$$y = 3\sin\left(\frac{2\pi}{3}\right) = 3\left(\frac{\sqrt{3}}{2}\right) = \frac{3\sqrt{3}}{2}$$

738. $\left(-2\sqrt{3}, 2\right)$

You find the x coordinate with $x = r\cos\theta$, so

$$x = 4\cos\left(\frac{5\pi}{6}\right) = 4\left(-\frac{\sqrt{3}}{2}\right) = -2\sqrt{3}$$

You find the y coordinate with $y = r\sin\theta$, so

$$y = 4\sin\left(\frac{5\pi}{6}\right) = 4\left(\frac{1}{2}\right) = 2$$

739. $\left(\frac{1}{2}, \frac{\sqrt{3}}{2}\right)$

You find the x coordinate with $x = r\cos\theta$, so

$$x = 1\cos\left(-\frac{5\pi}{3}\right)$$

An angle measuring $-\frac{5\pi}{3}$ is *coterminal* with the angle $\frac{\pi}{3}$ (they have the same terminal ray when graphed in standard position), so

$$x = 1\cos\left(-\frac{5\pi}{3}\right) = \cos\left(\frac{\pi}{3}\right) = \frac{1}{2}$$

You find the y coordinate with $y = r\sin\theta$, so

$$y = 1\sin\left(-\frac{5\pi}{3}\right) = \sin\left(\frac{\pi}{3}\right) = \frac{\sqrt{3}}{2}$$

740. (–2, 0)

You find the x coordinate with $x = r\cos\theta$, so

$$x = 2\cos\pi = 2(-1) = -2$$

You find the y coordinate with $y = r\sin\theta$, so

$$y = 2\sin\pi = 2(0) = 0$$

741. $\left(2, \frac{5\pi}{3}\right)$

You find the radius with $r = \sqrt{x^2 + y^2}$, so

$$r = \sqrt{1^2 + \left(-\sqrt{3}\right)^2} = \sqrt{1+3} = \sqrt{4} = 2$$

You find the angle θ with $\theta = \tan^{-1}\left(\frac{y}{x}\right)$, so

$$\theta = \tan^{-1}\left(\frac{-\sqrt{3}}{1}\right) = \tan^{-1}\left(-\sqrt{3}\right) = -\frac{\pi}{3}$$

The tangent of an angle is negative in the second and fourth quadrants. The rectangular coordinates are those of a point in the fourth quadrant, so you want an angle between $\frac{3\pi}{2}$ and 2π. Add 2π to $-\frac{\pi}{3}$ to get

$$\theta = \frac{5\pi}{3}$$

742. $\left(\sqrt{2}, \frac{3\pi}{4}\right)$

You find the radius with $r=\sqrt{x^2+y^2}$, so

$$r=\sqrt{(-1)^2+1^2}=\sqrt{1+1}=\sqrt{2}$$

You find the angle θ with $\theta=\tan^{-1}\left(\frac{y}{x}\right)$, so

$$\theta=\tan^{-1}\left(\frac{1}{-1}\right)=\tan^{-1}(-1)=-\frac{\pi}{4}$$

The tangent of an angle is negative in the second and fourth quadrants. The rectangular coordinates are those of a point in the second quadrant, so you want an angle between $\frac{\pi}{2}$ and $\frac{3\pi}{2}$. Add π to $-\frac{\pi}{4}$ to get

$$\theta=\frac{3\pi}{4}$$

743. $\left(2\sqrt{3}, \frac{\pi}{6}\right)$

You find the radius with $r=\sqrt{x^2+y^2}$, so

$$r=\sqrt{3^2+\left(\sqrt{3}\right)^2}=\sqrt{9+3}=\sqrt{12}=2\sqrt{3}$$

You find the angle θ with $\theta=\tan^{-1}\left(\frac{y}{x}\right)$, so

$$\theta=\tan^{-1}\left(\frac{\sqrt{3}}{3}\right)=\frac{\pi}{6}$$

The tangent of an angle is positive in the first and third quadrants. The rectangular coordinates are those of a point in the first quadrant, so

$$\theta=\frac{\pi}{6}$$

744. (3, 0)

You find the radius with $r=\sqrt{x^2+y^2}$, so

$$r=\sqrt{3^2+0^2}=\sqrt{9}=3$$

You find θ with $\theta=\tan^{-1}\left(\frac{y}{x}\right)$, so

$$\theta=\tan^{-1}\left(\frac{0}{3}\right)=\tan^{-1}(0)=0$$

The tangent of an angle is 0 on the x-axis. The rectangular coordinates are those of a point on the positive x-axis, so

$$\theta=0$$

745. (1, 4.429)

You find the radius with $r=\sqrt{x^2+y^2}$, so

$$r=\sqrt{\left(-\frac{7}{25}\right)^2+\left(-\frac{24}{25}\right)^2}=\sqrt{\frac{49}{625}+\frac{576}{625}}=\sqrt{\frac{625}{625}}=1$$

You find the angle θ with $\theta=\tan^{-1}\left(\frac{y}{x}\right)$, so

$$\theta=\tan^{-1}\left(\frac{-24/25}{-7/25}\right)=\tan^{-1}\left(\frac{24}{7}\right)\approx 1.287\text{ radians}$$

The tangent of an angle is positive in the first and third quadrants. The rectangular coordinates are those of a point in the third quadrant, so

$$\theta\approx 1.287+\pi\approx 4.429\text{ radians}$$

746. Archimedean spiral

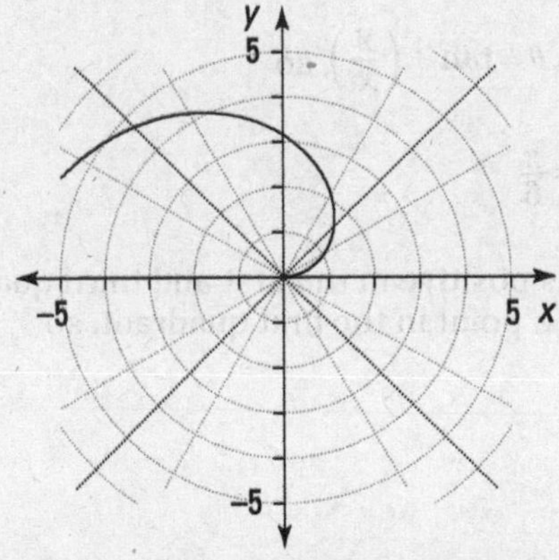

Illustration by Thomson Digital

The equation has the general form for an Archimedean spiral, $r=a+b\theta$. In this case, $a=0$ and $b=2$.

747. cardioid

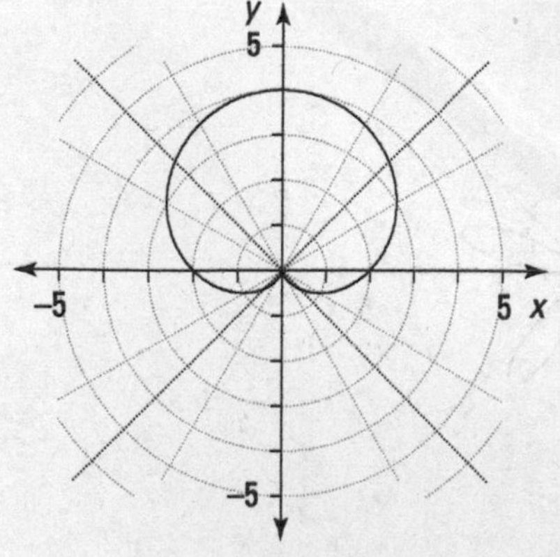

Illustration by Thomson Digital

The equation has the general form for a cardioid, $r = a + b\sin\theta$, where $\left|\frac{a}{b}\right| = 1$. In this case, $a = 2$ and $b = 2$.

748. lemniscate

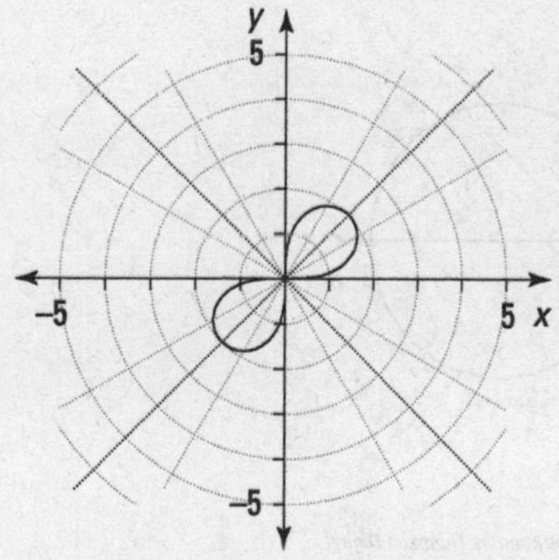

Illustration by Thomson Digital

The equation has the general form for a lemniscate, $r^2 = a^2 \sin 2\theta$. In this case, $a = 2$.

749. rose

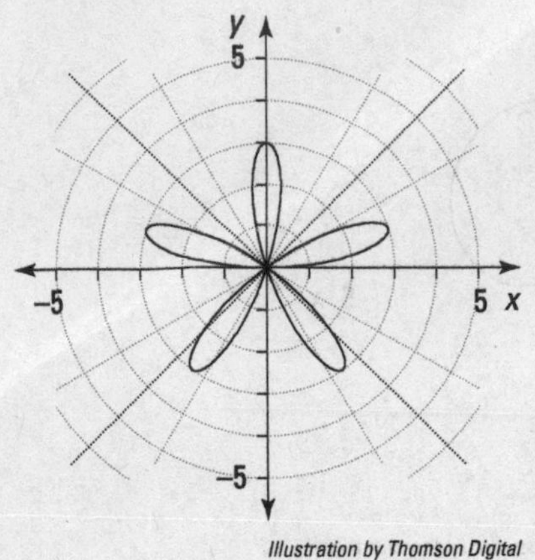

Illustration by Thomson Digital

The equation has the general form for a rose, $r = a \sin n\theta$, In this case, $a = 3$ and $n = 5$.

750. limaçon

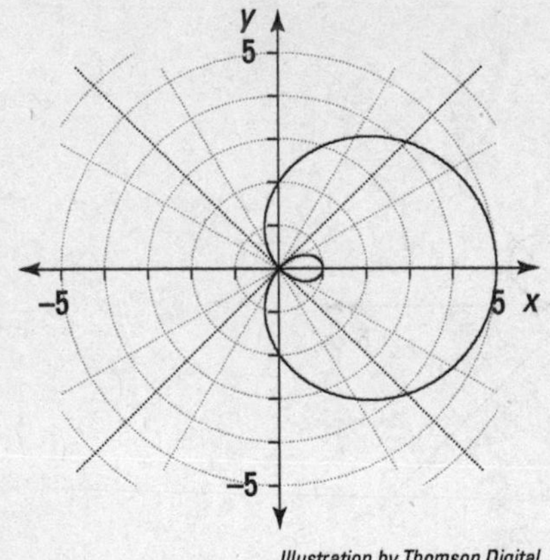

Illustration by Thomson Digital

The equation has the general form for a limaçon, $r = a + b\cos\theta$. In this case, $a = 2$ and $b = 3$.

751. hyperbola; center: (3, –2)

The standard form for the equation of a hyperbola with center (h, k) is

$$\frac{(x-h)^2}{a^2}+\frac{(y-k)^2}{b^2}=1$$

The given equation is already in this form, so you can identify the center coordinates by looking at the values substituted for h and k.

752. ellipse; center: (–4, 1)

The standard form for the equation of an ellipse with center (h, k) is

$$\frac{(x-h)^2}{a^2}+\frac{(y-k)^2}{b^2}=1$$

The given equation is already in this form, so you can identify the center coordinates by looking at the values substituted for h and k.

753. circle; center: (5, 0)

The standard form for the equation of a circle with center (h, k) is

$$(x-h)^2+(y-k)^2=r^2$$

The given equation is already in this form, with a bit of a twist: No k value is given, so you have to recognize that k is 0 (because $(y-0)^2=y^2$). You can identify the x coordinate more easily by looking at the values substituted for h.

754. ellipse; center: (0, –3)

The standard form for the equation of an ellipse with center (h, k) is

$$\frac{(x-h)^2}{a^2}+\frac{(y-k)^2}{b^2}=1$$

To put the given equation in the standard form, divide each term in the equation by 36:

$$\frac{9x^2}{36}+\frac{4(y+3)^2}{36}=\frac{36}{36}$$

$$\frac{x^2}{4}+\frac{(y+3)^2}{9}=1$$

Now you can identify the center coordinates by looking at the values substituted for h and k. In this case, no h value is given, so you have to recognize that h is 0 (because $(x-0)^2=x^2$).

755. hyperbola; center: (–3, 8)

The standard form for the equation of a hyperbola with center (h, k) is

$$\frac{(x-h)^2}{a^2}+\frac{(y-k)^2}{b^2}=1$$

To put the equation in the standard form, divide each term in the equation by 54:

$$\frac{6(y-8)^2}{54}-\frac{9(x+3)^2}{54}=\frac{54}{54}$$

$$\frac{(y-8)^2}{9}-\frac{(x+3)^2}{6}=1 \quad \text{or} \quad \frac{(x+3)^2}{6}-\frac{(y-8)^2}{9}=-1$$

This last form of the equation is an alternate that lets you see, from the –1 on the right, that the hyperbola will open upward and downward.

Now you can identify the center coordinates by looking at the values substituted for h and k.

756. $(x-3)^2+y^2=16$

Rewrite the equation by grouping the x terms and the y term and moving the constant to the right side:

$$x^2-6x+y^2=7$$

Complete the square with the x terms, adding 9 to both sides:

$$x^2-6x+9+y^2=7+9$$

Factor the perfect square trinomial and simplify the equation:

$$(x-3)^2+y^2=16$$

This is the equation of a circle with center (3, 0) and a radius of 4.

757. $\frac{(x+2)^2}{9}+\frac{(y-3)^2}{25}=1$

Rewrite the equation by grouping the x terms and the y terms and moving the constant to the right side:

$$25x^2+100x+9y^2-54y=44$$

Factor 25 from each of the x terms and 9 from each of the y terms:

$$25(x^2+4x)+9(y^2-6y)=44$$

Complete the square with the x terms by adding 4 in the parentheses; add 100 to the right side (because $25\cdot4=100$):

$$25(x^2+4x+4)+9(y^2-6y)=44+100$$

Complete the square with the y terms by adding 9 in the parentheses; add 81 to the right side (because $9\cdot9=81$).

$$25(x^2+4x+4)+9(y^2-6y+9)=44+100+81$$

Factor the perfect square trinomials and simplify the equation:

$$25(x+2)^2+9(y-3)^2=225$$

Divide each term by 225:

$$\frac{25(x+2)^2}{225}+\frac{9(y-3)^2}{225}=\frac{225}{225}$$

$$\frac{(x+2)^2}{9}+\frac{(y-3)^2}{25}=1$$

This is the equation of an ellipse with its center at (–2, 3).

758. $\frac{x^2}{25}-\frac{(y-3)^2}{1}=-1$

Rewrite the equation by grouping the y terms and moving the constant to the right side:

$$25y^2-150y-x^2=-200$$

Factor 25 from each of the y terms:

$$25\left(y^2-6y\right)-x^2=-200$$

Complete the square with the y terms by adding 9 in the parentheses; add 225 to the right side (because $25\cdot 9=225$):

$$25\left(y^2-6y+9\right)-x^2=-200+225$$

Factor the perfect square trinomial and simplify the equation:

$$25(y-3)^2-x^2=25$$

Divide each term by 25:

$$\frac{25(y-3)^2}{25}-\frac{x^2}{25}=\frac{25}{25}$$

$$\frac{(y-3)^2}{1}-\frac{x^2}{25}=1 \quad\text{or}\quad \frac{x^2}{25}-\frac{(y-3)^2}{1}=-1$$

This last form of the equation is an alternate that lets you see, from the –1 on the right, that the hyperbola will open upward and downward.

This is the equation of a hyperbola with its center at (0, 3).

759. $(x-4)^2+(y+4)^2=\frac{1}{4}$

Rewrite the equation by grouping the x terms and the y terms and moving the constant to the right side:

$$4x^2-32x+4y^2+32y=-127$$

Factor 4 from each of the x terms and 4 from each of the y terms:

$$4\left(x^2-8x\right)+4\left(y^2+8y\right)=-127$$

Complete the square with the x terms by adding 16 in the parentheses; add 64 to the right side (because $4\cdot 16=64$):

$$4\left(x^2-8x+16\right)+4\left(y^2+8y\right)=-127+64$$

Complete the square with the y terms by adding 16 in the parentheses; add another 64 to the right side (because $4 \cdot 16 = 64$):

$$4\left(x^2 - 8x + 16\right) + 4\left(y^2 + 8y + 16\right) = -127 + 64 + 64$$

Factor the perfect square trinomials and simplify the equation:

$$4(x-4)^2 + 4(y+4)^2 = 1$$

Divide each term by 4:

$$\frac{4(x-4)^2}{4} + \frac{4(y+4)^2}{4} = \frac{1}{4}$$

$$(x-4)^2 + (y+4)^2 = \frac{1}{4}$$

This is the equation of a circle with its center at (4, –4).

760. $\dfrac{(x-1)^2}{3/4} + \dfrac{(y-2)^2}{1} = 1$

Rewrite the equation by grouping the x terms and the y terms and moving the constant to the right side:

$$4x^2 - 8x + 3y^2 - 12y = -13$$

Factor 4 from each of the x terms and 3 from each of the y terms:

$$4\left(x^2 - 2x\right) + 3\left(y^2 - 4y\right) = -13$$

Complete the square with the x terms by adding 1 in the parentheses; add 4 to the right side (because $4 \cdot 1 = 4$):

$$4\left(x^2 - 2x + 1\right) + 3\left(y^2 - 4y\right) = -13 + 4$$

Complete the square with the y terms by adding 4 in the parentheses; add 12 to the right side (because $3 \cdot 4 = 12$):

$$4\left(x^2 - 2x + 1\right) + 3\left(y^2 - 4y + 4\right) = -13 + 4 + 12$$

Factor the perfect square trinomials and simplify the equation:

$$4(x-1)^2 + 3(y-2)^2 = 3$$

Divide each term by 3. Then multiply the first term by $\frac{1}{4}$ divided by $\frac{1}{4}$ to create a coefficient of 1 in the numerator:

$$\frac{4(x-1)^2}{3} + \frac{3(y-2)^2}{3} = \frac{3}{3}$$

$$\frac{4(x-1)^2}{3} + \frac{(y-2)^2}{1} = 1$$

$$\frac{1/4}{1/4} \cdot \frac{4(x-1)^2}{3} + \frac{(y-2)^2}{1} = 1 \cdot \frac{1/4}{1/4}$$

$$\frac{(x-1)^2}{3/4} + \frac{(y-2)^2}{1} = 1$$

This is the equation of an ellipse with its center at (1, 2).

761. $x^2+y^2=16$

The standard equation of a circle with radius (h, k) and radius r is $(x-h)^2+(y-k)^2=r^2$. Substitute the given point (0, 0) for the (h, k) and square the 4:

$$(x-0)^2+(y-0)^2=4^2$$
$$x^2+y^2=16$$

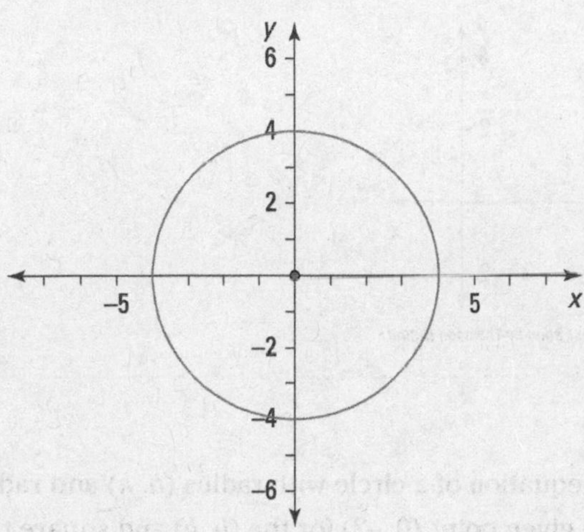

Illustration by Thomson Digital

762. $(x-4)^2+(y-3)^2=25$

The standard equation of a circle with radius (h, k) and radius r is $(x-h)^2+(y-k)^2=r^2$. Substitute the given point (4, 3) for the (h, k) and square the 5:

$$(x-4)^2+(y-3)^2=5^2$$
$$(x-4)^2+(y-3)^2=25$$

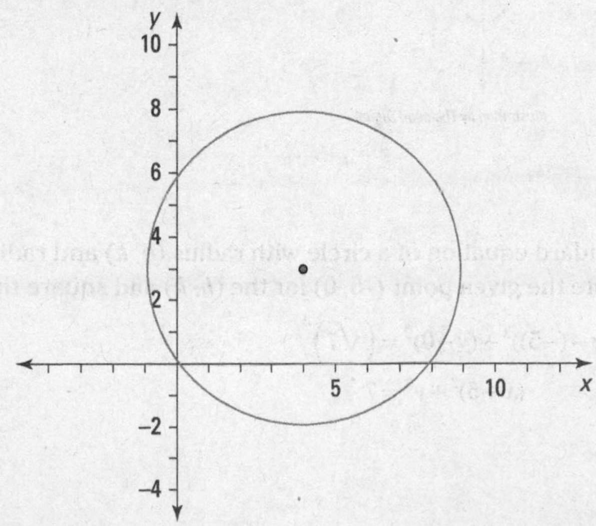

Illustration by Thomson Digital

763. $(x+2)^2+(y-1)^2=1$

The standard equation of a circle with radius (h, k) and radius r is $(x-h)^2+(y-k)^2=r^2$. Substitute the given point (–2, 1) for the (h, k) and square the 1:

$$(x+2)^2+(y-1)^2=1^2$$
$$(x+2)^2+(y-1)^2=1$$

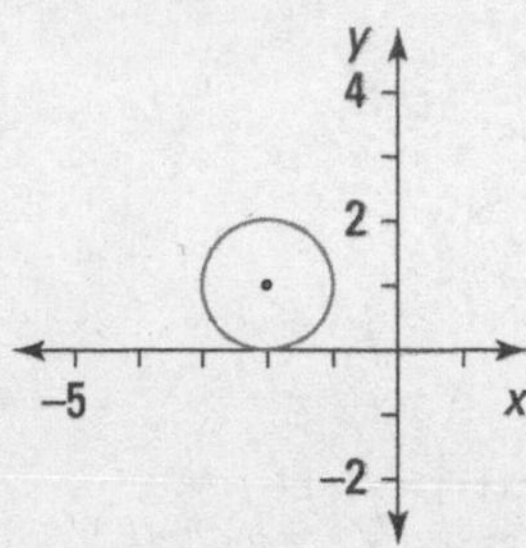

Illustration by Thomson Digital

764. $x^2+(y+2)^2=\frac{4}{9}$

The standard equation of a circle with radius (h, k) and radius r is $(x-h)^2+(y-k)^2=r^2$. Substitute the given point (0, –2) for the (h, k) and square the $\frac{2}{3}$.

$$(x-0)^2+(y-(-2))^2=\left(\frac{2}{3}\right)^2$$
$$x^2+(y+2)^2=\frac{4}{9}$$

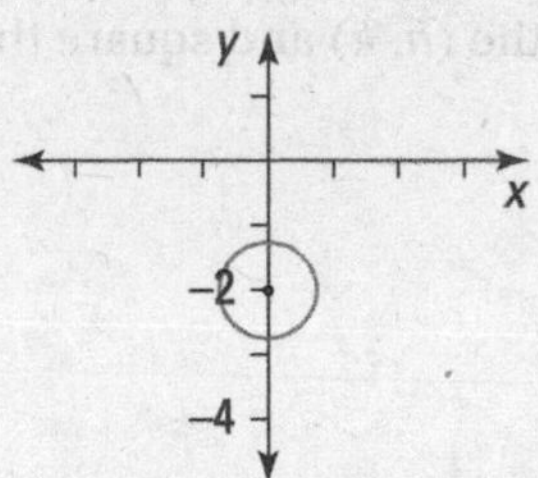

Illustration by Thomson Digital

Answers 701–800

765. $(x+5)^2+y^2=7$

The standard equation of a circle with radius (h, k) and radius r is $(x-h)^2+(y-k)^2=r^2$. Substitute the given point (–5, 0) for the (h, k) and square the $\sqrt{7}$:

$$(x-(-5))^2+(y-0)^2=\left(\sqrt{7}\right)^2$$
$$(x+5)^2+y^2=7$$

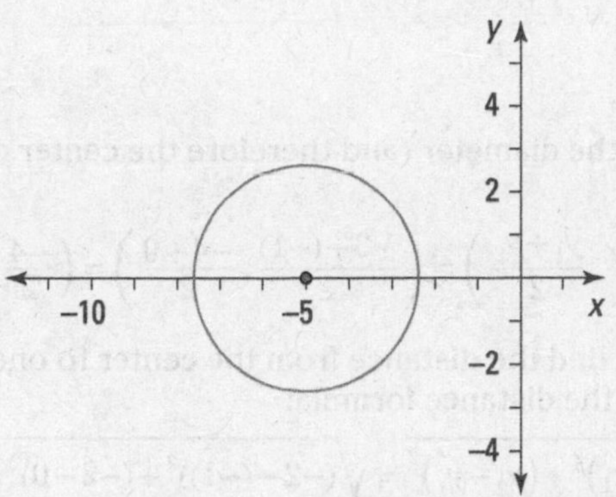

Illustration by Thomson Digital

766. $(x-1)^2+(y-3)^2=25$

Find the center of the diameter (and therefore the center of the circle) by using the midpoint formula:

$$M=\left(\frac{x_1+x_2}{2},\frac{y_1+y_2}{2}\right)=\left(\frac{4+(-2)}{2},\frac{7+(-1)}{2}\right)=\left(\frac{2}{2},\frac{6}{2}\right)=(1,3)$$

To find the radius, find the distance from the center to one of the endpoints of the diameter by using the distance formula:

$$d=\sqrt{(x_2-x_1)^2+(y_2-y_1)^2}=\sqrt{(1-4)^2+(3-7)^2}=\sqrt{9+16}=\sqrt{25}=5$$

The standard equation of a circle with radius (h, k) and radius r is $(x-h)^2+(y-k)^2=r^2$. Substitute the center point (1, 3) for the (h, k) and square the 5:

$$(x-1)^2+(y-3)^2=5^2$$
$$(x-1)^2+(y-3)^2=25$$

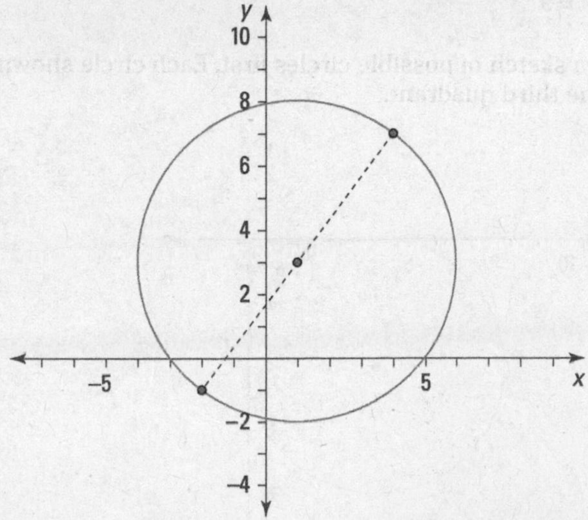

Illustration by Thomson Digital

Answers 701–800

767. $(x+2)^2+(y+2)^2=5$

Find the center of the diameter (and therefore the center of the circle) by using the midpoint formula:

$$M=\left(\frac{x_1+x_2}{2},\frac{y_1+y_2}{2}\right)=\left(\frac{-3+(-1)}{2},\frac{-4+0}{2}\right)=\left(\frac{-4}{2},\frac{-4}{2}\right)=(-2,-2)$$

To find the radius, find the distance from the center to one of the endpoints of the diameter by using the distance formula:

$$d=\sqrt{\left(x_2-x_1\right)^2+\left(y_2-y_1\right)^2}=\sqrt{(-2-(-1))^2+(-2-0)^2}=\sqrt{1+4}=\sqrt{5}$$

The standard equation of a circle with radius (h, k) and radius r is $(x-h)^2+(y-k)^2=r^2$. Substitute the center point (–2, –2) for the (h, k) and square the $\sqrt{5}$:

$$\begin{aligned}(x-(-2))^2+(y-(-2))^2&=\left(\sqrt{5}\right)^2\\(x+2)^2+(y+2)^2&=5\end{aligned}$$

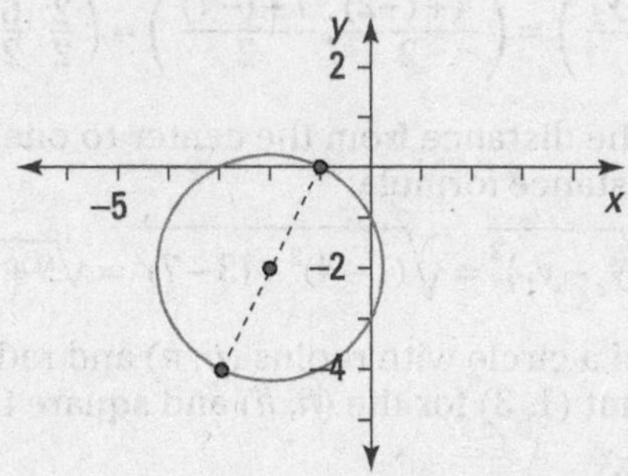

Illustration by Thomson Digital

768. $(x+3)^2+(y+3)^2=9$

Look at a sketch of possible circles first. Each circle shown is tangent to both axes and lies in the third quadrant.

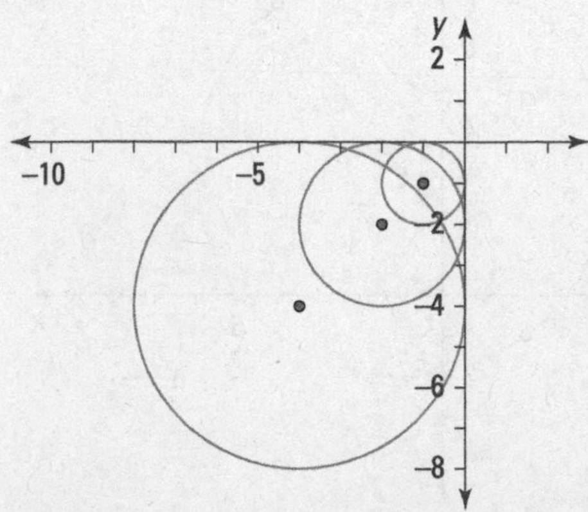

Illustration by Thomson Digital

For the circle you want, the points of tangency of the circle and axes are at (–3, 0) and (0, –3). The center of the circle is at (–3, –3).

The standard equation of a circle with radius (h, k) and radius r is $(x-h)^2+(y-k)^2=r^2$. Substitute the center point (–3, –3) for the (h, k) and square the 3:

$$(x-(-3))^2+(y-(-3))^2=9$$
$$(x+3)^2+(y+3)^2=9$$

Now you can sketch the exact graph:

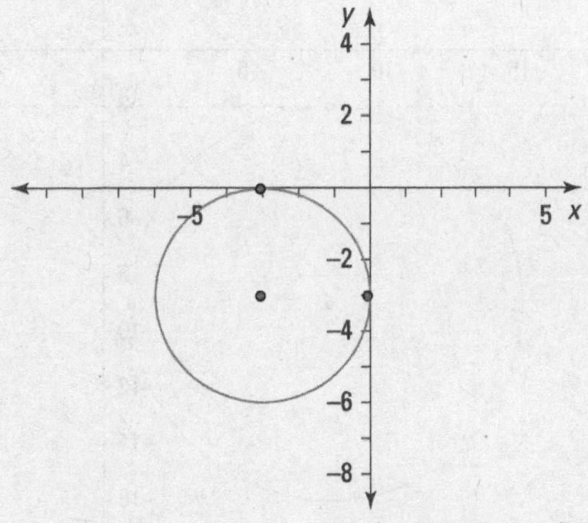

Illustration by Thomson Digital

769. $(x+8)^2+(y+9)^2=49$

Look at a sketch of some possible circles first. Each circle is tangent to the lines $x=-1$ and $y=-2$ and lies in the third quadrant.

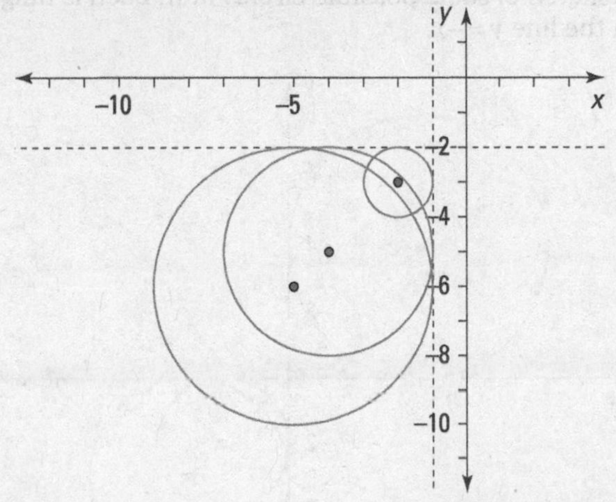

Illustration by Thomson Digital

The center of the circle is 7 units to the left of the vertical line $x=-1$, which means it's at $x=-8$. The center is also 7 units below the horizontal line $y=-2$, putting it at $y=-9$. So the center has coordinates (–8, –9).

The standard equation of a circle with radius (h, k) and radius r is $(x-h)^2+(y-k)^2=r^2$. Substitute the center point (–8, –9) for the (h, k) and square the 7:

$$(x-(-8))^2+(y-(-9))^2=49$$
$$(x+8)^2+(y+9)^2=49$$

Now you can sketch the exact graph:

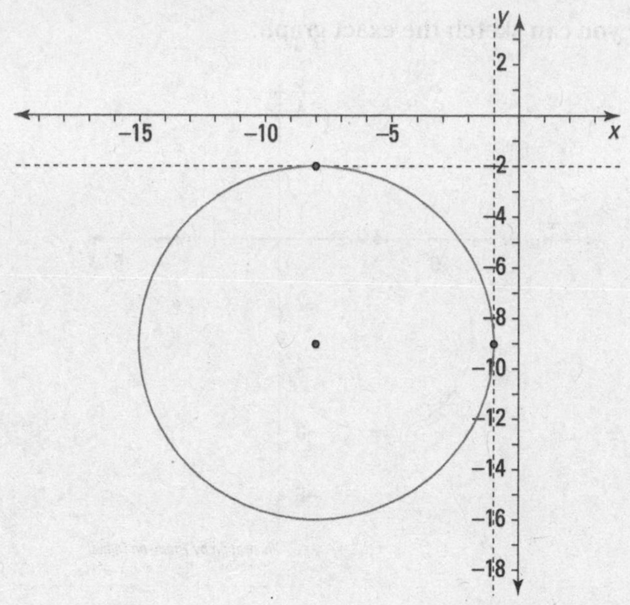

Illustration by Thomson Digital

770. $(x+4)^2+(y-4)^2=16$ and $(x-4)^2+(y+4)^2=16$

Look at a sketch of some possible circles first. Each is tangent to both axes and has its center on the line $y=-x$.

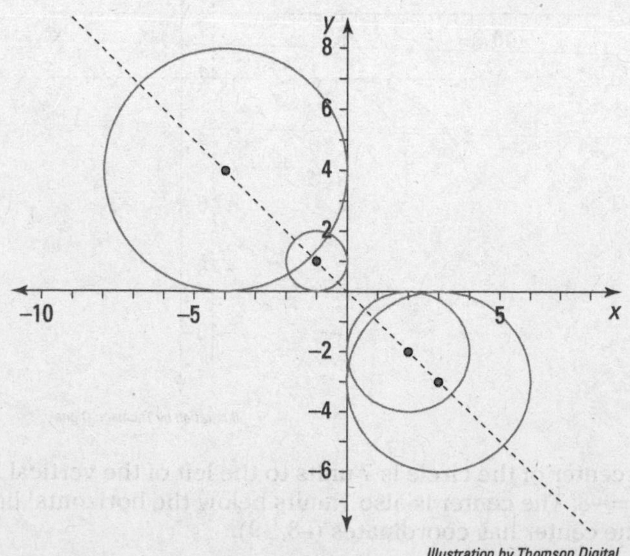

Illustration by Thomson Digital

Two circles fit this description: one in the second quadrant and the other in the fourth quadrant. With the centers on the line $y = -x$, the coordinates of the centers are opposites: (–4, 4) and (4, –4).

The standard equation of a circle with radius (h, k) and radius r is $(x-h)^2+(y-k)^2=r^2$. First, substitute one of the center points, (–4, 4), for the (h, k) and square the 4:

$$(x-(-4))^2+(y-4)^2=4^2$$
$$(x+4)^2+(y-4)^2=16$$

Now do the same for the second center point, (4, –4):

$$(x-4)^2+(y-(-4))^2=4^2$$
$$(x-4)^2+(y+4)^2=16$$

Now you can sketch the exact figures:

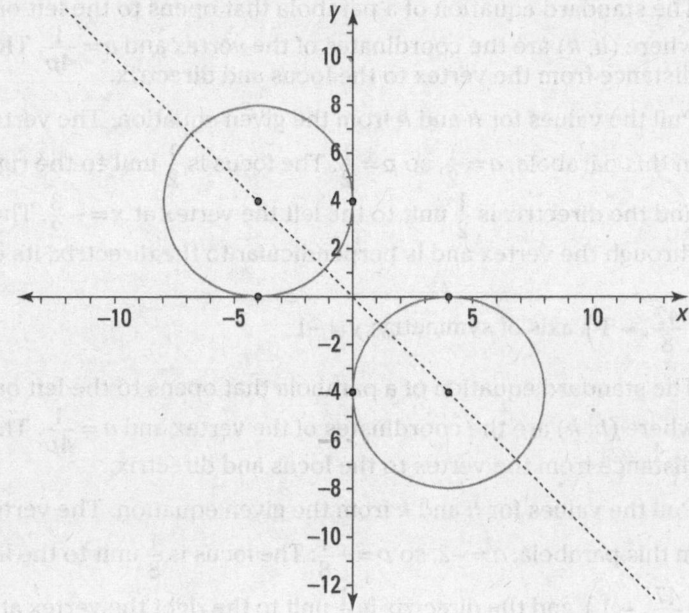

Illustration by Thomson Digital

771. focus: $\left(2, \frac{1}{4}\right)$; axis of symmetry: $x=2$

The standard equation of a parabola that opens upward or downward is $y-k=a(x-h)^2$, where (h, k) are the coordinates of the vertex and $a=\frac{1}{4p}$. The value of p gives you the distance from the vertex to the focus and directrix.

In this parabola, $a=1$, so $p=\frac{1}{4}$. The focus is $\frac{1}{4}$ unit above the vertex at $\left(2, \frac{1}{4}\right)$, and the directrix is $\frac{1}{4}$ unit below the vertex at $y=-\frac{1}{4}$. The axis of symmetry goes through the vertex and is perpendicular to the directrix; its equation is $x=2$.

772. focus: (–1, 1); axis of symmetry: $x=-1$

The standard equation of a parabola that opens upward or downward is $y-k=a(x-h)^2$, where (*h*, *k*) are the coordinates of the vertex and $a=\frac{1}{4p}$. The value of *p* gives you the distance from the vertex to the focus and directrix.

Pull the values for *h* and *k* from the given equation. The vertex is (–1, 2).

In this parabola, $a=-\frac{1}{4}$, so $p=-1$. The focus is 1 unit below the vertex at (–1, 1), and the directrix is 1 unit above the vertex at $y=3$. The axis of symmetry goes through the vertex and is perpendicular to the directrix; its equation is $x=-1$.

773. focus: $\left(-\frac{1}{2},4\right)$; axis of symmetry: $y=4$

The standard equation of a parabola that opens to the left or right is $x-h=a(y-k)^2$, where (*h*, *k*) are the coordinates of the vertex and $a=\frac{1}{4p}$. The value of *p* gives you the distance from the vertex to the focus and directrix.

Pull the values for *h* and *k* from the given equation. The vertex is (–1, 4).

In this parabola, $a=\frac{1}{2}$, so $p=\frac{1}{2}$. The focus is $\frac{1}{2}$ unit to the right of the vertex at $\left(-\frac{1}{2},4\right)$, and the directrix is $\frac{1}{2}$ unit to the left the vertex at $x=-\frac{3}{2}$. The axis of symmetry goes through the vertex and is perpendicular to the directrix; its equation is $y=4$.

774. focus: $\left(\frac{47}{8},-1\right)$; axis of symmetry: $y=-1$

The standard equation of a parabola that opens to the left or right is $x-h=a(y-k)^2$, where (*h*, *k*) are the coordinates of the vertex and $a=\frac{1}{4p}$. The value of *p* gives you the distance from the vertex to the focus and directrix.

Pull the values for *h* and *k* from the given equation. The vertex is (6, –1).

In this parabola, $a=-2$, so $p=-\frac{1}{8}$. The focus is $\frac{1}{8}$ unit to the left of the vertex at $\left(\frac{47}{8},-1\right)$, and the directrix is $\frac{1}{8}$ unit to the right the vertex at $x=\frac{49}{8}$. The axis of symmetry goes through the vertex and is perpendicular to the directrix; its equation is $y=-1$.

775. vertex: (0, –2); directrix: $y=-4$

The standard equation of a parabola that opens upward or downward is $y-k=a(x-h)^2$, where (*h*, *k*) are the coordinates of the vertex and $a=\frac{1}{4p}$. The value of *p* gives you the distance from the vertex to the focus and directrix.

Pull the values for *h* and *k* from the given equation. No *h* term is given, so you know the *x* coordinate is 0. The vertex is (0, –2).

In this parabola, $a=\frac{1}{8}$, so $p=2$. The focus is 2 units above the vertex at (0, 0), and the directrix is 2 units below the vertex at $y=-4$. The axis of symmetry goes through the vertex and is perpendicular to the directrix; its equation is $x=0$.

776. vertex: $(0,3)$; directrix: $x=\frac{3}{4}$

The standard equation of a parabola that opens to the left or right is $x-h=a(y-k)^2$, where (h, k) are the coordinates of the vertex and $a=\frac{1}{4p}$. The value of p gives you the distance from the vertex to the focus and directrix.

Pull the values for h and k from the given equation. No h term is given, so you know the x coordinate is 0. The vertex is (0, 3).

In this parabola, $a=-\frac{1}{3}$, so $p=-\frac{3}{4}$. The focus is $\frac{3}{4}$ unit to the left of the vertex at $\left(-\frac{3}{4},3\right)$, and the directrix is $\frac{3}{4}$ unit to the right the vertex at $x=\frac{3}{4}$. The axis of symmetry goes through the vertex and is perpendicular to the directrix; its equation is $y=3$.

777. vertex: (2, 2); directrix: $x=1$

The standard equation of a parabola that opens to the left or right is $x-h=a(y-k)^2$, where (h, k) are the coordinates of the vertex and $a=\frac{1}{4p}$.

To put the equation in the standard form, multiply each term by 2:

$$2\left(\frac{1}{2}x\right)-2(1)=2\left(\frac{1}{8}(y-2)^2\right)$$

$$x-2=\frac{1}{4}(y-2)^2$$

The value of p gives you the distance from the vertex to the focus and directrix.

Pull the values for h and k from the rewritten equation. The vertex is (2, 2).

In this parabola, $a=\frac{1}{4}$, so $p=1$. The focus is 1 unit to the right of the vertex at (3, 2), and the directrix is 1 unit to the left the vertex at $x=1$. The axis of symmetry goes through the vertex and is perpendicular to the directrix; its equation is $y=2$.

778. $y-2=2(x+3)^2$

The axis of symmetry is a vertical line, so the standard form of the parabola is $y-k=a(x-h)^2$. Replace the h and k in the equation with the coordinates of the vertex:

$$y-2=a(x-(-3))^2$$

$$y-2=a(x+3)^2$$

Solve for a by replacing the x and y in the equation with the coordinates of the given point, (0, 20):

$$20-2=a(0+3)^2$$

$$18=9a$$

$$a=2$$

The equation of the parabola is $y-2=2(x+3)^2$.

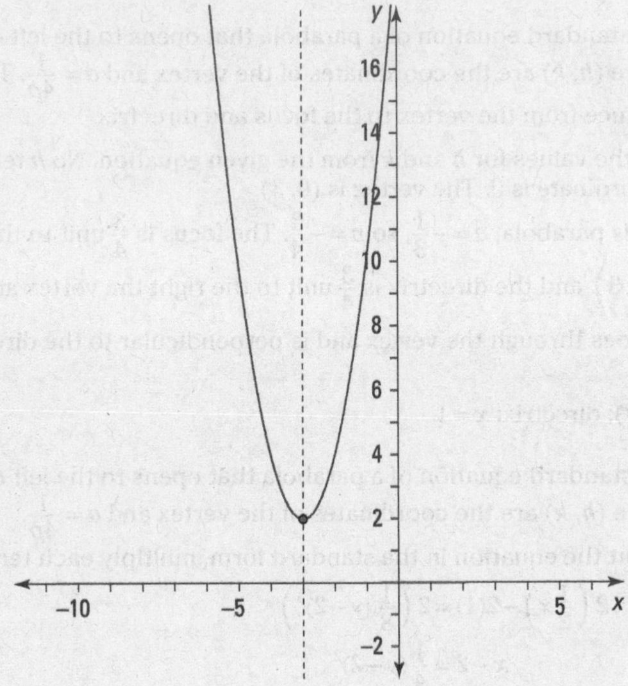

Illustration by Thomson Digital

779. $x-4=\frac{1}{2}(y-1)^2$

The axis of symmetry is a horizontal line, so the standard form of the parabola is $x-h=a(y-k)^2$. Replace the h and k in the equation with the coordinates of the vertex:

$$x-4=a(y-1)^2$$

Solve for a by replacing the x and y in the equation with the coordinates of the given point, (6, 3):

$$6-4=a(3-1)^2$$
$$2=4a$$
$$a=\frac{1}{2}$$

The equation of the parabola is $x-4=\frac{1}{2}(y-1)^2$.

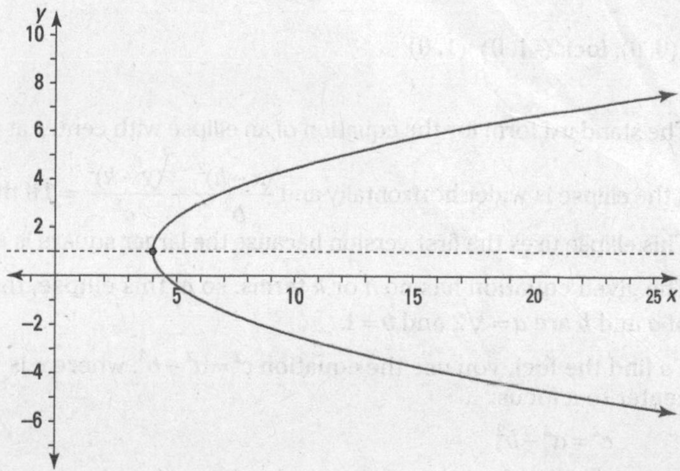

Illustration by Thomson Digital

780. $y = \frac{1}{4}(x-3)^2$

The axis of symmetry is a vertical line, so the standard form of the parabola is $y - k = a(x-h)^2$. Replace the h and k in the equation with the coordinates of the vertex:

$$y - 0 = a(x-3)^2$$

$$y = a(x-3)^2$$

The directrix, $y = -1$, is 1 unit below the vertex. Also, $a = \frac{1}{4p}$, where the value of p gives you the distance from the vertex to the focus and directrix. The value of p here is 1, so $a = \frac{1}{4}$.

The equation of the parabola is $y = \frac{1}{4}(x-3)^2$.

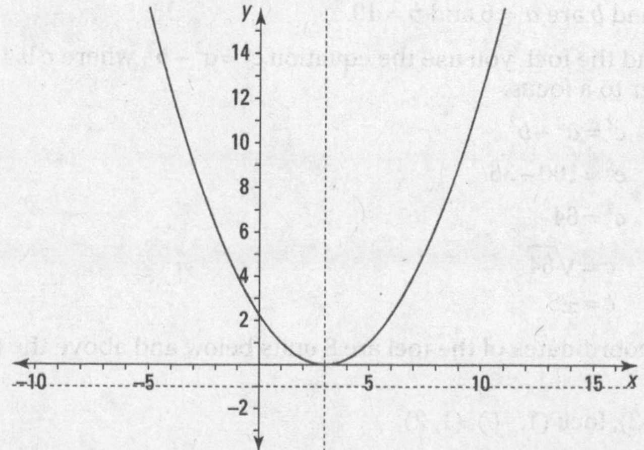

Illustration by Thomson Digital

781. center: (0, 0); foci: (–1, 0), (1, 0)

The standard form for the equation of an ellipse with center at (h, k) is $\frac{(x-h)^2}{a^2}+\frac{(y-k)^2}{b^2}=1$ if the ellipse is wider horizontally and $\frac{(x-h)^2}{b^2}+\frac{(y-k)^2}{a^2}=1$ if the ellipse is taller vertically. This ellipse uses the first version because the larger square is under the x term.

The given equation has no h or k terms, so in this ellipse, the center is (0, 0). The value of a and b are $a=\sqrt{2}$ and $b=1$.

To find the foci, you use the equation $c^2=a^2-b^2$, where c is the distance from the center to a focus:

$$c^2=a^2-b^2$$
$$c^2=2-1$$
$$c^2=1$$
$$c=\sqrt{1}$$
$$c=\pm 1$$

The coordinates of the foci are 1 unit to the left and the right of the center, at (–1, 0) and (1, 0).

782. center: (0, 0); foci: (0, –8), (0, 8).

The standard form for the equation of an ellipse with center at (h, k) is $\frac{(x-h)^2}{a^2}+\frac{(y-k)^2}{b^2}=1$ if the ellipse is wider horizontally and $\frac{(x-h)^2}{b^2}+\frac{(y-k)^2}{a^2}=1$ if the ellipse is taller vertically. This ellipse uses the second version because the larger square is under the y term.

The given equation has no h or k terms, so in this ellipse, the center is (0, 0). The value of a and b are $a = 6$ and $b = 10$.

To find the foci, you use the equation $c^2=a^2-b^2$, where c is the distance from the center to a focus:

$$c^2=a^2-b^2$$
$$c^2=100-36$$
$$c^2=64$$
$$c=\sqrt{64}$$
$$c=\pm 8$$

The coordinates of the foci are 8 units below and above the center, at (0, –8) and (0, 8).

783. center: (1, –2); foci: (1, –6), (1, 2).

The standard form for the equation of an ellipse with center at (h, k) is $\frac{(x-h)^2}{a^2}+\frac{(y-k)^2}{b^2}=1$ if the ellipse is wider horizontally and $\frac{(x-h)^2}{b^2}+\frac{(y-k)^2}{a^2}=1$ if the ellipse is taller vertically. This ellipse uses the second version because the larger square is under the y term.

In this ellipse, the center is (1, –2). Also, $a = 3$ and $b = 5$.

To find the foci, you use the equation $c^2 = a^2 - b^2$, where c is the distance from the center to a focus:

$$c^2 = a^2 - b^2$$
$$c^2 = 25 - 9$$
$$c^2 = 16$$
$$c = \sqrt{16}$$
$$c = \pm 4$$

The coordinates of the foci are 4 units below and above the center, at (1, –6) and (1, 2).

784. center: (–3, 0); foci: (–3, –12), (–3, 12).

The standard form for the equation of an ellipse with center at (h, k) is $\frac{(x-h)^2}{a^2} + \frac{(y-k)^2}{b^2} = 1$ if the ellipse is wider horizontally and $\frac{(x-h)^2}{b^2} + \frac{(y-k)^2}{a^2} = 1$ if the ellipse is taller vertically. This ellipse uses the second version because the larger square is under the y term.

In this ellipse, the center is (–3, 0). The values of a and b are $a = 5$ and $b = 13$.

To find the foci, you use the equation $c^2 = a^2 - b^2$, where c is the distance from the center to a focus:

$$c^2 = a^2 - b^2$$
$$c^2 = 169 - 25$$
$$c^2 = 144$$
$$c = \sqrt{144}$$
$$c = \pm 12$$

The coordinates of the foci are 12 units below and above the center, at (–3, –12) and (–3, 12).

785. center: (0, 1); foci: (–6, 1), (6, 1)

The standard form for the equation of an ellipse with center at (h, k) is $\frac{(x-h)^2}{a^2} + \frac{(y-k)^2}{b^2} = 1$ if the ellipse is wider horizontally and $\frac{(x-h)^2}{b^2} + \frac{(y-k)^2}{a^2} = 1$ if the ellipse is taller vertically. This ellipse uses the first version because the larger square is under the x term.

In this ellipse, the center is (0, 1). Also, $a = 10$ and $b = 8$.

To find the foci, you use the equation $c^2 = a^2 - b^2$, where c is the distance from the center to a focus:

$$c^2 = a^2 - b^2$$
$$c^2 = 100 - 64$$
$$c^2 = 36$$
$$c = \sqrt{36}$$
$$c = \pm 6$$

The coordinates of the foci are 6 units to the left and the right of the center, at (–6, 1) and (6, 1).

786. center: (7, –7); foci: (6, –7), (8, –7)

The standard form for the equation of an ellipse with center at (h, k) is $\frac{(x-h)^2}{a^2} + \frac{(y-k)^2}{b^2} = 1$ if the ellipse is wider horizontally and $\frac{(x-h)^2}{b^2} + \frac{(y-k)^2}{a^2} = 1$ if the ellipse is taller vertically. This ellipse uses the first version because the larger square is under the x term.

In this ellipse, the center is (7, –7). The values of a and b are $a = \sqrt{5}$ and $b = 2$.

To find the foci, you use the equation $c^2 = a^2 - b^2$, where c is the distance from the center to a focus:

$$c^2 = a^2 - b^2$$
$$c^2 = 5 - 4$$
$$c^2 = 1$$
$$c = \sqrt{1}$$
$$c = \pm 1$$

The coordinates of the foci are 1 unit to the left and the right of the center, at (6, –7) and (8, –7).

787. center: (2, 2); foci: (2, 0), (2, 4).

The standard form for the equation of an ellipse with center at (h, k) is $\frac{(x-h)^2}{a^2} + \frac{(y-k)^2}{b^2} = 1$ if the ellipse is wider horizontally and $\frac{(x-h)^2}{b^2} + \frac{(y-k)^2}{a^2} = 1$ if the ellipse is taller vertically. This ellipse uses the second version because the larger square is under the y term.

In this ellipse, the center is (2, 2). The values of a and b are $a = 4$ and $b = \sqrt{20}$.

To find the foci, you use the equation $c^2 = a^2 - b^2$, where c is the distance from the center to a focus:

$$c^2 = a^2 - b^2$$
$$c^2 = 20 - 16$$
$$c^2 = 4$$
$$c = \sqrt{4}$$
$$c = \pm 2$$

The coordinates of the foci are 2 units below and above the center, at (2, 0) and (2, 4).

788. $\frac{(x+3)^2}{25} + \frac{(y-2)^2}{1} = 1$

The distance between the center and the vertex is $2-(-3)=5$ units. This is a horizontal distance, so $a=5$. The distance between the center and the co-vertex is $3-2=1$; this is a vertical distance, so $b=1$.

The standard form for the equation of an ellipse with its vertices along a horizontal line and with center at (h, k) is $\frac{(x-h)^2}{a^2} + \frac{(y-k)^2}{b^2} = 1$. You know that this is a wide (rather than tall) ellipse because the given vertex and center both have y coordinates of 2. Replacing h with –3, k with 2, a with 5, and b with 1, you get the equation $\frac{(x+3)^2}{25} + \frac{(y-2)^2}{1} = 1$.

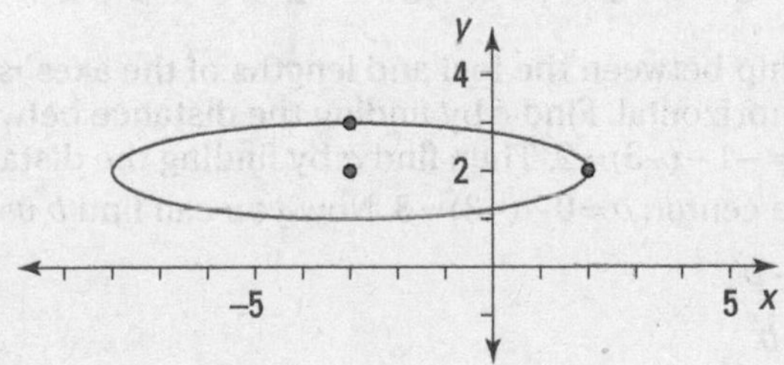

Illustration by Thomson Digital

789. $\frac{(x-1)^2}{16} + \frac{(y+4)^2}{25} = 1$

The center of the ellipse is at the midpoint, M, of the major axis between the vertices:

$$M = \left(\frac{x_1+x_2}{2}, \frac{y_1+y_2}{2}\right) = \left(\frac{1+1}{2}, \frac{1+(-9)}{2}\right) = \left(\frac{2}{2}, \frac{-8}{2}\right) = (1, -4)$$

The distance from the center to the vertex (1, 1) is $1-(-4)=5$. This is a vertical distance, so $b=5$. The distance between the center and the co-vertex (–3, –4) is $1-(-3)=4$; this is a horizontal distance, so $a=4$.

The standard form for the equation of an ellipse with its vertices along a vertical line and center at (h, k) is $\frac{(x-h)^2}{a^2} + \frac{(y-k)^2}{b^2} = 1$. You know that this is a tall (rather than wide) ellipse because the given vertices both have x coordinates of 1. Replacing h with 1, k with –4, a with 4, and b with 5, you get the equation $\frac{(x-1)^2}{16} + \frac{(y+4)^2}{25} = 1$.

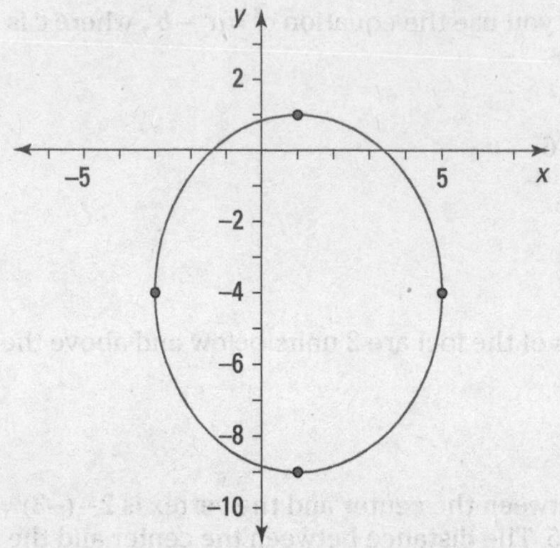

Illustration by Thomson Digital

790. $\frac{(x+3)^2}{9}+\frac{(y-5)^2}{5}=1$

The center of the ellipse is at the midpoint, M, of the major axis between the vertices:

$$M=\left(\frac{x_1+x_2}{2},\frac{y_1+y_2}{2}\right)=\left(\frac{-6+0}{2},\frac{5+5}{2}\right)=\left(\frac{-6}{2},\frac{10}{2}\right)=(-3,5)$$

The relationship between the foci and lengths of the axes is $c^2=a^2-b^2$ because the major axis is horizontal. Find c by finding the distance between the center and the right focus: $c=-1-(-3)=2$. Then find a by finding the distance between the right vertex and the center: $a=0-(-3)=3$. Now you can find b using $c^2=a^2-b^2$:

$$2^2=3^2-b^2$$
$$4=9-b^2$$
$$b^2=9-4$$
$$b^2=5$$

The standard form for the equation of an ellipse with its vertices along a horizontal line and center at (h, k) is $\frac{(x-h)^2}{a^2}+\frac{(y-k)^2}{b^2}=1$. You know that this is a wide (rather than tall) ellipse because the given vertices both have y coordinates of 5. Replacing h with -3, k with 5, a with 3 and b^2 with 5, you get the equation $\frac{(x+3)^2}{9}+\frac{(y-5)^2}{5}=1$.

Illustration by Thomson Digital

791. asymptotes: $y=x$, $y=-x$

The standard equation for this hyperbola is $\frac{(x-h)^2}{a^2}-\frac{(y-k)^2}{b^2}=1$ where (h, k) are the coordinates of the center. Because the given equation has no h or k values, the center of this hyperbola is (0, 0).

To find the foci, use $c^2=a^2+b^2$, where c is the distance from the center to a focus. Both a and b are 1, so

$$c^2=a^2+b^2$$
$$c^2=1+1$$
$$c^2=2$$
$$c=\pm\sqrt{2}$$

The foci are $\sqrt{2}$ units to the left and the right of the center, at $\left(-\sqrt{2},0\right)$ and $\left(\sqrt{2},0\right)$.

Solve for the asymptotes using the equation $\frac{(x-h)^2}{a^2}=\frac{(y-k)^2}{b^2}$:

$$\frac{x^2}{1}=\frac{y^2}{1}$$
$$x^2=y^2$$
$$y=\pm x$$

792. asymptotes: $y=\frac{3}{4}x$, $y=-\frac{3}{4}x$

The standard equation for this hyperbola is $\frac{(x-h)^2}{a^2}-\frac{(y-k)^2}{b^2}=1$, where (h, k) are the coordinates of the center. Because the given equation has no h or k values, the center of this hyperbola is (0, 0).

To find the foci, use $c^2 = a^2 + b^2$, where c is the distance from the center to a focus:

$$c^2 = a^2 + b^2$$
$$c^2 = 16 + 9$$
$$c^2 = 25$$
$$c = \pm 5$$

The foci are 5 units to the left and the right of the center, at (–5, 0) and (5, 0).

Solve for the asymptotes using the equation $\frac{(x-h)^2}{a^2} = \frac{(y-k)^2}{b^2}$:

$$\frac{x^2}{16} = \frac{y^2}{9}$$
$$9x^2 = 16y^2$$
$$y^2 = \frac{9}{16}x^2$$
$$y = \pm\frac{3}{4}x$$

793. asymptotes: $y = \frac{7}{24}x$, $y = -\frac{7}{24}x$

The standard equation for this hyperbola is $\frac{(y-k)^2}{b^2} - \frac{(x-h)^2}{a^2} = 1$, where (h, k) are the coordinates of the center. Because the given equation has no h or k values, the center of this hyperbola is (0, 0).

To find the foci, use $c^2 = a^2 + b^2$, where c is the distance from the center to a focus:

$$c^2 = a^2 + b^2$$
$$c^2 = 576 + 49$$
$$c^2 = 625$$
$$c = \pm 25$$

The foci are 25 units above and below the center, at (0, –25) and (0, 25).

Solve for the asymptotes using the equation $\frac{(y-k)^2}{b^2} = \frac{(x-h)^2}{a^2}$:

$$\frac{y^2}{49} = \frac{x^2}{576}$$
$$576y^2 = 49x^2$$
$$y^2 = \frac{49}{576}x^2$$
$$y = \pm\frac{7}{24}x$$

794. asymptotes: $y = x + 3$, $y = -x + 5$

The standard equation for this hyperbola is $\frac{(y-k)^2}{b^2} - \frac{(x-h)^2}{a^2} = 1$, where (h, k) are the coordinates of the center. The center of this hyperbola is (1, 4).

To find the foci, use $c^2 = a^2 + b^2$, where c is the distance from the center to a focus:

$$\begin{aligned} c^2 &= a^2 + b^2 \\ c^2 &= 8 + 8 \\ c^2 &= 16 \\ c &= \pm 4 \end{aligned}$$

The foci are 4 units above and below the center, at (1, 0) and (1, 8).

Solve for the asymptotes using the equation $\frac{(y-k)^2}{b^2} = \frac{(x-h)^2}{a^2}$:

$$\begin{aligned} \frac{(y-4)^2}{8} &= \frac{(x-1)^2}{8} \\ \cancel{8}(y-4)^2 &= \cancel{8}(x-1)^2 \\ \sqrt{(y-4)^2} &= \sqrt{(x-1)^2} \\ y-4 &= \pm(x-1) \\ y &= 4 \pm (x-1) \end{aligned}$$

The asymptotes are $y = x + 3$ and $y = 5 - x$.

795. asymptotes: $y = 2x + 10$, $y = -2x - 10$

The standard equation for this hyperbola is $\frac{(y-k)^2}{b^2} - \frac{(x-h)^2}{a^2} = 1$, where (h, k) are the coordinates of the center. The center of this hyperbola is (–5, 0).

To find the foci, use $c^2 = a^2 + b^2$, where c is the distance from the center to a focus:

$$\begin{aligned} c^2 &= a^2 + b^2 \\ c^2 &= 5 + 20 \\ c^2 &= 25 \\ c &= \pm 5 \end{aligned}$$

The foci are 5 units above and below the center, at (–5, –5) and (–5, 5).

Solve for the asymptotes using the equation $\frac{(y-k)^2}{b^2} = \frac{(x-h)^2}{a^2}$:

$$\begin{aligned} \frac{y^2}{20} &= \frac{(x+5)^2}{5} \\ \cancel{5}y^2 &= {}^{4}\cancel{20}(x+5)^2 \\ \sqrt{y^2} &= \sqrt{4(x+5)^2} \\ y &= \pm 2(x+5) \end{aligned}$$

The asymptotes are $y = 2x + 10$ and $y = -2x - 10$.

796. asymptotes: $y = \frac{1}{2}x + \frac{9}{2}$, $y = -\frac{1}{2}x + \frac{3}{2}$

The standard equation for this hyperbola is $\frac{(x-h)^2}{a^2} - \frac{(y-k)^2}{b^2} = 1$, where (h, k) are the coordinates of the center. The center of this hyperbola is (–3, 3).

To find the foci, use $c^2=a^2+b^2$, where c is the distance from the center to a focus:

$$c^2=a^2+b^2$$
$$c^2=36+9$$
$$c^2=45$$
$$c=\pm\sqrt{45}$$
$$c=\pm3\sqrt{5}$$

The foci are 5 units to the left and right of the center, at $(-3-3\sqrt{5},3)$ and $\left(-3+3\sqrt{5},3\right)$.

Solve for the asymptotes using the equation $\frac{(x-h)^2}{a^2}=\frac{(y-k)^2}{b^2}$:

$$\frac{(x+3)^2}{36}=\frac{(y-3)^2}{9}$$
$$\cancel{9}(x+3)^2={}^{4}\cancel{36}(y-3)^2$$
$$(y-3)^2=\frac{1}{4}(x+3)^2$$
$$\sqrt{(y-3)^2}=\pm\sqrt{\frac{1}{4}(x+3)^2}$$
$$y-3=\pm\frac{1}{2}(x+3)$$

Distribute the terms on the right side and then add 3 to both sides to isolate y. The equations of the asymptotes are $y=\frac{1}{2}x+\frac{9}{2}$ and $y=-\frac{1}{2}x+\frac{3}{2}$.

797. asymptotes: $y=\sqrt{3}x+4-4\sqrt{3}$, $y=-\sqrt{3}x+4+4\sqrt{3}$

The standard equation for this hyperbola is $\frac{(y-k)^2}{b^2}-\frac{(x-h)^2}{a^2}=1$, where (h, k) are the coordinates of the center. The center of this hyperbola is (4, 4).

To find the foci, use $c^2=a^2+b^2$, where c is the distance from the center to a focus:

$$c^2=a^2+b^2$$
$$c^2=6+18$$
$$c^2=24$$
$$c=\pm\sqrt{24}$$
$$c=\pm2\sqrt{6}$$

The foci are 5 units above and below the center, at $\left(4,4-2\sqrt{6}\right)$ and $\left(4,4+2\sqrt{6}\right)$.

Solve for the asymptotes using the equation $\frac{(y-k)^2}{b^2}=\frac{(x-h)^2}{a^2}$:

$$\frac{(y-4)^2}{18}=\frac{(x-4)^2}{6}$$
$$\cancel{6}(y-4)^2={}^{3}\cancel{18}(x-4)^2$$
$$\sqrt{(y-4)^2}=\pm\sqrt{3(x-4)^2}$$
$$y-4=\pm\sqrt{3}(x-4)$$

The asymptotes are $y=\sqrt{3}x+4-4\sqrt{3}$ and $y=-\sqrt{3}x+4+4\sqrt{3}$.

798. $\frac{(x-3)^2}{7}-\frac{(y+2)^2}{9}=1$

The two foci are to the left and right of the center, so the hyperbola has this standard form:

$$\frac{(x-h)^2}{a^2}-\frac{(y-k)^2}{b^2}=1$$

The foci are 4 units on either side of the center. You determine this by finding the difference between the x coordinate of the center and the x coordinate of a focus: $3-(-1)=4$.

The foci have the relation $c^2=a^2+b^2$, where c is the distance from the center to a focus. Substitute 4 for c and 3 for b:

$$c^2=a^2+b^2$$
$$4^2=a^2+3^2$$
$$16=a^2+9$$
$$a^2=7$$

Plug these values into the standard form equation, replacing h with 3, k with −2, a^2 with 7, and b^2 with 9:

$$\frac{(x-3)^2}{7}-\frac{(y+2)^2}{9}=1$$

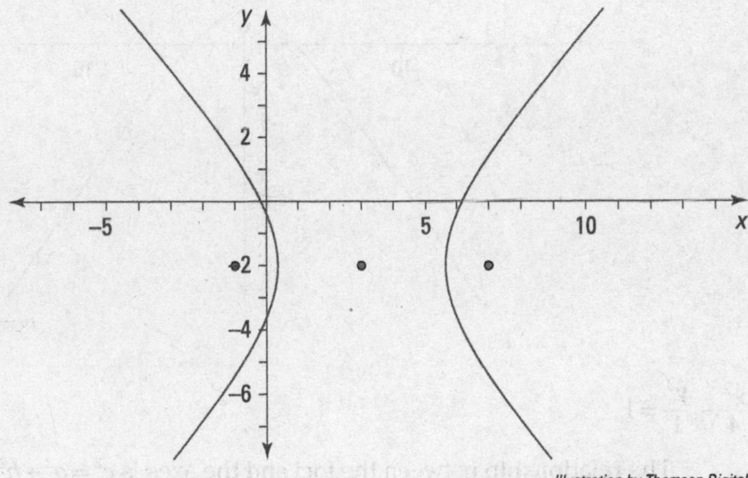

Illustration by Thomson Digital

799. $\frac{(y-7)^2}{25}-\frac{x^2}{11}=1$

The two foci are above and below the center, so the hyperbola has this standard form:

$$\frac{(y-k)^2}{b^2}-\frac{(x-h)^2}{a^2}=1$$

The foci are 6 units above and below the center. You determine this by finding the difference between the y coordinate of a focus and the y coordinate of the center: $13-7=6$.

The foci have the relation $c^2 = a^2 + b^2$, where c is the distance from the center to a focus. Substitute 6 for c and 5 for b:

$$c^2 = a^2 + b^2$$
$$6^2 = a^2 + 5^2$$
$$36 = a^2 + 25$$
$$a^2 = 11$$

Plug these values into the standard form equation, replacing h with 0, k with 7, a^2 with 11, and b^2 with 25:

$$\frac{(y-7)^2}{25} - \frac{x^2}{11} = 1$$

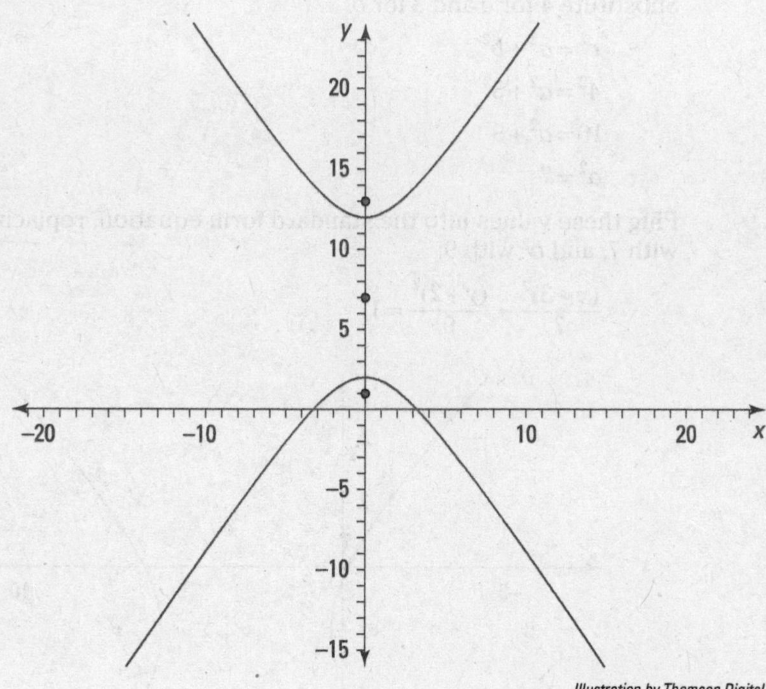

Illustration by Thomson Digital

800. $\frac{x^2}{4} - \frac{y^2}{1} = 1$

The relationship between the foci and the axes is $c^2 = a^2 + b^2$, where c is the distance from the hyperbola's center to a focus. The foci are on either side of the center, which lies on the x-axis (because the y coordinates are 0). Halfway between the foci is the origin, so the center of the hyperbola is (0, 0). Because the foci are $\left(-\sqrt{5}, 0\right)$ and $\left(\sqrt{5}, 0\right)$, you know that $c = \sqrt{5}$ and that $c^2 = 5$, giving you $5 = a^2 + b^2$.

The asymptotes are $y=\frac{1}{2}x$ and $y=-\frac{1}{2}x$. Work backward to start putting these equations into standard form:

$$y=\pm\frac{1}{2}x$$

$$y^2=\frac{1}{4}x^2$$

$$\frac{y^2}{1}=\frac{x^2}{4}$$

Because the foci lie along a horizontal line, the square under the x term is larger than that under the y term. Using $b^2=1$ and $a^2=4$, those values satisfy the equation $5=a^2+b^2$. The equation of the hyperbola has the form $\frac{(x-h)^2}{a^2}-\frac{(y-k)^2}{b^2}=1$, because the foci are along a horizontal axis, so $\frac{y^2}{1}=\frac{x^2}{4}$ becomes $\frac{x^2}{4}-\frac{y^2}{1}=1$.

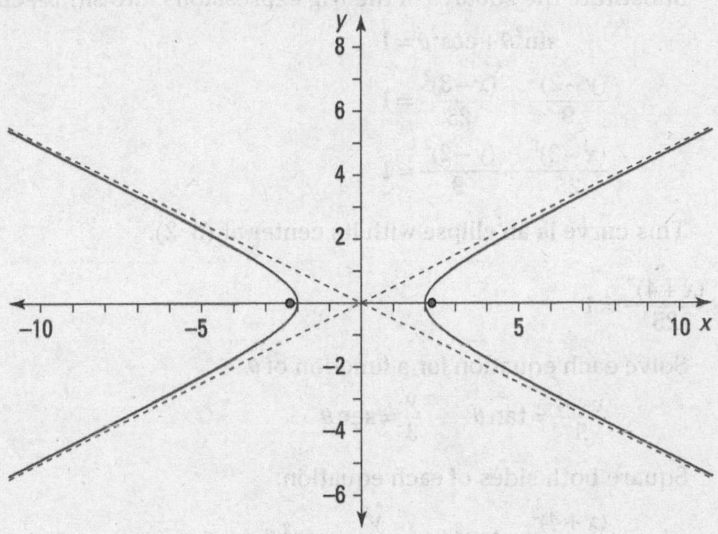

Illustration by Thomson Digital

801. $x^2+y^2=16$

Solve each equation for a function of θ:

$$\frac{x}{4}=\cos\theta \qquad \frac{y}{4}=\sin\theta$$

Square both sides of each equation:

$$\frac{x^2}{16}=\cos^2\theta \qquad \frac{y^2}{16}=\sin^2\theta$$

Substitute the squares of the trig expressions into $\sin^2\theta+\cos^2\theta=1$:

$$\sin^2\theta+\cos^2\theta=1$$

$$\frac{y^2}{16}+\frac{x^2}{16}=1$$

Multiply each side of the equation by 16 and simplify:

$$x^2+y^2=16$$

This curve is a circle with its center at the origin and a radius of 4.

802. $\frac{(x-3)^2}{25}+\frac{(y-2)^2}{9}=1$

Solve each equation for a function of θ:

$$\frac{x-3}{5}=\cos\theta \qquad \frac{y-2}{3}=\sin\theta$$

Square both sides of each equation:

$$\frac{(x-3)^2}{25}=\cos^2\theta \qquad \frac{(y-2)^2}{9}=\sin^2\theta$$

Substitute the squares of the trig expressions into $\sin^2\theta+\cos^2\theta=1$.

$$\sin^2\theta+\cos^2\theta=1$$

$$\frac{(y-2)^2}{9}+\frac{(x-3)^2}{25}=1$$

$$\frac{(x-3)^2}{25}+\frac{(y-2)^2}{9}=1$$

This curve is an ellipse with its center at (3, 2).

803. $\frac{y^2}{9}-\frac{(x+4)^2}{25}=1$

Solve each equation for a function of θ:

$$\frac{x+4}{5}=\tan\theta \qquad \frac{y}{3}=\sec\theta$$

Square both sides of each equation:

$$\frac{(x+4)^2}{25}=\tan^2\theta \qquad \frac{y^2}{9}=\sec^2\theta$$

Substitute the squares of the trig expressions into $\sec^2\theta-\tan^2\theta=1$:

$$\sec^2\theta-\tan^2\theta=1$$

$$\frac{y^2}{9}-\frac{(x+4)^2}{25}=1$$

This curve is a hyperbola with its center at (–4, 0).

804. $(x-5)^2+(y+2)^2=9$

Solve each equation for a function of θ:

$$\frac{x-5}{3}=\cos\theta \qquad \frac{y+2}{3}=\sin\theta$$

Square both sides of each equation.

$$\frac{(x-5)^2}{9}=\cos^2\theta \qquad \frac{(y+2)^2}{9}=\sin^2\theta$$

Substitute the squares of the trig expressions into $\sin^2\theta+\cos^2\theta=1$.

$$\sin^2\theta+\cos^2\theta=1$$

$$\frac{(y+2)^2}{9}+\frac{(x-5)^2}{9}=1$$

Multiply both sides of the equation by 9 and simplify:

$$(x-5)^2+(y+2)^2=9$$

This curve is a circle with its center at (5, –2) and radius 3.

805. $\dfrac{(y-6)^2}{1/16}-\dfrac{(x-3)^2}{1/4}=1$

Solve each equation for a function of θ:

$$2(x-3)=\tan\theta \qquad -4(y-6)=\sec\theta$$

Square both sides of each equation:

$$4(x-3)^2=\tan^2\theta \qquad 16(y-6)^2=\sec^2\theta$$

Substitute the squares of the trig expressions into $\sec^2\theta-\tan^2\theta=1$.

$$\sec^2\theta-\tan^2\theta=1$$

$$16(y-6)^2-4(x-3)^2=1$$

$$\frac{(y-6)^2}{1/16}-\frac{(x-3)^2}{1/4}=1$$

This curve is a hyperbola with its center at (3, 6).

806. $x=5\sin\theta,\ y=2\cos\theta$

Your goal is to substitute terms into a version of the Pythagorean theorem, $\sin^2\theta+\cos^2\theta=1$, so take advantage of the fact that each term is a perfect square. Divide each term by 100 to create a 1 on the right:

$$\frac{4x^2}{100}+\frac{25y^2}{100}=\frac{100}{100}$$

$$\frac{x^2}{25}+\frac{y^2}{4}=1$$

This equation looks similar to the identity. Let $\frac{x^2}{25}=\sin^2\theta$ and $\frac{y^2}{4}=\cos^2\theta$, corresponding to $\sin^2\theta+\cos^3\theta=1$. Now determine what x and y can be in this new format — how they work with the trig functions. Solving for x and y,

$$\frac{x^2}{25}=\sin^2\theta \qquad \frac{y^2}{4}=\cos^2\theta$$

$$\sqrt{\frac{x^2}{25}}=\sqrt{\sin^2\theta} \qquad \sqrt{\frac{y^2}{4}}=\sqrt{\cos^2\theta}$$

$$\frac{x}{5}=\sin\theta \qquad \frac{y}{2}=\cos\theta$$

$$x=5\sin\theta \qquad y=2\cos\theta$$

If you pick an angle θ and substitute it into the two equations for x and y, you have a point on the curve, now written in parametric form.

807. $x=3+3\sin\theta,\ y=-7+3\cos\theta$

First put the conic in standard form by completing the square:

$$x^2+y^2-6x+14y+49=0$$
$$x^2-6x+y^2+14y=-49$$
$$x^2-6x+9+y^2+14y+49=-49+9+49$$
$$(x-3)^2+(y+7)^2=9$$

Your goal is to substitute terms into a version of the Pythagorean theorem, $\sin^2\theta+\cos^2\theta=1$, so take advantage of the fact that each term is a perfect square. Divide each term by 9 to create a 1 on the right:

$$\frac{(x-3)^2}{9}+\frac{(y+7)^2}{9}=1$$

This equation looks similar to the identity. Let $\frac{(x-3)^2}{9}=\sin^2\theta$ and $\frac{(y+7)^2}{9}=\cos^2\theta$, corresponding to $\sin^2\theta+\cos^2\theta=1$. Now determine what x and y can be in this new format — how they work with the trig functions. Solving for x and y,

$$\frac{(x-3)^2}{9}=\sin^2\theta \qquad \frac{(y+7)^2}{9}=\cos^2\theta$$
$$\sqrt{\frac{(x-3)^2}{9}}=\sqrt{\sin^2\theta} \qquad \sqrt{\frac{(y+7)^2}{9}}=\sqrt{\cos^2\theta}$$
$$\frac{x-3}{3}=\sin\theta \qquad \frac{y+7}{3}=\cos\theta$$
$$x-3=3\sin\theta \qquad y+7=3\cos\theta$$
$$x=3+3\sin\theta \qquad y=-7+3\cos\theta$$

If you pick an angle θ and substitute it into the two equations for x and y, you have a point on the curve, now written in parametric form.

808. $x=1+\sqrt{11}\sec\theta,\ y=-2+\sqrt{7}\tan\theta$

First put the conic in standard form by completing the square:

$$7x^2-11y^2-14x-44y=114$$
$$7x^2-14x-11y^2-44y=114$$
$$7\left(x^2-2x+1\right)-11\left(y^2+4y+4\right)=114+7-44$$
$$7(x-1)^2-11(y+2)^2=77$$
$$\frac{7(x-1)^2}{77}-\frac{11(y+2)^2}{77}=\frac{77}{77}$$
$$\frac{(x-1)^2}{11}-\frac{(y+2)^2}{7}=1$$

Your goal is to substitute terms into a version of the Pythagorean theorem, $\sec^2\theta-\tan^2\theta=1$. Don't let the fact that the denominators aren't perfect squares throw you; radicals work just fine. Let $\frac{(x-1)^2}{11}=\sec^2\theta$ and $\frac{(y+2)^2}{7}=\tan^2\theta$, corresponding to $\sec^2\theta - \tan^2\theta = 1$. Now determine what x and y can be in this new format — how they work with the trig functions. Solving for x and y,

$$\frac{(x-1)^2}{11} = \sec^2\theta \qquad \frac{(y+2)^2}{7} = \tan^2\theta$$

$$\sqrt{\frac{(x-1)^2}{11}} = \sqrt{\sec^2\theta} \qquad \sqrt{\frac{(y+2)^2}{7}} = \sqrt{\tan^2\theta}$$

$$\frac{x-1}{\sqrt{11}} = \sec\theta \qquad \frac{y+2}{\sqrt{7}} = \tan\theta$$

$$x-1 = \sqrt{11}\sec\theta \qquad y+2 = \sqrt{7}\tan\theta$$

$$x = 1+\sqrt{11}\sec\theta \qquad y = -2+\sqrt{7}\tan\theta$$

If you pick an angle θ and substitute it into the two equations for x and y, you have a point on the curve, now written in parametric form.

809. $x = -2+4\sin\theta,\ y = 2+3\cos\theta$

First put the conic in standard form by using completing the square:

$$9x^2+16y^2+36x-64y = 44$$

$$9x^2+36x+16y^2-64y = 44$$

$$9(x^2+4x+4)+16(y^2-4y+4) = 44+36+64$$

$$9(x+2)^2+16(y-2)^2 = 144$$

$$\frac{9(x+2)^2}{144}+\frac{16(y-2)^2}{144} = \frac{144}{144}$$

$$\frac{(x+2)^2}{16}+\frac{(y-2)^2}{9} = 1$$

Your goal is to substitute terms into a version of the Pythagorean theorem, $\sin^2\theta+\cos^2\theta = 1$, so take advantage of the fact that each term is a perfect square. Let $\frac{(x+2)^2}{16} = \sin^2\theta$ and $\frac{(y-2)^2}{9} = \cos^2\theta$, corresponding to $\sin^2\theta+\cos^2\theta = 1$. Now determine what x and y can be in this new format — how they work with the trig functions. Solving for x and y,

$$\frac{(x+2)^2}{16} = \sin^2\theta \qquad \frac{(y-2)^2}{9} = \cos^2\theta$$

$$\sqrt{\frac{(x+2)^2}{16}} = \sqrt{\sin^2\theta} \qquad \sqrt{\frac{(y-2)^2}{9}} = \sqrt{\cos^2\theta}$$

$$\frac{x+2}{4} = \sin\theta \qquad \frac{y-2}{3} = \cos\theta$$

$$x+2 = 4\sin\theta \qquad y-2 = 3\cos\theta$$

$$x = -2+4\sin\theta \qquad y = 2+3\cos\theta$$

If you pick an angle θ and substitute it into the two equations for x and y, you have a point on the curve, now written in parametric form.

810. $y = 5 + 2\sqrt{2}\sec\theta,\ x = 1 + 2\tan\theta$

First put the conic in standard form by using completing the square:

$$4y^2 - 40y - 8x^2 + 16x + 60 = 0$$
$$4y^2 - 40y - 8x^2 + 16x = -60$$
$$4\left(y^2 - 10y + 25\right) - 8\left(x^2 - 2x + 1\right) = -60 + 100 - 8$$
$$4(y-5)^2 - 8(x-1)^2 = 32$$
$$\frac{4(y-5)^2}{32} - \frac{8(x-1)^2}{32} = \frac{32}{32}$$
$$\frac{(y-5)^2}{8} - \frac{(x-1)^2}{4} = 1$$

Your goal is to substitute terms into a version of the Pythagorean theorem, $\sec^2\theta - \tan^2\theta = 1$. Don't let the fact that the denominators aren't perfect squares throw you; radicals work just fine. Let $\frac{(y-5)^2}{8} = \sec^2\theta$ and $\frac{(x-1)^2}{4} = \tan^2\theta$, corresponding to $\sec^2\theta - \tan^2\theta = 1$. Now determine what x and y can be in this new format — how they work with the trig functions. Solving for x and y,

$$\frac{(y-5)^2}{8} = \sec^2\theta \qquad \frac{(x-1)^2}{4} = \tan^2\theta$$
$$\sqrt{\frac{(y-5)^2}{8}} = \sqrt{\sec^2\theta} \qquad \sqrt{\frac{(x-1)^2}{4}} = \sqrt{\tan^2\theta}$$
$$\frac{y-5}{\sqrt{8}} = \sec\theta \qquad \frac{x-1}{2} = \tan\theta$$
$$y - 5 = \sqrt{8}\sec\theta \qquad x - 1 = 2\tan\theta$$
$$x = 1 + 2\tan\theta$$
$$y = 5 + 2\sqrt{2}\sec\theta$$

If you pick an angle θ and substitute it into the two equations for x and y, you have a point on the curve, now written in parametric form.

811. (–1, 10)

Solve for y in the first equation, $2x + y = 8$:

$$y = 8 - 2x$$

Substitute the expression equal to y into the second equation; then solve for x:

$$3x + 2y = 17$$
$$3x + 2(8 - 2x) = 17$$
$$3x + 16 - 4x = 17$$
$$-x = 1$$
$$x = -1$$

Solve for y:

$$y = 8 - 2x = 8 - 2(-1) = 10$$

The answer is (–1, 10).

812. (4, 2)

Solve for x in the second equation, $x-3y=-2$:

$$x=3y-2$$

Substitute the expression equal to x into the first equation; then solve for y:

$$\begin{aligned}3x+5y&=22\\3(3y-2)+5y&=22\\9y-6+5y&=22\\14y&=28\\y&=2\end{aligned}$$

Solve for x:

$$x=3y-2=3(2)-2=4$$

The answer is (4, 2).

813. (–7, 1)

Solve for x in the first equation, $x+11y=4$:

$$x=4-11y$$

Substitute the expression equal to x into the second equation; then solve for y:

$$\begin{aligned}8y-3x&=29\\8y-3(4-11y)&=29\\8y-12+33y&=29\\41y&=41\\y&=1\end{aligned}$$

Solve for x:

$$x=4-11y=4-11(1)=-7$$

The answer is (–7, 1).

814. no solution

Solve for y in the first equation, $4x-y=6$:

$$y=4x-6$$

Substitute the expression equal to y into the second equation; then solve for x:

$$\begin{aligned}2y&=8x-9\\2(4x-6)&=8x-9\\8x-12&=8x-9\\-12&\neq-9\end{aligned}$$

The statement $-12=-9$ is never true, so the system has no solution.

815. $(k, 5k-9)$

Solve for y in the first equation, $5x-y=9$:

$$y=5x-9$$

Substitute the expression equal to y into the second equation; then solve for x:

$$\begin{aligned} 2y &= 10x-18 \\ 2(5x-9) &= 10x-18 \\ 10x-18 &= 10x-18 \\ 0 &= 0 \end{aligned}$$

This statement is always true, so the system has an infinite number of solutions. Choose the parameter k for x. You can write the solutions as the ordered pair $(k, 5k-9)$.

816. $(2, -3)$

Multiply the terms in the second equation $(2x+y=1)$ by 3 to eliminate the y term. Then add the two equations together:

$$\begin{array}{r} 4x-3y=17 \\ \underline{6x+3y=\ 3} \\ 10x \qquad =20 \end{array}$$

Divide by 10 to get $x = 2$.

To find y, substitute 2 for x in the second equation:

$$\begin{aligned} 2(2)+y &= 1 \\ y &= 1-4=-3 \end{aligned}$$

The answer is $(2, -3)$.

817. $(0, 4)$

Multiply the terms in the second equation $(3x+4y=16)$ by -1 to eliminate the x term. Then add the two equations together:

$$\begin{array}{r} 3x+5y=\ \ 20 \\ \underline{-3x-4y=-16} \\ y=\ \ \ 4 \end{array}$$

To find x, substitute 4 for y in the first equation:

$$\begin{aligned} 3x+5y &= 20 \\ 3x+5(4) &= 20 \\ 3x &= 20-20 \\ 3x &= 0 \\ x &= 0 \end{aligned}$$

The answer is $(0, 4)$.

818. (–2, 3)

Multiply the terms in the second equation ($3x+8y=18$) by –2 to eliminate the x term. Then add the two equations together:

$$\begin{aligned} 6x+\ 7y &= \ \ 9 \\ -6x-16y &= -36 \\ \hline -\ 9y &= -27 \end{aligned}$$

Divide both side of the equation by –9 to get $y=3$.

To find x, substitute 3 for y in the first equation:

$$\begin{aligned} 6x+7y &= 9 \\ 6x+7(3) &= 9 \\ 6x &= 9-21 \\ 6x &= -12 \\ x &= -2 \end{aligned}$$

The answer is (–2, 3).

819. no solution

Rearrange the terms in the second equation ($6y=10x-3$) so they align with those in the first equation:

$$\begin{aligned} 5x\ -3y &= 8 \\ -10x+6y &= -3 \end{aligned}$$

Multiply the terms in the first equation ($5x-3y=8$) by 2. Then add the two equations together:

$$\begin{aligned} 10x-6y &= 16 \\ -10x+6y &= -3 \\ \hline 0+\ 0 &= 13 \end{aligned}$$

This statement is false, so there's no solution.

820. $\left(2-\frac{4}{3}k, k\right)$

Rearrange the terms in the second equation ($8y=12-6x$) so they align with those in the first equation:

$$\begin{aligned} 3x+4y &= \ 6 \\ 6x+8y &= 12 \end{aligned}$$

Multiply the terms in the first equation ($3x+4y=6$) by –2. Then add the two equations together:

$$\begin{aligned} -6x-8y &= -12 \\ 6x+8y &= \ \ 12 \\ \hline 0+\ 0 &= \ \ \ 0 \end{aligned}$$

This statement is always true, so the system has an infinite number of solutions. Choose the parameter k for y. Then solve for x in the first equation:

$$3x = 6 - 4y$$
$$x = 2 - \frac{4}{3}y$$
$$x = 2 - \frac{4}{3}k$$

The answer is $\left(2 - \frac{4}{3}k, k\right)$.

821. (0, –2), (4, 50)

Using substitution, replace the y in the first equation with its equivalent in the second equation:

$$13x - 2 = 4x^2 - 3x - 2$$

Set this equation equal to 0 by adding $-13x$ and 2 to each side of the equation. Then factor and solve for x:

$$0 = 4x^2 - 16x$$
$$0 = 4x(x - 4)$$
$$x = 0, 4$$

Use the system's second equation, $y = 13x - 2$, to find the corresponding y value for each x:

$x = 0$	$x = 4$
$y = 13(0) - 2$	$y = 13(4) - 2$
$y = -2$	$y = 50$

822. (–3, –33), (1, –1)

Using substitution, replace the y in the first equation with its equivalent in the second equation:

$$8x - 9 = 3 - 4x^2$$

Set this equation equal to 0 by adding $-8x$ and 9 to each side of the equation. Then factor and solve for x:

$$0 = -4x^2 - 8x + 12$$
$$0 = -4\left(x^2 + 2x - 3\right)$$
$$0 = -4(x + 3)(x - 1)$$
$$x = -3, 1$$

Use the system's second equation, $y = 8x - 9$, to find the corresponding y value for each x:

$x = -3$	$x = 1$
$y = 8(-3) - 9$	$y = 8(1) - 9$
$y = -33$	$y = -1$

823. (0, 0), (1, 1)

Using substitution, replace the *y* in the first equation with its equivalent in the second equation:

$$x=\sqrt{x}$$

Square both sides of the equation:

$$(x)^2=\left(\sqrt{x}\right)^2$$
$$x^2=x$$

Set this equation equal to 0 by adding $-x$ to each side of the equation. Then factor and solve for *x*. (***Warning:*** Don't divide both sides of the equation by *x*; you lose a solution if you do that.)

$$x^2-x=0$$
$$x(x-1)=0$$
$$x=0,1$$

Use the system's second equation, $y=x$, to find the corresponding *y* value for each *x*:

$x=0$	$x=1$
$y=0$	$y=1$

824. (0, –4), (3, 2)

Using substitution, replace the *y* in the first equation with its equivalent in the second equation:

$$2x-4=x^3-3x^2+2x-4$$

Set this equation equal to 0 by adding 4 and $-2x$ to each side of the equation. Then factor and solve for *x*:

$$0=x^3-3x^2$$
$$0=x^2(x-3)$$
$$x=0,3$$

Use the system's second equation, $y=2x-4$, to find the corresponding *y* value for each *x*:

$x=0$	$x=3$
$y=2(0)-4$	$y=2(3)-4$
$y=-4$	$y=2$

825. (1, 0), (–2, –12), (3, 8)

Using substitution, replace the *y* in the first equation with its equivalent in the second equation:

$$4x-4=x^4-9x^2+8$$

Set this equation equal to 0 by adding 4 and $-4x$ to each side of the equation. Then factor. You can find the first two factorizations using synthetic division:

$$0=x^4-9x^2-4x+12$$

$$\begin{array}{r|rrrrr} 1 & 1 & 0 & -9 & -4 & 12 \\ & & 1 & 1 & -8 & -12 \\ \hline -2 & 1 & 1 & -8 & -12 & \\ & & -2 & 2 & 12 & \\ \hline & 1 & -1 & -6 & & \end{array}$$

$$0=(x-1)\left(x^3+x^2-8x-12\right)$$
$$0=(x-1)(x+2)\left(x^2-x-6\right)$$

Then factor the quadratic and solve for x:

$$0=(x-1)(x+2)\left(x^2-x-6\right)$$
$$0=(x-1)(x+2)(x+2)(x-3)$$
$$x=1,-2,-2,3$$

Use the system's second equation, $y=4x-4$, to find the corresponding y value for each x:

$x=1$	$x=-2$	$x=3$
$y=4(1)-4$	$y=4(-2)-4$	$y=4(3)-4$
$y=0$	$y=-12$	$y=8$

826. (1, –3), (–2, –6), (3, –11)

Using substitution, replace the y in the first equation with its equivalent in the second equation:

$$-x^2-2=x^3-3x^2-5x+4$$

Set this equation equal to 0 by adding x^2 and 2 to each side of the equation. Then factor. You can do the first factorization using synthetic division:

$$0=x^3-2x^2-5x+6$$

$$\begin{array}{r|rrrr} 1 & 1 & -2 & -5 & 6 \\ & & 1 & -1 & -6 \\ \hline & 1 & -1 & -6 & \end{array}$$

$$0=(x-1)\left(x^2-x-6\right)$$

Then factor the quadratic and solve for x:

$$0=(x-1)\left(x^2-x-6\right)$$
$$0=(x-1)(x+2)(x-3)$$
$$x=1,-2,3$$

Use the system's second equation, $y=-x^2-2$, to find the corresponding y value for each x:

$x=1$	$x=-2$	$x=3$
$y=-(1)^2-2$	$y=-(-2)^2-2$	$y=-(3)^2-2$
$y=-3$	$y=-6$	$y=-11$

827. (–3, 61), (5, –83), (1, 5)

Using substitution, replace the y in the first equation with its equivalent in the second equation:

$$-x^2-16x+22=-x^3+2x^2-3x+7$$

Set this equation equal to 0 by adding x^2, $16x$, and –22 to each side of the equation. Then factor. You can do the first factorization using synthetic division:

$$0=-x^3+3x^2+13x-15$$

$$\begin{array}{r|rrrr} -3 & -1 & 3 & 13 & -15 \\ & & 3 & -18 & 15 \\ \hline & -1 & 6 & -5 & \end{array}$$

$$0=(x+3)\left(-x^2+6x-5\right)$$

Then factor the quadratic and solve for x:

$$0=(x+3)\left(-x^2+6x-5\right)$$
$$0=-(x+3)(x-5)(x-1)$$
$$x=-3,5,1$$

Use the system's second equation to find the corresponding y value for each x:

$x=-3$	$x=5$	$x=1$
$y=-(-3)^2-16(-3)+22$	$y=-(5)^2-16(5)+22$	$y=-(1)^2-16(1)+22$
$y=-9+48+22$	$y=-25-80+22$	$y=-1-16+22$
$y=61$	$y=-83$	$y=5$

828. (8, 843), (–4, –237), (–1, –21)

Using substitution, replace the y in the first equation with its equivalent in the second equation:

$$2x^2+82x+59=2x^3-4x^2+10x-5$$

Set this equation equal to 0 by adding $-2x^2$, $-82x$, and –59 to each side of the equation; then factor. First take out the common factor of 2. You can then use synthetic division for the next factorization:

$$0=2x^3-6x^2-72x-64$$
$$0=2\left(x^3-3x^2-36x-32\right)$$

$$\begin{array}{r|rrrr} 8 & 1 & -3 & -36 & -32 \\ & & 8 & 40 & 32 \\ \hline & 1 & 5 & 4 & \end{array}$$

$$0=2(x-8)\left(x^2+5x+4\right)$$

Then factor the quadratic and solve for x:

$$0=2(x-8)\left(x^2+5x+4\right)$$
$$0=2(x-8)(x+4)(x+1)$$
$$x=8,-4,-1$$

Use the system's second equation, $y=2x^2+82x+59$, to find the corresponding y value for each x:

$x=8$	$x=-4$
$y=2(8)^2+82(8)+59$	$y=2(-4)^2+82(-4)+59$
$y=128+656+59$	$y=32-328+59$
$y=843$	$y=-237$

$x=-1$
$y=2(-1)^2+82(-1)+59$
$y=2-82+59$
$y=-21$

829. (–3, –6), (–5, 8)

Using substitution, replace the y in the first equation with its equivalent in the second equation:

$$-7x-27=\frac{x-3}{x+4}$$

Multiply each side of the equation by $(x+4)$, the denominator of the fraction on the right:

$$(x+4)(-7x-27)=\frac{x-3}{\cancel{x+4}}\cdot\cancel{(x+4)}$$
$$-7x^2-55x-108=x-3$$

Set this equation equal to 0 by adding $-x$ and 3 to each side of the equation. Then factor and solve for x:

$$-7x^2-56x-105=0$$
$$-7\left(x^2+8x+15\right)=0$$
$$-7(x+3)(x+5)=0$$
$$x=-3,\ -5$$

Use the system's first equation, $y=\frac{x-3}{x+4}$, to find the corresponding y value for each x:

$x=-3$	$x=-5$
$y=\frac{-3-3}{-3+4}$	$y=\frac{-5-3}{-5+4}$
$y=\frac{-6}{1}$	$y=\frac{-8}{-1}$
$y=-6$	$y=8$

830. (2, –2), $\left(\frac{1}{3},3\right)$, (–1, 7)

Using substitution, replace the y in the first equation with its equivalent in the second equation:

$$-3x+4=\frac{x-6}{x^2-2}$$

Multiply each side of the equation by (x^2-2), the denominator of the fraction on the right:

$$(x^2-2)(-3x+4)=\frac{x-6}{\cancel{x^2-2}}\cdot\cancel{(x^2-2)}$$

$$-3x^3+4x^2+6x-8=x-6$$

Set this equation equal to 0 by adding $-x$ and 6 to each side of the equation. Then factor. First take out the common factor of -1. You can then use synthetic division for the next factorization:

$$0=-3x^3+4x^2+5x-20$$

$$0=-1\left(3x^3-4x^2-5x+2\right)$$

$$\begin{array}{r|rrrr} 2 & 3 & -4 & -5 & 2 \\ & & 6 & 4 & -2 \\ \hline & 3 & 2 & -1 & \end{array}$$

$$0=-1(x-2)\left(3x^2+2x-1\right)$$

Then factor the quadratic and solve for x:

$$0=-1(x-2)\left(3x^2+2x-1\right)$$

$$0=-1(x-2)(3x-1)(x+1)$$

$$x=2,\frac{1}{3},-1$$

Use the system's second equation, $y=-3x+4$, to find the corresponding y value for each x:

$x=2$	$x=1/3$	$x=-1$
$y=-3(2)+4$ $y=-2$	$y=-3\left(\frac{1}{3}\right)+4$ $y=3$	$y=-3(-1)+4$ $y=7$

831. (2, –1, 3)

Eliminate z in the third equation. To do so, multiply the second equation $(2y+z=1)$ by 4 and add it to the third equation.

$$\begin{aligned} 8y+4z&=4 \\ y-4z&=-13 \\ \hline 9y&=-9 \end{aligned}$$

Divide each side of the equation by 9 to get $y=-1$.

Substitute -1 for y in the original second equation to solve for z:

$$2y+z=1$$

$$2(-1)+z=1$$

$$z=1+2=3$$

Now substitute -1 for y and 3 for z in the original first equation to solve for x:

$$x+3y-2z=-7$$

$$x+3(-1)-2(3)=-7$$

$$x=-7+3+6=2$$

In (x,y,z) form, the answer is $(2,-1,3)$.

832. (0, 4, 2)

Eliminate x in the first equation. To do so, multiply the second equation $(x-y-z=-6)$ by -4 and add it to the first equation:

$$\begin{array}{r} 4x + \quad\;\; 3z = \;\;6 \\ \underline{-4x + 4y + 4z = 24} \\ 4y + 7z = 30 \end{array}$$

Now use this new equation and the original third equation to eliminate y. Multiply the third equation $(y+2z=8)$ by -4 and add it to the new equation:

$$\begin{array}{r} -4y-8z=-32 \\ \underline{4y+7z=\;\;30} \\ -z=\;\;-2 \end{array}$$

Multiply each side of the equation by -1 to get $z=2$.

Substitute 2 for z in the original third equation to solve for y:

$$\begin{aligned} y+2z&=8 \\ y+2(2)&=8 \\ y&=4 \end{aligned}$$

To solve for x, substitute 2 for z in the original first equation:

$$\begin{aligned} 4x+3z&=6 \\ 4x+3(2)&=6 \\ 4x&=0 \\ x&=0 \end{aligned}$$

In (x,y,z) form, the answer is $(0,4,2)$.

833. (4, –1, –2)

Eliminate the x term in the second and third equations. To do so, multiply the first equation $(x-y+2z=1)$ by -3 and add it to the second equation:

$$\begin{array}{r} -3x+3y-6z=-3 \\ \underline{3x+\;y-\;\;z=13} \\ 4y-7z=10 \end{array}$$

Then multiply the original first equation by -2 and add it to the original third equation:

$$\begin{array}{r} -2x+2y-4z=-2 \\ \underline{2x-4y+3z=\;\;6} \\ -2y-\;\;z=\;\;4 \end{array}$$

Now eliminate the y term in the first new equation $(4y-7z=10)$ by adding it to twice the second new equation $(-2y-z=4)$:

$$\begin{array}{r} 4y-7z=10 \\ \underline{-4y-2z=\;\;8} \\ -9z=18 \end{array}$$

Divide each side of the equation by -9 to get $z=-2$.

Now substitute –2 for z in the second new equation to solve for y:

$$\begin{aligned} -2y-z&=4\\ -2y-(-2)&=4\\ -2y&=2\\ y&=-1 \end{aligned}$$

Now substitute –1 for y and –2 for z in the original first equation to solve for x:

$$\begin{aligned} x-y+2z&=1\\ x-(-1)+2(-2)&=1\\ x+1-4&=1\\ x&=4 \end{aligned}$$

In (x, y, z) form, the answer is $(4, -1, -2)$.

834. (–3, 0, 2)

Eliminate the z term in the second and third equations. To do so, add the second equation to the first equation:

$$\begin{array}{l} 5x+3y+z=-13\\ \underline{2x-4y-z=\ -8}\\ 7x-\ y\qquad =-21 \end{array}$$

Then multiply the original first equation $(5x+3y+z=-13)$ by –2 and add it to the third equation:

$$\begin{array}{l} -10x-6y-2z=26\\ \underline{-3x\ +\ y+2z=13}\\ -13x-5y\qquad =39 \end{array}$$

Now eliminate the y in the new equations. To do this, multiply the first new equation $(7x-y=-21)$ by –5 and add it to the second new equation:

$$\begin{array}{l} -35x+5y=105\\ \underline{-13x-5y=\ 39}\\ -48x\qquad =144 \end{array}$$

Divide each side of the equation by –48, and you have $x=-3$.

Substitute –3 for x in the first new equation to solve for y:

$$\begin{aligned} 7x-y&=-21\\ 7(-3)-y&=-21\\ -y&=0\\ y&=0 \end{aligned}$$

Now substitute 0 for y and –3 for x in the original first equation to solve for z:

$$\begin{aligned} 5x+3y+z&=-13\\ 5(-3)+3(0)+z&=-13\\ -15+z&=-13\\ z&=2\quad 5 \end{aligned}$$

In (x, y, z) form, the answer is $(-3, 0, 2)$.

835. (8, 4, 2)

Multiply the first equation ($2x-3y=4$) by –2 and add it to the second equation to eliminate the x term:

$$\begin{array}{r} -4x+6y \qquad =-8 \\ \underline{4x \qquad +5z=42} \\ 6y+5z=34 \end{array}$$

Eliminate the z term in this new equation and the original third equation. To do so, multiply the terms in the new equation ($6y+5z=34$) by 3 and the terms in the original third equation ($5y-3z=14$) by 5 before adding the equations:

$$\begin{array}{r} 18y+15z=102 \\ \underline{25y-15z=\ \ 70} \\ 43y \qquad =172 \end{array}$$

Divide each side of the equation by 43 to get $y=4$.

Substitute 4 for y in the original first equation to solve for x:

$$\begin{aligned} 2x-3y&=4 \\ 2x-3(4)&=4 \\ 2x&=16 \\ x&=8 \end{aligned}$$

Now substitute 4 for y in the original third equation to solve for z:

$$\begin{aligned} 5y-3z&=14 \\ 5(4)-3z&=14 \\ -3z&=-6 \\ z&=2 \end{aligned}$$

In (x, y, z) form, the answer is (8, 4, 2).

836. $\left(\frac{1}{2}, -\frac{1}{2}, -\frac{1}{2}\right)$

Add the first and second equations to eliminate the x and z terms:

$$\begin{array}{r} x+y+z=-\frac{1}{2} \\ \underline{-x \qquad -z=\ \ 0} \\ y \qquad =-\frac{1}{2} \end{array}$$

Substitute $-\frac{1}{2}$ for y in the original third equation to solve for z:

$$\begin{aligned} y+z&=-1 \\ -\frac{1}{2}+z&=-1 \\ z&=-\frac{1}{2} \end{aligned}$$

Substitute $-\frac{1}{2}$ for both y and z in the original first equation to solve for x:

$$x+y+z=-\frac{1}{2}$$
$$x-\frac{1}{2}-\frac{1}{2}=-\frac{1}{2}$$
$$x=\frac{1}{2}$$

In (x, y, z) form, the answer is $\left(\frac{1}{2}, -\frac{1}{2}, -\frac{1}{2}\right)$.

837. $\left(\frac{23+7k}{17}, \frac{9k+100}{17}, k\right)$

Eliminate the x term. To do so, first multiply the terms in the first equation $(x+3y-2z=19)$ by –4 and add them to the second equation:

$$\begin{aligned} -4x-12y+8z&=-76 \\ 4x-5y+z&=-24 \\ \hline -17y+9z&=-100 \end{aligned}$$

Now multiply the terms in the original first equation $(x+3y-2z=19)$ by –6 and add them to the third equation:

$$\begin{aligned} -6x-18y+12z&=-114 \\ 6x+y-3z&=14 \\ \hline -17y+9z&=-100 \end{aligned}$$

Next, eliminate the y term by multiplying the first new equation $(-17y+9z=-100)$ by –1 and adding it to the second new equation:

$$\begin{aligned} 17y-9z&=100 \\ -17y+9z&=-100 \\ \hline 0+0&=0 \end{aligned}$$

This statement is always true, so there are an infinite number of solutions for this system.

Let z be represented by the parameter k. Use the second new equation $(-17y+9z=-100)$ to solve for y in terms of k:

$$-17y+9k=-100$$
$$-17y=-9k-100$$
$$y=\frac{9k+100}{17}$$

Now solve for x in terms of k using the original first equation, the expression for y in terms of k, and $z=k$:

$$x+3y-2z=19$$
$$x+3\left(\frac{9k+100}{17}\right)-2k=19$$
$$x=19-\frac{27k+300}{17}+2k$$
$$x=\frac{323-27k-300+34k}{17}$$
$$x=\frac{23+7k}{17}$$

In (x, y, z) form, the answer is $\left(\frac{23+7k}{17}, \frac{9k+100}{17}, k\right)$.

838. $(k, -2k-11, -11k-59)$

Eliminate the z term in the second and third equations. To do so, first multiply the terms in the second equation ($x-5y+z=-4$) by –1 and add them to the third equation:

$$\begin{array}{r} -x+5y-z=4 \\ -3x-7y+z=18 \\ \hline -4x-2y\quad\;\; =22 \end{array}$$

Divide each term in this new equation by 2 and add it to the original first equation:

$$\begin{array}{r} -2x-y=11 \\ 2x+y=-11 \\ \hline 0+0=0 \end{array}$$

This statement is always true, so there are an infinite number of solutions for this system.

Let x be represented by the parameter k. Use the original first equation to solve for y in terms of k:

$$\begin{aligned} 2x+y&=-11 \\ 2k+y&=-11 \\ y&=-2k-11 \end{aligned}$$

Now substitute k for x and $(-2k-11)$ for y in the original second equation to solve for z in terms of k:

$$\begin{aligned} x-5y+z&=-4 \\ k-5(-2k-11)+z&=-4 \\ k+10k+55+z&=-4 \\ z&=-59-11k \end{aligned}$$

In (x,y,z) form, the answer is $(k, -2k-11, -11k-59)$.

839. $(-1, -3, 4, -2)$

Even though the coefficients on the three y terms offer a challenge, eliminating the y terms requires only two processes, whereas eliminating one of the other terms takes three. First, multiply the first equation ($x+2y-3z+w=-21$) by 3 and the second equation ($2x-3y+z-2w=15$) by 2 and add them together:

$$\begin{array}{r} 3x+6y-9z+3w=-63 \\ 4x-6y+2z-4w=30 \\ \hline 7x\qquad\;\; -7z-w=-33 \end{array}$$

Now multiply the original first equation ($x+2y-3z+w=-21$) by 2 and add it to the third equation:

$$\begin{array}{r} 2x+4y-6z+2w=-42 \\ x-4y-2z+3w=-3 \\ \hline 3x\qquad\;\; -8z+5w=-45 \end{array}$$

The new system equations, without the y term, consist of these two new equations and the original fourth equation:

$$\begin{cases} 7x-7z-w=-33 \\ 3x-8z+5w=-45 \\ 3x+2z-w=7 \end{cases}$$

The next steps involve eliminating the w term. First, multiply the top equation by 5 and add it to the middle equation:

$$\begin{array}{r} 35x-35z-5w=-165 \\ \underline{3x-8z+5w=-45} \\ 38x-43z=-210 \end{array}$$

Now multiply the bottom equation from that new system $(3x+2z-w=7)$ by 5 and add it to the middle equation:

$$\begin{array}{r} 15x+10z-5w=35 \\ \underline{3x-8z+5w=-45} \\ 18x+2z=-10 \end{array}$$

Each of these terms is divisible by 2:

$$9x+6z=-5$$

You now have a new system of two equations:

$$\begin{cases} 38x-43z=-210 \\ 9x+z=-5 \end{cases}$$

The numbers still aren't very pretty, but you can solve this newest system using elimination one more time. Multiply the terms in the second equation by 43 and add it to the first equation.

$$\begin{array}{r} 38x-43z=-210 \\ \underline{387x+43z=-215} \\ 425x=-425 \end{array}$$

Divide each side by 425, and you have $x=-1$.

Now back-substitute to find the values of the other variables:

$$\begin{aligned} 38x-43z&=-210 \\ 38(-1)-43z&=-210 \\ -43z&=-172 \\ z&=4 \\ 3x+2z-w&=7 \\ 3(-1)+2(4)-w&=7 \\ -w&=2 \\ w&=-2 \\ x-4y-2z+3w&=-3 \\ (-1)-4y-2(4)+3(-2)&=-3 \\ -4y&=12 \\ y&=-3 \end{aligned}$$

In (x,y,z,w) form, the answer is $(-1,-3,4,-2)$.

840. (1,1,0, – 2)

Start by eliminating the w term. Multiply the second equation $(2x-3y+w=-3)$ by 2 and add it to the third equation:

$$\begin{array}{l} 4x-6y \quad +2w=-6 \\ \quad\;\; 4y+z-2w=8 \\ \hline 4x-2y+z \quad\;\; =2 \end{array}$$

Next, multiply the fourth equation $(x-y+w=-2)$ by 2 and add it to the third equation:

$$\begin{array}{l} 2x-2y \quad +2w=-4 \\ \quad\;\; 4y+z-2w=8 \\ \hline 2x+2y+z \quad\;\; =4 \end{array}$$

The new system of equations, without the y term, consists of these two new equations and the original first equation:

$$\begin{cases} 4x-2y+\ z=2 \\ 2x+2y+\ z=4 \\ 5x \qquad +3z=5 \end{cases}$$

The next step involves eliminating the y term. Add the first two equations of the new system together:

$$\begin{array}{l} 4x-2y+\ z=2 \\ 2x+2y+\ z=4 \\ \hline 6x \qquad +2z=6 \end{array}$$

Each term in the new equation is divisible by 2, giving you $3x+z=3$. Multiply the terms in this equation by –3 and add it to the last equation in the new system:

$$\begin{array}{l} -9x-3z=-9 \\ \ 5x+3z=\ 5 \\ \hline -4x \qquad =-4 \end{array}$$

Dividing by –4, you have $x=1$. Now back-solve to find the values of the rest of the variables:

$$\begin{aligned} 6x+2z&=6 \\ 6(1)+2z&=6 \\ 2z&=0 \\ z&=0 \\ 2x+2y+z&=4 \\ 2(1)+2y+0&=4 \\ 2y&=2 \\ y&=1 \\ x-y+w&=-2 \\ 1-1+w&=-2 \\ w&=-2 \end{aligned}$$

In (x,y,z,w) form, the answer is $(1,1,0,-2)$.

841.

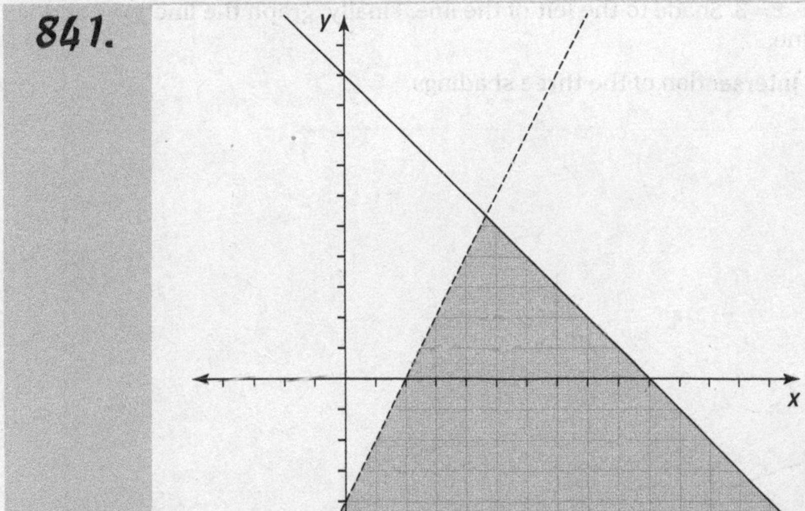

Illustration by Thomson Digital

To graph $x+y\leq 10$, first graph the line $x+y=10$ (or $y=-x+10$ in slope-intercept form). Choose a test point to determine which side of the line to shade. Using (0, 0) as a test point, $0+0\leq 10$, so shade the area below and to the left of the line.

Then graph the line $2x-y=4$ (or $y=2x-4$). Use a dashed line to show that the points on the line are not in the solution. Choose a test point. Using (0, 0) as a test point, $2(0)-0=0\not>4$. The point is not in the solution, so shade below and to the right of the line.

The solution is the intersection of the two shadings.

842.

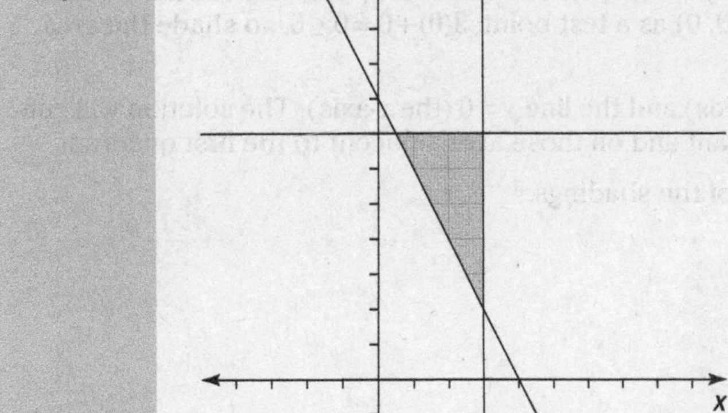

Illustration by Thomson Digital

To graph $2x+y\geq 8$, first graph the line $2x+y=8$ (or $y=-2x+8$ in slope-intercept form). Choose a test point to determine which side of the line to shade. Using (0, 0) as a test point, $2(0)+0=0\not\geq 8$, so shade the area above and to the right of the line.

Then graph the line $x=3$. Shade to the left of the line. Finally, graph the line $y=7$ and shade below that line.

The solution is the intersection of the three shadings.

843.

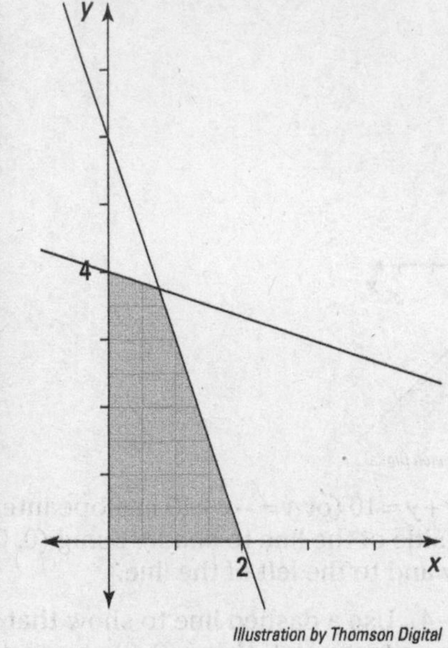

Illustration by Thomson Digital

To graph $x+3y\leq 12$, first graph the line $x+3y=12$ (or $y=-\frac{1}{3}x+4$ in slope-intercept form). Choose a test point to determine which side of the line to shade. Using (0, 0) as a test point, $0+3(0)=0\leq 12$, so shade the area below and to left of the line.

Next, graph the line $3x+y=6$ (or $y=-3x+6$). Choose a test point to determine which side of the line to shade. Using (0, 0) as a test point, $3(0)+0=0\leq 6$, so shade the area below and to left of the line.

Then graph the line $x=0$ (the y-axis) and the line $y=0$ (the x-axis). The solution will contain only points in the first quadrant and on those axes adjacent to the first quadrant.

The solution is the intersection of the shadings.

844.

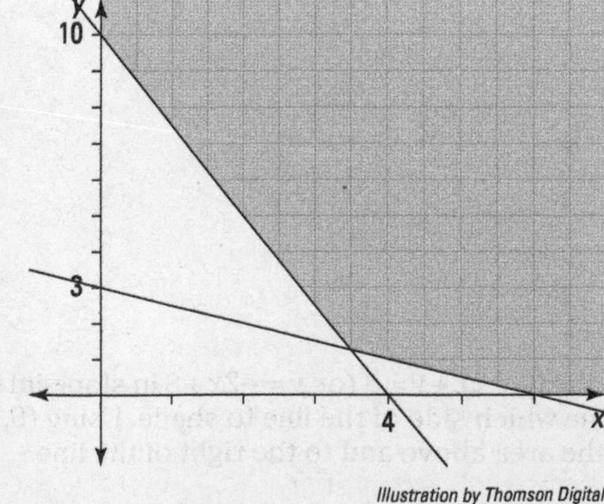

Illustration by Thomson Digital

First graph the line $2x+4y=12$ (or $y=-\frac{1}{2}x+3$ in slope-intercept form). Choose a test point to determine which side of the line to shade. Using (0, 0) as a test point, $2(0)+4(0)=0\ngeq 12$, so shade the area above and to right of the line.

Next, graph the line $5x+2y=20$ (or $y=-\frac{5}{2}x+10$). Choose a test point to determine which side of the line to shade. Using (0, 0) as a test point, $5(0)+2(0)=0\ngeq 20$, so shade the area above and to right of the line.

Then graph the line $x=0$ (the y-axis) and the line $y=0$ (the x-axis). The solution will contain only points in the first quadrant and on those axes adjacent to the first quadrant.

The solution is the intersection of the shadings.

845.

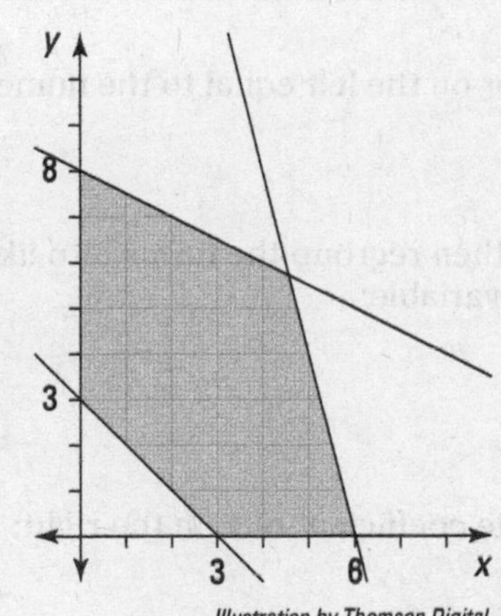

Illustration by Thomson Digital

First graph the line $x+y=3$ (or $y=-x+3$ in slope-intercept form). Choose a test point to determine which side of the line to shade. Using (0, 0) as a test point, $0+0\ngeq 3$, so shade the area above and to right of the line.

Next, graph the line $x+2y=16$ (or $y=-\frac{1}{2}x+8$). Choose a test point to determine which side of the line to shade. Using (0, 0) as a test point, $0+2(0)=0\leq 16$, so shade the area below and to the left of the line.

Then graph the line $4x+y=24$ (or $y=-4x+24$). Choose a test point to determine which side of the line to shade. Using (0, 0) as a test point, $4(0)+0=0\leq 24$, so shade the area below and to the left of the line.

Then graph the line $x=0$ (the y-axis) and the line $y=0$ (the x-axis). The solution will contain only points in the first quadrant and on those axes adjacent to the first quadrant.

The solution is the intersection of the shadings.

846. $\frac{5}{x-1}-\frac{2}{x+1}$

You want to decompose $\frac{3x+7}{x^2-1}$. Factor the denominator of the fraction:

$$x^2-1=(x-1)(x+1)$$

Next, set the original fraction equal to the sum of two fractions whose denominators are those factors you found. Let A and B represent the numerators of those fractions.

$$\frac{3x+7}{x^2-1}=\frac{3x+7}{(x-1)(x+1)}=\frac{A}{x-1}+\frac{B}{x+1}$$

The common denominator of the fractions on the right is, of course, the fraction you started with. Rewrite the fractions on the right so they all have that common denominator. You'll be multiplying the numerator and denominator by a missing factor, which is the denominator of the other fraction.

$$\frac{3x+7}{(x-1)(x+1)}=\frac{A}{x-1}\cdot\frac{x+1}{x+1}+\frac{B}{x+1}\cdot\frac{x-1}{x-1}$$

$$\frac{3x+7}{(x-1)(x+1)}=\frac{A(x+1)}{(x-1)(x+1)}+\frac{B(x-1)}{(x-1)(x+1)}$$

$$\frac{3x+7}{(x-1)(x+1)}=\frac{A(x+1)+B(x-1)}{(x-1)(x+1)}$$

Write a new equation, setting the numerator on the left equal to the numerator on the right:

$$3x+7=A(x+1)+B(x-1)$$

Distribute over the binomials on the right; then regroup the terms into like terms. Factor the x from the terms containing the variable:

$$3x+7=Ax+A+Bx-B$$

$$3x+7=Ax+Bx+A-B$$

$$3x+7=(A+B)x+(A-B)$$

The coefficient of x on the left is equal to the coefficient of x on the right:

$$3=A+B$$

The constant term on the left is equal to the constant term on the right:

$$7=A-B$$

The two equations in A and B form a system of equations whose solution gives you the numerators of the fractions.

Adding the two equations together, you get $10=2A$, or $A=5$. Substituting 5 for A in the first equation, you get $3=5+B$, or $B=-2$. Therefore, the decomposed fraction is

$$\frac{3x+7}{x^2-1}=\frac{5}{x-1}+\frac{-2}{x+1}$$

847. $\frac{6}{x-2}+\frac{5}{x+3}$

Factor the denominator of the fraction:

$$x^2+x-6=(x-2)(x+3)$$

Next, set the original fraction equal to the sum of two fractions whose denominators are those factors you found. Let A and B represent the numerators of those fractions.

$$\frac{11x+8}{x^2+x-6}=\frac{11x+8}{(x-2)(x+3)}=\frac{A}{x-2}+\frac{B}{x+3}$$

The common denominator of the fractions on the right is, of course, the fraction you started with. Rewrite the fractions on the right so they all have that common denominator. You'll be multiplying the numerator and denominator by a missing factor, which is the denominator of the other fraction.

$$\frac{11x+8}{(x-2)(x+3)}=\frac{A}{x-2}\cdot\frac{x+3}{x+3}+\frac{B}{x+3}\cdot\frac{x-2}{x-2}$$

$$\frac{11x+8}{(x-2)(x+3)}=\frac{A(x+3)}{(x-2)(x+3)}+\frac{B(x-2)}{(x-2)(x+3)}$$

$$\frac{11x+8}{(x-2)(x+3)}=\frac{A(x+3)+B(x-2)}{(x-2)(x+3)}$$

Write a new equation, setting the numerator on the left equal to the numerator on the right:

$$11x+8=A(x+3)+B(x-2)$$

Distribute over the binomials on the right; then regroup the terms into like terms. Factor the x from the terms containing the variable:

$$11x+8=Ax+3A+Bx-2B$$

$$11x+8=Ax+Bx+3A-2B$$

$$11x+8=(A+B)x+(3A-2B)$$

The coefficient of x on the left is equal to the coefficient of x on the right:

$$11=A+B$$

The constant term on the left is equal to the constant term on the right:

$$8=3A-2B$$

The two equations in A and B form a system of equations whose solution gives you the numerators of the fractions.

Multiply the terms in the first equation ($11=A+B$) by 2 and add them to the second equation:

$$22=2A+2B$$

$$\underline{8=3A-2B}$$

$$30=5A$$

Divide each side by 5, and you have $A=6$. Replace A with 6 in the original first equation, and you have $11=6+B$, so $B=5$. Therefore, the decomposed fraction is

$$\frac{11x+8}{x^2+x-6}=\frac{6}{x-2}+\frac{5}{x+3}$$

848. $\frac{4}{x}-\frac{7}{x+7}$

Factor the denominator of the fraction:

$$x^2+7x=x(x+7)$$

Next, set the original fraction equal to the sum of two fractions whose denominators are those factors you found. Let A and B represent the numerators of those fractions.

$$\frac{28-3x}{x^2+7x}=\frac{28-3x}{x(x+7)}=\frac{A}{x}+\frac{B}{x+7}$$

The common denominator of the fractions on the right is, of course, the fraction you started with. Rewrite the fractions on the right so they all have that common denominator. You'll be multiplying the numerator and denominator by a missing factor, which is the denominator of the other fraction.

$$\frac{28-3x}{x(x+7)}=\frac{A}{x}\cdot\frac{x+7}{x+7}+\frac{B}{x+7}\cdot\frac{x}{x}$$

$$\frac{28-3x}{x(x+7)}=\frac{A(x+7)}{x(x+7)}+\frac{Bx}{x(x+7)}$$

$$\frac{28-3x}{x(x+7)}=\frac{A(x+7)+Bx}{x(x+7)}$$

Write a new equation, setting the numerator on the left equal to the numerator on the right:

$$28-3x=A(x+7)+Bx$$

Distribute over the binomial on the right; then regroup the terms into like terms. Factor the x from the terms containing the variable:

$$\begin{aligned}28-3x&=Ax+7A+Bx\\28-3x&=Ax+Bx+7A\\-3x+28&=(A+B)x+7A\end{aligned}$$

The coefficient of x on the left is equal to the coefficient of x on the right:

$$-3=A+B$$

The constant term on the left is equal to the constant term on the right:

$$28=7A$$

Divide each side by 7, and you have $A=4$. Replace A with 4 in the original first equation, and you have $-3=4+B$, so $B=-7$. Therefore, the decomposed fraction is

$$\frac{28-3x}{x^2+7x}=\frac{4}{x}+\frac{-7}{x+7}$$

849. $\frac{3}{x}-\frac{2}{x+2}+\frac{5}{x-5}$

Factor the denominator of the fraction:

$$\begin{aligned}x^3-3x^2-10x&=x(x^2-3x-10)\\&=x(x+2)(x-5)\end{aligned}$$

Next, set the original fraction equal to the sum of three fractions whose denominators are those factors you found. Let A, B, and C represent the numerators of the fractions.

$$\frac{6x^2+11x-30}{x^3-3x^2-10x}=\frac{6x^2+11x-30}{x(x+2)(x-5)}=\frac{A}{x}+\frac{B}{x+2}+\frac{C}{x-5}$$

The common denominator of the fractions on the right is, of course, the fraction you started with. Rewrite the fractions on the right so they all have that common denominator. You'll be multiplying the numerator and the denominator by missing factors, which are the denominators of the other fractions.

$$\frac{6x^2+11x-30}{x(x+2)(x-5)}=\frac{A}{x}\cdot\frac{(x+2)(x-5)}{(x+2)(x-5)}+\frac{B}{x+2}\cdot\frac{x(x-5)}{x(x-5)}+\frac{C}{x-5}\cdot\frac{x(x+2)}{x(x+2)}$$

$$\frac{6x^2+11x-30}{x(x+2)(x-5)}=\frac{A(x^2-3x-10)}{x(x+2)(x-5)}+\frac{B(x^2-5x)}{x(x+2)(x-5)}+\frac{C(x^2+2x)}{x(x+2)(x-5)}$$

$$\frac{6x^2+11x-30}{x(x+2)(x-5)}=\frac{A(x^2-3x-10)+B(x^2-5x)+C(x^2+2x)}{x(x+2)(x-5)}$$

Write a new equation, setting the numerator on the left equal to the numerator on the right:

$$6x^2+11x-30=A\,(x^2-3x-10)+B\,(x^2-5x)+C\,(x^2+2x)$$

Distribute over the trinomial and binomials on the right; then regroup the terms into like terms. Factor the x from the terms containing the variable:

$$6x^2+11x-30=Ax^2-3Ax-10A+Bx^2-5Bx+Cx^2+2Cx$$

$$6x^2+11x-30=(A+B+C)x^2+(-3A-5B+2C)\,x-10A$$

The coefficient of x^2 on the left is equal to the coefficient of x^2 on the right:

$$6=A+B+C$$

The coefficient of x on the left is equal to the coefficient of x on the right:

$$11=-3A-5B+2C$$

The constant term on the left is equal to the constant term on the right:

$$-30=-10A$$

Divide each side by –10, and you have $A=3$. Replace A with 3 in the equations for x^2 and x:

$$\begin{aligned}6&=A+B+C\\6&=3+B+C\\3&=B+C\\11&=-3A-5B+2C\\11&=-9-5B+2C\\20&=-5B+2C\end{aligned}$$

You now have the following system of equations:

$$\begin{cases}3=\quad B+\ C\\20=-5B+2C\end{cases}$$

Solve the system by multiplying each term in the first equation by 5 and adding the two equations together:

$$\begin{aligned}15&=\ \ 5B+5C\\20&=-5B+2C\\ \hline 35&=\qquad\ 7C\end{aligned}$$

Divide each side of the equation by 7 to get $C=5$. Then solve for B using the first equation: $3=B+5$, so $B=-2$. Therefore, the decomposed fraction is

$$\frac{6x^2+11x-30}{x^3-3x^2-10x}=\frac{3}{x}+\frac{-2}{x+2}+\frac{5}{x-5}$$

850. $-\frac{4}{x}+\frac{2}{x-4}+\frac{5}{x+4}$

Factor the denominator of the fraction:

$$x^3-16x=x\,(x^2-16)=x(x-4)(x+4)$$

Next, set the original fraction equal to the sum of three fractions whose denominators are those factors you found. Let *A*, *B*, and *C* represent the numerators of those fractions.

$$\frac{3x^2-12x+64}{x^3-16x}=\frac{3x^2-12x+64}{x(x-4)(x+4)}=\frac{A}{x}+\frac{B}{x-4}+\frac{C}{x+4}$$

The common denominator of the fractions on the right is, of course, the fraction you started with. Rewrite the fractions on the right so they all have that common denominator. You'll be multiplying the numerator and the denominator by missing factors, which are the denominators of the other fractions.

$$\frac{3x^2-12x+64}{x(x-4)(x+4)}=\frac{A}{x}\cdot\frac{(x-4)(x+4)}{(x-4)(x+4)}+\frac{B}{x-4}\cdot\frac{x(x+4)}{x(x+4)}+\frac{C}{x+4}\cdot\frac{x(x-4)}{x(x-4)}$$

$$\frac{3x^2-12x+64}{x(x-4)(x+4)}=\frac{A(x^2-16)}{x(x-4)(x+4)}+\frac{B(x^2+4x)}{x(x-4)(x+4)}+\frac{C(x^2-4x)}{x(x-4)(x+4)}$$

$$\frac{3x^2-12x+64}{x(x-4)(x+4)}=\frac{A(x^2-16)+B(x^2+4x)+C(x^2-4x)}{x(x-4)(x+4)}$$

Write a new equation, setting the numerator on the left equal to the numerator on the right:

$$3x^2-12x+64=A(x^2-16)+B(x^2+4x)+C(x^2-4x)$$

Distribute over the trinomial and binomials on the right; then regroup the terms into like terms. Factor the *x* from the terms containing the variable:

$$3x^2-12x+64=Ax^2-16A+Bx^2+4Bx+Cx^2-4Cx$$

$$3x^2-12x+64=(A+B+C)x^2+(4B-4C)x-16A$$

The coefficient of x^2 on the left is equal to the coefficient of x^2 on the right:

$$3=A+B+C$$

The coefficient of *x* on the left is equal to the coefficient of *x* on the right:

$$-12=4B-4C$$

The constant term on the left is equal to the constant term on the right:

$$64=-16A$$

Divide each side by –16, and you have $A=-4$. Replace *A* with –4 in the equation for x^2:

$$3=A+B+C$$
$$3=-4+B+C$$
$$7=B+C$$

You now have the following system of equations:

$$\begin{cases}7=B+C\\-12=4B-4C\end{cases}$$

Solve the system by multiplying each term in the first equation by 4 and adding the two equations together:

$$\begin{array}{r}28=4B+4C\\ \underline{-12=4B-4C}\\ 16=8B\end{array}$$

Divide each side of the equation by 8 to get $B=2$. Then solve for *C* using the first equation: $7=2+C$, so $C=5$. Therefore, the decomposed fraction is

$$\frac{3x^2-12x+64}{x^3-16x}=-\frac{4}{x}+\frac{2}{x-4}+\frac{5}{x+4}$$

851. $\begin{bmatrix} 1 & -1 \\ 1 & -3 \end{bmatrix}$

Find the sum of each pair of corresponding elements:

$$A+B=\begin{bmatrix} 1 & 2 \\ -3 & 4 \end{bmatrix}+\begin{bmatrix} 0 & -3 \\ 4 & -7 \end{bmatrix}=\begin{bmatrix} 1+0 & 2+(-3) \\ -3+4 & 4+(-7) \end{bmatrix}=\begin{bmatrix} 1 & -1 \\ 1 & -3 \end{bmatrix}$$

852. $\begin{bmatrix} -1 & -5 \\ 7 & -11 \end{bmatrix}$

Find the difference between each pair of corresponding elements:

$$B-A=\begin{bmatrix} 0 & -3 \\ 4 & -7 \end{bmatrix}-\begin{bmatrix} 1 & 2 \\ -3 & 4 \end{bmatrix}=\begin{bmatrix} 0-1 & -3-2 \\ 4-(-3) & -7-4 \end{bmatrix}=\begin{bmatrix} -1 & -5 \\ 7 & -11 \end{bmatrix}$$

853. $\begin{bmatrix} 4 & 14 \\ -20 & 30 \end{bmatrix}$

You perform *scalar multiplication* by multiplying each element in the matrix by the multiplier outside the matrix. So first perform the scalar multiplication on A and B:

$$4A=4\cdot\begin{bmatrix} 1 & 2 \\ -3 & 4 \end{bmatrix}=\begin{bmatrix} 4 & 8 \\ -12 & 16 \end{bmatrix}$$

$$2B=2\cdot\begin{bmatrix} 0 & -3 \\ 4 & -7 \end{bmatrix}=\begin{bmatrix} 0 & -6 \\ 8 & -14 \end{bmatrix}$$

Subtract the results by subtracting the corresponding elements:

$$\begin{aligned} 4A-2B &= \begin{bmatrix} 4 & 8 \\ -12 & 16 \end{bmatrix}-\begin{bmatrix} 0 & -6 \\ 8 & -14 \end{bmatrix} \\ &= \begin{bmatrix} 4-0 & 8-(-6) \\ -12-8 & 16-(-14) \end{bmatrix} \\ &= \begin{bmatrix} 4 & 14 \\ -20 & 30 \end{bmatrix} \end{aligned}$$

854. $\begin{bmatrix} 8 & -17 \\ 16 & -19 \end{bmatrix}$

Multiply the elements in each row of matrix A by the elements in the columns of matrix B; then add the products:

$$\begin{aligned} A\cdot B &= \begin{bmatrix} 1 & 2 \\ -3 & 4 \end{bmatrix}\cdot\begin{bmatrix} 0 & -3 \\ 4 & -7 \end{bmatrix} \\ &= \begin{bmatrix} 1(0)+2(4) & 1(-3)+2(-7) \\ -3(0)+4(4) & -3(-3)+4(-7) \end{bmatrix} \\ &= \begin{bmatrix} 8 & -17 \\ 16 & -19 \end{bmatrix} \end{aligned}$$

855. $\begin{bmatrix} -12 & -3 & 12 & -6 \\ -24 & -7 & 36 & -2 \end{bmatrix}$

Multiply the elements in each row of matrix *B* by the elements in the columns of matrix *C*; then add the products:

$$B \cdot C = \begin{bmatrix} 0 & -3 \\ 4 & -7 \end{bmatrix} \cdot \begin{bmatrix} 1 & 0 & 2 & 3 \\ 4 & 1 & -4 & 2 \end{bmatrix}$$

$$= \begin{bmatrix} 0\cdot 1+(-3)\cdot 4 & 0\cdot 0+(-3)\cdot 1 & 0\cdot 2+(-3)(-4) & 0\cdot 3+(-3)\cdot 2 \\ 4\cdot 1+(-7)\cdot 4 & 4\cdot 0+(-7)\cdot 1 & 4\cdot 2+(-7)(-4) & 4\cdot 3+(-7)\cdot 2 \end{bmatrix}$$

$$= \begin{bmatrix} -12 & -3 & 12 & -6 \\ -24 & -7 & 36 & -2 \end{bmatrix}$$

856. $\begin{bmatrix} 1 & 3 & 2 \\ 0 & 1 & 1 \\ 0 & 0 & 1 \end{bmatrix}$

The *echelon form* of a matrix has only zeros as elements below the main diagonal and only ones as elements along that diagonal. You change a matrix to echelon form by performing various row operations: exchanging rows, multiplying a row by a constant, and adding a multiple of one row to another. You usually start with the first row, perform operations to change it to what you want, and then work your way down.

$$\begin{bmatrix} 1 & 3 & 2 \\ 4 & 0 & 3 \\ 2 & -1 & -3 \end{bmatrix}$$

$$\begin{matrix} -4R_1+R_2 \to R_2 \\ -2R_1+R_3 \to R_3 \end{matrix} \begin{bmatrix} 1 & 3 & 2 \\ 0 & -12 & -5 \\ 0 & -7 & -7 \end{bmatrix}$$

$$R_2 \leftrightarrow R_3 \begin{bmatrix} 1 & 3 & 2 \\ 0 & -7 & -7 \\ 0 & -12 & -5 \end{bmatrix}$$

$$-\frac{1}{7}R_2 \to R_2 \begin{bmatrix} 1 & 3 & 2 \\ 0 & 1 & 1 \\ 0 & -12 & -5 \end{bmatrix}$$

$$12R_2+R_3 \to R_3 \begin{bmatrix} 1 & 3 & 2 \\ 0 & 1 & 1 \\ 0 & 0 & 7 \end{bmatrix}$$

$$\frac{1}{7}R_3 \to R_3 \begin{bmatrix} 1 & 3 & 2 \\ 0 & 1 & 1 \\ 0 & 0 & 1 \end{bmatrix}$$

857. $\begin{bmatrix} 1 & 0 & 2 \\ 0 & 1 & 7 \\ 0 & 0 & 1 \end{bmatrix}$

$$\begin{bmatrix} 5 & -1 & 3 \\ 1 & 0 & 2 \\ 4 & 1 & -3 \end{bmatrix}$$

$$R_1 \leftrightarrow R_2 \begin{bmatrix} 1 & 0 & 2 \\ 5 & -1 & 3 \\ 4 & 1 & -3 \end{bmatrix}$$

$$\begin{matrix} -5R_1 + R_2 \rightarrow R_2 \\ -4R_1 + R_3 \rightarrow R_3 \end{matrix} \begin{bmatrix} 1 & 0 & 2 \\ 0 & -1 & -7 \\ 0 & 1 & -11 \end{bmatrix}$$

$$-1R_2 \rightarrow R_2 \begin{bmatrix} 1 & 0 & 2 \\ 0 & 1 & 7 \\ 0 & 1 & -11 \end{bmatrix}$$

$$-1R_2 + R_3 \rightarrow R_3 \begin{bmatrix} 1 & 0 & 2 \\ 0 & 1 & 7 \\ 0 & 0 & -18 \end{bmatrix}$$

$$-\frac{1}{18}R_3 \rightarrow R_3 \begin{bmatrix} 1 & 0 & 2 \\ 0 & 1 & 7 \\ 0 & 0 & 1 \end{bmatrix}$$

858. $\begin{bmatrix} 1 & -6 & -3 \\ 0 & 1 & 5/4 \\ 0 & 0 & 1 \end{bmatrix}$

$$\begin{bmatrix} 2 & -4 & 4 \\ -1 & 6 & 3 \\ 0 & -8 & 6 \end{bmatrix}$$

$$R_1 \leftrightarrow R_2 \begin{bmatrix} -1 & 6 & 3 \\ 2 & -4 & 4 \\ 0 & -8 & 6 \end{bmatrix}$$

$$-1R_1 \rightarrow R_1 \begin{bmatrix} 1 & -6 & -3 \\ 2 & -4 & 4 \\ 0 & -8 & 6 \end{bmatrix}$$

$$-2R_1 + R_2 \rightarrow R_2 \begin{bmatrix} 1 & -6 & -3 \\ 0 & 8 & 10 \\ 0 & -8 & 6 \end{bmatrix}$$

$$\frac{1}{8}R_2 \to R_2 \begin{bmatrix} 1 & -6 & -3 \\ 0 & 1 & 5/4 \\ 0 & -8 & 6 \end{bmatrix}$$

$$8R_2 + R_3 \to R_3 \begin{bmatrix} 1 & -6 & -3 \\ 0 & 1 & 5/4 \\ 0 & 0 & 16 \end{bmatrix}$$

$$\frac{1}{16}R_3 \to R_3 \begin{bmatrix} 1 & -6 & -3 \\ 0 & 1 & 5/4 \\ 0 & 0 & 1 \end{bmatrix}$$

859. $\begin{bmatrix} 1 & 0 & 4 & -2 \\ 0 & 1 & -14 & 11 \\ 0 & 0 & 1 & -5/4 \\ 0 & 0 & 0 & 1 \end{bmatrix}$

$$\begin{bmatrix} 1 & 0 & 4 & -2 \\ 3 & 1 & -2 & 5 \\ 4 & 3 & 2 & -10 \\ 0 & 1 & -6 & -3 \end{bmatrix}$$

$$\begin{matrix} -3R_1 + R_2 \to R_2 \\ -4R_1 + R_3 \to R_3 \end{matrix} \begin{bmatrix} 1 & 0 & 4 & -2 \\ 0 & 1 & -14 & 11 \\ 0 & 3 & -14 & -2 \\ 0 & 1 & -6 & -3 \end{bmatrix}$$

$$\begin{matrix} -3R_2 + R_3 \to R_3 \\ -1R_2 + R_4 \to R_4 \end{matrix} \begin{bmatrix} 1 & 0 & 4 & -2 \\ 0 & 1 & -14 & 11 \\ 0 & 0 & 28 & -35 \\ 0 & 0 & 8 & -14 \end{bmatrix}$$

$$\frac{1}{28}R_3 \to R_3 \begin{bmatrix} 1 & 0 & 4 & -2 \\ 0 & 1 & -14 & 11 \\ 0 & 0 & 1 & -5/4 \\ 0 & 0 & 8 & -14 \end{bmatrix}$$

$$-8R_3 + R_4 \to R_4 \begin{bmatrix} 1 & 0 & 4 & -2 \\ 0 & 1 & -14 & 11 \\ 0 & 0 & 1 & -5/4 \\ 0 & 0 & 0 & -4 \end{bmatrix}$$

$$-\frac{1}{4}R_4 \to R_4 \begin{bmatrix} 1 & 0 & 4 & -2 \\ 0 & 1 & -14 & 11 \\ 0 & 0 & 1 & -5/4 \\ 0 & 0 & 0 & 1 \end{bmatrix}$$

860. $\begin{bmatrix} 1 & -3 & 4 & 2 \\ 0 & 1 & -11 & -3 \\ 0 & 0 & 1 & 3/4 \\ 0 & 0 & 0 & 1 \end{bmatrix}$

$$\begin{bmatrix} 3 & -8 & 1 & 3 \\ 1 & -3 & 4 & 2 \\ 5 & -13 & -6 & 1 \\ -4 & 13 & -2 & 1 \end{bmatrix}$$

$$R_1 \leftrightarrow R_2 \begin{bmatrix} 1 & -3 & 4 & 2 \\ 3 & -8 & 1 & 3 \\ 5 & -13 & -6 & 1 \\ -4 & 13 & -2 & 1 \end{bmatrix}$$

$$\begin{matrix} -3R_1 + R_2 \to R_2 \\ -5R_1 + R_3 \to R_3 \\ 4R_1 + R_4 \to R_4 \end{matrix} \begin{bmatrix} 1 & -3 & 4 & 2 \\ 0 & 1 & -11 & -3 \\ 0 & 2 & -26 & -9 \\ 0 & 1 & 14 & 9 \end{bmatrix}$$

$$\begin{matrix} -2R_2 + R_3 \to R_3 \\ -1R_2 + R_4 \to R_4 \end{matrix} \begin{bmatrix} 1 & -3 & 4 & 2 \\ 0 & 1 & -11 & -3 \\ 0 & 0 & -4 & -3 \\ 0 & 0 & 25 & 12 \end{bmatrix}$$

$$-\frac{1}{4}R_3 \to R_3 \begin{bmatrix} 1 & -3 & 4 & 2 \\ 0 & 1 & -11 & -3 \\ 0 & 0 & 1 & 3/4 \\ 0 & 0 & 25 & 12 \end{bmatrix}$$

$$-25R_1 + R_4 \to R_4 \begin{bmatrix} 1 & -3 & 4 & 2 \\ 0 & 1 & -11 & -3 \\ 0 & 0 & 1 & 3/4 \\ 0 & 0 & 0 & -27/4 \end{bmatrix}$$

$$-\frac{4}{27}R_4 \to R_4 \begin{bmatrix} 1 & -3 & 4 & 2 \\ 0 & 1 & -11 & -3 \\ 0 & 0 & 1 & 3/4 \\ 0 & 0 & 0 & 1 \end{bmatrix}$$

861. (2, –3)

First write the augmented matrix for the system. The *augmented matrix* consists of the coefficient matrix separated from a column matrix of the constants by a bar:

$$\left[\begin{array}{cc|c} 2 & 1 & 1 \\ 1 & -2 & 8 \end{array}\right]$$

Perform row operations to put the matrix in echelon form. You perform row operations on the entire row — coefficients and constants. The ones along the main diagonal in the resulting echelon form correspond to the variables in the equation.

$$R_1 \leftrightarrow R_2 \left[\begin{array}{cc|c} 1 & -2 & 8 \\ 2 & 1 & 1 \end{array}\right]$$

$$-2R_1 + R_2 \rightarrow R_2 \left[\begin{array}{cc|c} 1 & -2 & 8 \\ 0 & 5 & -15 \end{array}\right]$$

$$\frac{1}{5}R_2 \rightarrow R_2 \left[\begin{array}{cc|c} 1 & -2 & 8 \\ 0 & 1 & -3 \end{array}\right]$$

From the bottom row, you have $y = -3$. Substitute –3 for y in the equation represented by the new first row to solve for x:

$$\begin{aligned} x - 2y &= 8 \\ x - 2(-3) &= 8 \\ x &= 2 \end{aligned}$$

862. (–1, –4)

First write the augmented matrix for the system. The *augmented matrix* consists of the coefficient matrix separated from a column matrix of the constants by a bar:

$$\left[\begin{array}{cc|c} 5 & -3 & 7 \\ 1 & -2 & 7 \end{array}\right]$$

Perform row operations to put the matrix in echelon form. You perform row operations on the entire row — coefficients and constants. The ones along the main diagonal in the resulting echelon form correspond to the variables in the equation.

$$R_1 \leftrightarrow R_2 \left[\begin{array}{cc|c} 1 & -2 & 7 \\ 5 & -3 & 7 \end{array}\right]$$

$$-5R_1 + R_2 \rightarrow R_2 \left[\begin{array}{cc|c} 1 & -2 & 7 \\ 0 & 7 & -28 \end{array}\right]$$

$$\frac{1}{7}R_2 \rightarrow R_2 \left[\begin{array}{cc|c} 1 & -2 & 7 \\ 0 & 1 & -4 \end{array}\right]$$

From the bottom row, you have $y = -4$. Substitute –4 for y in the equation represented by the new first row to solve for x:

$$\begin{aligned} x - 2y &= 7 \\ x - 2(-4) &= 7 \\ x &= -1 \end{aligned}$$

863. (2, 3)

First write the augmented matrix for the system. The *augmented matrix* consists of the coefficient matrix separated from a column matrix of the constants by a bar:

$$\left[\begin{array}{cc|c} 2 & -1 & 1 \\ 3 & 1 & 9 \end{array}\right]$$

Perform row operations to put the matrix in echelon form. You perform row operations on the entire row — coefficients and constants. The ones along the main diagonal in the resulting echelon form correspond to the variables in the equation.

$$\frac{1}{2}R_1 \to R_1 \left[\begin{array}{cc|c} 1 & -1/2 & 1/2 \\ 3 & 1 & 9 \end{array}\right]$$

$$-3R_1 + R_2 \to R_2 \left[\begin{array}{cc|c} 1 & -1/2 & 1/2 \\ 0 & 5/2 & 15/2 \end{array}\right]$$

$$\frac{2}{5}R_2 \to R_2 \left[\begin{array}{cc|c} 1 & -1/2 & 1/2 \\ 0 & 1 & 3 \end{array}\right]$$

From the bottom row, you have $y = 3$. Substitute 3 for y in the equation represented by the new first row to solve for x:

$$x - \frac{1}{2}y = \frac{1}{2}$$

$$x - \frac{1}{2}(3) = \frac{1}{2}$$

$$x = 2$$

864. $\left(\frac{1}{2}, -3\right)$

First write the augmented matrix for the system. The *augmented matrix* consists of the coefficient matrix separated from a column matrix of the constants by a bar:

$$\left[\begin{array}{cc|c} 2 & -3 & 10 \\ 4 & 1 & -1 \end{array}\right]$$

Perform row operations to put the matrix in echelon form. You perform row operations on the entire row — coefficients and constants. The ones along the main diagonal in the resulting echelon form correspond to the variables in the equation.

$$\frac{1}{2}R_1 \to R_1 \left[\begin{array}{cc|c} 1 & -3/2 & 5 \\ 4 & 1 & -1 \end{array}\right]$$

$$-4R_1 + R_2 \to R_2 \left[\begin{array}{cc|c} 1 & -3/2 & 5 \\ 0 & 7 & -21 \end{array}\right]$$

$$\frac{1}{7}R_2 \to R_2 \left[\begin{array}{cc|c} 1 & -3/2 & 5 \\ 0 & 1 & -3 \end{array}\right]$$

From the bottom row, you have $y=-3$. Substitute –3 for y in the equation represented by the new first row to solve for x:

$$x-\frac{3}{2}y=5$$
$$x-\frac{3}{2}(-3)=5$$
$$x=5-\frac{9}{2}=\frac{1}{2}$$

865. (–1, –2, 3)

First write the augmented matrix for the system. The *augmented matrix* consists of the coefficient matrix separated from a column matrix of the constants by a bar:

$$\left[\begin{array}{ccc|c} 1 & 3 & -2 & -13 \\ 2 & -1 & 3 & 9 \\ 1 & 8 & 6 & 1 \end{array}\right]$$

Perform row operations to put the matrix in echelon form. You perform row operations on the entire row — coefficients and constants. The ones along the main diagonal in the resulting echelon form correspond to the variables in the equation.

$$\begin{array}{l} -2R_1+R_2\to R_2 \\ -1R_1+R_3\to R_3 \end{array}\left[\begin{array}{ccc|c} 1 & 3 & -2 & -13 \\ 0 & -7 & 7 & 35 \\ 0 & 5 & 8 & 14 \end{array}\right]$$

$$-\frac{1}{7}R_2\to R_2\left[\begin{array}{ccc|c} 1 & 3 & -2 & -13 \\ 0 & 1 & -1 & -5 \\ 0 & 5 & 8 & 14 \end{array}\right]$$

$$-5R_2+R_3\to R_3\left[\begin{array}{ccc|c} 1 & 3 & -2 & -13 \\ 0 & 1 & -1 & -5 \\ 0 & 0 & 13 & 39 \end{array}\right]$$

$$\frac{1}{13}R_3\to R_3\left[\begin{array}{ccc|c} 1 & 3 & -2 & -13 \\ 0 & 1 & -1 & -5 \\ 0 & 0 & 1 & 3 \end{array}\right]$$

From the bottom row, you have $z=3$. Substitute 3 for z in the equation represented by the new second row to solve for y:

$$y-z=-5$$
$$y-3=-5$$
$$y=-2$$

Now substitute the values for y and z into the new first row to solve for x:

$$x+3y-2z=-13$$
$$x+3(-2)-2(3)=-13$$
$$x=-1$$

866. (2, –1, 0)

First write the augmented matrix for the system. The *augmented matrix* consists of the coefficient matrix separated from a column matrix of the constants by a bar:

$$\left[\begin{array}{ccc|c} 4 & -3 & -1 & 11 \\ 2 & 1 & 3 & 3 \\ 1 & 2 & 5 & 0 \end{array}\right]$$

Perform row operations to put the matrix in echelon form. You perform row operations on the entire row — coefficients and constants. The ones along the main diagonal in the resulting echelon form correspond to the variables in the equation.

$$R_1 \leftrightarrow R_3 \left[\begin{array}{ccc|c} 1 & 2 & 5 & 0 \\ 2 & 1 & 3 & 3 \\ 4 & -3 & -1 & 11 \end{array}\right]$$

$$\begin{array}{l} -2R_1+R_2 \to R_2 \\ -4R_1+R_3 \to R_3 \end{array} \left[\begin{array}{ccc|c} 1 & 2 & 5 & 0 \\ 0 & -3 & -7 & 3 \\ 0 & -11 & -21 & 11 \end{array}\right]$$

$$-\frac{1}{3}R_2 \to R_2 \left[\begin{array}{ccc|c} 1 & 2 & 5 & 0 \\ 0 & 1 & 7/3 & -1 \\ 0 & -11 & -21 & 11 \end{array}\right]$$

$$11R_2+R_3 \to R_3 \left[\begin{array}{ccc|c} 1 & 2 & 5 & 0 \\ 0 & 1 & 7/3 & -1 \\ 0 & 0 & 14/3 & 0 \end{array}\right]$$

$$\frac{3}{14}R_3 \to R_3 \left[\begin{array}{ccc|c} 1 & 2 & 5 & 0 \\ 0 & 1 & 7/3 & -1 \\ 0 & 0 & 1 & 0 \end{array}\right]$$

From the bottom row, you have $z=0$. Substitute 0 for z in the equation represented by the new second row to solve for y:

$$y+\frac{7}{3}z=-1$$
$$y+\frac{7}{3}(0)=-1$$
$$y=-1$$

Now substitute the values for y and z into the new first row to solve for x:

$$x+2y+5z=0$$
$$x+2(-1)+5(0)=0$$
$$x=2$$

867. (4, –3, 5)

First write the augmented matrix for the system. The *augmented matrix* consists of the coefficient matrix separated from a column matrix of the constants by a bar:

$$\left[\begin{array}{ccc|c} 1 & 3 & -4 & -25 \\ 2 & -1 & 11 & 66 \\ 3 & 4 & 1 & 5 \end{array}\right]$$

Perform row operations to put the matrix in echelon form. You perform row operations on the entire row — coefficients and constants. The ones along the main diagonal in the resulting echelon form correspond to the variables in the equation.

$$\begin{array}{l} -2R_1+R_2\to R_2 \\ -3R_1+R_3\to R_3 \end{array}\left[\begin{array}{ccc|c} 1 & 3 & -4 & -25 \\ 0 & -7 & 19 & 116 \\ 0 & -5 & 13 & 80 \end{array}\right]$$

$$R_2\leftrightarrow -\frac{1}{5}R_3\left[\begin{array}{ccc|c} 1 & 3 & -4 & -25 \\ 0 & 1 & -13/5 & -16 \\ 0 & -7 & 19 & 116 \end{array}\right]$$

$$7R_2+R_3\to R_3\left[\begin{array}{ccc|c} 1 & 3 & -4 & -25 \\ 0 & 1 & -13/5 & -16 \\ 0 & 0 & 4/5 & 4 \end{array}\right]$$

$$\frac{5}{4}R_3\to R_3\left[\begin{array}{ccc|c} 1 & 3 & -4 & -25 \\ 0 & 1 & -13/5 & -16 \\ 0 & 0 & 1 & 5 \end{array}\right]$$

From the bottom row, you have $z=5$. Substitute 5 for z in the equation represented by the new second row to solve for y:

$$\begin{aligned} y-\frac{13}{5}z&=-16 \\ y-\frac{13}{5}(5)&=-16 \\ y&=-3 \end{aligned}$$

Now substitute the values for y and z into the new first row to solve for x:

$$\begin{aligned} x+3y-4z&=-25 \\ x+3(-3)-4(5)&=-25 \\ x&=4 \end{aligned}$$

868. (7, –3, 2)

First write the augmented matrix for the system. The *augmented matrix* consists of the coefficient matrix separated from a column matrix of the constants by a bar:

$$\left[\begin{array}{ccc|c} 1 & 0 & 4 & 15 \\ 2 & -3 & 0 & 23 \\ 0 & 5 & -3 & -21 \end{array}\right]$$

Perform row operations to put the matrix in echelon form. You perform row operations on the entire row — coefficients and constants. The ones along the main diagonal in the resulting echelon form correspond to the variables in the equation.

$$-2R_1+R_2 \to R_2 \left[\begin{array}{ccc|c} 1 & 0 & 4 & 15 \\ 0 & -3 & -8 & -7 \\ 0 & 5 & -3 & -21 \end{array}\right]$$

$$-\frac{1}{3}R_2 \to R_2 \left[\begin{array}{ccc|c} 1 & 0 & 4 & 15 \\ 0 & 1 & 8/3 & 7/3 \\ 0 & 5 & -3 & -21 \end{array}\right]$$

$$-5R_2+R_3 \to R_3 \left[\begin{array}{ccc|c} 1 & 0 & 4 & 15 \\ 0 & 1 & 8/3 & 7/3 \\ 0 & 0 & -49/3 & -98/3 \end{array}\right]$$

$$-\frac{3}{49}R_3 \to R_3 \left[\begin{array}{ccc|c} 1 & 0 & 4 & 15 \\ 0 & 1 & 8/3 & 7/3 \\ 0 & 0 & 1 & 2 \end{array}\right]$$

From the bottom row, you have $z=2$. Substitute 2 for z in the equation represented by the new second row to solve for y:

$$y+\frac{8}{3}z=\frac{7}{3}$$

$$y+\frac{8}{3}(2)=\frac{7}{3}$$

$$y=\frac{7}{3}-\frac{16}{3}=-3$$

Now substitute the value for z into the first row to solve for x:

$$x+4z=15$$

$$x+4(2)=15$$

$$x=7$$

869. (1, –1, 2, –1)

First write the augmented matrix for the system. The *augmented matrix* consists of the coefficient matrix separated from a column matrix of the constants by a bar:

$$\left[\begin{array}{cccc|c} 1 & 2 & 1 & -1 & 2 \\ 2 & -3 & 0 & 2 & 3 \\ 4 & 1 & -1 & 0 & 1 \\ 0 & 2 & 4 & 1 & 5 \end{array}\right]$$

Perform row operations to put the matrix in echelon form. You perform row operations on the entire row — coefficients and constants. The ones along the main diagonal in the resulting echelon form correspond to the variables in the equation.

$$\begin{array}{l} \\ -2R_1+R_2 \to R_2 \\ -4R_1+R_3 \to R_3 \\ \\ \end{array} \left[\begin{array}{cccc|c} 1 & 2 & 1 & -1 & 2 \\ 0 & -7 & -2 & 4 & -1 \\ 0 & -7 & -9 & 4 & -7 \\ 0 & 2 & 4 & 1 & 5 \end{array}\right]$$

$$R_2 \leftrightarrow R_4 \left[\begin{array}{cccc|c} 1 & 2 & 1 & -1 & 2 \\ 0 & 2 & 4 & 1 & 5 \\ 0 & -7 & -9 & 4 & -7 \\ 0 & -7 & -2 & 4 & -1 \end{array}\right]$$

$$\frac{1}{2}R_2 \to R_2 \left[\begin{array}{cccc|c} 1 & 2 & 1 & -1 & 2 \\ 0 & 1 & 2 & 1/2 & 5/2 \\ 0 & -7 & -5 & 4 & -7 \\ 0 & -7 & -2 & 4 & -1 \end{array}\right]$$

$$\begin{array}{l} 7R_2 + R_3 \to R_3 \\ 7R_2 + R_4 \to R_4 \end{array} \left[\begin{array}{cccc|c} 1 & 2 & 1 & -1 & 2 \\ 0 & 1 & 2 & 1/2 & 5/2 \\ 0 & 0 & 9 & 15/2 & 21/2 \\ 0 & 0 & 12 & 15/2 & 33/2 \end{array}\right]$$

$$\frac{1}{9}R_3 \to R_3 \left[\begin{array}{cccc|c} 1 & 2 & 1 & -1 & 2 \\ 0 & 1 & 2 & 1/2 & 5/2 \\ 0 & 0 & 1 & 5/6 & 7/6 \\ 0 & 0 & 12 & 15/2 & 33/2 \end{array}\right]$$

$$-12R_3 + R_4 \to R_4 \left[\begin{array}{cccc|c} 1 & 2 & 1 & -1 & 2 \\ 0 & 1 & 2 & 1/2 & 5/2 \\ 0 & 0 & 1 & 5/6 & 7/6 \\ 0 & 0 & 0 & -5/2 & 5/2 \end{array}\right]$$

$$-\frac{2}{5}R_4 \to R_4 \left[\begin{array}{cccc|c} 1 & 2 & 1 & -1 & 2 \\ 0 & 1 & 2 & 1/2 & 5/2 \\ 0 & 0 & 1 & 5/6 & 7/6 \\ 0 & 0 & 0 & 1 & -1 \end{array}\right]$$

From the bottom row, you have $w = -1$. Substitute –1 for w in the equation represented by the new third row to solve for z:

$$z + \frac{5}{6}w = \frac{7}{6}$$

$$z + \frac{5}{6}(-1) = \frac{7}{6}$$

$$z = 2$$

Now substitute the values for z and w into the new second row to solve for y:

$$y + 2z + \frac{1}{2}w = \frac{5}{2}$$

$$y + 2(2) + \frac{1}{2}(-1) = \frac{5}{2}$$

$$y = \frac{5}{2} - \frac{7}{2} = -1$$

Finally, solve for x by substituting the known values into the new first row:

$$\begin{aligned} x+2y+z-w&=2 \\ x+2(-1)+2-(-1)&=2 \\ x&=1 \end{aligned}$$

870. (0, 3, –3, 2)

First write the augmented matrix for the system. The *augmented matrix* consists of the coefficient matrix separated from a column matrix of the constants by a bar:

$$\left[\begin{array}{cccc|c} 5 & 3 & -1 & 0 & 12 \\ 1 & -2 & 0 & 3 & 0 \\ 2 & 0 & 4 & -3 & -18 \\ 0 & 4 & 3 & -1 & 1 \end{array}\right]$$

Perform row operations to put the matrix in echelon form. You perform row operations on the entire row — coefficients and constants. The ones along the main diagonal in the resulting echelon form correspond to the variables in the equation.

$$R_1 \leftrightarrow R_2 \left[\begin{array}{cccc|c} 1 & -2 & 0 & 3 & 0 \\ 5 & 3 & -1 & 0 & 12 \\ 2 & 0 & 4 & -3 & -18 \\ 0 & 4 & 3 & -1 & 1 \end{array}\right]$$

$$\begin{array}{l} -5R_1+R_2 \to R_2 \\ -2R_1+R_3 \to R_3 \end{array} \left[\begin{array}{cccc|c} 1 & -2 & 0 & 3 & 0 \\ 0 & 13 & -1 & -15 & 12 \\ 0 & 4 & 4 & -9 & -18 \\ 0 & 4 & 3 & -1 & 1 \end{array}\right]$$

$$R_2 \leftrightarrow R_4 \left[\begin{array}{cccc|c} 1 & -2 & 0 & 3 & 0 \\ 0 & 4 & 3 & -1 & 1 \\ 0 & 4 & 4 & -9 & -18 \\ 0 & 13 & -1 & -15 & 12 \end{array}\right]$$

$$\frac{1}{4}R_2 \to R_2 \left[\begin{array}{cccc|c} 1 & -2 & 0 & 3 & 0 \\ 0 & 1 & 3/4 & -1/4 & 1/4 \\ 0 & 4 & 4 & -9 & -18 \\ 0 & 13 & -1 & -15 & 12 \end{array}\right]$$

$$\begin{array}{l} -4R_2+R_3 \to R_3 \\ -13R_2+R_4 \to R_4 \end{array} \left[\begin{array}{cccc|c} 1 & -2 & 0 & 3 & 0 \\ 0 & 1 & 3/4 & -1/4 & 1/4 \\ 0 & 0 & 1 & -8 & -19 \\ 0 & 0 & -43/4 & -47/4 & 35/4 \end{array}\right]$$

$$\frac{43}{4}R_3+R_4 \to R_4 \left[\begin{array}{cccc|c} 1 & -2 & 0 & 3 & 0 \\ 0 & 1 & 3/4 & -1/4 & 1/4 \\ 0 & 0 & 1 & -8 & -19 \\ 0 & 0 & 0 & -391/4 & -782/4 \end{array}\right]$$

$$-\frac{4}{391}R_4 \to R_4 \left[\begin{array}{cccc|c} 1 & -2 & 0 & 3 & 0 \\ 0 & 1 & 3/4 & -1/4 & 1/4 \\ 0 & 0 & 1 & -8 & -19 \\ 0 & 0 & 0 & 1 & 2 \end{array}\right]$$

From the bottom row, you have $w = 2$. Substitute 2 for w in the equation represented by the new third row to solve for z:

$$\begin{aligned} z - 8w &= -19 \\ z - 8(2) &= -19 \\ z &= -3 \end{aligned}$$

Now substitute the values for z and w into the second new row to solve for y:

$$\begin{aligned} y + \frac{3}{4}z - \frac{1}{4}w &= \frac{1}{4} \\ y + \frac{3}{4}(-3) - \frac{1}{4}(2) &= \frac{1}{4} \\ y &= \frac{1}{4} + \frac{11}{4} = 3 \end{aligned}$$

Finally, solve for x by substituting the known values into the first new row:

$$\begin{aligned} x - 2y + 3w &= 0 \\ x - 2(3) + 3(2) &= 0 \\ x &= 0 \end{aligned}$$

871. (–4, 5)

To determine the inverse of the coefficient matrix corresponding to the system of equations, use the formula for the inverse of a 2×2 matrix. (Recall that the determinant of a 2×2 matrix is $\det\left(\begin{bmatrix} a & b \\ c & d \end{bmatrix}\right) = ad - bc$.)

$$A^{-1} = \frac{\begin{bmatrix} a_{22} & -a_{12} \\ -a_{21} & a_{11} \end{bmatrix}}{\begin{vmatrix} a_{11} & a_{12} \\ a_{21} & a_{22} \end{vmatrix}}$$

The coefficient matrix for the system $\begin{cases} 2x + 3y = 7 \\ 3x + 4y = 8 \end{cases}$ is $\begin{bmatrix} 2 & 3 \\ 3 & 4 \end{bmatrix}$, so the inverse is

$$\frac{\begin{bmatrix} 4 & -3 \\ -3 & 2 \end{bmatrix}}{\begin{vmatrix} 2 & 3 \\ 3 & 4 \end{vmatrix}} = \frac{\begin{bmatrix} 4 & -3 \\ -3 & 2 \end{bmatrix}}{8-9} = \frac{\begin{bmatrix} 4 & -3 \\ -3 & 2 \end{bmatrix}}{-1} = \begin{bmatrix} -4 & 3 \\ 3 & -2 \end{bmatrix}$$

Multiply the inverse of the coefficient matrix by the constant matrix:

$$\begin{bmatrix} -4 & 3 \\ 3 & -2 \end{bmatrix} \cdot \begin{bmatrix} 7 \\ 8 \end{bmatrix} = \begin{bmatrix} -28 + 24 \\ 21 - 16 \end{bmatrix} = \begin{bmatrix} -4 \\ 5 \end{bmatrix}$$

The solution is $x = -4$, $y = 5$.

872. (2, –3)

To determine the inverse of the coefficient matrix corresponding to the system of equations, use the formula for the inverse of a 2×2 matrix. (Recall that the determinant of a 2×2 matrix is $\det\left(\begin{bmatrix} a & b \\ c & d \end{bmatrix}\right) = ad - bc$.)

$$A^{-1} = \frac{\begin{bmatrix} a_{22} & -a_{12} \\ -a_{21} & a_{11} \end{bmatrix}}{\begin{vmatrix} a_{11} & a_{12} \\ a_{21} & a_{22} \end{vmatrix}}$$

The coefficient matrix for the system $\begin{cases} 5x - 2y = 16 \\ 12x - 5y = 39 \end{cases}$ is $\begin{bmatrix} 5 & -2 \\ 12 & -5 \end{bmatrix}$, so the inverse is

$$\frac{\begin{bmatrix} -5 & 2 \\ -12 & 5 \end{bmatrix}}{\begin{vmatrix} 5 & -2 \\ 12 & -5 \end{vmatrix}} = \frac{\begin{bmatrix} -5 & 2 \\ -12 & 5 \end{bmatrix}}{-25-(-24)} = \frac{\begin{bmatrix} -5 & 2 \\ -12 & 5 \end{bmatrix}}{-1} = \begin{bmatrix} 5 & -2 \\ 12 & -5 \end{bmatrix}$$

This matrix is its own inverse! Multiply the inverse of the coefficient matrix by the constant matrix:

$$\begin{bmatrix} 5 & -2 \\ 12 & -5 \end{bmatrix} \cdot \begin{bmatrix} 16 \\ 39 \end{bmatrix} = \begin{bmatrix} 80-78 \\ 192-195 \end{bmatrix} = \begin{bmatrix} 2 \\ -3 \end{bmatrix}$$

The solution is $x = 2$, $y = -3$.

873. (–1, 3)

To determine the inverse of the coefficient matrix corresponding to the system of equations, use the formula for the inverse of a 2×2 matrix. (Recall that the determinant of a 2×2 matrix is $\det\left(\begin{bmatrix} a & b \\ c & d \end{bmatrix}\right) = ad - bc$.)

$$A^{-1} = \frac{\begin{bmatrix} a_{22} & -a_{12} \\ -a_{21} & a_{11} \end{bmatrix}}{\begin{vmatrix} a_{11} & a_{12} \\ a_{21} & a_{22} \end{vmatrix}}$$

The coefficient matrix for the system $\begin{cases} 3x + 5y = 12 \\ 7x + 12y = 29 \end{cases}$ is $\begin{bmatrix} 3 & 5 \\ 7 & 12 \end{bmatrix}$, so the inverse is

$$\frac{\begin{bmatrix} 12 & -5 \\ -7 & 3 \end{bmatrix}}{\begin{vmatrix} 3 & 5 \\ 7 & 12 \end{vmatrix}} = \frac{\begin{bmatrix} 12 & -5 \\ -7 & 3 \end{bmatrix}}{36-35} = \frac{\begin{bmatrix} 12 & -5 \\ -7 & 3 \end{bmatrix}}{1} = \begin{bmatrix} 12 & -5 \\ -7 & 3 \end{bmatrix}$$

Multiply the inverse of the coefficient matrix by the constant matrix:

$$\begin{bmatrix} 12 & -5 \\ -7 & 3 \end{bmatrix} \cdot \begin{bmatrix} 12 \\ 29 \end{bmatrix} = \begin{bmatrix} 144-145 \\ -84+87 \end{bmatrix} = \begin{bmatrix} -1 \\ 3 \end{bmatrix}$$

The solution is $x = -1$, $y = 3$.

874. (1, 0, –1)

The system is

$$\begin{cases} x+2y+3z=-2 \\ x-2y+4z=-3 \\ x+\ y+3z=-2 \end{cases}$$

Find the inverse of the coefficient matrix. To do so, perform row operations that change the matrix on the left of the bar to an identity matrix like the one on the right.

$$\left[\begin{array}{ccc|ccc} 1 & 2 & 3 & 1 & 0 & 0 \\ 1 & -2 & 4 & 0 & 1. & 0 \\ 1 & 1 & 3 & 0 & 0 & 1 \end{array}\right]=\left[\begin{array}{ccc|ccc} 1 & 2 & 3 & 1 & 0 & 0 \\ 0 & -4 & 1 & -1 & 1 & 0 \\ 0 & -1 & 0 & -1 & 0 & 1 \end{array}\right]=\left[\begin{array}{ccc|ccc} 1 & 2 & 3 & 1 & 0 & 0 \\ 0 & -1 & 0 & -1 & 0 & 1 \\ 0 & -4 & 1 & -1 & 1 & 0 \end{array}\right]$$

$$=\left[\begin{array}{ccc|ccc} 1 & 0 & 3 & -1 & 0 & 2 \\ 0 & 1 & 0 & 1 & 0 & -1 \\ 0 & 0 & 1 & 3 & 1 & -4 \end{array}\right]=\left[\begin{array}{ccc|ccc} 1 & 0 & 0 & -10 & -3 & 14 \\ 0 & 1 & 0 & 1 & 0 & -1 \\ 0 & 0 & 1 & 3 & 1 & -4 \end{array}\right]$$

Multiply the inverse matrix for the coefficients by the constant matrix:

$$\begin{bmatrix} -10 & -3 & 14 \\ 1 & 0 & -1 \\ 3 & 1 & -4 \end{bmatrix}\cdot\begin{bmatrix} -2 \\ -3 \\ -2 \end{bmatrix}=\begin{bmatrix} 20+9-28 \\ -2+0+2 \\ -6-3+8 \end{bmatrix}=\begin{bmatrix} 1 \\ 0 \\ -1 \end{bmatrix}$$

875. (–1, 2, –4)

The system is

$$\begin{cases} -7x+\ 2y+\ \ z= & 7 \\ 2x-\ 3y+\ 5z= & -28 \\ 4x\ \ \ \ \ -\ 3z= & 8 \end{cases}$$

Find the inverse of the coefficient matrix: To do so, perform row operations that change the matrix on the left of the bar to an identity matrix like the one on the right.

$$\left[\begin{array}{ccc|ccc} -7 & 2 & 1 & 1 & 0 & 0 \\ 2 & -3 & 5 & 0 & 1 & 0 \\ 4 & 0 & -3 & 0 & 0 & 1 \end{array}\right]=\left[\begin{array}{ccc|ccc} 1 & 0 & -3/4 & 0 & 0 & 1/4 \\ 2 & -3 & 5 & 0 & 1 & 0 \\ -7 & 2 & 1 & 1 & 0 & 0 \end{array}\right]$$

$$=\left[\begin{array}{ccc|ccc} 1 & 0 & -3/4 & 0 & 0 & 1/4 \\ 0 & 2 & -17/4 & 1 & 0 & 7/4 \\ 0 & -3 & 13/2 & 0 & 1 & -1/2 \end{array}\right]=\left[\begin{array}{ccc|ccc} 1 & 0 & -3/4 & 0 & 0 & 1/4 \\ 0 & 1 & -17/8 & 1/2 & 0 & 7/8 \\ 0 & -3 & 13/2 & 0 & 1 & -1/2 \end{array}\right]$$

$$=\left[\begin{array}{ccc|ccc} 1 & 0 & -3/4 & 0 & 0 & 1/4 \\ 0 & 1 & -17/8 & 1/2 & 0 & 7/8 \\ 0 & 0 & 1/8 & 3/2 & 1 & 17/8 \end{array}\right]=\left[\begin{array}{ccc|ccc} 1 & 0 & 0 & 9 & 6 & 13 \\ 0 & 1 & 0 & 26 & 17 & 37 \\ 0 & 0 & 1 & 12 & 8 & 17 \end{array}\right]$$

Multiply the inverse matrix for the coefficients by the constant matrix:

$$\begin{bmatrix} 9 & 6 & 13 \\ 26 & 17 & 37 \\ 12 & 8 & 17 \end{bmatrix}\cdot\begin{bmatrix} 7 \\ -28 \\ 8 \end{bmatrix}=\begin{bmatrix} -1 \\ 2 \\ -4 \end{bmatrix}$$

876. (–2, 4)

Cramer's Rule uses determinants to find x and y:

$$\begin{matrix} ax+by=e \\ cx+dy=f \end{matrix} \quad x=\frac{\begin{vmatrix} e & b \\ f & d \end{vmatrix}}{\begin{vmatrix} a & b \\ c & d \end{vmatrix}} \quad y=\frac{\begin{vmatrix} a & e \\ c & f \end{vmatrix}}{\begin{vmatrix} a & b \\ c & d \end{vmatrix}}$$

Recall that the determinant of a 2×2 matrix is

$$\det\left(\begin{bmatrix} a & b \\ c & d \end{bmatrix}\right)=ad-bc$$

With the system of $4x+5y=12$ and $5x+7y=18$, you have the following:

$$x=\frac{\begin{vmatrix} 12 & 5 \\ 18 & 7 \end{vmatrix}}{\begin{vmatrix} 4 & 5 \\ 5 & 7 \end{vmatrix}}=\frac{12(7)-5(18)}{4(7)-5(5)}=\frac{84-90}{28-25}=\frac{-6}{3}=-2$$

$$y=\frac{\begin{vmatrix} 4 & 12 \\ 5 & 18 \end{vmatrix}}{\begin{vmatrix} 4 & 5 \\ 5 & 7 \end{vmatrix}}=\frac{4(18)-5(12)}{4(7)-5(5)}=\frac{72-60}{28-25}=\frac{12}{3}=4$$

877. (5, 6)

Cramer's Rule uses determinants to find x and y:

$$\begin{matrix} ax+by=e \\ cx+dy=f \end{matrix} \quad x=\frac{\begin{vmatrix} e & b \\ f & d \end{vmatrix}}{\begin{vmatrix} a & b \\ c & d \end{vmatrix}} \quad y=\frac{\begin{vmatrix} a & e \\ c & f \end{vmatrix}}{\begin{vmatrix} a & b \\ c & d \end{vmatrix}}$$

Recall that the determinant of a 2×2 matrix is

$$\det\left(\begin{bmatrix} a & b \\ c & d \end{bmatrix}\right)=ad-bc$$

With the system of $9x-4y=21$ and $6x+7y=72$, you have the following:

$$x=\frac{\begin{vmatrix} 21 & -4 \\ 72 & 7 \end{vmatrix}}{\begin{vmatrix} 9 & -4 \\ 6 & 7 \end{vmatrix}}=\frac{21(7)-(-4)(72)}{9(7)-(-4)(6)}=\frac{147-(-288)}{63-(-24)}=\frac{435}{87}=5$$

$$y=\frac{\begin{vmatrix} 9 & 21 \\ 6 & 72 \end{vmatrix}}{\begin{vmatrix} 9 & -4 \\ 6 & 7 \end{vmatrix}}=\frac{9(72)-21(6)}{9(7)-(-4)(6)}=\frac{648-126}{63-(-24)}=\frac{522}{87}=6$$

878. $\left(\frac{4}{3},-3\right)$

Cramer's Rule uses determinants to find x and y:

$$\begin{matrix} ax+by=e \\ cx+dy=f \end{matrix} \quad x=\frac{\begin{vmatrix} e & b \\ f & d \end{vmatrix}}{\begin{vmatrix} a & b \\ c & d \end{vmatrix}} \quad y=\frac{\begin{vmatrix} a & e \\ c & f \end{vmatrix}}{\begin{vmatrix} a & b \\ c & d \end{vmatrix}}$$

Recall that the determinant of a 2×2 matrix is

$$\det\left(\begin{bmatrix} a & b \\ c & d \end{bmatrix}\right)=ad-bc$$

With the system of $3x+11y=-29$ and $12x-5y=31$, you have the following:

$$x=\frac{\begin{vmatrix} -29 & 11 \\ 31 & -5 \end{vmatrix}}{\begin{vmatrix} 3 & 11 \\ 12 & -5 \end{vmatrix}}=\frac{-29(-5)-(11)(31)}{3(-5)-11(12)}=\frac{145-341}{-15-132}=\frac{-196}{-147}=\frac{4}{3}$$

$$y=\frac{\begin{vmatrix} 3 & -29 \\ 12 & 31 \end{vmatrix}}{\begin{vmatrix} 3 & 11 \\ 12 & -5 \end{vmatrix}}=\frac{3(31)-(-29)(12)}{3(-5)-11(12)}=\frac{93-(-348)}{-15-132}=\frac{441}{-147}=-3$$

879. (7, –3)

Cramer's Rule uses determinants to find x and y:

$$\begin{matrix} ax+by=e \\ cx+dy=f \end{matrix} \quad x=\frac{\begin{vmatrix} e & b \\ f & d \end{vmatrix}}{\begin{vmatrix} a & b \\ c & d \end{vmatrix}} \quad y=\frac{\begin{vmatrix} a & e \\ c & f \end{vmatrix}}{\begin{vmatrix} a & b \\ c & d \end{vmatrix}}$$

Recall that the determinant of a 2×2 matrix is

$$\det\left(\begin{bmatrix} a & b \\ c & d \end{bmatrix}\right)=ad-bc$$

With the system of $5x+8y=11$ and $13x+4y=79$, you have the following:

$$x=\frac{\begin{vmatrix} 11 & 8 \\ 79 & 4 \end{vmatrix}}{\begin{vmatrix} 5 & 8 \\ 13 & 4 \end{vmatrix}}=\frac{(11)(4)-(8)(79)}{5(4)-8(13)}=\frac{44-623}{20-104}=\frac{-588}{-84}=7$$

$$y=\frac{\begin{vmatrix} 5 & 11 \\ 13 & 79 \end{vmatrix}}{\begin{vmatrix} 5 & 8 \\ 13 & 4 \end{vmatrix}}=\frac{5(79)-11(13)}{5(4)-8(13)}=\frac{395-143}{20-104}=\frac{252}{-84}=-3$$

880. (8, 2, 3)

Cramer's Rule uses determinants to find *x*, *y*, and *z*:

$$\begin{cases} ax+by+cz=p \\ dx+ey+fz=q \\ gx+hy+jz=r \end{cases} \quad x=\frac{\begin{vmatrix} p & b & c \\ q & e & f \\ r & h & j \end{vmatrix}}{\begin{vmatrix} a & b & c \\ d & e & f \\ g & h & j \end{vmatrix}} \quad y=\frac{\begin{vmatrix} a & p & c \\ d & q & f \\ g & r & j \end{vmatrix}}{\begin{vmatrix} a & b & c \\ d & e & f \\ g & h & j \end{vmatrix}} \quad z=\frac{\begin{vmatrix} a & b & p \\ d & e & q \\ g & h & r \end{vmatrix}}{\begin{vmatrix} a & b & c \\ d & e & f \\ g & h & j \end{vmatrix}}$$

Recall that you find the determinant of a 3×3 matrix by multiplying along the diagonals and finding a difference:

$$\det\left(\begin{bmatrix} a & b & c \\ d & e & f \\ g & h & i \end{bmatrix}\right)=aei+bfg+cdh-(ceg+fha+ibd)$$

With the system of $x+3y-2z=8$, $4x+3y-5z=23$, and $2x+5y+4z=38$, you have the following:

$$x=\frac{\begin{vmatrix} 8 & 3 & -2 \\ 3 & 3 & -5 \\ 38 & 5 & 4 \end{vmatrix}}{\begin{vmatrix} 1 & 3 & -2 \\ 4 & 3 & -5 \\ 2 & 5 & 4 \end{vmatrix}}=\frac{(96-570-230)-(-228-200+276)}{(12-30-40)-(-12-25+48)}=\frac{-552}{-69}=8$$

$$y=\frac{\begin{vmatrix} 1 & 8 & -2 \\ 4 & 23 & -5 \\ 2 & 38 & 4 \end{vmatrix}}{\begin{vmatrix} 1 & 3 & -2 \\ 4 & 3 & -5 \\ 2 & 5 & 4 \end{vmatrix}}=\frac{(92-80-304)-(-92-190+128)}{(12-30-40)-(-12-25+48)}=\frac{-138}{-69}=2$$

$$z=\frac{\begin{vmatrix} 1 & 3 & 8 \\ 4 & 3 & 23 \\ 2 & 5 & 38 \end{vmatrix}}{\begin{vmatrix} 1 & 3 & -2 \\ 4 & 3 & -5 \\ 2 & 5 & 4 \end{vmatrix}}=\frac{(114+138+160)-(48+115+456)}{(12-30-40)-(-12-25+48)}=\frac{-207}{-69}=3$$

881. 31

You find the fifth term, a_5, by replacing the *n* with 5 in the general term and simplifying:

$$a_5=2^5-1=32-1=31$$

882. $\frac{7}{3}$

You find the fifth term, a_5, by replacing each n with 5 in the general term and simplifying:

$$a_5 = \frac{4(5)+1}{2(5)-1} = \frac{21}{9} = \frac{7}{3}$$

883. 10

You find the fifth term, a_5, by replacing each n with 5 in the general term and simplifying:

$$a_5 = 5^2 - 3(5) = 25 - 15 = 10$$

884. $-\frac{1}{26}$

You find the fifth term, a_5, by replacing each n with 5 in the general term and simplifying:

$$a_5 = \frac{(-1)^5}{5^2+1} = \frac{-1}{26}$$

885. 3

You find the fifth term, a_5, by replacing the n with 5 in the general term and simplifying:

$$a_5 = 3\sin\left(\frac{5\pi}{2}\right) = 3\sin\left(\frac{5\pi}{2} - 2\pi\right) = 3\sin\left(\frac{\pi}{2}\right) = 3(1) = 3$$

886. $a_n = 3^{n-1}$

Each term is a power of 3. The number 1 is 3^0, so you subtract 1 from n to create a 0 exponent for the first term.

887. $a_n = n!$

Each term is the result of performing the factorial operation on the number of the term.

888. $a_n = 2ne^n$

The coefficient of each term is twice the number of the term, and the power of the term is the number of the term.

889. $a_n = n^2 + 1$

Each term is one more than a perfect square. The number of the term is squared.

890. $a_n = \frac{n+3}{n^2}$

The pattern is easier to see if you write the first term as a fraction and the third term with a denominator of 9:

$$\frac{4}{1}, \frac{5}{4}, \frac{6}{9}, \frac{7}{16}, \frac{8}{25}, \dots$$

The numerators are all 3 more than the number of the term, and the denominators are the number of the term squared.

891. –1, 5

A sequence has a *recursively defined* general term if the rule involves one or more previous terms in the computation of a given term. In this recursively defined sequence, the next two terms are as follows:

$$a_3 = 2a_{3-2} - 3a_{3-1} = 2a_1 - 3a_2 = 2(1) - 3(1) = -1$$
$$a_4 = 2a_{4-2} - 3a_{4-1} = 2a_2 - 3a_3 = 2(1) - 3(1) = 2 + 3 = 5$$

892. $\frac{3}{2}, \frac{7}{3}$

A sequence has a *recursively defined* general term if the rule involves one or more previous terms in the computation of a given term. In this recursively defined sequence, the next two terms are as follows:

$$a_3 = \frac{a_{3-2} + a_{3-1}}{a_{3-1}} = \frac{a_1 + a_2}{a_2} = \frac{1+2}{2} = \frac{3}{2}$$

$$a_4 = \frac{a_{4-2} + a_{4-1}}{a_{4-1}} = \frac{a_2 + a_3}{a_3} = \frac{2 + \frac{3}{2}}{\frac{3}{2}} = \frac{\frac{7}{2}}{\frac{3}{2}} = \frac{7}{3}$$

893. 3, 10

A sequence has a *recursively defined* general term if the rule involves one or more previous terms in the computation of a given term. In this recursively defined sequence the next two terms are as follows:

$$a_3 = 2^{a_{3-2}} + 2^{a_{3-1}} = 2^{a_1} + 2^{a_2} = 2^0 + 2^1 = 1 + 2 = 3$$
$$a_4 = 2^{a_{4-2}} + 2^{a_{4-1}} = 2^{a_2} + 2^{a_3} = 2^1 + 2^3 = 2 + 8 = 10$$

894. $1, \frac{1}{e}$

A sequence has a *recursively defined* general term if the rule involves one or more previous terms in the computation of a given term. In this recursively defined sequence, the next two terms are as follows:

$$a_3 = (a_{3-2})(a_{3-1}) = (a_1)(a_2) = (e)\left(\frac{1}{e}\right) = 1$$
$$a_4 = (a_{4-2})(a_{4-1}) = (a_2)(a_3) = \left(\frac{1}{e}\right)(1) = \frac{1}{e}$$

895. 2, 3

A sequence has a *recursively defined* general term if the rule involves one or more previous terms in the computation of a given term. In this recursively defined sequence, the next two terms are as follows:

$$a_3 = a_{3-2} + a_{3-1} = a_1 + a_2 = 1 + 1 = 2$$
$$a_4 = a_{4-2} + a_{4-1} = a_2 + a_3 = 1 + 2 = 3$$

This sequence is, of course, the famous Fibonacci sequence.

896. 205

You find the sum of an arithmetic series with the formula $\sum_{i=1}^{n} a_i = \frac{n}{2}(a_1 + a_n)$, where you're averaging the first and last terms and multiplying by *n*, the number of terms. In this series, $n = 10$, the first term is $4 + 3 = 7$, and the last term is $4 + 30 = 34$:

$$\sum_{i=1}^{10}(4+3i) = \frac{10}{2}(7+34) = 5(41) = 205$$

897. –24

You find the sum of an arithmetic series with the formula $\sum_{i=1}^{n} a_i = \frac{n}{2}(a_1 + a_n)$, where you're averaging the first and last terms and multiplying by *n*, the number of terms. In this series, $n = 8$, the first term is $6 - 2 = 4$, and the last term is $6 - 16 = -10$:

$$\sum_{i=1}^{10}(6-2i) = \frac{8}{2}(4+(-10)) = 4(-6) = -24$$

898. 190

You find the sum of an arithmetic series with the formula $\sum_{i=1}^{n} a_i = \frac{n}{2}(a_1 + a_n)$, where you're averaging the first and last terms and multiplying by *n*, the number of terms. In this series, you're actually going from the eighth to the eleventh term, which means you're adding four terms. Using the formula, $n = 4$, the eighth term is 40, and the eleventh term is 55:

$$\sum_{i=8}^{11}(5i) = \frac{4}{2}(40+55) = 2(95) = 190$$

Answers 801–900

899. 28

You find the sum of an arithmetic series with the formula $\sum_{i=1}^{n} a_i = \frac{n}{2}(a_1 + a_n)$, where you're averaging the first and last terms and multiplying by *n*, the number of terms.

In this series, you're actually going from the third to the ninth term, which means you're adding seven terms. Using the formula, $n = 7$, the third term is $\frac{5}{2}$, and the ninth term is $\frac{11}{2}$:

$$\sum_{i=3}^{9}\left(\frac{i}{2}+1\right)=\frac{7}{2}\left(\frac{5}{2}+\frac{11}{2}\right)=\frac{7}{2}(8)=28$$

900. $\frac{35}{3}$

You find the sum of an arithmetic series with the formula $\sum_{i=1}^{n} a_i = \frac{n}{2}(a_1 + a_n)$, where you're averaging the first and last terms and multiplying by n, the number of terms.

In this series, you're actually going from the zeroth to the sixth term, which means you're adding seven terms. (It's often more convenient to number the terms starting with 0 rather than 1; you can write the formula more simply or combine terms of different series more easily.) Using the formula, $n = 7$, the zeroth term is $\frac{2}{3}$, and the sixth term is $\frac{8}{3}$:

$$\sum_{i=0}^{6}\left(\frac{1}{3}i+\frac{2}{3}\right)=\frac{7}{2}\left(\frac{2}{3}+\frac{8}{3}\right)=\frac{7}{2}\left(\frac{10}{3}\right)=\frac{35}{3}$$

901. 37

Write out the four terms and add them:

$$\sum_{i=1}^{4}\left(3+\frac{12}{i}\right)=\left(3+\frac{12}{1}\right)+\left(3+\frac{12}{2}\right)+\left(3+\frac{12}{3}\right)+\left(3+\frac{12}{4}\right)$$
$$=15+9+7+6=37$$

902. 11

Write out the five terms, simplify, and add them. The signs will alternate with the powers of –1.

$$\sum_{i=1}^{5}(-1)^{i+1}2^{i-1}=(-1)^2(2^0)+(-1)^3(2^1)+(-1)^4(2^2)+(-1)^5(2^3)+(-1)^6(2^4)$$
$$=1-2+4-8+16=11$$

903. 0

Write out the six terms, compute the value of the trig function, and then add the terms:

$$\sum_{i=0}^{5}(\cos i\pi)=\cos 0\pi+\cos 1\pi+\cos 2\pi+\cos 3\pi+\cos 4\pi+\cos 5\pi$$
$$=1+(-1)+1+(-1)+1+(-1)=0$$

904. $\frac{10}{11}$

Write out the ten terms and add them together:

$$\sum_{i=1}^{10}\left(\frac{1}{i}-\frac{1}{i+1}\right)=\left(\frac{1}{1}-\frac{1}{1+1}\right)+\left(\frac{1}{2}-\frac{1}{2+1}\right)+\left(\frac{1}{3}-\frac{1}{3+1}\right)+\left(\frac{1}{4}-\frac{1}{4+1}\right)+\left(\frac{1}{5}-\frac{1}{5+1}\right)$$

$$+\left(\frac{1}{6}-\frac{1}{6+1}\right)+\left(\frac{1}{7}-\frac{1}{7+1}\right)+\left(\frac{1}{8}-\frac{1}{8+1}\right)+\left(\frac{1}{9}-\frac{1}{9+1}\right)\left(\frac{1}{10}-\frac{1}{10+1}\right)$$

$$=1-\frac{1}{2}+\frac{1}{2}-\frac{1}{3}+\frac{1}{3}-\frac{1}{4}+\frac{1}{4}-\frac{1}{5}+\frac{1}{5}-\frac{1}{6}+\frac{1}{6}$$

$$-\frac{1}{7}+\frac{1}{7}-\frac{1}{8}+\frac{1}{8}-\frac{1}{9}+\frac{1}{9}-\frac{1}{10}+\frac{1}{10}-\frac{1}{11}$$

$$=1-\frac{1}{11}=\frac{10}{11}$$

905. $\frac{364}{243}$

Write out the six terms and add them:

$$\sum_{i=0}^{5}\left(\frac{1}{3}\right)^i=\left(\frac{1}{3}\right)^0+\left(\frac{1}{3}\right)^1+\left(\frac{1}{3}\right)^2+\left(\frac{1}{3}\right)^3+\left(\frac{1}{3}\right)^4+\left(\frac{1}{3}\right)^5$$

$$=1+\frac{1}{3}+\frac{1}{9}+\frac{1}{27}+\frac{1}{81}+\frac{1}{243}$$

$$=\frac{243+81+27+9+3+1}{243}=\frac{364}{243}$$

906. 100

This series is an arithmetic series with $a_1=1$ and a difference of 2. First, find the tenth term by using the formula $a_n=a_1+(n-1)d$:

$$a_n=a_1+(n-1)d$$
$$a_{10}=1+(10-1)\cdot 2$$
$$=1+18=19$$

Now find the sum of the first ten terms by using the following formula:

$$\sum_{i=1}^{n}a_i=\frac{n}{2}\left(a_1+a_n\right)$$

$$\sum_{i=1}^{10}a_i=\frac{10}{2}(1+19)=5(20)=100$$

907. 145

This series is an arithmetic series with $a_1=-8$ and a difference of 5. First, find the tenth term by using the formula $a_n=a_1+(n-1)d$:

$$a_n=a_1+(n-1)d$$
$$a_{10}=-8+(10-1)(5)=-8+45=37$$

Now find the sum of the first ten terms by using the following formula:

$$\sum_{i=1}^{n} a_i = \frac{n}{2}(a_1 + a_n)$$

$$\sum_{i=1}^{10} a_i = \frac{10}{2}(-8+37) = 5(29) = 145$$

908. 25

This series is an arithmetic series with $a_1 = 1$ and a difference of $\frac{1}{3}$. First, find the tenth term by using the formula $a_n = a_1 + (n-1)d$:

$$a_n = a_1 + (n-1)d$$

$$a_{10} = 1 + (10-1)\left(\frac{1}{3}\right) = 1 + 3 = 4$$

Now find the sum of the first ten terms by using the following formula:

$$\sum_{i=1}^{n} a_i = \frac{n}{2}(a_1 + a_n)$$

$$\sum_{i=1}^{10} a_i = \frac{10}{2}(1+4) = 5(5) = 25$$

909. –15

This series is an arithmetic series with $a_1 = 12$ and a difference of –3. First, find the tenth term by using the formula $a_n = a_1 + (n-1)d$:

$$a_n = a_1 + (n-1)d$$

$$a_{10} = 12 + (10-1)(-3) = 12 + (-27) = -15$$

Now find the sum of the first ten terms by using the following formula:

$$\sum_{i=1}^{n} a_i = \frac{n}{2}(a_1 + a_n)$$

$$\sum_{i=1}^{10} a_i = \frac{10}{2}(12 + (-15)) = 5(-3) = -15$$

910. –380

This series is an arithmetic series with $a_1 = -2$ and a difference of –8. Find the tenth term by using the formula $a_n = a_1 + (n-1)d$:

$$a_n = a_1 + (n-1)d$$

$$a_{10} = -2 + (10-1)(-8) = -2 + (-72) = -74$$

Now find the sum of the first ten terms by using the following formula:

$$\sum_{i=1}^{n} a_i = \frac{n}{2}(a_1 + a_n)$$

$$\sum_{i=1}^{10} a_i = \frac{10}{2}(-2 + (-74)) = 5(-76) = -380$$

911. 63

You find the sum of the first n terms of a geometric series by using the following formula:

$$\sum_{i=0}^{n} ar^i = a\frac{1-r^{n+1}}{1-r}$$

The sum of the terms in this series with $r=2$ is

$$\sum_{i=0}^{5} 2^i = \frac{1-2^{5+1}}{1-2} = \frac{1-2^6}{-1} = \frac{1-64}{-1} = 63$$

912. 2,188

You find the sum of the first n terms of a geometric series by using the following formula:

$$\sum_{i=0}^{n} ar^i = a\frac{1-r^{n+1}}{1-r}$$

The sum of the terms in this series with $a=4$ and $r=-3$ is

$$\sum_{i=0}^{6} 4(-3)^i = 4\cdot\frac{1-(-3)^{6+1}}{1-(-3)} = 4\cdot\frac{1-(-3)^7}{1+3} = \cancel{4}\cdot\frac{1-(-2,187)}{\cancel{4}} = 1+2,187 = 2,188$$

913. $\frac{12,610}{2,187}$

You find the sum of the first n terms of a geometric series by using the following formula:

$$\sum_{i=0}^{n} ar^i = a\frac{1-r^{n+1}}{1-r}$$

The sum of the terms in this series with $a=2$ and $r=\frac{2}{3}$ is

$$\sum_{i=0}^{7} 2\left(\frac{2}{3}\right)^i = 2\cdot\frac{1-\left(\frac{2}{3}\right)^{7+1}}{1-\left(\frac{2}{3}\right)} = 2\cdot\frac{1-\left(\frac{2}{3}\right)^8}{\frac{1}{3}} = 2\cdot\frac{1-\frac{256}{6,561}}{\frac{1}{3}}$$

$$= 2\cdot\frac{3}{1}\cdot\frac{6,561-256}{6,561} = \cancel{6}^2\cdot\frac{6,305}{_{2,187}\cancel{6,561}} = \frac{12,610}{2,187}$$

914. $\frac{511}{2}$

You find the sum of the first n terms of a geometric series by using the following formula:

$$\sum_{i=0}^{n} ar^i = a\frac{1-r^{n+1}}{1-r}$$

To use the formula for this series, you need to rewrite the general term with an exponent of i:

$$2^{i-1}=2^i\cdot 2^{-1}=\frac{1}{2}\cdot 2^i$$

You can now rewrite the problem as

$$\sum_{i=0}^{8}\frac{1}{2}(2)^i$$

This version has the correct format for the formula, letting $a=\frac{1}{2}$ and $r=2$:

$$\sum_{i=0}^{8}\frac{1}{2}(2)^i=\frac{1}{2}\cdot\frac{1-(2)^{8+1}}{1-2}=\frac{1}{2}\cdot\frac{1-2^9}{-1}=\frac{1}{2}\cdot\frac{1-512}{-1}=\frac{1}{2}\cdot\frac{-511}{-1}=\frac{511}{2}$$

915. $\frac{364}{729}$

You find the sum of the first n terms of a geometric series by using the following formula:

$$\sum_{i=0}^{n}ar^i=a\frac{1-r^{n+1}}{1-r}$$

To use the formula for this series, you need to rewrite the problem from 0 to n. The first term in this series is $\frac{1}{3}$, so to begin the summation with $i=0$, write the general term as $\frac{1}{3}\left(\frac{1}{3}\right)^i$ and change n from 6 to 5. You can now rewrite the problem as

$$\sum_{i=0}^{5}\frac{1}{3}\left(\frac{1}{3}\right)^i$$

This version has the correct format for the formula, letting $a=\frac{1}{3}$ and $r=\frac{1}{3}$:

$$\sum_{i=0}^{5}\frac{1}{3}\left(\frac{1}{3}\right)^i=\frac{1}{3}\cdot\frac{1-\left(\frac{1}{3}\right)^{5+1}}{1-\frac{1}{3}}=\frac{1}{3}\cdot\frac{1-\left(\frac{1}{3}\right)^6}{\frac{2}{3}}=\frac{1}{3}\cdot\frac{1-\frac{1}{729}}{\frac{2}{3}}$$

$$=\frac{1}{\cancel{3}}\cdot\frac{\cancel{3}}{2}\cdot\frac{729-1}{729}=\frac{1}{\cancel{2}}\cdot\frac{\overset{364}{\cancel{728}}}{729}=\frac{364}{729}$$

916. 32

You find the sum of the terms of an infinite geometric series with $|r|<1$ by using the formula $\sum_{i=0}^{\infty}ar^i=\frac{a}{1-r}$. In this series, the first term, a, is 16, and the ratio, r, is $\frac{1}{2}$. You find the ratio by dividing any term by the term immediately preceding it.

$$r=\frac{8}{16}=\frac{1}{2}$$

$$\sum_{i=0}^{\infty}16\left(\frac{1}{2}\right)^i=\frac{16}{1-\frac{1}{2}}=\frac{16}{\frac{1}{2}}=16\cdot\frac{2}{1}=32$$

917. 81

You find the sum of the terms of an infinite geometric series with $|r|<1$ by using the formula $\sum_{i=0}^{\infty} ar^i = \frac{a}{1-r}$. In this series, the first term, a, is 27, and the ratio, r, is $\frac{2}{3}$. You find the ratio by dividing any term by the term immediately preceding it.

$$r = \frac{8}{12} = \frac{2}{3}$$

$$\sum_{i=0}^{\infty} 27\left(\frac{2}{3}\right)^i = \frac{27}{1-\frac{2}{3}} = \frac{27}{\frac{1}{3}} = 27 \cdot \frac{3}{1} = 81$$

918. $\frac{256}{27}$

You find the sum of the terms of an infinite geometric series with $|r|<1$ by using the formula $\sum_{i=0}^{\infty} ar^i = \frac{a}{1-r}$. In this series, the first term, a, is $\frac{64}{27}$, and the ratio, r, is $\frac{3}{4}$. You find the ratio by dividing any term by the term immediately preceding it.

$$r = \frac{\frac{4}{3}}{\frac{16}{9}} = \frac{\cancel{4}}{\cancel{3}} \cdot \frac{^3\cancel{9}}{_4\cancel{16}} = \frac{3}{4}$$

$$\sum_{i=0}^{\infty} \frac{64}{27}\left(\frac{3}{4}\right)^i = \frac{\frac{64}{27}}{1-\frac{3}{4}} = \frac{\frac{64}{27}}{\frac{1}{4}} = \frac{64}{27} \cdot \frac{4}{1} = \frac{256}{27}$$

919. $\frac{256}{5}$

You find the sum of the terms of an infinite geometric series with $|r|<1$ by using the formula $\sum_{i=0}^{\infty} ar^i = \frac{a}{1-r}$. In this series, the first term, a, is 64, and the ratio, r, is $-\frac{1}{4}$. You find the ratio by dividing any term by the term immediately preceding it.

$$r = \frac{4}{-16} = -\frac{1}{4}$$

$$\sum_{i=0}^{\infty} 64\left(-\frac{1}{4}\right)^i = \frac{64}{1-\left(-\frac{1}{4}\right)} = \frac{64}{1+\frac{1}{4}} = \frac{64}{\frac{5}{4}} = \frac{64}{1} \cdot \frac{4}{5} = \frac{256}{5}$$

920. $\frac{e^4}{e-1}$

You find the sum of the terms of an infinite geometric series with $|r|<1$ by using the following formula: $\sum_{i=0}^{\infty} ar^i = \frac{a}{1-r}$.

In this series, the first term, a, is e^3 and the ratio, r, is $\frac{1}{e}$. You find the ratio by dividing any term by the term immediately preceding it.

$$\frac{e^2}{e^3} = \frac{1}{e}$$

$$\sum_{i=0}^{\infty} e^3\left(\frac{1}{e}\right)^i = \frac{e^3}{1-\frac{1}{e}} = \frac{e^3}{\frac{e}{e}-\frac{1}{e}} = \frac{e^3}{\frac{e-1}{e}} = \frac{e^3}{1}\cdot\frac{e}{e-1} = \frac{e^4}{e-1}$$

921. 15

You find the combination of n things taken r at a time by using the formula ${}_nC_r = \frac{n!}{r!(n-r)!}$, so

$${}_6C_2 = \frac{6!}{2!(6-2)!} = \frac{6!}{2!4!}$$

Instead of writing 6! as $6\cdot5\cdot4\cdot3\cdot2\cdot1$, you can use 4! for the last four factors to simplify the fraction reduction:

$$\frac{6!}{2!4!} = \frac{6\cdot5\cdot4!}{2\cdot1\cdot4!} = \frac{\cancel{6}^{3}\cdot5\cdot\cancel{4!}}{\cancel{2}\cdot1\cdot\cancel{4!}} = \frac{15}{1} = 15$$

922. 9

You find the combination of n things taken r at a time by using the formula ${}_nC_r = \frac{n!}{r!(n-r)!}$, so

$${}_9C_8 = \frac{9!}{8!(9-8)!} = \frac{9!}{8!1!}$$

Instead of writing 9! as $9\cdot8\cdot7\cdot6\cdot5\cdot4\cdot3\cdot2\cdot1$, you can use 8! for the last eight factors to simplify the fraction reduction:

$$\frac{9\cdot8!}{1\cdot8!} = \frac{9\cdot\cancel{8!}}{1\cdot\cancel{8!}} = \frac{9}{1} = 9$$

923. 1

You find the combination of n things taken r at a time by using the formula ${}_nC_r = \frac{n!}{r!(n-r)!}$, so

$${}_4C_4 = \frac{4!}{4!(4-4)!} = \frac{4!}{4!0!}$$

By special definition, 0! = 1:

$$\frac{4!}{1\cdot4!} = \frac{1}{1} = 1$$

924. 1

You find the combination of n things taken r at a time by using the formula ${}_nC_r = \frac{n!}{r!(n-r)!}$, so

$${}_5C_0 = \frac{5!}{0!(5-0)!} = \frac{5!}{0!5!} = \frac{5!}{1\cdot5!} = \frac{1}{1} = 1$$

925. 25,827,165

You find the combination of *n* things taken *r* at a time by using the formula $_nC_r = \frac{n!}{r!(n-r)!}$, so

$$_{54}C_6 = \frac{54!}{6!(54-6)!} = \frac{54!}{6!48!}$$

Instead of writing 54! as $54 \cdot 53 \cdot 52$ and so on, you can use 48! for the last 48 factors to simplify the fraction reduction:

$$\frac{54 \cdot 53 \cdot 52 \cdot 51 \cdot 50 \cdot 49 \cdot \cancel{48!}}{6 \cdot 5 \cdot 4 \cdot 3 \cdot 2 \cdot 1 \cdot \cancel{48!}} = \frac{{}^{9}\cancel{54} \cdot 53 \cdot {}^{13}\cancel{52} \cdot 51 \cdot 50 \cdot 49}{\cancel{6} \cdot 5 \cdot \cancel{4} \cdot 3 \cdot 2 \cdot 1} = \frac{{}^{3}\cancel{9} \cdot 53 \cdot 13 \cdot 51 \cdot {}^{5}\cancel{50} \cdot 49}{\cancel{5} \cdot \cancel{3} \cdot \cancel{2} \cdot 1}$$
$$= \frac{3 \cdot 53 \cdot 13 \cdot 51 \cdot 5 \cdot 49}{1} = 25{,}827{,}165$$

926. bottom row is 1 2 1

You form each new row by placing the sums of the two adjacent numbers from the row above diagonally below. The beginning and ending numbers of each row are 1.

1
1 1
1 2 1

927. bottom row is 1 4 6 4 1

You form each new row by placing the sums of the two adjacent numbers from the row above diagonally below. The beginning and ending numbers of each row are 1.

1
1 1
1 2 1
1 3 3 1
1 4 6 4 1

928. bottom row is 1 5 10 10 5 1

You form each new row by placing the sums of the two adjacent numbers from the row above diagonally below. The beginning and ending numbers of each row are 1.

1
1 1
1 2 1
1 3 3 1
1 4 6 4 1
1 5 10 10 5 1

929. bottom row is 1 6 15 20 15 6 1

You form each new row by placing the sums of the two adjacent numbers from the row above diagonally below. The beginning and ending numbers of each row are 1.

1

1 1

1 2 1

1 3 3 1

1 4 6 4 1

1 5 10 10 5 1

1 6 15 20 15 6 1

930. bottom row is 1 1

You form each new row by placing the sums of the two adjacent numbers from the row above diagonally below. The beginning and ending numbers of each row are 1.

1

1 1

931. $x^3+3x^2y+3xy^2+y^3$

First, write the coefficients from the fourth row of the triangle:

1 3 3 1

Next, write in decreasing powers of x:

$1x^3 \quad 3x^2 \quad 3x^1 \quad 1x^0$

Now write in increasing powers of y:

$1x^3y^0 \quad 3x^2y^1 \quad 3x^1y^2 \quad 1x^0y^3$

Simplify and write the sum of the terms:

$x^3+3x^2y+3xy^2+y^3$

932. $x^4+8x^3+24x^2+32x+16$

First, write the coefficients from the fifth row of the triangle:

1 4 6 4 1

Next, write in decreasing powers of x:

$1x^4 \quad 4x^3 \quad 6x^2 \quad 4x^1 \quad 1x^0$

Now write in increasing powers of 2:

$1x^4 2^0 \quad 4x^3 2^1 \quad 6x^2 2^2 \quad 4x^1 2^3 \quad 1x^0 2^4$

Simplify and write the sum of the terms:

$x^4+8x^3+24x^2+32x+16$

933. $y^5 - 15y^4 + 90y^3 - 270y^2 + 405y - 243$

First, write the coefficients from the sixth row of the triangle:

1 5 10 10 5 1

Next, write in decreasing powers of y:

$1y^5 \quad 5y^4 \quad 10y^3 \quad 10y^2 \quad 5y^1 \quad 1y^0$

Now write in increasing powers of –3:

$1y^5(-3)^0 \quad 5y^4(-3)^1 \quad 10y^3(-3)^2 \quad 10y^2(-3)^3 \quad 5y^1(-3)^4 \quad 1y^0(-3)^5$

Simplify and write the sum of the terms:

$y^5 - 15y^4 + 90y^3 - 270y^2 + 405y - 243$

934. $8x^3 - 60x^2 + 150x - 125$

First, write the coefficients from the fourth row of the triangle:

1 3 3 1

Next, write in decreasing powers of $2x$:

$1(2x)^3 \quad 3(2x)^2 \quad 3(2x)^1 \quad 1(2x)^0$

Now write in increasing powers of –5:

$1(2x)^3(-5)^0 \quad 3(2x)^2(-5)^1 \quad 3(2x)^1(-5)^2 \quad 1(2x)^0(-5)^3$

Simplify and write the sum of the terms:

$8x^3 - 60x^2 + 150x - 125$

935. $256x^4 - 768x^3y + 864x^2y^2 - 432xy^3 + 81y^4$

First, write the coefficients from the fifth row of the triangle:

1 4 6 4 1

Next, write in decreasing powers of $4x$:

$1(4x)^4 \quad 4(4x)^3 \quad 6(4x)^2 \quad 4(4x)^1 \quad 1(4x)^0$

Now write in increasing powers of $-3y$:

$1(4x)^4(-3y)^0 \quad 4(4x)^3(-3y)^1 \quad 6(4x)^2(-3y)^2 \quad 4(4x)^1(-3y)^3 \quad 1(4x)^0(-3y)^4$

Simplify and write the sum of the terms:

$256x^4 - 768x^3y + 864x^2y^2 - 432xy^3 + 81y^4$

936. 36

The binomial theorem produces the expansion of the binomial $(x+y)^n$ by using the following formula:

$$\sum_{k=0}^{n} \binom{n}{k} x^{n-k} y^k$$

You can find the x^7y^2 term in $(x+y)^9$ by plugging in the given information:

$$\binom{9}{2}x^{9-2}y^2=\frac{9!}{(9-2)!2!}x^7y^2=\frac{9!}{7!2!}x^7y^2$$

Instead of writing 9! as $9\cdot8\cdot7\cdot6\cdot5\cdot4\cdot3\cdot2\cdot1$, you can use 7! for the last seven factors to simplify the fraction reduction:

$$\frac{9\cdot8\cdot7!}{7!2\cdot1}x^7y^2=\frac{9\cdot8\cdot\cancel{7!}}{\cancel{7!}2\cdot1}x^7y^2=\frac{9\cdot\cancel{8}^4}{\cancel{2}\cdot1}x^7y^2=36x^7y^2$$

937. −56

The binomial theorem produces the expansion of the binomial $(x+y)^n$ by using the following formula:

$$\sum_{k=0}^{n}\binom{n}{k}x^{n-k}y^k$$

You can find the x^3y^5 term in $(x-y)^8$ by plugging in the given information:

$$\binom{8}{5}x^{8-5}(-y)^5=\frac{8!}{(8-5)!5!}x^3(-y)^5=\frac{8!}{3!5!}x^3(-y)^5$$

Instead of writing 8! as $8\cdot7\cdot6\cdot5\cdot4\cdot3\cdot2\cdot1$, you can use 5! for the last five factors to simplify the fraction reduction:

$$=\frac{8\cdot7\cdot6\cdot5!}{3\cdot2\cdot1\cdot5!}x^3(-y)^5=\frac{8\cdot7\cdot6\cdot\cancel{5!}}{3\cdot2\cdot1\cdot\cancel{5!}}x^3(-y)^5=\frac{8\cdot7\cdot\cancel{6}}{\cancel{3}\cdot\cancel{2}\cdot1}x^3(-y)^5=-56x^3y^5$$

938. 1,280

The binomial theorem produces the expansion of the binomial $(x+y)^n$ by using the following formula:

$$\sum_{k=0}^{n}\binom{n}{k}x^{n-k}y^k$$

You can find the x^3 term in $(x+4)^6$ by plugging in the given information:

$$\binom{6}{3}x^{6-3}4^3=\frac{6!}{(6-3)!3!}x^34^3=\frac{6!}{3!3!}x^34^3$$

Instead of writing 6! as $6\cdot5\cdot4\cdot3\cdot2\cdot1$, you can use 3! for the last three factors to simplify the fraction reduction:

$$\frac{6\cdot5\cdot4\cdot\cancel{3!}}{3\cdot2\cdot1\cdot\cancel{3!}}x^34^3=\frac{\cancel{6}\cdot5\cdot4}{\cancel{3}\cdot\cancel{2}\cdot1}x^34^3=20\cdot4^3x^3=1{,}280x^3$$

939. −2,835

The binomial theorem produces the expansion of the binomial $(x+y)^n$ by using the following formula:

$$\sum_{k=0}^{n}\binom{n}{k}x^{n-k}y^k$$

You can find the x^4y^3 term in $(3x-y)^7$ by plugging in the given information:

$$\binom{7}{3}(3x)^{7-3}(-y)^3 = \frac{7!}{(7-3)!3!}(3x)^4(-y)^3$$
$$= \frac{7!}{4!3!}(3x)^4(-y)^3$$

Instead of writing 7! as $7\cdot6\cdot5\cdot4\cdot3\cdot2\cdot1$, you can use 4! for the last four factors to simplify the fraction reduction:

$$= \frac{7\cdot6\cdot5\cdot\cancel{4!}}{\cancel{4!}\cdot3\cdot2\cdot1}(3x)^4(-y)^3 = \frac{7\cdot\cancel{6}\cdot5}{\cancel{3\cdot2}\cdot1}(3x)^4(-y)^3 = 35\left(81x^4\right)\left(-y^3\right) = -2{,}835x^4y^3$$

940. 2,000

The binomial theorem produces the expansion of the binomial $(x+y)^n$ by using the following formula:

$$\sum_{k=0}^{n}\binom{n}{k}x^{n-k}y^k$$

You can find the x^3y^2 term in $(2x+5y)^5$ by plugging in the given information:

$$\binom{5}{2}(2x)^{5-2}(5y)^2 = \frac{5!}{(5-2)!2!}(2x)^3(5y)^2$$
$$= \frac{5!}{3!2!}(2x)^3(5y)^2$$

Instead of writing 5! as $5\cdot4\cdot3\cdot2\cdot1$, you can use 3! for the last three factors to simplify the fraction reduction:

$$\frac{5\cdot4\cdot\cancel{3!}}{\cancel{3!}2\cdot1}(2x)^3(5y)^2 = \frac{5\cdot4}{2\cdot1}(2x)^3(5y)^2 = 10\cdot8x^3\cdot25y^2 = 2{,}000x^3y^2$$

941. 3

The function is continuous, and it has y values that approach 3 when x is approaching 2.

942. 3

The function has a hole at (2, 3). The limit as x approaches 2 from the left is 3, and the limit as x approaches 2 from the right is 3.

943. does not exist

The limit as x approaches 2 from the left is 3, and the limit as x approaches 2 from the right is –2. They aren't the same, so the limit doesn't exist.

944. 3

The function has a hole at (2, 3). The limit as x approaches 2 from the left is 3, and the limit as x approaches 2 from the right is 3. The asymptotes aren't involved in this limit.

945. does not exist

The function has a vertical asymptote when $x = 2$. The function approaches $-\infty$ when x approaches 2 from the left and $+\infty$ when x approaches 2 from the right. The left- and right-hand limits don't match, so the limit doesn't exist.

946. –3

The function is approaching a y value of –3 as x moves toward –1 from the right.

947. 0

The function is approaching a y value of 0 as x moves toward –3 from the right.

948. $+\infty$

The function is rising and approaching $+\infty$ as x gets closer and closer to –3 from the left.

949. –2

The function is approaching a y value of –2 as x moves toward 2 from the right.

950. $-\infty$

The function is dropping and approaching $-\infty$ as x gets closer and closer to 3 from the left.

951. 8

The y values are getting closer and closer to 8 as x approaches 3 from the left and from the right.

952. 728

The y values are getting closer and closer to 728 as x approaches 9 from the left and from the right.

953. –9

The y values are getting closer and closer to –9 as x approaches –2 from the left and from the right.

954. does not exist

The y values are getting closer and closer to $-\infty$ as x approaches 3 from the left and to $+\infty$ as x approaches 3 from the right.

955. $+\infty$

The y values are getting closer and closer to $+\infty$ as x approaches 5 from the left and from the right.

956. 1

Replace each x in the function rule with 3; then simplify:

$$\lim_{x\to 3}\frac{x+5}{x^2-x+2}=\frac{3+5}{(3)^2-3+2}=\frac{8}{9-1}=\frac{8}{8}=1$$

957. 0

Replace each x in the function rule with –2; then simplify:

$$\lim_{x\to -2}\frac{x^2-4}{x+3}=\frac{(-2)^2-4}{-2+3}=\frac{4-4}{1}=\frac{0}{1}=0$$

958. $-\frac{3}{2}$

Replace each x in the function rule with 1; then simplify:

$$\lim_{x\to 1}\frac{x^2-5x+4}{x^2-1}=\frac{1^2-5(1)+4}{(1)^2-1}=\frac{1-5+4}{1-1}=\frac{0}{0}$$

Division by 0 is undefined, so go back to the original problem. Factor and reduce the rational expression before replacing each x with 1:

$$\lim_{x\to 1}\frac{x^2-5x+4}{x^2-1}=\lim_{x\to 1}\frac{(x-4)\cancel{(x-1)}}{\cancel{(x-1)}(x+1)}=\lim_{x\to 1}\frac{x-4}{x+1}=\frac{1-4}{1+1}=\frac{-3}{2}$$

The function has a removable discontinuity when $x=1$. A *removable discontinuity* occurs when you can rewrite a function rule by factoring out a common factor. There's still a discontinuity at the point identified by the removed factor, but this discontinuity behaves differently from other discontinuities.

959. $\frac{11}{6}$

Replace each x in the function rule with –3; then simplify:

$$\lim_{x\to -3}\frac{2x^2+x-15}{x^2-9}=\frac{2(-3)^2+(-3)-15}{(-3)^2-9}=\frac{18-18}{9-9}=\frac{0}{0}$$

Division by 0 is undefined, so go back to the original problem. Factor and reduce the rational expression before replacing each x with –3:

$$\lim_{x\to -3}\frac{2x^2+x-15}{x^2-9}=\lim_{x\to -3}\frac{(2x-5)\cancel{(x+3)}}{(x-3)\cancel{(x+3)}}=\lim_{x\to -3}\frac{2x-5}{x-3}$$
$$=\frac{2(-3)-5}{(-3)-3}=\frac{-11}{-6}=\frac{11}{6}$$

The function has a removable discontinuity when $x=-3$.

960. does not exist

Replace each x in the function rule with 5; then simplify:

$$\lim_{x\to 5}\frac{x^2+5x}{x^2-25}=\frac{5^2+5(5)}{5^2-25}=\frac{25+25}{25-25}=\frac{50}{0}$$

Division by 0 is undefined, so go back to the original problem. Factor and reduce the rational expression before replacing each x with 5:

$$\lim_{x\to 5}\frac{x^2+5x}{x^2-25}=\lim_{x\to 5}\frac{x\cancel{(x+5)}}{(x-5)\cancel{(x+5)}}=\lim_{x\to 5}\frac{x}{x-5}=\frac{5}{5-5}=\frac{5}{0}$$

The resulting expression still has no limit.

961. does not exist

Replace each x in the function rule with 5; then simplify:

$$\lim_{x\to 5}\frac{3x^2+16x+5}{x^2-3x-10}=\frac{3(5)^2+16(5)+5}{(5)^2-3(5)-10}=\frac{75+80+5}{25-15-10}=\frac{160}{0}$$

Division by 0 is undefined, so go back to the original problem. Factor the rational expression before replacing each x with 5:

$$\lim_{x\to 5}\frac{3x^2+16x+5}{x^2-3x-10}=\lim_{x\to 5}\frac{(3x+1)(x+5)}{(x+2)(x-5)}$$

The resulting expression still has no limit.

962. $\frac{1}{4}$

Replace each x in the function rule with 3; then simplify:

$$\lim_{x\to 3}\frac{x^3-27}{x^4-81}=\frac{3^3-27}{3^4-81}=\frac{27-27}{81-81}=\frac{0}{0}$$

Division by 0 is undefined, so go back to the original problem. Factor and reduce the rational expression before replacing each x with 3:

$$\lim_{x\to 3}\frac{x^3-27}{x^4-81}=\lim_{x\to 3}\frac{\cancel{(x-3)}(x^2+3x+9)}{\cancel{(x-3)}(x+3)(x^2+9)}=\lim_{x\to 3}\frac{x^2+3x+9}{(x+3)(x^2+9)}$$

$$=\frac{3^2+3(3)+9}{(3+3)(3^2+9)}=\frac{9+9+9}{6(9+9)}=\frac{27}{6(18)}=\frac{27}{108}=\frac{1}{4}$$

The function has a removable discontinuity when $x=3$.

963. –3

Replace each x in the function rule with –2; then simplify:

$$\lim_{x\to -2}\frac{x^3+8}{x^2-4}=\frac{(-2)^3+8}{(-2)^2-4}=\frac{-8+8}{4-4}=\frac{0}{0}$$

Division by 0 is undefined, so go back to the original problem. Factor and reduce the rational expression before replacing each x with –2:

$$\lim_{x\to -2}\frac{x^3+8}{x^2-4}=\lim_{x\to -2}\frac{\cancel{(x+2)}(x^2-2x+4)}{(x-2)\cancel{(x+2)}}=\lim_{x\to -2}\frac{x^2-2x+4}{x-2}$$
$$=\frac{(-2)^2-2(-2)+4}{-2-2}=\frac{4+4+4}{-4}=\frac{12}{-4}=-3$$

The function has a removable discontinuity when $x=-2$.

964. $-\frac{12}{5}$

Replace each x in the function rule with –3; then simplify:

$$\lim_{x\to -3}\frac{x^3+5x^2-9x-45}{x^3+3x^2-4x-12}=\lim_{x\to -3}\frac{(-3)^3+5(-3)^2-9(-3)-45}{(-3)^3+3(-3)^2-4(-3)-12}$$
$$=\frac{-27+45+27-45}{27+27+12-12}=\frac{0}{0}$$

Division by 0 is undefined, so go back to the original problem. Factor and reduce the rational expression before replacing each x with –3:

$$\lim_{x\to -3}\frac{x^3+5x^2-9x-45}{x^3+3x^2-4x-12}=\lim_{x\to -3}\frac{x^2(x+5)-9(x+5)}{x^2(x+3)-4(x+3)}=\lim_{x\to -3}\frac{(x+5)(x^2-9)}{(x+3)(x^2-4)}$$
$$=\lim_{x\to -3}\frac{(x+5)(x-3)\cancel{(x+3)}}{\cancel{(x+3)}(x-2)(x+2)}=\lim_{x\to -3}\frac{(x+5)(x-3)}{(x-2)(x+2)}$$
$$=\frac{(-3+5)(-3-3)}{(-3-2)(-3+2)}=\frac{(2)(-6)}{(-5)(-1)}=\frac{-12}{5}$$

The function has a removable discontinuity when $x=-3$.

965. does not exist

Replace each x in the function rule with –2; then simplify:

$$\lim_{x\to -2}\frac{x^3+x^2-25x-25}{4x^3+8x^2-9x-18}=\frac{(-2)^3+(-2)^2-25(-2)-25}{4(-2)^3+8(-2)^2-9(-2)-18}$$
$$=\frac{-8+4+50-25}{-32+32+18-18}=\frac{21}{0}$$

Division by 0 is undefined, so go back to the original problem. Factor the rational expression before replacing each x with –2:

$$\lim_{x\to -2}\frac{x^2(x+1)-25(x+1)}{4x^2(x+2)-9(x+2)}=\lim_{x\to -2}\frac{(x+1)(x^2-25)}{(x+2)(4x^2-9)}$$
$$=\lim_{x\to -2}\frac{(x+1)(x-5)(x+5)}{(x+2)(2x-3)(2x+3)}$$

The resulting expression still has no limit.

966. 1

Replace each x in the function rule with $\frac{\pi}{4}$; then simplify:

$$\lim_{x\to\frac{\pi}{4}}\frac{\sin x}{\cos x}=\frac{\sin\frac{\pi}{4}}{\cos\frac{\pi}{4}}=\frac{\frac{\sqrt{2}}{2}}{\frac{\sqrt{2}}{2}}=1$$

967. does not exist

Replace each x in the function rule with $\frac{\pi}{2}$; then simplify:

$$\lim_{x\to\frac{\pi}{2}}\frac{\tan x+1}{\cos x}=\frac{\tan\frac{\pi}{2}+1}{\cos\frac{\pi}{2}}=\frac{0+1}{0}=\frac{1}{0}$$

Division by 0 is undefined, so the limit doesn't exist.

968. $-\frac{1}{2}$

Replace each x in the function rule with $\frac{\pi}{3}$; then simplify:

$$\lim_{x\to\frac{\pi}{3}}\frac{\sin^2 x-1}{\cos x}=\frac{\left(\sin\frac{\pi}{3}\right)^2-1}{\cos\frac{\pi}{3}}=\frac{\left(\frac{\sqrt{3}}{2}\right)^2-1}{\frac{1}{2}}=\frac{\frac{3}{4}-1}{\frac{1}{2}}=\frac{-\frac{1}{4}}{\frac{1}{2}}=-\frac{1}{4}\cdot\frac{2}{1}=-\frac{1}{2}$$

969. –2

Replace each x in the function rule with 0; then simplify:

$$\lim_{x\to 0}\frac{1-\sec^2 x}{\sec x-1}=\frac{1-(\sec 0)^2}{\sec 0-1}=\frac{1-1}{1-1}=\frac{0}{0}$$

Division by 0 is undefined, so go back to the original problem. Factor the numerator and reduce the rational expression before replacing each x with 0:

$$\lim_{x\to 0}\frac{1-\sec^2 x}{\sec x-1}=\lim_{x\to 0}\frac{(1-\sec x)(1+\sec x)}{\sec x-1}=\lim_{x\to 0}\frac{-1\cancel{(\sec x-1)}(1+\sec x)}{\cancel{\sec x-1}}$$
$$=\lim_{x\to 0}\frac{-1(1+\sec x)}{1}=\lim_{x\to 0}\left[-1(1+\sec x)\right]$$
$$=-1(1+\sec 0)=-1(1+1)=-2$$

The function has a removable discontinuity when $x=0$.

970. does not exist

Replace each x in the function rule with $-\frac{\pi}{6}$; then simplify:

$$\lim_{x\to-\frac{\pi}{6}} \frac{\cos x}{1+2\sin x} = \frac{\cos\left(-\frac{\pi}{6}\right)}{1+2\sin\left(-\frac{\pi}{6}\right)} = \frac{\frac{\sqrt{3}}{2}}{1+2\left(-\frac{1}{2}\right)} = \frac{\frac{\sqrt{3}}{2}}{1+(-1)} = \frac{\frac{\sqrt{3}}{2}}{0}$$

Division by 0 is undefined, so the limit doesn't exist.

971. $\frac{1}{6}$

Replace each x in the function rule with 9; then simplify:

$$\lim_{x\to 9} \frac{\sqrt{x}-3}{x-9} = \frac{\sqrt{9}-3}{9-9} = \frac{3-3}{9-9} = \frac{0}{0}$$

Division by 0 is undefined, so go back to the original problem. Multiply both the numerator and the denominator of the rational expression by the conjugate of the numerator before replacing each x with 9:

$$\begin{aligned}\lim_{x\to 9} \frac{\sqrt{x}-3}{x-9} \cdot \frac{\sqrt{x}+3}{\sqrt{x}+3} &= \lim_{x\to 9} \frac{\left(\sqrt{x}-3\right)\left(\sqrt{x}+3\right)}{(x-9)\left(\sqrt{x}+3\right)} = \lim_{x\to 9} \frac{\cancel{x-9}}{\cancel{(x-9)}\left(\sqrt{x}+3\right)} \\ &= \lim_{x\to 9} \frac{1}{\sqrt{x}+3} = \frac{1}{\sqrt{9}+3} = \frac{1}{6}\end{aligned}$$

The function has a removable discontinuity when $x=9$.

972. does not exist

Replace each x in the function rule with 4; then simplify:

$$\lim_{x\to 4} \frac{5-x}{2-\sqrt{x}} = \frac{5-4}{2-\sqrt{4}} = \frac{1}{2-2} = \frac{1}{0}$$

The result is not the fraction $\frac{0}{0}$, so rationalizing won't change the value. The limit doesn't exist.

973. $\frac{1}{2}$

Replace each x in the function rule with 3; then simplify:

$$\lim_{x\to 3} \frac{\sqrt{x-2}-1}{x-3} = \frac{\sqrt{3-2}-1}{3-3} = \frac{\sqrt{1}-1}{0} = \frac{0}{0}$$

The fraction $\frac{0}{0}$ is undefined, so go back to the original problem. Multiply both the numerator and the denominator of the rational expression by the conjugate of the numerator before replacing each x with 3:

$$\begin{aligned}\lim_{x\to3}\frac{\sqrt{x-2}-1}{x-3} &= \lim_{x\to3}\frac{\sqrt{x-2}-1}{x-3}\cdot\frac{\sqrt{x-2}+1}{\sqrt{x-2}+1}\\ &= \lim_{x\to3}\frac{\left(\sqrt{x-2}-1\right)\left(\sqrt{x-2}+1\right)}{(x-3)\left(\sqrt{x-2}+1\right)} = \lim_{x\to3}\frac{x-2-1}{(x-3)\left(\sqrt{x-2}+1\right)}\\ &= \lim_{x\to3}\frac{\cancel{x-3}}{\cancel{(x-3)}\left(\sqrt{x-2}+1\right)} = \lim_{x\to3}\frac{1}{\sqrt{x-2}+1}\\ &= \frac{1}{\sqrt{3-2}+1} = \frac{1}{\sqrt{1}+1} = \frac{1}{2}\end{aligned}$$

The function has a removable discontinuity when $x = 3$.

974. $-\frac{1}{25}$

Replace each x in the function rule with 0; then simplify:

$$\lim_{x\to0}\frac{\frac{1}{x-5}+\frac{1}{5}}{x} = \frac{\frac{1}{0-5}+\frac{1}{5}}{0} = \frac{-\frac{1}{5}+\frac{1}{5}}{0} = \frac{0}{0}$$

The fraction $\frac{0}{0}$ is undefined, so go back to the original problem. Add the two fractions in the numerator:

$$\begin{aligned}\lim_{x\to0}\frac{\frac{1}{x-5}+\frac{1}{5}}{x} &= \lim_{x\to0}\frac{\frac{1}{x-5}\cdot\frac{5}{5}+\frac{1}{5}\cdot\frac{x-5}{x-5}}{x} = \lim_{x\to0}\frac{\frac{5}{5(x-5)}+\frac{x-5}{5(x-5)}}{x}\\ &= \lim_{x\to0}\frac{\frac{5+x-5}{5(x-5)}}{x} = \lim_{x\to0}\frac{\frac{x}{5(x-5)}}{x}\end{aligned}$$

Simplify the complex fraction by multiplying the numerator by the reciprocal of the denominator; then replace each x with 0:

$$\lim_{x\to0}\frac{\frac{x}{5(x-5)}}{\frac{x}{1}} = \lim_{x\to0}\frac{\cancel{x}}{5(x-5)}\cdot\frac{1}{\cancel{x}} = \lim_{x\to0}\frac{1}{5(x-5)} = \frac{1}{5(0-5)} = \frac{1}{-25}$$

The function has a removable discontinuity when $x = 0$.

975. $\frac{1}{2}$

Replace each x in the function rule with $\frac{\pi}{4}$; then simplify:

$$\lim_{x\to\frac{\pi}{4}} \frac{\sin^2 x - \frac{1}{2}}{\tan x - 1} = \frac{\left(\sin\frac{\pi}{4}\right)^2 - \frac{1}{2}}{\tan\frac{\pi}{4} - 1} = \frac{\left(\frac{\sqrt{2}}{2}\right)^2 - \frac{1}{2}}{1-1} = \frac{\frac{1}{2} - \frac{1}{2}}{1-1} = \frac{0}{0}$$

The fraction $\frac{0}{0}$ is undefined, so go back to the original problem. Rewrite the expression using the ratio identity for the tangent. Combine the fractions in the numerator and denominator.

$$\lim_{x\to\frac{\pi}{4}} \frac{\sin^2 x - \frac{1}{2}}{\tan x - 1} = \lim_{x\to\frac{\pi}{4}} \frac{\sin^2 x - \frac{1}{2}}{\frac{\sin x}{\cos x} - 1} = \lim_{x\to\frac{\pi}{4}} \frac{\frac{2\sin^2 x}{2} - \frac{1}{2}}{\frac{\sin x}{\cos x} - \frac{\cos x}{\cos x}}$$

$$= \lim_{x\to\frac{\pi}{4}} \frac{\frac{2\sin^2 x - 1}{2}}{\frac{\sin x - \cos x}{\cos x}}$$

Replace the 1 in the expression $2\sin^2 x - 1$ with $\sin^2 x + \cos^2 x$ from the Pythagorean identity. Simplify. Then multiply the numerator of the rational expression by the reciprocal of the denominator to simplify the complex fraction.

$$\lim_{x\to\frac{\pi}{4}} \frac{\frac{2\sin^2 x - (\sin^2 x + \cos^2 x)}{2}}{\frac{\sin x - \cos x}{\cos x}} = \lim_{x\to\frac{\pi}{4}} \frac{\frac{\sin^2 x - \cos^2 x}{2}}{\frac{\sin x - \cos x}{\cos x}}$$

$$= \lim_{x\to\frac{\pi}{4}} \frac{\sin^2 x - \cos^2 x}{2} \cdot \frac{\cos x}{\sin x - \cos x}$$

Factor the numerator of the first fraction and reduce before replacing each x with $\frac{\pi}{4}$:

$$\lim_{x\to\frac{\pi}{4}} \frac{\cancel{(\sin x - \cos x)}(\sin x + \cos x)}{2} \cdot \frac{\cos x}{\cancel{\sin x - \cos x}}$$

$$= \lim_{x\to\frac{\pi}{4}} \frac{\sin x + \cos x}{2} \cdot \frac{\cos x}{1} = \lim_{x\to\frac{\pi}{4}} \frac{(\sin x + \cos x)\cos x}{2}$$

$$= \frac{\left(\sin\frac{\pi}{4} + \cos\frac{\pi}{4}\right)\cos\frac{\pi}{4}}{2} = \frac{\left(\frac{\sqrt{2}}{2} + \frac{\sqrt{2}}{2}\right)\cdot\frac{\sqrt{2}}{2}}{2} = \frac{\left(\frac{2\sqrt{2}}{2}\right)\cdot\frac{\sqrt{2}}{2}}{2}$$

$$= \frac{4/4}{2} = \frac{1}{2}$$

The function has a removable discontinuity when $x = \frac{\pi}{4}$.

976. 1

$f(0)=2(0)+1=1$, and as x approaches 0 from the left, $x^2+1 \to 0^2+1=1$.

Remember: When working with piece-wise functions, you determine the function value for a given x by using the rule corresponding to that x value. When evaluating a limit at an x value that occurs where the rule changes, you find the function values for both rules — coming from both the left and the right — to determine whether they give you the same y value.

977. –4

$g(-2)=(-2)^2-8=4-8=-4$, and as x approaches –2 from the left, $2x \to 2(-2)=-4$.

978. –3

$h(0)=4(0)-3=-3$, and as x approaches 0 from the left, $x^2-3 \to 0^2-3=-3$.

979. does not exist

$k(-1)=\sqrt{-1+1}=\sqrt{0}=0$, and as x approaches –1 from the left, $\frac{1}{x-1} \to \frac{1}{-1-1}=-\frac{1}{2}$.

980. 2

$k(3)=\frac{4}{\sqrt{3+1}}=\frac{4}{\sqrt{4}}=\frac{4}{2}=2$, and as x approaches 3 from the left, $\sqrt{x+1} \to \sqrt{3+1}=\sqrt{4}=2$.

981. 13

The addition law states that if $\lim\limits_{x\to k} f(x)=A$ and $\lim\limits_{x\to k} g(x)=B$, then $\lim\limits_{x\to k}\left[f(x)+g(x)\right]=A+B$. Using the law,

$$\lim_{x\to 1}\left[f(x)+g(x)\right]=4+9=13$$

982. –144

The multiplication law states that if $\lim\limits_{x\to k} f(x)=A$ and $\lim\limits_{x\to k} g(x)=B$, then $\lim\limits_{x\to k}(f(x)\cdot g(x))=A\cdot B$. Using the law,

$$\lim_{x\to 0}\left[f(x)\cdot g(x)\right]=12(-12)=-144$$

983. does not exist

The division law states that if $\lim\limits_{x\to k} f(x)=A$ and $\lim\limits_{x\to k} g(x)=B$, then $\lim\limits_{x\to k}\left[\frac{f(x)}{g(x)}\right]=\frac{A}{B}$ as long as $B\neq 0$. In $\lim\limits_{x\to -2}\left[\frac{f(x)}{g(x)}\right]$, the limit doesn't exist because $\lim\limits_{x\to -2} g(x)=0$.

984. –27

The power law states that if $\lim_{x\to k} f(x)=A$, then $\lim_{x\to k}[f(x)]^p=A^p$. Using the law,

$$\lim_{x\to -4}[f(x)]^{-3}=\left(-\frac{1}{3}\right)^{-3}=(-3)^3=-27$$

985. –5

The constant factor law states that if $\lim_{x\to k} f(x)=A$, then $\lim_{x\to k}[n\cdot f(x)]=n\cdot A$. Using the law,

$$\lim_{x\to -2}[5\cdot g(x)]=5(-1)=-5$$

986. –21

The constant factor law states that if $\lim_{x\to k} f(x)=A$, then $\lim_{x\to k}[n\cdot f(x)]=n\cdot A$. The power law states that if $\lim_{x\to k} f(x)=A$, then $\lim_{x\to k}[f(x)]^p=A^p$. The subtraction law states that if $\lim_{x\to k} f(x)=A$ and $\lim_{x\to k} g(x)=B$, then $\lim_{x\to k}[f(x)-g(x)]=A-B$. Using these laws,

$$\lim_{x\to 5}\left[2f(x)-(g(x))^2\right]=2(-6)-(-3)^2=-12-9=-21$$

987. $-\frac{124}{5}$

The constant factor law states that if $\lim_{x\to k} f(x)=A$, then $\lim_{x\to k}[n\cdot f(x)]=n\cdot A$. The subtraction law states that if $\lim_{x\to k} f(x)=A$ and $\lim_{x\to k} g(x)=B$, then $\lim_{x\to k}[f(x)-g(x)]=A-B$. The power law states that if $\lim_{x\to k} f(x)=A$, then $\lim_{x\to k}[f(x)]^p=A^p$. The division law states that if $\lim_{x\to k} f(x)=A$ and $\lim_{x\to k} g(x)=B$, then $\lim_{x\to k}\left[\frac{f(x)}{g(x)}\right]=\frac{A}{B}$ as long as $B\neq 0$. Using these laws,

$$\lim_{x\to -3}\left[\frac{3f(x)-4g(x)}{\sqrt{g(x)}}\right]=\frac{3(-8)-4(25)}{\sqrt{25}}=\frac{-24-100}{5}=\frac{-124}{5}$$

988. $\frac{97}{5}$

The power law states that if $\lim_{x\to k} f(x)=A$, then $\lim_{x\to k}[f(x)]^p=A^p$. The constant factor law states that if $\lim_{x\to k} f(x)=A$, then $\lim_{x\to k}[n\cdot f(x)]=n\cdot A$. The addition law states that if $\lim_{x\to k} f(x)=A$ and $\lim_{x\to k} g(x)=B$, then $\lim_{x\to k}[f(x)+g(x)]=A+B$. The division law states that if $\lim_{x\to k} f(x)=A$ and $\lim_{x\to k} g(x)=B$, then $\lim_{x\to k}\left[\frac{f(x)}{g(x)}\right]=\frac{A}{B}$ as long as $B\neq 0$. Using these laws,

$$\lim_{x\to 6}\left[\frac{(f(x))^2+3(g(x))^3}{2g(x)-1}\right]=\frac{(4)^2+3(3)^3}{2\cdot 3-1}=\frac{16+81}{6-1}=\frac{97}{5}$$

989. −3

The power law states that if $\lim_{x\to k} f(x)=A$, then $\lim_{x\to k}\left[f(x)\right]^p=A^p$. The subtraction law states that if $\lim_{x\to k} f(x)=A$ and $\lim_{x\to k} g(x)=B$, then $\lim_{x\to k}\left[f(x)-g(x)\right]=A-B$. The division law states that if $\lim_{x\to k} f(x)=A$ and $\lim_{x\to k} g(x)=B$, then $\lim_{x\to k}\left[\frac{f(x)}{g(x)}\right]=\frac{A}{B}$ as long as $B\neq 0$. Using these laws,

$$\lim_{x\to -2}\left[\frac{(f(x))^2-(g(x))^2}{g(x)+1}\right]=\frac{(5)^2-(-4)^2}{-4+1}=\frac{25-16}{-3}=\frac{9}{-3}=-3$$

990. −5

The constant factor law states that if $\lim_{x\to k} f(x)=A$, then $\lim_{x\to k}\left[n\cdot f(x)\right]=n\cdot A$. The power law states that if $\lim_{x\to k} f(x)=A$, then $\lim_{x\to k}\left[f(x)\right]^p=A^p$. The addition law states that if $\lim_{x\to k} f(x)=A$ and $\lim_{x\to k} g(x)=B$, then $\lim_{x\to k}\left[f(x)+g(x)\right]=A+B$. The subtraction law states that if $\lim_{x\to k} f(x)=A$ and $\lim_{x\to k} g(x)=B$, then $\lim_{x\to k}\left[f(x)-g(x)\right]=A-B$. The division law states that if $\lim_{x\to k} f(x)=A$ and $\lim_{x\to k} g(x)=B$, then $\lim_{x\to k}\left[\frac{f(x)}{g(x)}\right]=\frac{A}{B}$ as long as $B\neq 0$. Using these laws,

$$\lim_{x\to 0}\left[\frac{4\sqrt{f(x)}+\sqrt{g(x)}}{\sqrt{f(x)}-\sqrt{g(x)}}\right]=\frac{4\sqrt{16}+\sqrt{81}}{\sqrt{16}-\sqrt{81}}=\frac{4(4)+9}{4-9}=\frac{16+9}{-5}=\frac{25}{-5}=-5$$

991. 3

The function $f(x)=\frac{x-4}{x-3}$ is not continuous when $x=3$, because $f(3)=\frac{3-4}{3-3}=\frac{-1}{0}$. When working with rational functions, you determine discontinuities by setting the denominator equal to 0 and solving for x. The domain consists of all real numbers except 3.

992. 4

The function $f(x)=\frac{x^2-16}{x-4}$ is not continuous when $x=4$, because $f(4)=\frac{4^2-16}{4-4}=\frac{0}{0}$. When working with rational functions, you determine discontinuities by setting the denominator equal to 0 and solving for x. The domain consists of all real numbers except 4.

993. $x<-4$ or $x>4$

The function $f(x)=\sqrt{16-x^2}$ is not continuous when x is greater than 4 or x is less than −4. When the root is an even number, you can't have a negative value under the radical. Solving the inequality $16-x^2\geq 0$, you have

$$(4-x)(4+x)\geq 0$$

This is true when $-4\leq x\leq 4$, so the domain consists of all x such that $-4\leq x\leq 4$. Therefore, x values smaller than −4 or larger than 4 are *not* in the domain.

994. 3, –3

The function $f(x)=\dfrac{x+3}{x^2-9}$ is not continuous when $x=-3$, because $f(-3)=\dfrac{-3+3}{(-3)^2-9}=\dfrac{0}{9-9}=\dfrac{0}{0}$. Also, $f(3)=\dfrac{6}{0}$, which is undefined. When working with rational functions, you determine discontinuities by setting the denominator equal to 0 and solving for x. The domain consists of all real numbers except 3 and –3.

995. π

The function $f(x)=\dfrac{\sin x}{\cos x+1}$ is not continuous when $x=\pi$, because $f(\pi)=\dfrac{\sin(\pi)}{\cos(\pi)+1}=\dfrac{0}{-1+1}=\dfrac{0}{0}$. The domain consists of all real numbers except those that create a 0 in the denominator, so set the denominator equal to 0 and solve for x:

$$\cos x+1=0$$

$$\cos x=-1$$

This equation is true for all x such that $x=(2k+1)\pi$ — all odd multiples of π.

996. $-\dfrac{\pi}{2}$

The function $f(x)=\tan x$ is not continuous when $x=-\dfrac{\pi}{2}$, because $f\left(-\dfrac{\pi}{2}\right)=\tan\left(-\dfrac{\pi}{2}\right)$, and $-\dfrac{\pi}{2}$ is not in the domain of the tangent. The domain consists of all real numbers except those that are odd multiples of $\dfrac{\pi}{2}$.

997. $\dfrac{3\pi}{2}$

The function $f(x)=\dfrac{\sin 2x}{1+\cos 2x}$ is not continuous when $x=\dfrac{3\pi}{2}$, because

$$f\left(\frac{3\pi}{2}\right)=\frac{\sin 2\left(\frac{3\pi}{2}\right)}{1+\cos 2\left(\frac{3\pi}{2}\right)}=\frac{\sin(3\pi)}{1+\cos(3\pi)}=\frac{0}{1+(-1)}=\frac{0}{0}.$$

The domain consists of all real numbers except those that create a 0 in the denominator, so set the denominator equal to 0 and solve for x:

$$\cos 2x+1=0$$

$$\cos 2x=-1$$

This is true for all x such that $2x=(2k+1)\pi$, or $x=\dfrac{(2k+1)}{2}\pi$.

998. 0

The function $f(x)=\dfrac{e^x+1}{e^x-1}$ is not continuous when $x=0$, because $f(0)=\dfrac{e^0+1}{e^0-1}=\dfrac{1+1}{1-1}=\dfrac{2}{0}$. When working with rational functions, you determine discontinuities by setting the denominator equal to 0 and solving for x. In this case, $e^x-1=0$ when $e^x=1$, which is when $x=0$. The domain consists of all real numbers except 0.

999. 1

The function $f(x)=\frac{2e^{x-1}+2}{2e^{x-1}-2}$ is not continuous when $x=1$, because $f(1)=\frac{2e^{1-1}+2}{2e^{1-1}-2}=\frac{2(1)+2}{2(1)-2}=\frac{4}{0}$. When working with rational functions, you determine discontinuities by setting the denominator equal to 0 and solving for x. In this case, $2e^{x-1}-2=0$ when $e^{x-1}=1$, which is when $x=1$. The domain consists of all real numbers except 1.

1,000. 0

$f(0)=\cos 0=1$, and as x approaches 0 from the left, $\sin x \to \sin 0=0$. The sine and cosine functions are continuous for all x, so the function is continuous everywhere except at 0.

1,001. 1

$f(1)=|1-1|=0$, and as x approaches 1 from the right, $e^x-1 \to e^1-1 \approx 1.718$. All the function rules used in the piecewise function are continuous for all real numbers, but the values when $x=1$ aren't the same.

The function $f(x)=\frac{2e^{x-1}+2}{2e^{x-1}-2}$ is not continuous when $x=1$ because $f(1)=\frac{2e^{1-1}+2}{2e^{1-1}-2}=\frac{2(1)+2}{2(1)-2}=\frac{4}{0}$. When working with rational functions, you determine discontinuities by setting the denominator equal to 0 and solving for x. In this case, $2e^{x-1}-2=0$ when $e^{x-1}=1$, which is when $x=1$. The domain consists of all real numbers except 1.

$f(0)=\cos 0=1$, and as x approaches 0 from the left, $\sin x \to \sin 0=0$. The sine and cosine functions are continuous for all x, so the function is continuous everywhere except at 0.

$f(1)=(1-1)=0$, and as x approaches 1 from the right, $e^{x}+1 \to e+1 \approx 3.718$. All the function rules used in the piecewise function are continuous for all real numbers, but the values when $x=1$ aren't the same.

Index

• A •

• B •

• C •

• D •

• E •

• F •

• G •

• H •

• I •

• Q •

• R •

• S •

• U •

• V •

• W •

Workspace

Workspace

Workspace

About the Author

Mary Jane Sterling is the author of six *For Dummies* titles: *Algebra I For Dummies, Algebra II For Dummies, Trigonometry For Dummies, Math Word Problems For Dummies, Business Math For Dummies,* and *Linear Algebra For Dummies* (all published by Wiley). She has also written many supplements — workbooks and study aids.

Mary Jane has been teaching at Bradley University in Peoria, Illinois, for 35 years and loves hearing from former students. She still remembers her favorite student evaluation remark: "Mrs. Sterling is way too excited about mathematics." Yes!

Dedication

I dedicate this book to my friends and colleagues — current and former — at Bradley University. The feeling of community at this institution is unique and has made my tenure there a delight.

Author's Acknowledgments

I issue a big thank you to project editor Georgette Beatty, who has taken on the huge challenge of pulling together this project. She has been a delight to work with — always upbeat and helpful. Thank you so much for your hard work and patience. Also, another big salute with sincere appreciation goes to the copy editors, Megan Knoll and Danielle Voirol. Their thoroughness and attention to detail help make for a polished product.

And, of course, a heartfelt thank you to the technical editors, Mark Kannowski and Becky Moening. As much as I try to check the problems carefully, there is always that chance of a silly error. The editors keep me honest!

As always, a grateful thank you to acquisitions editor Lindsay Lefevere, who again found me another interesting project.

Publisher's Acknowledgments

Executive Editor: Lindsay Sandman Lefevere

Senior Project Editor: Georgette Beatty

Senior Copy Editor: Danielle Voirol

Copy Editor: Megan Knoll

Technical Editors: Mark Kannowski, Becky Moening

Art Coordinator: Alicia B. South

Project Coordinator: Lauren Buroker

Cover Image: ©iStock.com/ngkaki